New Operators and Their Applications in Geometric Function Theory

SERIES ON CONCRETE AND APPLICABLE MATHEMATICS

ISSN: 1793-1142

Series Editor: Professor George A. Anastassiou
Department of Mathematical Sciences
University of Memphis
Memphis, TN 38152, USA

---

*Published**

Vol. 27 *New Operators and Their Applications in Geometric Function Theory*
by Alina Alb Lupaş

Vol. 26 *Rainbow and Properly Colored Subgraphs*
by Jie Hu & Xueliang Li

Vol. 25 *Advanced Topics on Semilinear Evolution Equations*
by Mouffak Benchohra, Gaston M. N'Guérékata & Abdelkrim Salim

Vol. 24 *Trigonometric and Hyperbolic Generated Approximation Theory*
by George A. Anastassiou

Vol. 23 *Introduction to Matrix Theory: with Applications in Economics and Engineering*
Second Edition
by Ferenc Szidarovszky, Sandor Molnar & Mark Molnar

Vol. 22 *Bloch-Type Periodic Functions: Theory and Applications to Evolution Equations*
by Yong-Kui Chang, Gaston M. N'Guérékata & Rodrigo Ponce

Vol. 21 *Functional Equations and Inequalities: Solutions and Stability Results*
by John Michael Rassias, E. Thandapani, K. Ravi & B. V. Senthil Kumar

Vol. 20 *Discrete Approximation Theory*
by George A. Anastassiou & Merve Kester

Vol. 19 *Stochastic Models with Applications to Genetics, Cancers, AIDS and Other Biomedical Systems*
Second Edition
by Wai-Yuan Tan

*To view the complete list of the published volumes in the series, please visit:
https://www.worldscientific.com/series/scam

Series on Concrete and Applicable Mathematics – Vol. 27

New Operators and Their Applications in Geometric Function Theory

Alina Alb Lupaş
University of Oradea, Romania

World Scientific

W JERSEY • LONDON • SINGAPORE • BEIJING • SHANGHAI • HONG KONG • TAIPEI • CHENNAI • TOKYO

Published by

World Scientific Publishing Co. Pte. Ltd.
5 Toh Tuck Link, Singapore 596224
USA office: 27 Warren Street, Suite 401-402, Hackensack, NJ 07601
UK office: 57 Shelton Street, Covent Garden, London WC2H 9HE

Library of Congress Control Number: 2025035019

British Library Cataloguing-in-Publication Data
A catalogue record for this book is available from the British Library.

Series on Concrete and Applicable Mathematics — Vol. 27
NEW OPERATORS AND THEIR APPLICATIONS IN GEOMETRIC FUNCTION THEORY

ISBN 978-981-98-1878-5 (hardcover)
ISBN 978-981-98-1879-2 (ebook for institutions)
ISBN 978-981-98-1880-8 (ebook for individuals)

For any available supplementary material, please visit
https://www.worldscientific.com/worldscibooks/10.1142/14467#t=suppl

Desk Editors: Nambirajan Karuppiah/Lai Fun Kwong

Typeset by Stallion Press
Email: enquiries@stallionpress.com

Dedicated to all those I love.

Preface

The Romanian Mathematical School has an important contribution in the development of the field of Complex Analysis. A part of Complex Analysis studies the geometric properties of functions of one complex variable, known as "The geometric theory of functions of one complex variable".

The theory of functions of one complex variable, founded at the middle of the past century, is one of the mathematical fields, in which the rigor of the analytic judgment is tightly linked with the geometric intuition. The analytic functions of one complex variable are the ideal model of geometric transformations in the plan.

An important area of the theory of the functions of one complex variable is the geometric theory of analytic functions, whose aim consists of the geometric interpretation of some purely analytically expressed conditions and the transposal of some geometric properties in the analytical form. The geometric interpretation of some mathematical properties only in analytic form and, on the other side, the expression in analytic language of some geometric properties constitute some of the main objectives of the geometric theory of analytic functions. The geometric theory of analytic functions is based on the notion of conformal representation and in this theory the univalent functions have an essential role.

The univalent functions realize conformal representations, having applications in areas such as fluid mechanics, electrotechnics, nuclear physics and others. The first study of univalent functions is due to P. Koebe and was published as a paper in 1907. In these 100 years that passed until today, the univalent functions theory has developed considerably.

Many more papers have been published regarding the univalent functions theory. Among these, we mention here: Z. Nehari, *Conformal mappings*, 1952; L.V. Ahlfors, *Conformal invariants, Topics in Geometric Function Theory*, 1973; Ch. Pommerenke, *Univalent functions*, 1975; A.W. Goodman, *Univalent functions*, 1984; S.S. Miller, P.T. Mocanu, *Differential Subordinations, Theory and Applications*, 2000.

An important problem considered in the field of Geometric Function Theory is the study of differential and integral operators and of their properties on different classes of analytic functions. These operators were studied by many mathematicians, such as: St. Ruscheweyh, G.S. Sălăgean, N.E. Cho, H.M. Srivastava, T.H. Kim, F.M. Al-Oboudi, M. Darus, B.A. Uralegaddi, C. Somanatha. It is worth mentioning that the Romanian School of Geometric Function Theory was imposed worldwide by the remarkable results of many researchers.

Fractional calculus is being used nowadays in a growing variety of scientific domains as researchers develop and discover novel applications. Additionally, q-calculus is used in physics, engineering, and a number of mathematical fields. Moreover, combining fractional and q-calculus in Geometric Function Theory revealed a variety of noteworthy findings that Srivastava emphasized in a review paper that elucidates the relevance of the advancements and encourages more research on the topic.

In 1910, Jackson used the q-calculus in Mathematical Analysis to define the concepts of q-derivative and q-integral. Research incorporating quantum calculus features into Geometric Function Theory studies began the development after Srivastava provided a general framework for these types of investigations in a chapter of a book released in 1989. The foundations laid by Srivastava generated continued research in this context leading to the development of numerous concepts among which q-analogue operators play an important part. Noteworthy operators generating numerous important results in Geometric Function Theory can be listed as the q-analogue of Sălăgean differential operator, the q-analogue of Ruscheweyh differential operator, the q-analogue of multiplier transformation, q-hypergeometric function.

The current research in this domain aims at obtaining general results and tries to establish a substantial relation with the different ways of approaching the problems such that their applicability to be the best for the concrete problems that arise in practice.

The theme of this book concerns the Geometric Function Theory. A particular chapter in this field is represented by the theory of differential subordinations and superordinations. The differential subordinations theory is also known as the admissible functions method and it was initiated by S.S. Miller and P.T. Mocanu in two publications in 1978 and 1981. The same two reputed researchers have introduced the notion of differential superordination as a dual notion of differential subordination in 2003. Generalizing the differential subordinations, J.A. Antonino and S. Romaguera introduced the notion of strong differential subordination in 1994. G.I. Oros has introduced the notion of strong differential superordination as a dual notion of strong differential subordination in 2009. Using the differential subordinations and superordinations, strong differential subordinations and superordinations methods it was easier to prove some classical results from this domain, and also many extensions or generalizations of these, at the same time new results being obtained. The famous notion of subordination was interpreted in the fuzzy context in 2011 by G.I. Oros and G. Oros and fuzzy differential subordination was introduced in 2012 as an extension of the classical notion of differential subordination due to Miller and Mocanu. The theory started to develop in the next years and the dual notion of fuzzy differential superordination was introduced in 2017 by W.G. Atshan and K.O. Hussain. Later, studies continued to emerge using both theories, different types of operators and various notions familiar to the classical theory of differential subordinations.

A useful tool in the study of different types of operators is offered by the theory of classical differential subordinations and superordinations, the theory of strong differential subordinations and superordinations, and the theory of fuzzy differential subordinations and superordinations. Some remarkable results in the differential and integral operators theory were successfully proved and new results were also obtained using those theories. A particular chapter in the operators' study is occupied by the study of the different geometric properties of some classes of analytic functions defined using these operators. The differential subordinations and superordinations as well as the strong differential subordinations and superordinations play an important role in defining some new classes of holomorphic functions that are obtained using some generalizations of known operators.

In this book, we use the classical and strong differential subordinations and superordinations theories to obtain certain subclasses of analytic functions defined by different types of operators. Using several differential and integral operators, we introduce new subclasses of analytic functions and we study some properties of these classes (coefficient estimates, distortion theorems, extreme points, closure theorems, neighborhood properties and the radii of starlikeness, convexity and close-to-convexity).

In the first chapter, we present the already established results that are required for the studies. The definitions of the main classes of analytic functions are presented and also the basic notions regarding the theories of differential subordinations and superordinations: classical differential subordinations and superordinations, strong differential subordinations and superordinations and fuzzy differential subordinations and superordinations. Next, we present several classical differential subordination, superordination, respectively sandwich-type results involving few of the defined operators.

In Chapter 2, we remember the definitions of certain very well known differential and fractional integral operators that were used by the author and then we describe the differential and fractional integral operators introduced and studied by the author.

In Chapter 3, we expose some of the results obtained by the author using the classical differential subordinations and superordinations theory regarding some of the most interesting operators presented in the previous chapter. We define several subclasses of analytical functions using the classical differential subordinations theory and we describe some of the geometrical properties of the functions belonging to the defined subclasses regarding distortion bounds, extreme points, radii of close-to-convexity, starlikeness and convexity and neighborhood properties.

In Chapter 4, we expose applications of differential and fractional operators in the theory of strong differential subordinations and superordinations.

In Chapter 5, we give applications of differential and fractional integral operators in the theory of fuzzy differential subordinations and superordinations.

This book's results are expected to find applications in the many areas of Pure and Applied Mathematics and Engineering. All in all what is presented here is a valuable tool for a large range of applications.

This monograph is addressed to researchers and graduate students for research, seminars and advanced graduate courses in complex analysis and it is the first manuscript of the habilitation thesis of the author.

Overall, the topic covered in the monograph entitled New Operators and their Applications in Geometric Function Theory is an exclusive one, and it has been able to be developed with theoretical foundations through fractional calculus in the recent years particularly. Since there is not an adequate number of books regarding this topic, this monograph could prove to be a resource and reference to be resorted to for models that could be developed on the foundation of theoretical aspects. Thus, it has potential appeal.

I would like to thank my family for their support to me.

Alina Alb Lupaş
February 1, 2025

About the Author

Alina Alb Lupaş was born in Oradea, Romania in 1977. She received her BSc degree in Mathematics-Informatics from University of Oradea, Romania in 1999 and her MA in Analysis and topologico-algebraic systems from the Faculty of Sciences, University of Oradea, Romania in 2000. She was awarded her PhD in Mathematics from Institute of Mathematics and Informatics of Academy of Science of Moldova, State University from Tiraspol, Faculty of Mathematics, with the thesis "Categorial aspects of topological groups and rings theory". Since 1999, she is a faculty member at Department of Mathematics and Computer Science, Faculty of Informatics and Sciences, University of Oradea, Romania. She is currently an Associate Professor of Mathematics since 2016. She changed her research domain in 2008 when she started to work in Complex Analysis, conducting studies related to the Geometric Function Theory. Romanian Mathematical School has an important contribution in the development of this theory and University of Oradea is an important center of research in this field. She has published over 230 research articles in international mathematical journals and 22 monographs, proceedings and textbooks in well-known publishing houses. She is associate editor in many international mathematical journals and has been guest editor and assistant editor in 3 journals.

Contents

Chapter 1

Preliminaries on Differential Subordinations and Superordinations Theories

In this chapter, we present the already known results that are required for the studies. The definitions of the main classes of analytic functions are presented and also the basic notions regarding the theories of differential subordinations and superordinations: classical differential subordinations and superordinations, strong differential subordinations and superordinations and fuzzy differential subordinations and superordinations.

1.1 Differential Subordinations and Superordinations

Denote by U the unit disc of the complex plane, $U = \{z \in \mathbb{C} : |z| < 1\}$ and $\mathcal{H}(U)$ the space of holomorphic functions in U.

Let $\mathcal{A}_n = \{f \in \mathcal{H}(U) : f(z) = z + a_{n+1}z^{n+1} + \cdots,\ z \in U\}$ with $\mathcal{A}_1 = \mathcal{A}$ and $\mathcal{H}[a, n] = \{f \in \mathcal{H}(U) : f(z) = a + a_n z^n + a_{n+1}z^{n+1} + \cdots,\ z \in U\}$ for $a \in \mathbb{C}$ and $n \in \mathbb{N}$.

We present here the notion of differential subordination introduced by Miller and Mocanu (1985).

Definition 1.1 (Miller and Mocanu, 1985). If f and g are analytic functions in U, we say that f is subordinate to g, written $f \prec g$, if there is a function w analytic in U, with $w(0) = 0$, $|w(z)| < 1$, $z \in U$, such that $f(z) = g(w(z))$, $z \in U$. If g is univalent, then $f \prec g$ if and only if $f(0) = g(0)$ and $f(U) \subseteq g(U)$.

Let $\psi : \mathbb{C}^3 \times U \to \mathbb{C}$ and h an univalent function in U. If p is analytic in U and satisfies the (second-order) differential subordination

$$\psi(p(z), zp'(z), z^2p''(z); z) \prec h(z), \quad z \in U, \tag{1.1}$$

then p is called a solution of the differential subordination. The univalent function q is called a dominant of the solutions of the differential subordination, or more simply a dominant, if $p \prec q$ for all p satisfying (1.1).

A dominant $\widetilde{q}$ that satisfies $\widetilde{q} \prec q$ for all dominants q of (1.1) is said to be the best dominant of (1.1). The best dominant is unique up to a rotation of U.

We used the following lemmas in the research:

Lemma 1.1 (Hallenbeck and Ruscheweyh, 1975). *Let h be a convex function with $h(0) = a$, and let $\gamma \in \mathbb{C}\backslash\{0\}$ be a complex number with $\operatorname{Re} \gamma \geq 0$. If $p \in \mathcal{H}[a, n]$ and*

$$p(z) + \frac{1}{\gamma} zp'(z) \prec h(z), \quad z \in U,$$

then

$$p(z) \prec g(z) \prec h(z), \quad z \in U,$$

where $g(z) = \frac{\gamma}{nz^{\frac{\gamma}{n}}} \int_0^z h(t) t^{\frac{\gamma}{n}-1} dt$, $z \in U$.

Lemma 1.2 (Miller and Mocanu, 2000). *Let g be a convex function in U and let $h(z) = g(z) + n\alpha zg'(z)$, $z \in U$, where $\alpha > 0$ and n be a positive integer.*

If $p(z) = g(0) + p_n z^n + p_{n+1} z^{n+1} + \cdots$, $z \in U$, is holomorphic in U and

$$p(z) + \alpha zp'(z) \prec h(z), \quad z \in U,$$

then

$$p(z) \prec g(z), \quad z \in U,$$

and this result is sharp.

Lemma 1.3 (Miller and Mocanu, 2000). *Let the function g be univalent in the unit disc U and θ and ϕ be analytic in a domain D containing $g(U)$ with $\phi(w) \neq 0$ when $w \in g(U)$. Set $H(z) = zg'(z)\phi(g(z))$ and $h(z) = \theta(g(z)) + H(z)$. Suppose that*

1. *H is starlike univalent in U and*
2. $\operatorname{Re}\left(\frac{zh'(z)}{H(z)}\right) > 0$ *for $z \in U$.*

If p is analytic with $p(0) = g(0)$, $p(U) \subseteq D$ and

$$\theta(p(z)) + zp'(z)\,\phi(p(z)) \prec \theta(g(z)) + zg'(z)\,\phi(g(z)),$$

then

$$p(z) \prec g(z)$$

and g is the best dominant.

As a dual notion of differential subordination appears the notion of differential superordination introduced by Miller and Mocanu (2003).

Definition 1.2 (Miller and Mocanu, 2003). If f and g are analytic functions in U, we say that f is superordinate to g, written $g \prec f$, if there is a function w analytic in U, with $w(0) = 0$, $|w(z)| < 1$, $z \in U$, such that $g(z) = f(w(z))$, $z \in U$. If f is univalent, then $g \prec f$ if and only if $f(0) = g(0)$ and $g(U) \subseteq f(U)$.

Let $\psi : \mathbb{C}^2 \times U \to \mathbb{C}$ and h analytic in U. If p and $\psi(p(z), zp'(z); z)$ are univalent in U and satisfies the (first-order) differential superordination

$$h(z) \prec \psi(p(z), zp'(z); z), \quad z \in U, \tag{1.2}$$

then p is called a solution of the differential superordination. The analytic function q is called a subordinant of the solutions of the differential superordination, or more simply a subordinant, if $q \prec p$ for all p satisfying (1.2).

A univalent subordinant $\widetilde{q}$ that satisfies $q \prec \widetilde{q}$ for all subordinants q of (1.2) is said to be the best subordinant of (1.2). The best subordinant is unique up to a rotation of U.

Definition 1.3 (Miller and Mocanu, 2003). We denote by Q the set of functions analytic and injective on $\overline{U}\backslash E(f)$, where $E(f) = \{\zeta \in \partial U : \lim_{z\to\zeta} f(z) = \infty\}$, and are such that $f'(\zeta) \neq 0$ for $\zeta \in \partial U \backslash E(f)$. The subclass of Q for which $f(0) = a$ is denoted by $Q(a)$.

Lemma 1.4 (Miller and Mocanu, 2003). *Let h be a convex function with $h(0) = a$ and let $\gamma \in \mathbb{C}\backslash\{0\}$ be a complex number with $\operatorname{Re}\gamma \geq 0$.*

If $p \in \mathcal{H}[a,n] \cap Q$, $p(z) + \frac{1}{\gamma}zp'(z)$ is univalent in U and

$$h(z) \prec p(z) + \frac{1}{\gamma}zp'(z), \quad z \in U,$$

then

$$g(z) \prec p(z), \quad z \in U,$$

where $g(z) = \frac{\gamma}{nz^{\frac{\gamma}{n}}} \int_0^z h(t) t^{\frac{\gamma}{n}-1} dt$, $z \in U$. *The function g is convex and is the best subordinant.*

Lemma 1.5 (Miller and Mocanu, 2003). *Let g be a convex function in U and let* $h(z) = g(z) + \frac{1}{\gamma} zg'(z)$, $z \in U$, *where* $\mathrm{Re}\,\gamma \geq 0$.

If $p \in \mathcal{H}[a,n] \cap Q$, $p(z) + \frac{1}{\gamma} zp'(z)$ *is univalent in U and*

$$g(z) + \frac{1}{\gamma} zg'(z) \prec p(z) + \frac{1}{\gamma} zp'(z), \quad z \in U,$$

then

$$g(z) \prec p(z), \quad z \in U,$$

where $g(z) = \frac{\gamma}{nz^{\frac{\gamma}{n}}} \int_0^z h(t) t^{\frac{\gamma}{n}-1} dt$, $z \in U$. *The function g is the best subordinant.*

Lemma 1.6 (Bulboacă, 2002). *Let the function g be convex univalent in the open unit disc U and* θ *and* ϕ *be analytic in a domain D containing* $g(U)$. *Suppose that*

1. $\mathrm{Re}\left(\frac{\theta'(g(z))}{\phi(g(z))}\right) > 0$ *for* $z \in U$ *and*
2. $H(z) = zg'(z)\phi(g(z))$ *is starlike univalent in U.*

If $p(z) \in \mathcal{H}[g(0),1] \cap Q$, *with* $p(U) \subseteq D$ *and* $\theta(p(z)) + zp'(z)\phi(p(z))$ *is univalent in U and*

$$\theta(g(z)) + zg'(z)\phi(g(z)) \prec \theta(p(z)) + zp'(z)\phi(p(z)),$$

then

$$g(z) \prec p(z)$$

and g is the best subordinant.

1.2 Strong Differential Subordinations and Superordinations

Extending the well known classes of functions $\mathcal{A}_n$ and $\mathcal{H}[a,n]$, Oros has introduced in Oros (2012) new classes of analytic functions $\mathcal{A}^*_{n\zeta}$ and $\mathcal{H}^*[a,n,\zeta]$.

Denote by $\overline{U} = \{z \in \mathbb{C} : |z| \leq 1\}$ the closed unit disc of the complex plane and $\mathcal{H}(U \times \overline{U})$ the class of analytic functions in $U \times \overline{U}$.

Let

$$\mathcal{A}^*_{n\zeta} = \{f \in \mathcal{H}(U \times \overline{U}),\ f(z,\zeta) = z + a_{n+1}(\zeta) z^{n+1} + \cdots,\ z \in U,\ \zeta \in \overline{U}\},$$

where $a_k(\zeta)$ are holomorphic functions in $\overline{U}$ for $k \geq 2$, and

$$\mathcal{H}^*[a,n,\zeta] = \{f \in \mathcal{H}(U \times \overline{U}),\ f(z,\zeta) = a + a_n(\zeta) z^n + a_{n+1}(\zeta) z^{n+1} + \cdots,\\ z \in U,\ \zeta \in \overline{U}\},$$

for $a \in \mathbb{C}$, $n \in \mathbb{N}$, $a_k(\zeta)$ are holomorphic functions in $\overline{U}$ for $k \geq n$.

Generalizing the notion of differential subordination, Antonino and Romaguera have introduced in Antonino and Romaguera (1994) the notion of strong differential subordination, which was developed by Oros and Oros (2009) and Oros (2012).

Definition 1.4 (Oros and Oros, 2009). Let $f(z,\zeta)$, $g(z,\zeta)$ analytic in $U \times \overline{U}$. The function $f(z,\zeta)$ is said to be strongly subordinate to $g(z,\zeta)$ if there exists a function w analytic in U, with $w(0) = 0$ and $|w(z)| < 1$ such that $f(z,\zeta) = g(w(z),\zeta)$ for all $\zeta \in \overline{U}$. In such a case we write $f(z,\zeta) \prec\prec g(z,\zeta)$, $z \in U$, $\zeta \in \overline{U}$.

Remark 1.1.

(i) Since $f(z,\zeta)$ is analytic in $U \times \overline{U}$, for all $\zeta \in \overline{U}$, and univalent in U, for all $\zeta \in \overline{U}$, as in mentioned in Oros and Oros (2009), Definition 1.4 is equivalent to $f(0,\zeta) = g(0,\zeta)$, for all $\zeta \in \overline{U}$, and $f(U \times \overline{U}) \subset g(U \times \overline{U})$.

(ii) If $g(z,\zeta) \equiv g(z)$ and $f(z,\zeta) \equiv f(z)$, the strong subordination becomes the usual notion of subordination.

For studying the strong differential subordinations, we need the following lemmas.

Lemma 1.7 (Alb Lupaş *et al.*, 2012). *Let $h(z,\zeta)$ be a convex function with $h(0,\zeta) = a$ for every $\zeta \in \overline{U}$ and let $\gamma \in \mathbb{C}^*$ be a complex number with* Re $\gamma \geq 0$. *If $p \in \mathcal{H}^*[a,n,\zeta]$ and*

$$p(z,\zeta) + \frac{1}{\gamma} z p'_z(z,\zeta) \prec\prec h(z,\zeta),$$

then

$$p(z,\zeta) \prec\prec g(z,\zeta) \prec\prec h(z,\zeta),$$

where $g(z,\zeta) = \frac{\gamma}{n z^{\frac{\gamma}{n}}} \int_0^z h(t,\zeta) t^{\frac{\gamma}{n}-1} dt$ is convex and it is the best dominant.

Lemma 1.8 (Alb Lupaş *et al.*, 2012). *Let $g(z,\zeta)$ be a convex function in $U \times \overline{U}$, for all $\zeta \in \overline{U}$, and let*

$$h(z,\zeta) = g(z,\zeta) + n\alpha z g'_z(z,\zeta), \quad z \in U, \ \zeta \in \overline{U},$$

where $\alpha > 0$ and n is a positive integer. If

$$p(z,\zeta) = g(0,\zeta) + p_n(\zeta) z^n + p_{n+1}(\zeta) z^{n+1} + \cdots, \quad z \in U, \ \zeta \in \overline{U},$$

is holomorphic in $U \times \overline{U}$ and

$$p(z,\zeta) + \alpha z p'_z(z,\zeta) \prec\prec h(z,\zeta), \quad z \in U, \ \zeta \in \overline{U},$$

then

$$p(z,\zeta) \prec\prec g(z,\zeta)$$

and this result is sharp.

Lemma 1.9 (Miller and Mocanu, 2007). *Consider the univalent function g in $U \times \overline{U}$ and the analytic functions θ and ϕ in a domain $D \supset g(U \times \overline{U})$ such that $\phi(w) \neq 0$ for $w \in g(U \times \overline{U})$. Assume that function $H(z,\zeta) = z g'_z(z,\zeta)\phi(g(z,\zeta))$ is starlike univalent in $U \times \overline{U}$ and function $h(z,\zeta) = \theta(g(z,\zeta)) + H(z,\zeta)$ has the property $\operatorname{Re}\left(\frac{z h'_z(z,\zeta)}{H(z,\zeta)}\right) > 0$ for $(z,\zeta) \in U \times \overline{U}$.*

If analytic function p with properties $p(0,\zeta) = g(0,\zeta)$ and $p(U \times \overline{U}) \subseteq D$ satisfies the strong differential subordination

$$\theta(p(z,\zeta)) + z p'_z(z,\zeta)\phi(p(z,\zeta)) \prec\prec \theta(g(z,\zeta)) + z g'_z(z,\zeta)\phi(g(z,\zeta)),$$

then

$$p(z,\zeta) \prec\prec g(z,\zeta), \quad (z,\zeta) \in U \times \overline{U},$$

and g is the best dominant.

As a dual notion of strong differential subordination, Oros has introduced and developed the notion of strong differential superordination in Oros (2009).

Definition 1.5 (Oros, 2009). Let $f(z,\zeta)$, $g(z,\zeta)$ analytic in $U \times \overline{U}$. The function $f(z,\zeta)$ is said to be strongly superordinate to $g(z,\zeta)$ if there exists a function w analytic in U, with $w(0) = 0$ and $|w(z)| < 1$, such that $g(z,\zeta) = f(w(z),\zeta)$, for all $\zeta \in \overline{U}$. In such a case we write $g(z,\zeta) \prec\prec f(z,\zeta)$, $z \in U$, $\zeta \in \overline{U}$.

Remark 1.2.

(i) Since $f(z,\zeta)$ is analytic in $U \times \overline{U}$, for all $\zeta \in \overline{U}$, and univalent in U, for all $\zeta \in \overline{U}$, as in mentioned in Oros (2009), Definition 1.5 is equivalent to $g(0,\zeta) = f(0,\zeta)$, for all $\zeta \in \overline{U}$, and $g\left(U \times \overline{U}\right) \subset f\left(U \times \overline{U}\right)$.
(ii) If $g(z,\zeta) \equiv g(z)$ and $f(z,\zeta) \equiv f(z)$, the strong superordination becomes the usual notion of superordination.

Definition 1.6 (Alb Lupaş, 2014f). We denote by Q^* the set of functions that are analytic and injective on $\overline{U} \times \overline{U} \backslash E(f,\zeta)$, where $E(f,\zeta) = \{y \in \partial U : \lim\limits_{z \to y} f(z,\zeta) = \infty\}$, and are such that $f'_z(y,\zeta) \neq 0$ for $y \in \partial U \times \overline{U} \backslash E(f,\zeta)$. The subclass of Q^* for which $f(0,\zeta) = a$ is denoted by $Q^*(a)$.

For studying the strong differential superordinations, we need the following lemmas.

Lemma 1.10 (Alb Lupaş, 2014f). *Let $h(z,\zeta)$ be a convex function with $h(0,\zeta) = a$ and let $\gamma \in \mathbb{C}^*$ be a complex number with* Re $\gamma \geq 0$. *If $p \in \mathcal{H}^*[a,n,\zeta] \cap Q^*$, $p(z,\zeta) + \frac{1}{\gamma} z p'_z(z,\zeta)$ is univalent in $U \times \overline{U}$ and*

$$h(z,\zeta) \prec\prec p(z,\zeta) + \frac{1}{\gamma} z p'_z(z,\zeta), \quad z \in U, \ \zeta \in \overline{U},$$

then

$$g(z,\zeta) \prec\prec p(z,\zeta), \quad z \in U, \ \zeta \in \overline{U},$$

where $g(z,\zeta) = \frac{\gamma}{n z^{\frac{\gamma}{n}}} \int_0^z h(t,\zeta)\, t^{\frac{\gamma}{n}-1} dt$, $z \in U$, $\zeta \in \overline{U}$. The function g is convex and is the best subordinant.

Lemma 1.11 (Alb Lupaş, 2014f). *Let $g(z,\zeta)$ be a convex function in $U \times \overline{U}$ and let $h(z,\zeta) = g(z,\zeta) + \frac{1}{\gamma} z g'_z(z,\zeta)$, $z \in U$, $\zeta \in \overline{U}$, where* Re $\gamma \geq 0$.

If $p \in \mathcal{H}^*[a,n,\zeta] \cap Q^*$, $p(z,\zeta) + \frac{1}{\gamma} z p'_z(z,\zeta)$ is univalent in $U \times \overline{U}$ and

$$g(z,\zeta) + \frac{1}{\gamma} z g'_z(z,\zeta) \prec\prec p(z,\zeta) + \frac{1}{\gamma} z p'_z(z,\zeta), \quad z \in U, \ \zeta \in \overline{U},$$

then

$$g(z,\zeta) \prec\prec p(z,\zeta), \quad z \in U, \ \zeta \in \overline{U},$$

where $g(z,\zeta) = \frac{\gamma}{n z^{\frac{\gamma}{n}}} \int_0^z h(t,\zeta)\, t^{\frac{\gamma}{n}-1} dt$, $z \in U$, $\zeta \in \overline{U}$. The function g is the best subordinant.

Lemma 1.12 (Miller and Mocanu, 2007). *Consider the convex univalent function g in $U \times \overline{U}$ and the analytic functions θ and ϕ in a domain $D \supset g\left(U \times \overline{U}\right)$ such that $\operatorname{Re}\left(\frac{\theta_z'(g(z,\zeta))}{\phi(g(z,\zeta))}\right) > 0$ for $(z,\zeta) \in U \times \overline{U}$ and $H(z,\zeta) = zg_z'(z,\zeta)\,\phi(g(z,\zeta))$ is starlike univalent in $U \times \overline{U}$.*

If function $p(z,\zeta) \in \mathcal{H}^*[g(0,\zeta),1,\zeta] \cap Q^*$, with properties $p\left(U \times \overline{U}\right) \subseteq D$ and $\theta(p(z,\zeta)) + zp_z'(z)\,\phi(p(z,\zeta))$ is univalent in $U \times \overline{U}$, satisfies the strong differential superordination

$$\theta(g(z,\zeta)) + zg_z'(z,\zeta)\,\phi(g(z,\zeta)) \prec\prec \theta(p(z,\zeta)) + zp_z'(z,\zeta)\,\phi(p(z,\zeta)),$$

then

$$g(z,\zeta) \prec\prec p(z,\zeta), \quad (z,\zeta) \in U \times \overline{U},$$

and g is the best subordinant.

1.3 Fuzzy Differential Subordinations and Superordinations

Fuzzy sets theory was connected to geometric function theory in 2011 when the notion of fuzzy subordination was introduced in Oros and Oros (2011) having as inspiration the theory of differential subordinations initiated by Miller and Mocanu (1978, 1981). The core of the theory of differential subordinations was gradually adapted to fuzzy sets notions in the next years in Oros and Oros (2012a, 2012b, 2012c) following the main lines of research as they can be found in the monograph published in Miller and Mocanu (2000).

In order to use the concept of fuzzy differential subordination, we remember the following definitions:

Definition 1.7 (Gal and Ban, 1996). A pair (A, F_A), where $F_A : X \to [0,1]$ and $A = \{x \in X : 0 < F_A(x) \leq 1\}$ is called fuzzy subset of X. The set A is called the support of the fuzzy set (A, F_A) and F_A is called the membership function of the fuzzy set (A, F_A). One can also denote $A = \operatorname{supp}(A, F_A)$.

Remark 1.3. Throughout my thesis, we use the following notations.

$$F_{f(D)}(f(z)) = \operatorname{supp}\left(f(D), F_{f(D)}\cdot\right) = \{z \in D : 0 < F_{f(D)}f(z) \leq 1\},$$
$$F_{g(D)}(g(z)) = \operatorname{supp}\left(g(D), F_{g(D)}\cdot\right) = \{z \in D : 0 < F_{g(D)}g(z) \leq 1\},$$
$$p(U) = \operatorname{supp}\left(p(U), F_{p(U)}\cdot\right) = \{z \in U : 0 < F_{p(U)}(p(z)) \leq 1\},$$

$q(U) = \operatorname{supp}\big(q(U), F_{q(U)}\cdot\big) = \{z \in U : 0 < F_{q(U)}(q(z)) \le 1\},$
$h(U) = \operatorname{supp}\big(h(U), F_{h(U)}\cdot\big) = \{z \in U : 0 < F_{h(U)}(h(z)) \le 1\}.$

Remark 1.4. We call the subset $F_{\mathbb{C}}(\mathbb{C}) = \{z \in \mathbb{C} : 0 < |F(z)| \le 1\} = U_{\mathcal{F}}(0,1)$ the fuzzy unit disk.

The membership functions can be defined otherwise as follows:

Definition 1.8 (Alb Lupaş and Oros, 2015). Let $f(D) = \operatorname{supp}\big(f(D), F_{f(D)}\big) = \{z \in D : 0 < F_{f(D)} f(z) \le 1\}$, where $F_{f(D)}\cdot$ is the membership function of the fuzzy set $f(D)$ associated to the function f.

The membership function of the fuzzy set $(\mu f)(D)$ associated to the function μf coincide with the membership function of the fuzzy set $f(D)$ associated to the function f, i.e. $F_{(\mu f)(D)}((\mu f)(z)) = F_{f(D)} f(z)$, $z \in D$.

The membership function of the fuzzy set $(f+g)(D)$ associated to the function $f+g$ coincide with the half of the sum of the membership functions of the fuzzy sets $f(D)$, respectively $g(D)$, associated to the function f, respectively g, i.e. $F_{(f+g)(D)}((f+g)(z)) = \frac{F_{f(D)} f(z) + F_{g(D)} g(z)}{2}$, $z \in D$.

Remark 1.5. $F_{(f+g)(D)}((f+g)(z))$ can be defined in other ways.

Since $0 < F_{f(D)} f(z) \le 1$ and $0 < F_{g(D)} g(z) \le 1$, it is evidently that $0 < F_{(f+g)(D)}((f+g)(z)) \le 1$, $z \in D$.

Definition 1.9 (Oros and Oros, 2011). Let $D \subset \mathbb{C}$, $z_0 \in D$ be a fixed point and let the functions $f, g \in \mathcal{H}(D)$. The function f is said to be fuzzy subordinate to g and write $f \prec_{\mathcal{F}} g$ or $f(z) \prec_{\mathcal{F}} g(z)$, if are satisfied the conditions:

(1) $f(z_0) = g(z_0)$,
(2) $F_{f(D)} f(z) \le F_{g(D)} g(z)$, $z \in D$.

Definition 1.10 (Oros and Oros, 2012a, Definition 2.2). Let $\psi : \mathbb{C}^3 \times U \to \mathbb{C}$ and h univalent in U, with $\psi(a, 0; 0) = h(0) = a$. If p is analytic in U, with $p(0) = a$ and satisfies the (second-order) fuzzy differential subordination

$$F_{\psi(\mathbb{C}^3 \times U)} \psi(p(z), zp'(z), z^2 p''(z); z) \le F_{h(U)} h(z), \quad z \in U, \tag{1.3}$$

then p is called a fuzzy solution of the fuzzy differential subordination. The univalent function q is called a fuzzy dominant of the fuzzy solutions of the fuzzy differential subordination, or more simple a fuzzy dominant, if $F_{p(U)} p(z) \le F_{q(U)} q(z)$, $z \in U$, for all p satisfying (1.3). A fuzzy dominant

$\widetilde{q}$ that satisfies $F_{\widetilde{q}(U)}\widetilde{q}(z) \leq F_{q(U)}q(z)$, $z \in U$, for all fuzzy dominants q of (1.3) is said to be the fuzzy best dominant of (1.3).

Lemma 1.13 (Miller and Mocanu, 2000, Corollary 2.6g.2, p. 66). *Let $h \in \mathcal{A}$ and*

$$G(z) = \frac{1}{z}\int_0^z h(t)\,dt, \ \ z \in U.$$

If

$$\operatorname{Re}\left(\frac{zh''(z)}{h'(z)} + 1\right) > -\frac{1}{2}, \ z \in U,$$

then $G \in \mathcal{K}$.

Lemma 1.14 (Oros and Oros, 2012b). *Let h be a convex function with $h(0) = a$, and let $\gamma \in \mathbb{C}^*$ be a complex number with* $\operatorname{Re}\gamma \geq 0$. *If $p \in \mathcal{H}[a,n]$ with $p(0) = a$, $\psi : \mathbb{C}^2 \times U \to \mathbb{C}$, $\psi(p(z), zp'(z); z) = p(z) + \frac{1}{\gamma}zp'(z)$ an analytic function in U and*

$$F_{\psi(\mathbb{C}^2\times U)}\left(p(z) + \frac{1}{\gamma}zp'(z)\right) \leq F_{h(U)}h(z), \text{ i.e. } p(z) + \frac{1}{\gamma}zp'(z) \prec_{\mathcal{F}} h(z),$$

$z \in U$, then

$$F_{p(U)}p(z) \leq F_{g(U)}g(z) \leq F_{h(U)}h(z), \text{ i.e. } p(z) \prec_{\mathcal{F}} g(z) \prec_{\mathcal{F}} h(z), \quad z \in U,$$

where $g(z) = \frac{\gamma}{nz^{\frac{\gamma}{n}}}\int_0^z h(t)t^{\frac{\gamma}{n}-1}dt$, $z \in U$. The function g is convex and is the fuzzy best dominant.

Lemma 1.15 (Oros and Oros, 2012b). *Let g be a convex function in U and let $h(z) = g(z) + n\alpha zg'(z)$, $z \in U$, where $\alpha > 0$ and n is a positive integer.*

If $p(z) = g(0) + p_n z^n + p_{n+1}z^{n+1} + \cdots$, $z \in U$, is holomorphic in U and

$$F_{p(U)}\left(p(z) + \alpha zp'(z)\right) \leq F_{h(U)}h(z), \text{ i.e. } p(z) + \alpha zp'(z) \prec_{\mathcal{F}} h(z), \quad z \in U,$$

then

$$F_{p(U)}p(z) \leq F_{g(U)}g(z), \text{ i.e. } p(z) \prec_{\mathcal{F}} g(z), \quad z \in U,$$

and this result is sharp.

Lemma 1.16 (Oros and Oros, 2012b). *Let the function g be univalent in the unit disc U and θ and ϕ be analytic in a domain D containing $g(U)$ with $\phi(w) \neq 0$ when $w \in g(U)$. Set $H(z) = zg'(z)\phi(g(z))$ and $h(z) = \theta(g(z)) + H(z)$. Suppose that*

1. *H is starlike univalent in U and*
2. $\operatorname{Re}\left(\frac{zh'(z)}{H(z)}\right) > 0$ *for $z \in U$.*

If p is analytic with $p(0) = g(0)$, $p(U) \subseteq D$ and

$$F_{p(U)}\theta(p(z)) + zp'(z)\phi(p(z)) \leq F_{h(U)}\theta(g(z)) + zg'(z)\phi(g(z)),$$

then

$$F_{p(U)}p(z) \leq F_{g(U)}g(z)$$

and g is the fuzzy best dominant.

The dual notion of fuzzy differential superordination was introduced in Atshan and Hussain (2017).

Definition 1.11 (Atshan and Hussain, 2017). Let $\psi : \mathbb{C}^3 \times U \to \mathbb{C}$ and let h be analytic in U. If p and $\psi(p(z), zp'(z), z^2p''(z); z)$ are univalent in U and satisfy the (second-order) fuzzy differential superordination

$$F_{h(U)}h(z) \leq F_{\psi(\mathbb{C}^3\times U)}\psi(p(z), zp'(z), z^2p''(z); z), \quad z \in U, \tag{1.4}$$

i.e.

$$h(z) \prec_{\mathcal{F}} \psi(p(z), zp'(z), z^2p''(z); z), \quad z \in U,$$

then p is called a fuzzy solution of the fuzzy differential superordination. An analytic function q is called a fuzzy subordinant of the fuzzy differential superordination (1.4), or more simply a fuzzy sub-ordination if

$$F_{q(U)}q(z) \leq F_{p(U)}p(z), \quad z \in U,$$

for all p satisfying (1.4). A univalent fuzzy subordination $\widetilde{q}$ that satisfies $F_{q(U)}q \leq F_{q(U)}\widetilde{q}$ for all fuzzy subordinate q of (1.4) is said to be the fuzzy best subordinate of (1.4). Note that the fuzzy best subordinant is unique up to a relation of U.

Definition 1.12 (Oros and Oros, 2012b). Denote by Q the set of all functions f that are analytic and injective on $\overline{U}\backslash E(f)$, where $E(f) = \{\zeta \in \partial U : \lim_{z\to\zeta} f(z) = \infty\}$, and for which $f'(\zeta) \neq 0$ for $\zeta \in \partial U\backslash E(f)$.

Lemma 1.17 (Alb Lupaş, 2013d, Corollary 2.6g.2, p. 66). *Consider $\gamma \in \mathbb{C}^*$ such that* Re $\gamma \geq 0$ *and h a convex function with the property $h(0) = a$. If $p \in Q \cap \mathcal{H}[a, n]$ and $\frac{zp'(z)}{\gamma} + p(z)$ verifies for any $z \in U$ the fuzzy differential superordination*

$$\mathcal{F}_{h(U)}h(z) \leq \mathcal{F}_{p(U)}\left(\frac{zp'(z)}{\gamma} + p(z)\right),$$

and it is univalent in U, then the fuzzy differential superordination

$$\mathcal{F}_{g(U)}g(z) \leq \mathcal{F}_{p(U)}p(z), \quad z \in U,$$

is satisfied for any $z \in U$ by the convex function $g(z) = \frac{\gamma}{nz^{\frac{\gamma}{n}}} \int_0^z h(t)t^{\frac{\gamma}{n}-1}dt$, which is the fuzzy best subordinant.

Lemma 1.18 (Alb Lupaş, 2013d, Corollary 2.6g.2, p. 66). *Set for any $z \in U$ the function*

$$h(z) = \frac{zg'(z)}{\gamma} + g(z),$$

for g a convex function in U, $\gamma \in \mathbb{C}^*$ such that Re $\gamma \geq 0$. *When $p \in Q \cap \mathcal{H}[a, n]$ and $\frac{zp'(z)}{\gamma} + p(z)$ verifies for any $z \in U$ the fuzzy differential superordination*

$$\mathcal{F}_{g(U)}\left(\frac{zg'(z)}{\gamma} + g(z)\right) \leq \mathcal{F}_{p(U)}\left(\frac{zp'(z)}{\gamma} + p(z)\right),$$

and it is univalent in U, then the fuzzy differential superordination

$$\mathcal{F}_{g(U)}g(z) \leq \mathcal{F}_{p(U)}p(z),$$

is satisfies for any $z \in U$ by the function $g(z) = \frac{\gamma}{nz^{\frac{\gamma}{n}}} \int_0^z h(t)t^{\frac{\gamma}{n}-1}dt$, which is the fuzzy best subordinant.

Lemma 1.19 (Atshan and Hussain, 2017). *Let g be a convex function, univalent in the open unit disc U and θ and ϕ be analytic in a domain D containing $g(U)$. Suppose that*

1. Re $\left(\frac{\theta'(g(z))}{\phi(g(z))}\right) > 0$ *for $z \in U$ and*
2. *$H(z) = zg'(z)\phi(g(z))$ is starlike univalent in U.*

If $p(z) \in \mathcal{H}[g(0), 1] \cap Q$*, with* $p(U) \subseteq D$ *and* $\theta(p(z)) + zp'(z)\phi(p(z))$ *is univalent in* U *and*

$$F_{g(U)}\theta(g(z)) + zg'(z)\phi(g(z)) \leq F_{p(U)}\theta(p(z)) + zp'(z)\phi(p(z)),$$

then

$$F_{g(U)}g(z) \leq F_{p(U)}p(z)$$

and g *is the fuzzy best subordinant.*

Chapter 2

Defining New Operators

In this chapter, we remember the definitions of certain very well-known differential and fractional integral operators that were used by the author and then we describe the differential and fractional integral operators introduced and studied by the author.

2.1 Preliminaries

In this chapter, we use the following well-known differential operators.

Definition 2.1 (Sălăgean, 1983). For $f \in \mathcal{A}_n$, $m \in \mathbb{N}$, the operator S^m is defined by $S^m : \mathcal{A}_n \to \mathcal{A}_n$,

$$\begin{aligned} S^0 f(z) &= f(z), \\ S^1 f(z) &= z f'(z), \ldots, \\ S^{m+1} f(z) &= z \left(S^m f(z)\right)', \quad z \in U. \end{aligned}$$

Remark 2.1. If $f \in \mathcal{A}_n$, $f(z) = z + \sum_{j=n+1}^{\infty} a_j z^j$, then $S^m f(z) = z + \sum_{j=n+1}^{\infty} j^m a_j z^j$, $z \in U$.

If $f \in \mathcal{A}$, $f(z) = z + \sum_{j=2}^{\infty} a_j z^j$, then $S^n f(z) = z + \sum_{j=2}^{\infty} j^n a_j z^j$, $z \in U$.

Definition 2.2 (Ruscheweyh, 1975). For $f \in \mathcal{A}_n$, $m \in \mathbb{N}$, the operator R^m is defined by $R^m : \mathcal{A}_n \to \mathcal{A}_n$,

$$\begin{aligned} R^0 f(z) &= f(z), \\ R^1 f(z) &= z f'(z), \ldots, \\ (m+1) R^{m+1} f(z) &= z \left(R^m f(z)\right)' + m R^m f(z), \quad z \in U. \end{aligned}$$

Remark 2.2. If $f \in \mathcal{A}_n$, $f(z) = z + \sum_{j=n+1}^{\infty} a_j z^j$, then $R^m f(z) = z + \sum_{j=n+1}^{\infty} \frac{(m+j-1)!}{m!(j-1)!} a_j z^j$, $z \in U$.

If $f \in \mathcal{A}$, $f(z) = z + \sum_{j=2}^{\infty} a_j z^j$, then $R^n f(z) = z + \sum_{j=2}^{\infty} \frac{(n+j-1)!}{n!(j-1)!} a_j z^j$, $z \in U$.

Definition 2.3 (Al Oboudi, 2004). For $f \in \mathcal{A}_n$, $\lambda \geq 0$ and $n, m \in \mathbb{N}$, the operator D_λ^m is defined by $D_\lambda^m : \mathcal{A}_n \to \mathcal{A}_n$,

$$\begin{aligned} D_\lambda^0 f(z) &= f(z), \\ D_\lambda^1 f(z) &= (1-\lambda) f(z) + \lambda z f'(z) = D_\lambda f(z), \ldots, \\ D_\lambda^{m+1} f(z) &= (1-\lambda) D_\lambda^m f(z) + \lambda z \left(D_\lambda^m f(z)\right)' = D_\lambda \left(D_\lambda^m f(z)\right), \quad z \in U. \end{aligned}$$

Remark 2.3. If $f \in \mathcal{A}_n$ and $f(z) = z + \sum_{j=n+1}^{\infty} a_j z^j$, then $D_\lambda^m f(z) = z + \sum_{j=n+1}^{\infty} \left[1 + (j-1)\lambda\right]^m a_j z^j$, $z \in U$.

If $f \in \mathcal{A}$ and $f(z) = z + \sum_{j=2}^{\infty} a_j z^j$, then $D_\lambda^n f(z) = z + \sum_{j=2}^{\infty} \left[1 + (j-1)\lambda\right]^n a_j z^j$, $z \in U$.

Remark 2.4. For $\lambda = 1$ in the above definition, we obtain the Sălăgean differential operator.

Definition 2.4 (Cătaş, 2007). For $f \in \mathcal{A}_n$, $n \in \mathbb{N}$, $m \in \mathbb{N} \cup \{0\}$, $\lambda, l \geq 0$, the operator $I(m, \lambda, l) f(z)$ is defined by the following infinite series:

$$I(m, \lambda, l) f(z) = z + \sum_{j=n+1}^{\infty} \left(\frac{\lambda(j-1) + l + 1}{l+1}\right)^m a_j z^j.$$

Remark 2.5. It follows from the above definition that

$$I(0, \lambda, l) f(z) = f(z),$$
$$(l+1) I(m+1, \lambda, l) f(z) = (l+1-\lambda) I(m, \lambda, l) f(z) + \lambda z (I(m, \lambda, l) f(z))',$$

$z \in U$.

If $f \in \mathcal{T}$, $f(z) = z - \sum_{j=2}^{\infty} a_j z^j$, then $I(n, \lambda, l) f(z) = z - \sum_{j=2}^{\infty} \left(\frac{\lambda(j-1)+l+1}{l+1}\right)^n a_j z^j$.

Remark 2.6. For $l = 0$, $\lambda \geq 0$, the operator $D_\lambda^m = I(m, \lambda, 0)$ was introduced and studied by Al-Oboudi (2004), which is reduced to the Sălăgean differential operator for $\lambda = 1$. The operator $I(m, 1, l)$ was studied by Cho and Srivastava (2003) and Cho and Kim (2003). The operator $I(m, 1, 1)$ was studied by Uralegaddi and Somanatha (1992) and the operator $I(\alpha, \lambda, 0)$ was introduced by Acu and Owa (2006). Cătaş (2007) has studied the operator $I_p(m, \lambda, l)$ which generalizes the operator $I(m, \lambda, l)$.

The author published 12 papers studying the operator described in Definition 2.4 (e.g. Alb Lupaş, 2010a, 2011c, 2012b, 2014a, 2015a, 2017a).

2.2 Differential Operators

In this section, we present nine differential operators defined and studied by the author.

Using the Sălăgean differential operator and Ruscheweyh derivative, we have defined a new operator as a linear combination of these 2 operators, publishing 18 papers regarding this differential operator (e.g. Alb Lupaş, 2009, 2012, 2014c, 2014d, 2014f, 2015c, 2015e, 2016a, 2016b; Alb Lupaş and Breaz, 2010; Alb Lupaş and Oros, 2015).

Definition 2.5 (Alb Lupaş, 2009). Let $\alpha \geq 0$, $n \in \mathbb{N}$. Denote by L_α^n the operator given by $L_\alpha^n : \mathcal{A} \to \mathcal{A}$,

$$L_\alpha^n f(z) = (1-\alpha)R^n f(z) + \alpha S^n f(z), \quad \text{for } z \in U.$$

Remark 2.7 (Alb Lupaş, 2009). L_α^n is a linear operator and if $f \in \mathcal{A}$, $f(z) = z + \sum_{j=2}^{\infty} a_j z^j$, then $L_\alpha^n f(z) = z + \sum_{j=2}^{\infty} (\alpha j^n + (1-\alpha) C_{n+j-1}^n) a_j z^j$, for $z \in U$.

Next, we have generalized this operator and we have defined a new operator as a linear combination between the generalized Sălăgean operator and Ruscheweyh derivative. The author has published 15 papers regarding this differential operator (e.g. Alb Lupaş, 2011a, 2011d, 2012c, 2012e, 2013c, 2014b, 2015a; Alb Lupaş and Andrei, 2014a, 2015b, 2015c).

Definition 2.6 (Alb Lupaş, 2011a). Let $\alpha, \lambda \geq 0$, $n \in \mathbb{N}$. Denote by $RD_{\lambda,\alpha}^n$ the operator given by $RD_{\lambda,\alpha}^n : \mathcal{A} \to \mathcal{A}$,

$$RD_{\lambda,\alpha}^n f(z) = (1-\alpha)R^n f(z) + \alpha D_\lambda^n f(z), \quad \text{for } z \in U.$$

Remark 2.8 (Alb Lupaş, 2011a). If $f \in \mathcal{A}$, $f(z) = z + \sum_{j=2}^{\infty} a_j z^j$, then $RD_{\lambda,\alpha}^n f(z) = z + \sum_{j=2}^{\infty} \{\alpha[1+(j-1)\lambda]^n + (1-\alpha) C_{n+j-1}^n\} a_j z^j$, for $z \in U$.

Remark 2.9 (Alb Lupaş, 2011a). For $\alpha = 0$, $RD_{\lambda,0}^n f(z) = R^n f(z)$, where $z \in U$ and for $\alpha = 1$, $RD_{\lambda,1}^n f(z) = D_\lambda^n f(z)$, where $z \in U$.

For $\lambda = 1$, we obtain $RD_{1,\alpha}^n f(z) = L_\alpha^n f(z)$ which was defined in the work of Alb Lupaş (2009).

For $n = 0$, $RD_{\lambda,\alpha}^0 f(z) = (1-\alpha)R^0 f(z) + \alpha D_\lambda^0 f(z) = f(z) = R^0 f(z) = D_\lambda^0 f(z)$, where $z \in U$.

Generalizing again, we have defined a new operator as a linear combination of multiplier transformation and Ruscheweyh derivative, publishing 17 papers regarding this differential operator (e.g. Alb Lupaş, 2010d, 2011e, 2012d, 2012f, 2014b, 2014e, 2015b, 2015d, 2015f, 2015g, 2018a).

Definition 2.7 (Alb Lupaş, 2010d). Let $\alpha, \lambda, l \geq 0$, $n \in \mathbb{N}$. Denote by $RI^{\alpha}_{n,\lambda,l}$ the operator given by $RI^{\alpha}_{n,\lambda,l} : \mathcal{A} \to \mathcal{A}$,

$$RI^{\alpha}_{n,\lambda,l} f(z) = (1-\alpha) R^n f(z) + \alpha I(n,\lambda,l) f(z), \quad \text{for } z \in U.$$

Remark 2.10 (Alb Lupaş, 2010d). If $f \in \mathcal{A}$, $f(z) = z + \sum_{j=2}^{\infty} a_j z^j$, then $RI^{\alpha}_{n,\lambda,l} f(z) = z + \sum_{j=2}^{\infty} \left\{ \alpha \left(\frac{1+\lambda(j-1)+l}{l+1} \right)^n + (1-\alpha) C^n_{n+j-1} \right\} a_j z^j$, for $z \in U$.

Remark 2.11 (Alb Lupaş, 2010d). For $\alpha = 0$, $RI^0_{n,\lambda,l} f(z) = R^n f(z)$, where $z \in U$ and for $\alpha = 1$, $RI^1_{n,\lambda,l} f(z) = I(n,\lambda,l) f(z)$, where $z \in U$, which was studied in the work of Alb Lupaş (2010a).

For $l = 0$, we obtain $RI^{\alpha}_{n,\lambda,0} f(z) = RD^n_{1,\alpha} f(z)$ which was studied in the work of Alb Lupaş (2011a) and for $l = 0$ and $\lambda = 1$, we obtain $RI^{\alpha}_{n,1,0} f(z) = L^n_{\alpha} f(z)$ which was studied in the work of Alb Lupaş (2009).

For $n = 0$, $RI^{\alpha}_{0,\lambda,l} f(z) = (1-\alpha) R^0 f(z) + \alpha I(0,\lambda,l) f(z) = f(z) = R^0 f(z) = I(0,\lambda,l) f(z)$, where $z \in U$.

Using Sălăgean differential operator and Ruscheweyh derivative, we have defined a new operator as a convolution product of these two operators and we have published 9 papers regarding this operator (e.g. Alb Lupaş, 2010e, 2011f, 2012a, 2012g, 2013d).

Definition 2.8 (Alb Lupaş, 2012a). Let $n \in \mathbb{N}$. Denote by SR^n the operator given by the Hadamard product of the Sălăgean operator S^n and the Ruscheweyh operator R^n, $SR^n : \mathcal{A} \to \mathcal{A}$,

$$SR^n f(z) = (S^n * R^n) f(z).$$

Remark 2.12 (Alb Lupaş, 2012a). If $f \in \mathcal{A}$, $f(z) = z + \sum_{j=2}^{\infty} a_j z^j$, then $SR^n f(z) = z + \sum_{j=2}^{\infty} C^n_{n+j-1} j^n a_j^2 z^j$.

Generalizing this operator, we introduced a new operator as a convolution product between the generalized Sălăgean operator and Ruscheweyh derivative and we have published 9 papers regarding this operator (e.g. Alb Lupaş, 2010b, 2011b, 2012h, 2013b, 2023a).

Definition 2.9 (Alb Lupaş, 2010b). Let $\lambda \geq 0$ and $n \in \mathbb{N}$. Denote by $DR_\lambda^n : \mathcal{A} \to \mathcal{A}$ the operator given by the Hadamard product (the convolution product) of the generalized Sălăgean operator D_λ^n and the Ruscheweyh operator R^n:

$$DR_\lambda^n f(z) = (D_\lambda^n * R^n) f(z),$$

for any $z \in U$ and each non-negative integer n.

Remark 2.13 (Alb Lupaş, 2010b). If $f \in \mathcal{A}$ and $f(z) = z + \sum_{j=2}^{\infty} a_j z^j$, then $DR_\lambda^n f(z) = z + \sum_{j=2}^{\infty} C_{n+j-1}^n [1 + (j-1)\lambda]^n a_j^2 z^j$, for $z \in U$.

Remark 2.14 (Alb Lupaş, 2010b). For $\lambda = 1$, we obtain the Hadamard product SR^n (Alb Lupaş, 2012a) of the Sălăgean operator S^n and Ruscheweyh operator R^n.

A new operator defined as a convolution product of Ruscheweyh derivative and multiplier transformation was introduced and studied by the author in 19 papers (e.g. Alb Lupaş, 2010c, 2011g, 2012i, 2013a, 2015h, 2015i, 2015j, 2016c, 2016d, 2017b, 2017c, 2017d, 2018b, 2022a, 2023b).

Definition 2.10 (Alb Lupaş, 2010c). Let $m, n, \lambda, l \in \mathbb{N}$. Denote by $IR_{\lambda,l}^{m,n}$ the operator given by the Hadamard product (the convolution product) of the operator $I(m, \lambda, l)$ and the Ruscheweyh operator R^n, $IR_{\lambda,l}^{m,n} : \mathcal{A} \to \mathcal{A}$,

$$IR_{\lambda,l}^{m,n} f(z) = (I(m, \lambda, l) * R^n) f(z).$$

Remark 2.15 (Alb Lupaş, 2010c). If $f \in \mathcal{A}$, $f(z) = z + \sum_{j=2}^{\infty} a_j z^j$, then

$$IR_{\lambda,l}^{m,n} f(z) = z + \sum_{j=2}^{\infty} \left(\frac{1 + \lambda (j-1) + l}{l+1} \right)^m \frac{(n+j-1)!}{n! \, (j-1)!} a_j^2 z^j, \text{ for } z \in U.$$

Remark 2.16 (Alb Lupaş, 2010c). For $l = 0$, $\lambda \geq 0$, we obtain the Hadamard product DR_λ^n (Alb Lupaş, 2010b) of the generalized Sălăgean operator D_λ^n and Ruscheweyh operator R^n.

For $l = 0$ and $\lambda = 1$, we obtain the Hadamard product SR^n (Alb Lupaş, 2012a) of the Sălăgean operator S^n and Ruscheweyh operator R^n.

Introducing q-calculus in the study of the Sălăgean differential operator, the q-analogue of Sălăgean differential operator was defined in the work of

Govindaraj and Sivasubramanian (2017) and studied by the author in two published papers (Alb Lupaş, 2022e, 2022f).

Definition 2.11 (Govindaraj and Sivasubramanian, 2017). Denote by $\mathcal{S}_q^m$ the q-analogue of Sălăgean differential operator

$$\mathcal{S}_q^m f(z) = z + \sum_{k=2}^{\infty} [k]_q^m a_k z^k,$$

when $f(z) = z + \sum_{k=2}^{\infty} a_k z^k \in \mathcal{A}$, $z \in U$.

We note that $\lim_{q\to 1} \mathcal{S}_q^m f(z) = \lim_{q\to 1} \left(z + \sum_{k=2}^{\infty} [k]_q^m a_k z^k \right) = z + \sum_{k=2}^{\infty} k^m a_k z^k = S^m f(z)$.

We can write $\mathcal{D}_q \left(\mathcal{S}_q^m f(z) \right) = 1 + \sum_{k=2}^{\infty} [k]_q^{m+1} a_k z^{k-1}$ and $z\mathcal{D}_q \left(\mathcal{S}_q^m f(z) \right) = z + \sum_{k=2}^{\infty} [k]_q^{m+1} a_k z^k$; therefore, the following identity holds for the operator $\mathcal{S}_q^m$:

$$z\mathcal{D}_q \left(\mathcal{S}_q^m f(z) \right) = z + \sum_{k=2}^{\infty} [k]_q^{m+1} a_k z^k = \mathcal{S}_q^{m+1} f(z).$$

The q-analogue of the multiplier transformation was defined in the work of Shah and Noor (2019) and studied by the author in three published papers (Alb Lupaş and Cătas, 2023; Alb Lupaş and Ghanim, 2023; Alb Lupaş *et al.*, 2023).

Definition 2.12 (Shah and Noor, 2019). Denote by $\mathcal{I}_q^{m,l}$ the q-analogue of multiplier transformation

$$\mathcal{I}_q^{m,l} f(z) := z + \sum_{k=2}^{\infty} \left(\frac{[k+l]_q}{[1+l]_q} \right)^m a_k z^k,$$

with $l > -1$, $q \in (0,1)$, m a real number, and $f(z) = z + \sum_{k=2}^{\infty} a_k z^k \in \mathcal{A}$, $z \in U$.

Using the properties of q-calculus, we obtain

$$z\mathcal{D}_q \left(\mathcal{I}_q^{m,l} f(z) \right) = \left(1 + \frac{[l]_q}{q^l} \right) \mathcal{I}_q^{m+1,l} f(z) - \frac{[l]_q}{q^l} \mathcal{I}_q^{m,l} f(z).$$

We extend the q-analogue of the multiplier transformation to the new class of analytic functions $\mathcal{A}_\zeta^*$.

Definition 2.13 (Alb Lupaş and Ghanim, 2023). The extended q-analogue of multiplier transformation has the following form:

$$\mathcal{I}_q^{m,l} f(z,\zeta) = z + \sum_{k=2}^{\infty} \left(\frac{[k+l]_q}{[1+l]_q} \right)^m a_k(\zeta) z^k, \quad z \in U,\ \zeta \in \overline{U},$$

with $l > -1$, $q \in (0,1)$, m a real number, and $f(z,\zeta) = z + \sum_{k=2}^{\infty} a_k(\zeta) z^k \in \mathcal{A}_{\zeta}^{*}$.

Using the properties of q-calculus, we obtain

$$z\mathcal{D}_q\left(\mathcal{I}_q^{m,l} f(z,\zeta)\right) = \left(1 + \frac{[l]_q}{q^l}\right)\mathcal{I}_q^{m+1,l} f(z,\zeta) - \frac{[l]_q}{q^l}\mathcal{I}_q^{m,l} f(z,\zeta).$$

2.3 Fractional Integral Operators

In this section, we describe seven fractional integral operators defined and studied by the author.

First, we remind the definition of Riemann–Liouville fractional integral.

Definition 2.14 (Cho and Aouf, 1996). For a function f, analytic in a simply connected region of the z-plane containing the origin, the fractional integral of order λ ($\lambda > 0$) is defined by $D_z^{-\lambda} f(z) = \frac{1}{\Gamma(\lambda)} \int_0^z \frac{f(t)}{(z-t)^{1-\lambda}} dt$.

Applying fractional integral to the previously defined operators, we have introduced and studied seven new fractional integral operators.

Definition 2.15 (Alb Lupaş and Oros, 2021a). Let $\alpha \geq 0$, $\lambda > 0$ and $n \in \mathbb{N}$. The fractional integral associated with the linear differential operator $L_\alpha^n f$ is defined by

$$D_z^{-\lambda} L_\alpha^n f(z) = \frac{1}{\Gamma(\lambda)} \int_0^z \frac{L_\alpha^n f(t)}{(z-t)^{1-\lambda}} dt = \frac{1}{\Gamma(\lambda)} \int_0^z \frac{t}{(z-t)^{1-\lambda}} dt$$
$$+ \sum_{j=2}^{\infty} \left(\alpha j^n + (1-\alpha)\frac{\Gamma(n+j)}{\Gamma(n+1)\Gamma(j)}\right) a_j \int_0^z \frac{t^j}{(z-t)^{1-\lambda}} dt,$$

which has the following form, after a simple calculation:

$$D_z^{-\lambda} L_\alpha^n f(z) = \frac{1}{\Gamma(\lambda+2)} z^{\lambda+1}$$
$$+ \sum_{j=2}^{\infty} \left[\frac{\alpha j^m \Gamma(j+1)}{\Gamma(j+\lambda+1)} + \frac{(1-\alpha) j \Gamma(m+j)}{\Gamma(m+1)\Gamma(j+\lambda+1)}\right] a_j z^{j+\lambda},$$

for the function $f(z) = z + \sum_{j=2}^{\infty} a_j z^j \in \mathcal{A}$. We note that $D_z^{-\lambda} L_\alpha^n f(z) \in \mathcal{H}[0, \lambda+1]$.

This new operator was introduced and studied by the author in two papers (Alb Lupaş and Oros, 2021a; Alb Lupaş, 2022c).

Applying fractional integral to the operator defined as convolution product of Sălăgean operator and Ruscheweyh derivative, we defined a new fractional differential operator, defined and studied by the author in three published papers (Alb Lupaş, 2019a, 2019b, 2022b).

Definition 2.16 (Alb Lupaş, 2022b). Let $\lambda > 0$ and $m, n \in \mathbb{N}$. The fractional integral associated with the linear differential operator $SR^{m,n}f$ is defined by

$$\begin{aligned} D_z^{-\lambda} SR^{m,n} f(z) &= \frac{1}{\Gamma(\lambda)} \int_0^z \frac{SR^{m,n} f(t)}{(z-t)^{1-\lambda}} dt \\ &= \frac{1}{\Gamma(\lambda)} \left(\int_0^z \frac{t}{(z-t)^{1-\lambda}} dt + \sum_{j=l+1}^{\infty} \left(j^m \frac{\Gamma(n+j)}{\Gamma(n+1)\Gamma(j)} \right) \right. \\ &\quad \left. \times\ a_j^2 \int_0^z \frac{t^j}{(z-t)^{1-\lambda}} dt \right), \end{aligned}$$

which has the following form, after a simple calculation:

$$D_z^{-\lambda} SR^{m,n} f(z) = \frac{1}{\Gamma(\lambda+2)} z^{\lambda+1} + \sum_{j=l+1}^{\infty} \frac{j^{m+1}\Gamma(n+j)}{\Gamma(n+1)\Gamma(j+\lambda+1)} a_j^2 z^{j+\lambda},$$

for a function $f(z) = z + \sum_{j=l+1}^{\infty} a_j z^j \in \mathcal{A}_l$. We note that $D_z^{-\lambda} SR^{m,n} f(z) \in \mathcal{A}(\lambda+1, l)$.

After a simple calculation, we obtain the relation

$$z\left(D_z^{-\lambda} SR^{m,n} f(z)\right)' = \lambda D_z^{-\lambda} SR^{m,n} f(z) + D_z^{-\lambda} SR^{m+1,n} f(z), \quad z \in U.$$

Another fractional differential operator defined and studied by the author in two published papers (Alb Lupaş, 2018c; Alb Lupaş and Cioban, 2018) is the following.

Definition 2.17 (Alb Lupaş, 2018c). Let $\lambda > 0$, $\alpha \geq 0$ and $m, n \in \mathbb{N}$. The fractional integral associated with the convolution product of generalized Sălăgean operator and Ruscheweyh derivative is defined by

$$
\begin{aligned}
D_z^{-\lambda} DR_\alpha^{m,n} f(z) &= \frac{1}{\Gamma(\lambda)} \int_0^z \frac{DR_\alpha^{m,n} f(t)}{(z-t)^{1-\lambda}} dt \\
&= \frac{1}{\Gamma(\lambda)} \int_0^z \frac{t}{(z-t)^{1-\lambda}} dt \\
&\quad + \sum_{j=2}^{\infty} [1+(j-1)\alpha]^m \frac{\Gamma(n+j)}{\Gamma(\lambda)\Gamma(n+1)\Gamma(j)} \\
&\quad \times a_j^2 \int_0^z \frac{t^j}{(z-t)^{1-\lambda}} dt,
\end{aligned}
$$

which has the following form, after a simple calculation:

$$
D_z^{-\lambda} DR_\alpha^{m,n} f(z) = \frac{1}{\Gamma(\lambda+2)} z^{\lambda+1} + \sum_{j=2}^{\infty} \frac{[1+(j-1)\alpha]^m j\Gamma(n+j)}{\Gamma(n+1)\Gamma(j+\lambda+1)} a_j^2 z^{j+\lambda},
$$

for the function $f(z) = z + \sum_{j=2}^{\infty} a_j z^j \in \mathcal{A}$. We note that $D_z^{-\lambda} DR_\alpha^{m,n} f(z) \in \mathcal{A}(\lambda+1, 1)$.

Applying fractional integral to the convolution product of multiplier transformation and Ruscheweyh derivative, the author defined and studied a new fractional differential operator in two published papers (Alb Lupaş and Acu, 2023; Alb Lupaş, 2022d).

Definition 2.18 (Alb Lupaş and Acu, 2023). The Riemann–Liouville fractional integral applied to the convolution product of multiplier transformation and Ruscheweyh derivative is defined by

$$
\begin{aligned}
D_z^{-\lambda} IR_{\alpha,l}^{m,n} f(z) &= \frac{1}{\Gamma(\lambda)} \int_0^z \frac{IR_{\alpha,l}^{m,n} f(t)}{(z-t)^{1-\lambda}} dt = \frac{1}{\Gamma(\lambda)} \int_0^z \frac{t}{(z-t)^{1-\lambda}} dt \\
&\quad + \frac{1}{\Gamma(\lambda)\Gamma(n+1)} \sum_{k=2}^{\infty} \frac{\Gamma(n+k)}{\Gamma(k)} \left(\frac{1+\alpha(k-1)+l}{l+1} \right)^m \\
&\quad \times a_k^2 \int_0^z \frac{t^k}{(z-t)^{1-\lambda}} dt,
\end{aligned}
$$

which can be written making a simple calculation as follows:

$$D_z^{-\lambda} IR_{\alpha,l}^{m,n} f(z) = \frac{1}{\Gamma(\lambda+2)} z^{\lambda+1} + \frac{1}{\Gamma(n+1)} \sum_{k=2}^{\infty} \frac{k\Gamma(n+k)}{\Gamma(k+\lambda+1)} \times \left(\frac{1+\alpha(k-1)+l}{l+1}\right)^m a_k^2 z^{k+\lambda},$$

considering the function $f(z) = z + \sum_{k=2}^{\infty} a_k z^k \in \mathcal{A}$. We observe that $D_z^{-\lambda} IR_{\alpha,l}^{m,n} f(z) \in \mathcal{H}(\lambda+1,1)$.

Another newly defined fractional operator that implies the confluent (Kummer) hypergeometric function of the first kind is given in the following:

Definition 2.19 (Miller and Mocanu, 2000, p. 5). Let a and c be complex numbers with $c \neq 0, -1, -2, \ldots$ and consider

$$\phi(a,c;z) = {}_1F_1(a,c;z) = 1 + \frac{a}{c}\frac{z}{1!} + \frac{a(a+1)}{c(c+1)}\frac{z^2}{2!} + \cdots, \quad z \in U.$$

The function $\phi(a,c;\cdot)$ is called confluent (Kummer) hypergeometric function, and it is analytic in $\mathbb{C}$ and satisfies Kummer's differential equation

$$zw''(z) + (c-z)w'(z) - aw(z) = 0.$$

If we let

$$(d)_k = \frac{\Gamma(d+k)}{\Gamma(d)} = d(d+1)(d+2)\cdots(d+k-1) \quad \text{and} \quad (d)_0 = 1,$$

then the confluent (Kummer) hypergeometric function can be written in the form

$$\phi(a,c;z) = \sum_{k=0}^{\infty} \frac{(a)_k}{(c)_k}\frac{z^k}{k!} = \frac{\Gamma(c)}{\Gamma(a)} \sum_{k=0}^{\infty} \frac{\Gamma(a+k)}{\Gamma(c+k)}\frac{z^k}{k!}.$$

The author introduced and studied fractional integral of confluent hypergeometric function in six published papers (Alb Lupaş, 2021a, 2021c, 2021b; Alb Lupaş and Oros, 2021b, 2022a, 2022b).

Definition 2.20 (Alb Lupaş and Oros, 2021b). Let a and c be complex numbers with $c \neq 0, -1, -2, \ldots$ and $\lambda > 0$. We define the fractional integral of confluent hypergeometric function

$$D_z^{-\lambda}\phi(a,c;z) = \frac{1}{\Gamma(\lambda)}\int_0^z \frac{\phi(a,c;t)}{(z-t)^{1-\lambda}}dt$$
$$= \frac{1}{\Gamma(\lambda)}\frac{\Gamma(c)}{\Gamma(a)}\sum_{k=0}^{\infty}\frac{\Gamma(a+k)}{\Gamma(c+k)\Gamma(k+1)}\int_0^z \frac{t^k}{(z-t)^{1-\lambda}}dt.$$

After a simple calculation, the fractional integral of confluent hypergeometric function has the following form:

$$D_z^{-\lambda}\phi(a,c;z) = \frac{\Gamma(c)}{\Gamma(a)}\sum_{k=0}^{\infty}\frac{\Gamma(a+k)}{\Gamma(c+k)\Gamma(\lambda+k+1)}z^{k+\lambda}.$$

We note that $D_z^{-\lambda}\phi(a,c;z) \in \mathcal{H}[0,\lambda]$.

After a simple calculation, the Riemann–Liouville fractional integral of q-confluent hypergeometric function has the following form:

$$D_z^{-\lambda}\phi(a,b;q,z) = \sum_{k=0}^{\infty}\frac{(a,q)_k}{(q,q)_k(b,q)_k(k+1)_\lambda}z^{\lambda+k}.$$

We note that $D_z^{-\lambda}\phi(a,b;q,z) \in \mathcal{H}[0,\lambda]$.

Extending this operator to the class of normalized analytic functions $\mathcal{A}_{n\zeta}^* = \{f \in \mathcal{H}(U\times\overline{U}),\ f(z,\zeta) = z + a_{n+1}(\zeta)z^{n+1} + \cdots,\ z \in U,\ \zeta \in \overline{U}\}$, with $\mathcal{A}_{1\zeta}^* = \mathcal{A}_\zeta^*$, we obtained a new operator studied in the work of Alb Lupaş (2023a, 2024):

Definition 2.21 (Alb Lupaş and Oros, 2023a). The Riemann–Liouville fractional integral of the extended q-confluent hypergeometric function is

$$D_z^{-\alpha}\phi(m(\zeta),n(\zeta);q,z,\zeta) = \frac{1}{\Gamma(\alpha)}\int_0^z \frac{\phi(m(\zeta),n(\zeta);q,t,\zeta)}{(z-t)^{1-\alpha}}dt$$
$$= \frac{1}{\Gamma(\alpha)}\sum_{j=0}^{\infty}\frac{(m(\zeta),q)_j}{(q,q)_j(n(\zeta),q)_j}\int_0^z \frac{t^j}{(z-t)^{1-\alpha}}dt,$$

where the q-hypergeometric function $\phi(m(\zeta),n(\zeta);q,z,\zeta)$ is defined by

$$\phi(m(\zeta),n(\zeta);q,z,\zeta) = \sum_{j=0}^{\infty}\frac{(m(\zeta),q)_j}{(q,q)_j(n(\zeta),q)_j}z^j,$$

with

$$(m(\zeta), q)_j = \begin{cases} 1, & j = 0, \\ (1 - m(\zeta))(1 - m(\zeta)q)(1 - m(\zeta)q^2) \cdots (1 - m(\zeta)q^{j-1}), \\ \quad j \in \mathbb{N}, \end{cases}$$

$$(d)_k = \frac{\Gamma(d+k)}{\Gamma(d)} = d(d+1)(d+2)\cdots(d+k-1) \text{ and } (d)_0 = 1,$$

and $m(\zeta), n(\zeta)$ are holomorphic functions depending on the parameter $\zeta \in \overline{U}$, $\alpha > 0$, $0 < q < 1$.

After some calculations, it can be written using the following form:

$$D_z^{-\alpha}\phi(m(\zeta), n(\zeta); q, z, \zeta) = \sum_{j=0}^{\infty} \frac{(m(\zeta), q)_j}{(q, q)_j (n(\zeta), q)_j (j+1)_\alpha} z^{\alpha+j}$$

and $D_z^{-\alpha}\phi(m(\zeta), n(\zeta); q, z, \zeta) \in \mathcal{H}[0, \alpha, \zeta]$.

Applying Riemann–Liouville fractional integral to q-analogue of multiplier transformation, we define in the work of Alhily and Alb Lupaş (2024) a new fractional differential operator:

Definition 2.22 (Alhily and Alb Lupaş, 2024). Let q, m, l, λ be real numbers, $q \in (0,1)$, $l > -1$ and $\lambda > 0$. The fractional integral applied to the q-analogue of multiplier transformation is defined by

$$D_z^{-\lambda}\mathcal{I}_q^{m,l} f(z) = \frac{1}{\Gamma(\lambda)} \int_0^z \frac{\mathcal{I}_q^{m,l} f(t)}{(z-t)^{1-\lambda}} dt$$

$$= \frac{1}{\Gamma(\lambda)} \int_0^z \frac{t}{(z-t)^{1-\lambda}} dt + \sum_{j=2}^{\infty} \left(\frac{[l+j]_q}{[l+1]_q}\right)^m a_j \int_0^z \frac{t^j}{(z-t)^{1-\lambda}} dt.$$

After a laborious calculation, we get that the fractional integral applied to the q-analogue of multiplier transformation takes the following form:

$$D_z^{-\lambda}\mathcal{I}_q^{m,l} f(z) = \frac{1}{\Gamma(\lambda+2)} z^{\lambda+1} + \sum_{j=2}^{\infty} \left(\frac{[l+j]_q}{[l+1]_q}\right)^m \frac{\Gamma(j+1)}{\Gamma(j+\lambda+1)} a_j z^{j+\lambda},$$

when $f(z) = z + \sum_{j=2}^{\infty} a_j z^j \in \mathcal{A}$. We note that $D_z^{-\lambda}\mathcal{I}_q^{m,l} f(z) \in \mathcal{H}[0, \lambda+1]$.

Chapter 3

Recent Developments in the Classical Differential Subordinations and Superordinations Theory

In this chapter, we expose some of the results obtained by the author regarding few operators presented in the previous chapter using the classical differential subordinations and superordinations theory.

In the first section, we define several subclasses of analytical functions using the classical differential subordinations theory and we describe some of the geometrical properties of the functions belonging to the defined subclasses regarding distortion bounds, extreme points, radii of close-to-convexity, starlikeness and convexity and neighborhood properties.

In the second section of this chapter, we present several classical differential subordination, superordination, respectively sandwich-type results involving few of the defined operators.

3.1 Applications of Differential Operators in the Study of Special Classes of Analytical Functions

3.1.1 *Aspects of Univalent Holomorphic Functions Involving Multiplier Transformation and Ruscheweyh Derivative*

Making use of multiplier transformation and Ruscheweyh derivative, we introduced in the paper Alb Lupaş (2014b) a new subclass of analytic functions $\mathcal{RI}(\gamma, \lambda, l, \alpha, \beta)$ defined on the open unit disc, and investigate its various characteristics. Further we obtained distortion bounds, extreme points and radii of close-to-convexity, starlikeness and convexity and neighborhood property for functions belonging to the subclass $\mathcal{RI}(\gamma, \lambda, l, \alpha, \beta)$.

Let $\mathcal{A}$ denote the class of functions of the form $f(z) = z + \sum_{j=2}^{\infty} a_j z^j$, which are analytic and univalent in the open unit disc $U = \{z : z \in \mathbb{C} : |z| < 1\}$. $\mathcal{T}$ is a subclass of $\mathcal{A}$ consisting of the functions of the form $f(z) = z - \sum_{j=2}^{\infty} |a_j| z^j$.

Let $\gamma, \lambda, l \geq 0$, $n \in \mathbb{N}$. The operator from Definition 2.7 is given by $RI_{n,\lambda,l}^{\gamma} : \mathcal{A} \to \mathcal{A}$,

$$RI_{n,\lambda,l}^{\gamma} f(z) = (1-\gamma) R^n f(z) + \gamma I(n,\lambda,l) f(z), \quad z \in U.$$

If $f \in \mathcal{A}$, $f(z) = z + \sum_{j=2}^{\infty} a_j z^j$, then

$$RI_{n,\lambda,l}^{\gamma} f(z) = z + \sum_{j=2}^{\infty} \left\{ \gamma \left(\frac{1+\lambda(j-1)+l}{l+1} \right)^n + (1-\gamma) \frac{(n+j-1)!}{n!\,(j-1)!} \right\} a_j z^j,$$

$z \in U$.

If $f \in \mathcal{T}$, $f(z) = z - \sum_{j=2}^{\infty} a_j z^j$, then

$$RI_{n,\lambda,l}^{\gamma} f(z) = z - \sum_{j=2}^{\infty} \left\{ \gamma \left(\frac{1+\lambda(j-1)+l}{l+1} \right)^n + (1-\gamma) \frac{(n+j-1)!}{n!\,(j-1)!} \right\} a_j z^j,$$

$z \in U$.

We define the subclass $\mathcal{RI}(\gamma, \lambda, l, \alpha, \beta)$ as follows.

Definition 3.1 (Alb Lupaş, 2014b). For $\gamma, \lambda, l \geq 0$, $0 \leq \alpha < 1$ and $0 < \beta \leq 1$, let $\mathcal{RI}(\gamma, \lambda, l, \alpha, \beta)$ be the subclass of $\mathcal{T}$ consisting of functions that satisfy the inequality

$$\left| \frac{RI_{n,\lambda,l}^{\mu,\gamma} f(z) - 1}{2\nu(RI_{n,\lambda,l}^{\mu,\gamma} f(z) - \alpha) - (RI_{n,\lambda,l}^{\mu,\gamma} f(z) - 1)} \right| < \beta \tag{3.1.1.1}$$

where

$$RI_{n,\lambda,l}^{\mu,\gamma} f(z) = (1-\mu) \frac{RI_{n,\lambda,l}^{\gamma} f(z)}{z} + \mu (RI_{n,\lambda,l}^{\gamma} f(z))', \tag{3.1.1.2}$$

$0 < \mu \leq 1$.

Remark 3.1 (Alb Lupaş, 2014b). If $f \in \mathcal{T}$, $f(z) = z - \sum_{j=2}^{\infty} a_j z^j$, then

$$RI_{n,\lambda,l}^{\mu,\gamma} f(z) = 1 - \sum_{j=t+1}^{\infty} [1+\mu(j-1)] \left\{ \gamma \left(\frac{1+\lambda(j-1)+l}{l+1} \right)^n + (1-\gamma) \frac{(n+j-1)!}{n!\,(j-1)!} \right\} a_j z^{j-1}, \quad z \in U.$$

Coefficient Bounds

We obtain coefficient bounds and extreme points for functions in $\mathcal{RI}(\gamma, \lambda, l, \alpha, \beta)$.

Theorem 3.1 (Alb Lupaş, 2014b). *Let the function $f \in \mathcal{T}$. Then $f \in \mathcal{RI}(\gamma, \lambda, l, \alpha, \beta)$ if and only if*

$$\sum_{j=t+1}^{\infty} (1+\mu(j-1))[1+\beta(2\nu-1)]$$

$$\times \left\{\gamma\left(\frac{1+\lambda(j-1)+l}{l+1}\right)^n + (1-\gamma)\frac{(n+j-1)!}{n!\,(j-1)!}\right\} a_j$$

$$< 2\beta\nu(1-\alpha). \tag{3.1.1.3}$$

The result is sharp for the function $F(z)$ defined by

$$F(z) = z - \frac{2\beta\nu(1-\alpha)}{(1+\mu(j-1))[1+\beta(2\nu-1)]\left\{\gamma\left(\frac{1+\lambda(j-1)+l}{l+1}\right)^n + (1-\gamma)\frac{(n+j-1)!}{n!(j-1)!}\right\}} z^j,$$

$j \geq t+1$.

Proof. Suppose f satisfies (3.1.1.3). Then for $|z| < 1$, we have

$$\left|RI_{n,\lambda,l}^{\mu,\gamma} f(z) - 1\right| - \beta\left|2\nu(RI_{n,\lambda,l}^{\mu,\gamma} f(z) - \alpha) - (RI_{n,\lambda,l}^{\mu,\gamma} f(z) - 1)\right|$$

$$= \left| - \sum_{j=t+1}^{\infty} (1+\mu(j-1))\left\{\gamma\left(\frac{1+\lambda(j-1)+l}{l+1}\right)^n \right.\right.$$

$$\left.\left. + (1-\gamma)\frac{(n+j-1)!}{n!\,(j-1)!}\right\} a_j z^{j-1}\right|$$

$$- \beta\left|2\nu(1-\alpha) - (2\nu-1)\sum_{j=t+1}^{\infty}\left\{\gamma\left(\frac{1+\lambda(j-1)+l}{l+1}\right)^n \right.\right.$$

$$\left.\left. + (1-\gamma)\frac{(n+j-1)!}{n!\,(j-1)!}\right\}[1+\mu(j-1)]\, a_j z^{j-1}\right|$$

$$\leq \sum_{j=t+1}^{\infty} [1+\mu(j-1)] \left\{ \gamma \left(\frac{1+\lambda(j-1)+l}{l+1} \right)^n \right.$$

$$\left. + (1-\gamma) \frac{(n+j-1)!}{n!\,(j-1)!} \right\} a_k - 2\beta\nu(1-\alpha)$$

$$+ \sum_{j=t+1}^{\infty} \beta(2\nu-1)(1+\mu(j-1)) \left\{ \gamma \left(\frac{1+\lambda(j-1)+l}{l+1} \right)^n \right.$$

$$\left. + (1-\gamma) \frac{(n+j-1)!}{n!\,(j-1)!} \right\} a_j$$

$$= \sum_{j=t+1}^{\infty} [1+\mu(j-1)]\,[1+\beta(2\nu-1)] \left\{ \gamma \left(\frac{1+\lambda(j-1)+l}{l+1} \right)^n \right.$$

$$\left. + (1-\gamma) \frac{(n+j-1)!}{n!\,(j-1)!} \right\} a_j - 2\beta\nu(1-\alpha) < 0.$$

Hence, by using the maximum modulus theorem and (3.1.1.1), $f \in \mathcal{RI}(\gamma, \lambda, l, \alpha, \beta)$.

Conversely, assume that

$$\left| \frac{RI_{n,\lambda,l}^{\mu,\gamma} f(z) - 1}{2\nu(RI_{n,\lambda,l}^{\mu,\gamma} f(z) - \alpha) - (RI_{n,\lambda,l}^{\mu,\gamma} f(z) - 1)} \right|$$

$$= \left| \frac{-\sum_{j=t+1}^{\infty}[1+\mu(j-1)]\left\{\gamma\left(\frac{1+\lambda(j-1)+l}{l+1}\right)^n+(1-\gamma)\frac{(n+j-1)!}{n!(j-1)!}\right\}a_j z^{j-1}}{2\nu(1-\alpha)-\sum_{j=t+1}^{\infty}[1+\mu(j-1)](2\nu-1)\left\{\gamma\left(\frac{1+\lambda(j-1)+l}{l+1}\right)^n+(1-\gamma)\frac{(n+j-1)!}{n!(j-1)!}\right\}a_j z^{j-1}} \right| < \beta,$$

$z \in U$.

Since $\mathrm{Re}(z) \leq |z|$ for all $z \in U$, we have

$$\mathrm{Re}\left\{ \frac{\sum_{j=t+1}^{\infty}[1+\mu(j-1)]\left\{\gamma\left(\frac{1+\lambda(j-1)+l}{l+1}\right)^n+(1-\gamma)\frac{(n+j-1)!}{n!(j-1)!}\right\}a_j z^{j-1}}{2\nu(1-\alpha)-\sum_{j=t+1}^{\infty}[1+\mu(j-1)](2\nu-1)\left\{\gamma\left(\frac{1+\lambda(j-1)+l}{l+1}\right)^n+(1-\gamma)\frac{(n+j-1)!}{n!(j-1)!}\right\}a_j z^{j-1}} \right\} < \beta. \tag{3.1.1.4}$$

By choosing choose values of z on the real axis so that $RI_{n,\lambda,l}^{\mu,\gamma} f(z)$ is real and letting $z \to 1$ through real values, we obtain the desired inequality (3.1.1.3). □

Corollary 3.1 (Alb Lupaş, 2014b). *If $f \in \mathcal{RI}(\gamma, \lambda, l, \alpha, \beta)$, then*

$$a_j \leq \frac{2\beta\nu(1-\alpha)}{[1+\mu(j-1)]\,[1+\beta(2\nu-1)]\left\{\gamma\left(\frac{1+\lambda(j-1)+l}{l+1}\right)^n + (1-\gamma)\frac{(n+j-1)!}{n!(j-1)!}\right\}}, \tag{3.1.1.5}$$

$j \geq t+1$, with equality only for $F(z)$.

Theorem 3.2 (Alb Lupaş, 2014b). *Let $f_1(z) = z$ and*

$$f_j(z) = z - \frac{2\beta\nu(1-\alpha)}{[1+\mu(j-1)]\,[1+\beta(2\nu-1)]\left\{\gamma\left(\frac{1+\lambda(j-1)+l}{l+1}\right)^n + (1-\gamma)\frac{(n+j-1)!}{n!(j-1)!}\right\}} z^j, \tag{3.1.1.6}$$

$j \geq t+1$, for $0 \leq \alpha < 1$, $0 < \beta \leq 1$, $\gamma, \lambda, l \geq 0$ and $0 < \nu \leq 1$.

Then $f(z)$ is in the subclass $\mathcal{RI}(\gamma, \lambda, l, \alpha, \beta)$ if and only if it can be expressed in the form

$$f(z) = \sum_{j=t}^{\infty} \omega_j f_j(z), \tag{3.1.1.7}$$

where $\omega_j \geq 0$ and $\sum_{j=1}^{\infty} \omega_j = 1$.

Proof. Suppose $f(z)$ can be written as in (3.1.1.7). Then

$$f(z) = z - \sum_{j=t+1}^{\infty} \omega_j \frac{2\beta\nu(1-\alpha)}{[1+\mu(j-1)][1+\beta(2\nu-1)]\left\{\gamma\left(\frac{1+\lambda(j-1)+l}{l+1}\right)^n + (1-\gamma)\frac{(n+j-1)!}{n!(j-1)!}\right\}} z^j.$$

Now,

$$\sum_{j=t+1}^{\infty} \frac{[1+\mu(j-1)]\,[1+\beta(2\nu-1)]\left\{\gamma\left(\frac{1+\lambda(j-1)+l}{l+1}\right)^n + (1-\gamma)\frac{(n+j-1)!}{n!(j-1)!}\right\}}{2\beta\nu(1-\alpha)} \omega_j$$

$$\times \frac{2\beta\nu(1-\alpha)}{[1+\mu(j-1)]\,[1+\beta(2\nu-1)]\left\{\gamma\left(\frac{1+\lambda(j-1)+l}{l+1}\right)^n + (1-\gamma)\frac{(n+j-1)!}{n!(j-1)!}\right\}}$$

$$= \sum_{j=t+1}^{\infty} \omega_j = 1 - \omega_1 \leq 1.$$

Thus, $f \in \mathcal{RI}(\gamma, \lambda, l, \alpha, \beta)$.

Conversely, let $f \in \mathcal{RI}(\gamma, \lambda, l, \alpha, \beta)$. Then by using (3.1.1.5), setting

$$\omega_j = \frac{[1+\mu(j-1)]\,[1+\beta(2\nu-1)]\left\{\gamma\left(\frac{1+\lambda(j-1)+l}{l+1}\right)^n + (1-\gamma)\frac{(n+j-1)!}{n!(j-1)!}\right\}}{2\beta\nu(1-\alpha)} a_j,$$

$j \geq t+1$, and $\omega_1 = 1 - \sum_{j=2}^{\infty} \omega_j$, we have $f(z) = \sum_{j=t}^{\infty} \omega_j f_j(z)$. This completes the proof of Theorem. □

Distortion Bounds

We obtain distortion bounds for the subclass $\mathcal{RI}(\gamma, \lambda, l, \alpha, \beta)$.

Theorem 3.3 (Alb Lupaş, 2014b). *If $f \in \mathcal{RI}(\gamma, \lambda, l, \alpha, \beta)$, then*

$$r - \frac{2\beta\nu(1-\alpha)}{(1+\mu t)[1+\beta(2\nu-1)]\left\{\gamma\left(\frac{1+\lambda t+l}{l+1}\right)^n + (1-\gamma)\frac{(n+t)!}{n!t!}\right\}} r^{t+1}$$

$$\leq |f(z)| \leq r + \frac{2\beta\nu(1-\alpha)}{(1+\mu t)[1+\beta(2\nu-1)]\left\{\gamma\left(\frac{1+\lambda t+l}{l+1}\right)^n + (1-\gamma)\frac{(n+t)!}{n!t!}\right\}} r^{t+1} \tag{3.1.1.8}$$

holds if the sequence $\{\sigma_j(\gamma, \lambda, l, \beta, \nu)\}_{j=t+1}^{\infty}$ is non-decreasing, and

$$1 - \frac{2\beta\nu(1-\alpha)(t+1)}{(1+\mu t)[1+\beta(2\nu-1)]\left\{\gamma\left(\frac{1+\lambda t+l}{l+1}\right)^n + (1-\gamma)\frac{(n+t)!}{n!t!}\right\}} r^{t}$$

$$\leq |f'(z)| \leq 1 + \frac{2\beta\nu(1-\alpha)(t+1)}{(1+\mu t)[1+\beta(2\nu-1)]\left\{\gamma\left(\frac{1+\lambda t+l}{l+1}\right)^n + (1-\gamma)\frac{(n+t)!}{n!t!}\right\}} r^{t} \tag{3.1.1.9}$$

holds if the sequence $\{\frac{\sigma_j(\gamma, \lambda, l, \beta, \nu)}{j}\}_{j=t+1}^{\infty}$ is non-decreasing, where

$$\sigma_j(\gamma, \beta, \nu) = [1+\mu(j-1)]\,[1+\beta(2\nu-1)]$$

$$\times \left\{\gamma\left(\frac{1+\lambda\,(j-1)+l}{l+1}\right)^n + (1-\gamma)\frac{(n+j-1)!}{n!\,(j-1)!}\right\}.$$

The bounds in (3.1.1.8) *and* (3.1.1.9) *are sharp, for* $f(z)$ *given by*

$$f(z) = z - \frac{2\beta\nu(1-\alpha)}{(1+\mu t)[1+\beta(2\nu-1)]\left\{\gamma\left(\frac{1+\lambda t+l}{l+1}\right)^n + (1-\gamma)\frac{(n+t)!}{n!t!}\right\}} z^{t+1}, \tag{3.1.1.10}$$

$z = \pm r$.

Proof. In view of Theorem 3.1, we have

$$\sum_{j=t+1}^{\infty} a_j \leq \frac{2\beta\nu(1-\alpha)}{(1+\mu t)[1+\beta(2\nu-1)]\left\{\gamma\left(\frac{1+\lambda t+l}{l+1}\right)^n + (1-\gamma)\frac{(n+t)!}{n!t!}\right\}}. \tag{3.1.1.11}$$

We obtain

$$|z| - |z|^{t+1}\sum_{j=t+1}^{\infty} a_j \leq |f(z)| \leq |z| + |z|^{t+1}\sum_{j=t+1}^{\infty} a_j.$$

Thus,

$$\begin{aligned} r &- \frac{2\beta\nu(1-\alpha)}{(1+\mu t)[1+\beta(2\nu-1)]\left\{\gamma\left(\frac{1+\lambda t+l}{l+1}\right)^n + (1-\gamma)\frac{(n+t)!}{n!t!}\right\}} r^{t+1} \\ &\leq |f(z)| \leq r + \frac{2\beta\nu(1-\alpha)}{(1+\mu t)[1+\beta(2\nu-1)]\left\{\gamma\left(\frac{1+\lambda t+l}{l+1}\right)^n + (1-\gamma)\frac{(n+t)!}{n!t!}\right\}} r^{t+1}. \end{aligned} \tag{3.1.1.12}$$

Hence, (3.1.1.8) follows from (3.1.1.12). Further,

$$\sum_{j=t+1}^{\infty} j a_j \leq \frac{2\beta\nu(1-\alpha)}{(1+\mu t)[1+\beta(2\nu-1)]\left\{\gamma\left(\frac{1+\lambda t+l}{l+1}\right)^n + (1-\gamma)\frac{(n+t)!}{n!t!}\right\}}.$$

Hence, (3.1.1.9) follows from

$$1 - r^t\sum_{j=t+1}^{\infty} j a_j \leq |f'(z)| \leq 1 + r^t\sum_{j=t+1}^{\infty} j a_j.$$

□

Radius of Starlikeness and Convexity

The radii of close-to-convexity, starlikeness and convexity for the subclass $\mathcal{RI}(\gamma,\lambda,l,\alpha,\beta)$ are given in the following.

Theorem 3.4 (Alb Lupaş, 2014b). *Let the function $f \in \mathcal{RI}(\gamma,\lambda,l,\alpha,\beta)$, Then $f(z)$ is close-to-convex of order δ, $0 \leq \delta < 1$, in the disc $|z| < r$, where*

$$r := \inf_{j \geq t+1} \left[\frac{(1-\delta)[1+\mu(j-1)][1+\beta(2\nu-1)]\left\{\gamma\left(\frac{1+\lambda(j-1)+l}{l+1}\right)^n + (1-\gamma)\frac{(n+j-1)!}{n!(j-1)!}\right\}}{2\beta\nu j(1-\alpha)} \right]^{\frac{1}{t}}. \tag{3.1.1.13}$$

The result is sharp, with extremal function $f(z)$ given by (3.1.1.5).

Proof. For given $f \in \mathcal{T}$, we must show that

$$|f'(z) - 1| < 1 - \delta. \tag{3.1.1.14}$$

By a simple calculation we have

$$|f'(z) - 1| \leq \sum_{j=t+1}^{\infty} j a_j |z|^t.$$

The last expression is less than $1 - \delta$ if

$$\sum_{j=t+1}^{\infty} \frac{j}{1-\delta} a_j |z|^t < 1.$$

We use the fact that $f \in \mathcal{RI}(\gamma,\lambda,l,\alpha,\beta)$ if and only if

$$\sum_{j=t+1}^{\infty} \frac{[1+\mu(j-1)][1+\beta(2\nu-1)]\left\{\gamma\left(\frac{1+\lambda(j-1)+l}{l+1}\right)^n + (1-\gamma)\frac{(n+j-1)!}{n!(j-1)!}\right\}}{2\beta\nu(1-\alpha)} a_j \leq 1.$$

Equation (3.1.1.14) holds true if

$$\frac{j}{1-\delta}|z|^t \leq \sum_{j=t+1}^{\infty} \frac{[1+\mu(j-1)][1+\beta(2\nu-1)]\left\{\gamma\left(\frac{1+\lambda(j-1)+l}{l+1}\right)^n + (1-\gamma)\frac{(n+j-1)!}{n!(j-1)!}\right\}}{2\beta\nu(1-\alpha)}.$$

Or, equivalently,

$$|z|^t \leq \sum_{j=t+1}^{\infty} \frac{(1-\delta)[1+\mu(j-1)][1+\beta(2\nu-1)]\left\{\gamma\left(\frac{1+\lambda(j-1)+l}{l+1}\right)^n+(1-\gamma)\frac{(n+j-1)!}{n!(j-1)!}\right\}}{2\beta\nu j(1-\alpha)},$$

which completes the proof. □

Theorem 3.5 (Alb Lupaş, 2014b). *Let $f \in \mathcal{RI}(\gamma, \lambda, l, \alpha, \beta)$. Then*

1. *f is starlike of order δ, $0 \leq \delta < 1$, in the disc $|z| < r_1$, where*

$$r_1 = \inf_{j \geq t+1} \left\{ \frac{(1-\delta)[1+\mu(j-1)][1+\beta(2\nu-1)]\left\{\gamma\left(\frac{1+\lambda(j-1)+l}{l+1}\right)^n+(1-\gamma)\frac{(n+j-1)!}{n!(j-1)!}\right\}}{2\beta\nu(1-\alpha)(j-\delta)} \right\}^{\frac{1}{t}}.$$

2. *f is convex of order δ, $0 \leq \delta < 1$, in the disc $|z| < r_2$ where,*

$$r_2 = \inf_{j \geq t+1} \left\{ \frac{(1-\delta)[1+\mu(j-1)][1+\beta(2\nu-1)]\left\{\gamma\left(\frac{1+\lambda(j-1)+l}{l+1}\right)^n+(1-\gamma)\frac{(n+j-1)!}{n!(j-1)!}\right\}}{2\beta\nu j(j-1)(1-\alpha)} \right\}^{\frac{1}{t}}.$$

Each of these results is sharp for the extremal function $f(z)$ given by (3.1.1.5).

Proof. 1. For $0 \leq \delta < 1$, we need to show that

$$\left|\frac{zf'(z)}{f(z)} - 1\right| < 1 - \delta. \tag{3.1.1.15}$$

We have

$$\left|\frac{zf'(z)}{f(z)} - 1\right| \leq \left|\frac{\sum_{j=t+1}^{\infty}(j-1)a_j|z|^t}{1-\sum_{j=t+1}^{\infty}a_j|z|^t}\right|.$$

The last expression is less than $1 - \delta$ if

$$\sum_{j=t+1}^{\infty} \frac{(j-\delta)}{1-\delta} a_j |z|^t < 1.$$

We use the fact that $f \in \mathcal{RI}(\gamma, \lambda, l, \alpha, \beta)$ if and only if

$$\sum_{j=t+1}^{\infty} \frac{[1+\mu(j-1)]\,[1+\beta(2\nu-1)]\left\{\gamma\left(\frac{1+\lambda(j-1)+l}{l+1}\right)^n + (1-\gamma)\frac{(n+j-1)!}{n!(j-1)!}\right\}}{2\beta\nu(1-\alpha)} a_j < 1.$$

Equation (3.1.1.15) holds true if

$$\frac{j-\delta}{1-\delta}|z|^t < \frac{[1+\mu(j-1)][1+\beta(2\nu-1)]\left\{\gamma\left(\frac{1+\lambda(j-1)+l}{l+1}\right)^n + (1-\gamma)\frac{(n+j-1)!}{n!(j-1)!}\right\}}{2\beta\nu(1-\alpha)}.$$

Or, equivalently,

$$|z|^t < \frac{(1-\delta)[1+\mu(j-1)][1+\beta(2\nu-1)]\left\{\gamma\left(\frac{1+\lambda(j-1)+l}{l+1}\right)^n + (1-\gamma)\frac{(n+j-1)!}{n!(j-1)!}\right\}}{2\beta\nu(1-\alpha)(j-\delta)},$$

which yields the starlikeness of the family.

2. Using the fact that f is convex if and only zf' is starlike, we can prove (2) with a similar way of the proof of (1). The function f is convex if and only if

$$|zf''(z)| < 1-\delta. \tag{3.1.1.16}$$

We have

$$|zf''(z)| \leq \left|\sum_{j=t+1}^{\infty} j(j-1)a_j |z|^{t-1}\right|$$

$$< 1-\delta \sum_{j=t+1}^{\infty} \frac{j(j-1)}{1-\delta} a_j |z|^{t-1} < 1.$$

We use the fact that $f \in \mathcal{RI}(\gamma, \lambda, l, \alpha, \beta)$ if and only if

$$\sum_{j=t+1}^{\infty} \frac{[1+\mu(j-1)][1+\beta(2\nu-1)]\left\{\gamma\left(\frac{1+\lambda(j-1)+l}{l+1}\right)^n + (1-\gamma)\frac{(n+j-1)!}{n!(j-1)!}\right\}}{2\beta\nu(1-\alpha)} a_j < 1.$$

Equation (3.1.1.16) holds true if

$$\frac{j(j-1)}{1-\delta}|z|^{t-1}$$
$$< \frac{[1+\mu(j-1)][1+\beta(2\nu-1)]\left\{\gamma\left(\frac{1+\lambda(j-1)+l}{l+1}\right)^n + (1-\gamma)\frac{(n+j-1)!}{n!(j-1)!}\right\}}{2\beta\nu(1-\alpha)},$$

or, equivalently,

$$|z|^{t-1} < \frac{(1-\delta)[1+\mu(j-1)][1+\beta(2\nu-1)]\left\{\gamma\left(\frac{1+\lambda(j-1)+l}{l+1}\right)^n + (1-\gamma)\frac{(n+j-1)!}{n!(j-1)!}\right\}}{2\beta\nu j(j-1)(1-\alpha)},$$

which yields the convexity of the family. □

Neighborhood Property

We study neighborhood property for functions in the subclass $\mathcal{RI}(\gamma, \lambda, l, \alpha, \beta)$.

Definition 3.2 (Alb Lupaş, 2014b). For functions f belonging to $\mathcal{A}$ and $\varepsilon \geq 0$, we define $(\eta - \varepsilon)$-neighborhood of f by

$$N_{\varepsilon}^{\eta}(f) = \left\{ g(z) \in \mathcal{A} : g(z) = z + \sum_{j=2}^{\infty} b_j z^j, \quad \sum_{j=2}^{\infty} j^{\eta+1} \left| a_j - b_j \right| \leq \varepsilon \right\},$$

where η is a fixed positive integer.

By using the following lemmas, we will investigate the $(\eta - \varepsilon)$-neighborhood of function in $\mathcal{RI}(\gamma, \lambda, l, \alpha, \beta)$.

Lemma 3.1 (Alb Lupaş, 2014b). *Let* $-1 \leq \beta < 1$, *if* $g(z) = z + \sum_{j=2}^{\infty} b_j z^j$ *satisfies*

$$\sum_{j=2}^{\infty} j^{\rho+1} \left| b_j \right| \leq \frac{2\beta\nu(1-\alpha)}{1+\beta(2\nu-1)},$$

then $g(z) \in \mathcal{RI}(\gamma, \lambda, l, \alpha, \beta)$.

Proof. By using of Theorem 3.1, it is sufficient to show that

$$\frac{[1+\mu(j-1)]\,[1+\beta(2\nu-1)] \left\{ \gamma \left(\frac{1+\lambda(j-1)+l}{l+1} \right)^{\rho} + (1-\gamma) \frac{(\rho+j-1)!}{\rho!(j-1)!} \right\}}{2\beta\nu(1-\alpha)}$$

$$= \frac{j^{\rho+1}}{2\beta\nu(1-\alpha)} [1+\beta(2\nu-1)].$$

But

$$\frac{[1+\beta(2\nu-1)] \left\{ \gamma \left(\frac{1+\lambda(j-1)+l}{l+1} \right)^{\rho} + (1-\gamma) \frac{(\rho+j-1)!}{\rho!(j-1)!} \right\}}{2\beta\nu(1-\alpha)}$$

$$\leq \frac{j^{\rho+1}}{2\beta\nu(1-\alpha)} [1+\beta(2\nu-1)].$$

Therefore, it is enough to prove that

$$Q(j,\rho) = \frac{\gamma \left(\frac{1+\lambda(j-1)+l}{l+1} \right)^{\rho} + (1-\gamma) \frac{(\rho+j-1)!}{\rho!(j-1)!}}{j^{\rho+1}} \leq 1,$$

the result follows because the last inequality holds for all $j \geq t+1$. □

Lemma 3.2 (Alb Lupaş, 2014b). *Let* $f(z) = z - \sum_{k=2}^{\infty} a_k z^k \in \mathcal{T}$, $\gamma, \lambda, l \geq 0$, $0 \leq \alpha < 1$, $0 < \beta \leq 1$ *and* $\varepsilon \geq 0$. *If* $\frac{f(z)+\epsilon z}{1+\epsilon} \in \mathcal{RI}(\gamma, \lambda, l, \alpha, \beta)$, *then*

$$\sum_{j=t+1}^{\infty} j^{\rho+1} a_j \leq \frac{2\beta\nu(1-\alpha)(1+\epsilon)(t+1)^{\rho+1}}{[1+\mu(j-1)]\,[1+\beta(2\nu-1)]\left\{\gamma\left(\frac{1+\lambda t+l}{l+1}\right)^n + (1-\gamma)\frac{(n+t)!}{n!t!}\right\}},$$

where either $\rho = 0$ *or* $\rho = 1$. *The result is sharp with the extremal function*

$$f(z) = z - \frac{2\beta\nu(1-\alpha)(1+\epsilon)}{[1+\mu(j-1)]\,[1+\beta(2\nu-1)]\left\{\gamma\left(\frac{1+\lambda t+l}{l+1}\right)^n + (1-\gamma)\frac{(n+t)!}{n!t!}\right\}} z^{t+1},$$

$z \in U$.

Proof. Letting $g(z) = \frac{f(z)+\epsilon z}{1+\epsilon}$, we have

$$g(z) = z - \sum_{j=t+1}^{\infty} \frac{a_j}{1+\epsilon} z^j, \quad z \in U.$$

In view of Theorem 3.2, $g(z) = \sum_{j=1}^{\infty} \eta_j g_j(z)$ where $\eta_j \geq 0$, $\sum_{j=1}^{\infty} \eta_j = 1$,

$$g_1(z) = z$$

and

$$g_j(z) = z - \frac{2\beta\nu(1-\alpha)(1+\epsilon)}{[1+\mu(j-1)]\,[1+\beta(2\nu-1)]\left\{\gamma\left(\frac{1+\lambda(j-1)+l}{l+1}\right)^n + (1-\gamma)\frac{(n+j-1)!}{n!(j-1)!}\right\}} z^j,$$

$j \geq t+1$.

So we obtain,

$$\begin{aligned} g(z) &= \eta_1 z + \sum_{j=t+1}^{\infty} \eta_j \\ &\quad \times \left[z - \frac{2\beta\nu(1-\alpha)(1+\epsilon)}{[1+\mu(j-1)]\,[1+\beta(2\nu-1)]\left\{\gamma\left(\frac{1+\lambda(j-1)+l}{l+1}\right)^n + (1-\gamma)\frac{(n+j-1)!}{n!(j-1)!}\right\}} z^j \right] \\ &= z - \sum_{j=t+1}^{\infty} \eta_j \frac{2\beta\nu(1-\alpha)(1+\epsilon)}{[1+\mu(j-1)][1+\beta(2\nu-1)]\left\{\gamma\left(\frac{1+\lambda(j-1)+l}{l+1}\right)^n + (1-\gamma)\frac{(n+j-1)!}{n!(j-1)!}\right\}} z^j. \end{aligned}$$

Since $\eta_j \geq 0$ and $\sum_{j=2}^{\infty} \eta_j \leq 1$, it follows that

$$\sum_{j=t+1}^{\infty} a_k \leq \sup_{j \geq t+1} j^{\rho+1}$$

$$\times \frac{2\beta\nu(1-\alpha)(1+\epsilon)}{[1+\mu(j-1)]\,[1+\beta(2\nu-1)]\left\{\gamma\left(\frac{1+\lambda(j-1)+l}{l+1}\right)^n + (1-\gamma)\frac{(n+j-1)!}{n!(j-1)!}\right\}}.$$

Since whenever $\rho = 0$ or $\rho = 1$, we conclude that

$$W(j,\rho,\gamma,\alpha,\beta,\epsilon)$$

$$= j^{\rho+1}\frac{2\beta\nu(1-\alpha)(1+\epsilon)}{[1+\mu(j-1)]\,[1+\beta(2\nu-1)]\left\{\gamma\left(\frac{1+\lambda(j-1)+l}{l+1}\right)^n + (1-\gamma)\frac{(n+j-1)!}{n!(j-1)!}\right\}},$$

is a decreasing function of j, the result will follow. The proof is complete. □

Theorem 3.6 (Alb Lupaş, 2014b). *Let* $\rho = 0$ *or* $\rho = 1$ *and suppose* $0 \leq \beta < 1$ *and*

$$-1 \leq \theta < \frac{[1+\mu(t-1)][1+\beta(2\nu-1)]\left\{\gamma\left(\frac{1+\lambda t+l}{l+1}\right)^n+(1-\gamma)\frac{(n+t)!}{n!t!}\right\}-2\beta\nu(1-\alpha)(1+\epsilon)(t+1)^{\eta+1}}{[1+\mu(t-1)][1+\beta(2\nu-1)]\left\{\gamma\left(\frac{1+\lambda t+l}{l+1}\right)^n+(1-\gamma)\frac{(n+t)!}{n!t!}\right\}},$$

$f(z) \in \mathcal{T}$ *and* $\frac{f(z)+\epsilon z}{1+\epsilon} \in \mathcal{RI}(\gamma,\lambda,l,\alpha,\beta)$. *Then the* $(\eta-\varepsilon)$*-neighborhood of* f *is the subset of* $\mathcal{RI}(\lambda,\lambda,l,\alpha,\beta)$, *where*

$$\varepsilon \leq \frac{\begin{array}{c}2(1-\alpha)\left\{\theta\gamma\,[1+\mu(t-1)]\,[1+\beta(2\nu-1)]\left\{\gamma\left(\frac{1+\lambda t+l}{l+1}\right)^n+(1-\gamma)\frac{(n+t)!}{n!t!}\right\}\right.\\ \left.-\,\beta\gamma\,[1+\theta(2\nu-1)]\,(1+\epsilon)(t+1)^{\eta+1}\right\}\end{array}}{[1+\theta(2\nu-1)][1+\mu(t-1)][1+\beta(2\nu-1)]\left\{\gamma\left(\frac{1+\lambda t+l}{l+1}\right)^n+(1-\gamma)\frac{(n+t)!}{n!t!}\right\}}.$$

The result is sharp.

Proof. For $f(z) = z - \sum_{j=2}^{\infty}|a_j|\,z^j$, let $g(z) = z + \sum_{j=2}^{\infty} b_j z^j$ be in $N_{\varepsilon}^{\eta}(f)$. So by Lemma 3.2, we have

$$\sum_{j=2}^{\infty} j^{\eta+1}|b_j| = \sum_{j=2}^{\infty} j^{\eta+1}|a_j - b_j - a_j|$$

$$\leq \varepsilon + \frac{2\beta\nu(1-\alpha)(1+\epsilon)(t+1)^{\eta+1}}{[1+\mu(t-1)]\,[1+\beta(2\nu-1)]\left\{\gamma\left(\frac{1+\lambda t+l}{l+1}\right)^n + (1-\gamma)\frac{(n+t)!}{n!t!}\right\}}.$$

By using Lemma 3.1, $g(z) \in \mathcal{L}(\gamma, \alpha, \beta)$ if

$$\varepsilon + \frac{2\beta\nu(1-\alpha)(1+\epsilon)(t+1)^{\eta+1}}{[1+\mu(t-1)]\,[1+\beta(2\nu-1)]\left\{\gamma\left(\frac{1+\lambda t+l}{l+1}\right)^n + (1-\gamma)\frac{(n+t)!}{n!t!}\right\}} \leq \frac{2\theta\nu(1-\alpha)}{1+\theta(2\nu-1)},$$

that is,

$$\varepsilon \leq \frac{2(1-\alpha)\left\{\theta\gamma\,[1+\mu(t-1)]\,[1+\beta(2\nu-1)]\left\{\gamma\left(\frac{1+\lambda t+l}{l+1}\right)^n + (1-\gamma)\frac{(n+t)!}{n!t!}\right\} - \beta\gamma\,[1+\theta(2\nu-1)]\,(1+\epsilon)(t+1)^{\eta+1}\right\}}{[1+\theta(2\nu-1)][1+\mu(t-1)][1+\beta(2\nu-1)]\left\{\gamma\left(\frac{1+\lambda t+l}{l+1}\right)^n+(1-\gamma)\frac{(n+t)!}{n!t!}\right\}}$$

and the proof is complete. □

3.1.2 *Properties of a Subclass of Univalent Functions Defined by Using Multiplier Transformation and Ruscheweyh Derivative*

In the paper Alb Lupaş (2015d), we have introduced and studied the subclass $\mathcal{RI}(d, \alpha, \beta)$ of univalent functions defined by the linear operator $RI^{\gamma}_{n,\lambda,l}f(z)$ from Definition 2.7 defined by using the Ruscheweyh derivative $R^n f(z)$ and multiplier transformation $I\,(n, \lambda, l)\, f(z)$, as $RI^{\gamma}_{n,\lambda,l} : \mathcal{A} \to \mathcal{A}$,

$$RI^{\gamma}_{n,\lambda,l}f(z) = (1-\gamma)R^n f(z) + \gamma I\,(n, \lambda, l)\, f(z), z \in U,$$

where $\mathcal{A}_n = \{f \in \mathcal{H}(U) : f(z) = z + a_{n+1}z^{n+1} + \cdots,\ z \in U\}$ is the class of normalized analytic functions with $\mathcal{A}_1 = \mathcal{A}$.

The main object is to investigate several properties such as coefficient estimates, distortion theorems, closure theorems, neighborhoods and the radii of starlikeness, convexity and close-to-convexity of functions belonging to the subclass $\mathcal{RI}(d, \alpha, \beta)$.

Definition 3.3 (Alb Lupaş, 2015d). Let the function $f \in \mathcal{A}$. Then $f(z)$ is said to be in the subclass $\mathcal{RI}(d, \alpha, \beta)$ if it satisfies the following criterion:

$$\left|\frac{1}{d}\left(\frac{z(RI^{\gamma}_{n,\lambda,l}f(z))' + \alpha z^2 RI^{\gamma}_{n,\lambda,l}f(z))''}{(1-\alpha)RI^{\gamma}_{n,\lambda,l}f(z) + \alpha z(RI^{\gamma}_{n,\lambda,l}f(z))'} - 1\right)\right| < \beta, \qquad (3.1.2.1)$$

where $d \in \mathbb{C} - \{0\}$, $0 \leq \alpha \leq 1$, $0 < \beta \leq 1$, $z \in U$.

Coefficient Inequality

Theorem 3.7 (Alb Lupaş, 2015d). *Let the function $f \in \mathcal{A}$. Then $f(z)$ is in the subclass $\mathcal{RI}(d,\alpha,\beta)$ if and only if*

$$\sum_{j=2}^{\infty}(1+\alpha(j-1))(j-1+\beta|d|)\left\{\gamma\left(\frac{1+\lambda(j-1)+l}{l+1}\right)^{n}\right.$$
$$\left.+(1-\gamma)\frac{(n+j-1)!}{n!\,(j-1)!}\right\}a_j$$
$$\leq \beta|d|, \tag{3.1.2.2}$$

where $d \in \mathbb{C}-\{0\}$, $0 \leq \alpha \leq 1$, $0 < \beta \leq 1$, $z \in U$.

Proof. Let $f(z) \in \mathcal{RI}(d,\alpha,\beta)$. Assume that inequality (3.1.2.2) holds true. Then we find that

$$\left|\frac{z(RI^{\gamma}_{n,\lambda,l}f(z))' + \alpha z^2 (RI^{\gamma}_{n,\lambda,l}f(z))''}{(1-\alpha)RI^{\gamma}_{n,\lambda,l}f(z) + \alpha z (RI^{\gamma}_{n,\lambda,l}f(z))'} - 1\right|$$
$$= \left|\frac{\sum_{j=2}^{\infty}(1+\alpha(j-1))(j-1)\left\{\gamma\left(\frac{1+\lambda(j-1)+l}{l+1}\right)^{n} + (1-\gamma)\frac{(n+j-1)!}{n!(j-1)!}\right\}a_j z^j}{z+\sum_{j=2}^{\infty}(1+\alpha(j-1))\left\{\gamma\left(\frac{1+\lambda(j-1)+l}{l+1}\right)^{n} + (1-\gamma)\frac{(n+j-1)!}{n!(j-1)!}\right\}a_j z^j}\right|$$
$$\leq \frac{\sum_{j=2}^{\infty}(1+\alpha(j-1))(j-1)\left\{\gamma\left(\frac{1+\lambda(j-1)+l}{l+1}\right)^{n} + (1-\gamma)\frac{(n+j-1)!}{n!(j-1)!}\right\}a_j |z|^{j-1}}{1-\sum_{j=2}^{\infty}(1+\alpha(j-1))\left\{\gamma\left(\frac{1+\lambda(j-1)+l}{l+1}\right)^{n} + (1-\gamma)\frac{(n+j-1)!}{n!(j-1)!}\right\}a_j |z|^{j-1}}$$
$$< \beta|d|.$$

Choosing values of z on real axis and letting $z \to 1^-$, we have

$$\sum_{j=2}^{\infty}(1+\alpha(j-1))(j-1+\beta|d|)$$
$$\times\left\{\gamma\left(\frac{1+\lambda(j-1)+l}{l+1}\right)^{n} + (1-\gamma)\frac{(n+j-1)!}{n!\,(j-1)!}\right\}a_j \leq \beta|d|.$$

Conversely, assume that $f(z) \in \mathcal{RI}(d,\alpha,\beta)$, then we get the following inequality

$$\mathrm{Re}\left\{\frac{z(RI^{\gamma}_{n,\lambda,l}f(z))' + \alpha z^2(RI^{\gamma}_{n,\lambda,l}f(z))''}{(1-\alpha)RI^{\gamma}_{n,\lambda,l}f(z) + \alpha z(RI^{\gamma}_{n,\lambda,l}f(z))'} - 1\right\} > -\beta|d|$$

$$\mathrm{Re}\left\{\frac{z+\sum_{j=2}^{\infty} j(1+\alpha(j-1))\{\gamma(\frac{1+\lambda(j-1)+l}{l+1})^n+(1-\gamma)\frac{(n+j-1)!}{n!(j-1)!}\}a_j z^j}{z+\sum_{j=2}^{\infty}(1+\alpha(j-1))\{\gamma(\frac{1+\lambda(j-1)+l}{l+1})^n+(1-\gamma)\frac{(n+j-1)!}{n!(j-1)!}\}a_j z^j} - 1 + \beta|d|\right\} > 0$$

$$\mathrm{Re}\frac{\beta|d|z+\sum_{j=2}^{\infty}(1+\alpha(j-1))(j-1+\beta|d|)\{\gamma(\frac{1+\lambda(j-1)+l}{l+1})^n+(1-\gamma)\frac{(n+j-1)!}{n!(j-1)!}\}a_j z^j}{z+\sum_{j=2}^{\infty}(1+\alpha(j-1))\{\gamma(\frac{1+\lambda(j-1)+l}{l+1})^n+(1-\gamma)\frac{(n+j-1)!}{n!(j-1)!}\}a_j z^j} > 0.$$

Since $\mathrm{Re}(-e^{i\theta}) \geq -|e^{i\theta}| = -1$, the above inequality reduces to

$$\frac{\beta|d|r-\sum_{j=2}^{\infty}(1+\alpha(j-1))(j-1+\beta|d|)\{\gamma(\frac{1+\lambda(j-1)+l}{l+1})^n+(1-\gamma)\frac{(n+j-1)!}{n!(j-1)!}\}a_j r^j}{r-\sum_{j=2}^{\infty}(1+\alpha(j-1))\{\gamma(\frac{1+\lambda(j-1)+l}{l+1})^n+(1-\gamma)\frac{(n+j-1)!}{n!(j-1)!}\}a_j r^j} > 0.$$

Letting $r \to 1^-$ and by the mean value theorem, we have desired inequality (3.1.2.2).

This completes the proof of theorem. □

Corollary 3.2 (Alb Lupaş, 2015d). *Let the function $f \in \mathcal{A}$ be in the subclass $\mathcal{RI}(d,\alpha,\beta)$. Then*

$$a_j \leq \frac{\beta|d|}{(1+\alpha(j-1))(j-1+\beta|d|)\left\{\gamma\left(\frac{1+\lambda(j-1)+l}{l+1}\right)^n + (1-\gamma)\frac{(n+j-1)!}{n!(j-1)!}\right\}},$$

$j \geq 2.$

Distortion Theorems

Theorem 3.8 (Alb Lupaş, 2015d). *Let the function $f \in \mathcal{A}$ be in the subclass $\mathcal{RI}(d,\alpha,\beta)$. Then for $|z| = r < 1$, we have*

$$r - \frac{\beta|d|}{(1+\alpha)(1+\beta|d|)\left[\gamma\left(\frac{1+\lambda+l}{l+1}\right)^n + (1-\gamma)(n+1)\right]}r^2 \leq |f(z)|$$

$$\leq r + \frac{\beta|d|}{(1+\alpha)(1+\beta|d|)\left[\gamma\left(\frac{1+\lambda+l}{l+1}\right)^n + (1-\gamma)(n+1)\right]}r^2.$$

The result is sharp for the function $f(z)$ given by

$$f(z) = z + \frac{\beta|d|}{(1+\alpha)(1+\beta|d|)\left[\gamma\left(\frac{1+\lambda+l}{l+1}\right)^n + (1-\gamma)(n+1)\right]} z^2.$$

Proof. Given that $f(z) \in \mathcal{RI}(d, \alpha, \beta)$, from the equation (3.1.2.2) and since

$$(1+\alpha)(1+\beta|d|)\left[\gamma\left(\frac{1+\lambda+l}{l+1}\right)^n + (1-\gamma)(n+1)\right]$$

is non decreasing and positive for $j \geq 2$, then we have

$$(1+\alpha)(1+\beta|d|)\left[\gamma\left(\frac{1+\lambda+l}{l+1}\right)^n + (1-\gamma)(n+1)\right]\sum_{j=2}^{\infty} a_j$$

$$\leq \sum_{j=2}^{\infty}(1+\alpha(j-1))(j-1+\beta|d|)$$

$$\times\left\{\gamma\left(\frac{1+\lambda(j-1)+l}{l+1}\right)^n + (1-\gamma)\frac{(n+j-1)!}{n!\,(j-1)!}\right\} a_j \leq \beta|d|,$$

which is equivalent to,

$$\sum_{j=2}^{\infty} a_j \leq \frac{\beta|d|}{(1+\alpha)(1+\beta|d|)\left[\gamma\left(\frac{1+\lambda+l}{l+1}\right)^n + (1-\gamma)(n+1)\right]}. \qquad (3.1.2.3)$$

Using (3.1.2.3), we obtain for

$$f(z) = z + \sum_{j=2}^{\infty} a_j z^j$$

that

$$|f(z)| \leq |z| + \sum_{j=2}^{\infty} a_j |z|^j \leq r + \sum_{j=2}^{\infty} a_j r^j \leq r + r^2 \sum_{j=2}^{\infty} a_j$$

$$\leq r + \frac{\beta|d|}{(1+\alpha)(1+\beta|d|)\left[\gamma\left(\frac{1+\lambda+l}{l+1}\right)^n + (1-\gamma)(n+1)\right]} r^2.$$

Similarly,

$$|f(z)| \geq r^2 - \frac{\beta|d|}{(1+\alpha)(1+\beta|d|)\left[\gamma\left(\frac{1+\lambda+l}{l+1}\right)^n + (1-\gamma)(n+1)\right]} r^2.$$

This completes the proof of theorem. □

Theorem 3.9 (Alb Lupaş, 2015d). *Let the function $f \in \mathcal{A}$ be in the subclass $\mathcal{RI}(d, \alpha, \beta)$. Then for $|z| = r < 1$, we have*

$$-\frac{2\beta|d|}{(1+\alpha)(1+\beta|d|)\left[\gamma\left(\frac{1+\lambda+l}{l+1}\right)^n + (1-\gamma)(n+1)\right]}r \leq |f'(z)|$$

$$\leq \frac{2\beta|d|}{(1+\alpha)(1+\beta|d|)\left[\gamma\left(\frac{1+\lambda+l}{l+1}\right)^n + (1-\gamma)(n+1)\right]}r.$$

The result is sharp for the function $f(z)$ given by

$$f(z) = z + \frac{\beta|d|}{(1+\alpha)(1+\beta|d|)\left[\gamma\left(\frac{1+\lambda+l}{l+1}\right)^n + (1-\gamma)(n+1)\right]}z^2.$$

Proof. From (3.1.2.3)

$$f'(z) = 1 + \sum_{j=2}^{\infty} ja_j z^{j-1}$$

and

$$|f'(z)| \leq 1 - \sum_{j=2}^{\infty} ja_j|z|^{j-1} \leq 1 + \sum_{j=2}^{\infty} ja_j r^{j-1}$$

$$\leq 1 + \frac{2\beta|d|}{(1+\alpha)(1+\beta|d|)\left[\gamma\left(\frac{1+\lambda+l}{l+1}\right)^n + (1-\gamma)(n+1)\right]}r.$$

Similarly,

$$|f'(z)| \geq 1 - \frac{2\beta|d|}{(1+\alpha)(1+\beta|d|)\left[\gamma\left(\frac{1+\lambda+l}{l+1}\right)^n + (1-\gamma)(n+1)\right]}r.$$

This completes the proof of theorem. □

Closure Theorems

Theorem 3.10 (Alb Lupaş, 2015d). *Let the functions f_k, $k = 1, 2, \ldots, m$, defined by*

$$f_k(z) = z + \sum_{j=2}^{\infty} a_{j,k} z^j, \quad a_{j,k} \geq 0, \tag{3.1.2.4}$$

be in the subclass $\mathcal{RI}(d,\alpha,\beta)$. Then the function $h(z)$ defined by

$$h(z)=\sum_{k=1}^{m}\mu_k f_k(z),\quad \mu_k\geq 0,$$

is also in the subclass $\mathcal{RI}(d,\alpha,\beta)$, where

$$\sum_{k=1}^{m}\mu_k=1.$$

Proof. We can write

$$h(z)=\sum_{k=1}^{m}\mu_m z+\sum_{k=1}^{m}\sum_{j=2}^{\infty}\mu_k a_{j,k}z^j=z+\sum_{j=2}^{\infty}\sum_{k=1}^{m}\mu_k a_{j,k}z^j.$$

Furthermore, since the functions $f_k(z)$, $k=1,2,\ldots,m$, are in the subclass $\mathcal{RI}(d,\alpha,\beta)$, then from Theorem 3.7, we have

$$\sum_{j=2}^{\infty}(1+\alpha(j-1))(j-1+\beta|d|)\\ \times\left\{\gamma\left(\frac{1+\lambda(j-1)+l}{l+1}\right)^n+(1-\gamma)\frac{(n+j-1)!}{n!\,(j-1)!}\right\}a_{j,k}\leq\beta|d|.$$

Thus, it is enough to prove that

$$\begin{aligned}&\sum_{j=2}^{\infty}(1+\alpha(j-1))(j-1+\beta|d|)\\ &\quad\times\left\{\gamma\left(\frac{1+\lambda(j-1)+l}{l+1}\right)^n+(1-\gamma)\frac{(n+j-1)!}{n!\,(j-1)!}\right\}\left(\sum_{k=1}^{m}\mu_k a_{j,k}\right)\\ &=\sum_{k=1}^{m}\mu_k\sum_{j=2}^{\infty}(1+\alpha(j-1))(j-1+\beta|d|)\\ &\quad\times\left\{\gamma\left(\frac{1+\lambda(j-1)+l}{l+1}\right)^n+(1-\gamma)\frac{(n+j-1)!}{n!\,(j-1)!}\right\}a_{j,k}\\ &\leq\sum_{k=1}^{m}\mu_k\beta|d|=\beta|d|.\end{aligned}$$

Hence, the proof is complete. □

Corollary 3.3 (Alb Lupaş, 2015d). *Let the functions f_k, $k = 1, 2$, defined by* (3.1.2.4) *be in the subclass $\mathcal{RI}(d, \alpha, \beta)$. Then the function $h(z)$ defined by*

$$h(z) = (1 - \zeta) f_1(z) + \zeta f_2(z), \quad 0 \leq \zeta \leq 1,$$

is also in the subclass $\mathcal{RI}(d, \alpha, \beta)$.

Theorem 3.11 (Alb Lupaş, 2015d). *Let*

$$f_1(z) = z,$$

and

$$f_j(z) = z + \frac{\beta|d|}{(1 + \alpha(j-1))(j - 1 + \beta|d|)\left\{\gamma\left(\frac{1+\lambda(j-1)+l}{l+1}\right)^n + (1-\gamma)\frac{(n+j-1)!}{n!(j-1)!}\right\}} z^j,$$

$j \geq 2$.

Then the function $f(z)$ is in the subclass $\mathcal{RI}(d, \alpha, \beta)$ if and only if it can be expressed in the form:

$$f(z) = \mu_1 f_1(z) + \sum_{j=2}^{\infty} \mu_j f_j(z),$$

where $\mu_1 \geq 0$, $\mu_j \geq 0$, $j \geq 2$ and $\mu_1 + \sum_{j=2}^{\infty} \mu_j = 1$.

Proof. Assume that $f(z)$ can be expressed in the form

$$f(z) = \mu_1 f_1(z) + \sum_{j=2}^{\infty} \mu_j f_j(z) = z$$

$$+ \sum_{j=2}^{\infty} \frac{\beta|d|}{(1 + \alpha(j-1))(j - 1 + \beta|d|)\left\{\gamma\left(\frac{1+\lambda(j-1)+l}{l+1}\right)^n + (1-\gamma)\frac{(n+j-1)!}{n!(j-1)!}\right\}} \mu_j z^j.$$

Thus,

$$\sum_{j=2}^{\infty} \frac{(1 + \alpha(j-1))(j - 1 + \beta|d|)\left\{\gamma\left(\frac{1+\lambda(j-1)+l}{l+1}\right)^n + (1-\gamma)\frac{(n+j-1)!}{n!(j-1)!}\right\}}{\beta|d|}$$

$$\cdot \frac{\beta|d|}{(1 + \alpha(j-1))(j - 1 + \beta|d|)\left\{\gamma\left(\frac{1+\lambda(j-1)+l}{l+1}\right)^n + (1-\gamma)\frac{(n+j-1)!}{n!(j-1)!}\right\}} \mu_j$$

$$= \sum_{j=2}^{\infty} \mu_j = 1 - \mu_1 \leq 1.$$

Hence, $f(z) \in \mathcal{RI}(d, \alpha, \beta)$.

Conversely, assume that $f(z) \in \mathcal{RI}(d, \alpha, \beta)$.

Setting

$$\mu_j = \frac{(1+\alpha(j-1))(j-1+\beta|d|)\left\{\gamma\left(\frac{1+\lambda(j-1)+l}{l+1}\right)^n + (1-\gamma)\frac{(n+j-1)!}{n!(j-1)!}\right\}}{\beta|d|} a_j,$$

since

$$\mu_1 = 1 - \sum_{j=2}^{\infty} \mu_j.$$

Thus,

$$f(z) = \mu_1 f_1(z) + \sum_{j=2}^{\infty} \mu_j f_j(z).$$

Hence, the proof is complete. □

Corollary 3.4 (Alb Lupaş, 2015d). *The extreme points of the subclass $\mathcal{RI}(d, \alpha, \beta)$ are the functions*

$$f_1(z) = z,$$

and

$$f_j(z) = z + \frac{\beta|d|}{(1+\alpha(j-1))(j-1+\beta|d|)\left\{\gamma\left(\frac{1+\lambda(j-1)+l}{l+1}\right)^n + (1-\gamma)\frac{(n+j-1)!}{n!(j-1)!}\right\}} z^j,$$

$j \geq 2$.

Inclusion and Neighborhood Results

We define the δ-neighborhood of a function $f(z) \in \mathcal{A}$ by

$$N_\delta(f) = \left\{ g \in \mathcal{A} : g(z) = z + \sum_{j=2}^{\infty} b_j z^j \text{ and } \sum_{j=2}^{\infty} j|a_j - b_j| \leq \delta \right\}. \quad (3.1.2.5)$$

In particular, for $e(z) = z$

$$N_\delta(e) = \left\{ g \in \mathcal{A} : g(z) = z + \sum_{j=2}^{\infty} b_j z^j \text{ and } \sum_{j=2}^{\infty} j|b_j| \leq \delta \right\}. \quad (3.1.2.6)$$

Furthermore, a function $f \in \mathcal{A}$ is said to be in the subclass $\mathcal{RI}^{\xi}(d,\alpha,\beta)$ if there exists a function $h(z) \in \mathcal{RI}(d,\alpha,\beta)$ such that

$$\left|\frac{f(z)}{h(z)} - 1\right| < 1-\xi, \quad z \in U, \ 0 \le \xi < 1. \tag{3.1.2.7}$$

Theorem 3.12 (Alb Lupaş, 2015d). *If*

$$\begin{aligned}&\left\{\gamma\left(\frac{1+\lambda(j-1)+l}{l+1}\right)^n + (1-\gamma)\frac{(n+j-1)!}{n!\,(j-1)!}\right\}\\ &\quad \ge \left[\gamma\left(\frac{1+\lambda+l}{l+1}\right)^n + (1-\gamma)(n+1)\right], \quad j \ge 2\end{aligned}$$

and

$$\delta = \frac{2\beta|d|}{(1+\alpha)(1+\beta|d|)\left[\gamma\left(\frac{1+\lambda+l}{l+1}\right)^n + (1-\gamma)(n+1)\right]},$$

then

$$\mathcal{RI}(d,\alpha,\beta) \subset N_\delta(e).$$

Proof. Let $f \in \mathcal{RI}(d,\alpha,\beta)$. Then in view of assertion (3.1.2.2) of Theorem 3.7 and the condition

$$\begin{aligned}&\left\{\gamma\left(\frac{1+\lambda(j-1)+l}{l+1}\right)^n + (1-\gamma)\frac{(n+j-1)!}{n!\,(j-1)!}\right\}\\ &\quad \ge \left[\gamma\left(\frac{1+\lambda+l}{l+1}\right)^n + (1-\gamma)(n+1)\right]\end{aligned}$$

for $j \ge 2$, we get

$$\begin{aligned}&(1+\alpha)(1+\beta|d|)\left[\gamma\left(\frac{1+\lambda+l}{l+1}\right)^n + (1-\gamma)(n+1)\right]\sum_{j=2}^{\infty} a_j\\ &\le \sum_{j=2}^{\infty}(1+\alpha(j-1))(j-1+\beta|d|)\\ &\quad \times \left\{\gamma\left(\frac{1+\lambda(j-1)+l}{l+1}\right)^n + (1-\gamma)\frac{(n+j-1)!}{n!\,(j-1)!}\right\} a_j\\ &\le \beta|d|,\end{aligned}$$

which implies

$$\sum_{j=2}^{\infty} a_j \leq \frac{\beta|d|}{(1+\alpha)(1+\beta|d|)\left[\gamma\left(\frac{1+\lambda+l}{l+1}\right)^n + (1-\gamma)(n+1)\right]}. \tag{3.1.2.8}$$

Applying assertion (3.1.2.2) of Theorem 3.7 in conjunction with (3.1.2.8), we obtain

$$(1+\alpha)(1+\beta|d|)\left[\gamma\left(\frac{1+\lambda+l}{l+1}\right)^n + (1-\gamma)(n+1)\right]\sum_{j=2}^{\infty} a_j \leq \beta|d|,$$

$$2(1+\alpha)(1+\beta|d|)\left[\gamma\left(\frac{1+\lambda+l}{l+1}\right)^n + (1-\gamma)(n+1)\right]\sum_{j=2}^{\infty} a_j \leq 2\beta|d|$$

$$\sum_{j=2}^{\infty} ja_j \leq \frac{2\beta|d|}{(1+\alpha)(1+\beta|d|)\left[\gamma\left(\frac{1+\lambda+l}{l+1}\right)^n + (1-\gamma)(n+1)\right]} = \delta,$$

by virtue of (3.1.2.5), we have $f \in N_\delta(e)$.

This completes the proof of the theorem. □

Theorem 3.13 (Alb Lupaş, 2015d). *If* $h \in \mathcal{RI}(d, \alpha, \beta)$ *and*

$$\xi = 1 - \frac{\delta}{2}\frac{(1+\alpha)(1+\beta|d|)\left[\gamma\left(\frac{1+\lambda+l}{l+1}\right)^n + (1-\gamma)(n+1)\right]}{(1+\alpha)(1+\beta|d|)\left[\gamma\left(\frac{1+\lambda+l}{l+1}\right)^n + (1-\gamma)(n+1)\right] - \beta|d|}, \tag{3.1.2.9}$$

then

$$N_\delta(h) \subset \mathcal{RI}^{\xi}(d, \alpha, \beta).$$

Proof. Suppose that $f \in N_\delta(h)$, we then find from (3.1.2.5) that

$$\sum_{j=2}^{\infty} j|a_j - b_j| \leq \delta,$$

which readily implies the following coefficient inequality

$$\sum_{j=2}^{\infty} |a_j - b_j| \leq \frac{\delta}{2}. \tag{3.1.2.10}$$

Next, since $h \in \mathcal{RI}(d,\alpha,\beta)$ in the view of (3.1.2.8), we have

$$\sum_{j=2}^{\infty} b_j \le \frac{\beta|d|}{(1+\alpha)(1+\beta|d|)\left[\gamma\left(\frac{1+\lambda+l}{l+1}\right)^n + (1-\gamma)(n+1)\right]}. \tag{3.1.2.11}$$

Using (3.1.2.10) and (3.1.2.11), we get

$$\begin{aligned}\left|\frac{f(z)}{h(z)}-1\right| &\le \frac{\sum_{j=2}^{\infty}|a_j-b_j|}{1-\sum_{j=2}^{\infty}b_j} \le \frac{\delta}{2\left(1-\frac{\beta|d|}{(1+\alpha)(1+\beta|d|)\left[\gamma\left(\frac{1+\lambda+l}{l+1}\right)^n+(1-\gamma)(n+1)\right]}\right)}\\ &\le \frac{\delta}{2}\frac{(1+\alpha)(1+\beta|d|)\left[\gamma\left(\frac{1+\lambda+l}{l+1}\right)^n+(1-\gamma)(n+1)\right]}{(1+\alpha)(1+\beta|d|)\left[\gamma\left(\frac{1+\lambda+l}{l+1}\right)^n+(1-\gamma)(n+1)\right]-\beta|d|}\\ &= 1-\xi,\end{aligned}$$

provided that ξ is given by (3.1.2.9), thus by condition (3.1.2.7), $f \in \mathcal{RI}^{\xi}(d,\alpha,\beta)$, where ξ is given by (3.1.2.9). □

Radii of Starlikeness, Convexity and Close-to-Convexity

Theorem 3.14 (Alb Lupaş, 2015d). *Let the function $f \in \mathcal{A}$ be in the subclass $\mathcal{RI}(d,\alpha,\beta)$. Then f is univalent starlike of order δ, $0 \le \delta < 1$, in $|z| < r_1$, where*

$$r_1 = \inf_j \left\{\frac{(1-\delta)(1+\alpha(j-1))(j-1+\beta|d|)\left\{\gamma\left(\frac{1+\lambda(j-1)+l}{l+1}\right)^n+(1-\gamma)\frac{(n+j-1)!}{n!(j-1)!}\right\}}{\beta|d|(1-\delta)}\right\}^{\frac{1}{j-1}}.$$

The result is sharp for the function $f(z)$ given by

$$f_j(z) = z + \frac{\beta|d|}{(1+\alpha(j-1))(j-1+\beta|d|)\left\{\gamma\left(\frac{1+\lambda(j-1)+l}{l+1}\right)^n+(1-\gamma)\frac{(n+j-1)!}{n!(j-1)!}\right\}}z^j,$$

$j \ge 2$.

Proof. It suffices to show that

$$\left|\frac{zf'(z)}{f(z)}-1\right| \le 1-\delta, \quad |z| < r_1.$$

Since

$$\left|\frac{zf'(z)}{f(z)}-1\right| = \left|\frac{\sum_{j=2}^{\infty}(j-1)a_j z^{j-1}}{1+\sum_{j=2}^{\infty}a_j z^{j-1}}\right| \le \frac{\sum_{j=2}^{\infty}(j-1)a_j|z|^{j-1}}{1-\sum_{j=2}^{\infty}a_j|z|^{j-1}},$$

to prove the theorem, we must show that

$$\frac{\sum_{j=2}^{\infty}(j-1)a_j|z|^{j-1}}{1-\sum_{j=2}^{\infty}a_j|z|^{j-1}} \leq 1-\delta.$$

This is equivalent to

$$\sum_{j=2}^{\infty}(j-\delta)a_j|z|^{j-1} \leq 1-\delta,$$

using Theorem 3.7, we obtain

$$|z| \leq \left\{ \frac{\begin{array}{c}(1-\delta)(1+\alpha(j-1))(j-1+\beta|d|) \\ \times\left\{\gamma\left(\frac{1+\lambda(j-1)+l}{l+1}\right)^n+(1-\gamma)\frac{(n+j-1)!}{n!(j-1)!}\right\}\end{array}}{\beta|d|(1-\delta)} \right\}^{\frac{1}{j-1}}.$$

Hence, the proof is complete. □

Theorem 3.15 (Alb Lupaş, 2015d). *Let the function $f \in \mathcal{A}$ be in the subclass $\mathcal{RI}(d,\alpha,\beta)$. Then f is univalent convex of order δ, $0 \leq \delta \leq 1$, in $|z| < r_2$, where*

$$r_2 = \inf_j \left\{ \frac{(1-\delta)(1+\alpha(j-1))(j-1+\beta|d|)\left\{\gamma\left(\frac{1+\lambda(j-1)+l}{l+1}\right)^n+(1-\gamma)\frac{(n+j-1)!}{n!(j-1)!}\right\}}{2(j-\delta)\beta|d|} \right\}^{\frac{1}{k-p}}.$$

The result is sharp for the function $f(z)$ given by

$$f_j(z) = z + \frac{\beta|d|}{(1+\alpha(j-1))(j-1+\beta|d|)\left\{\gamma\left(\frac{1+\lambda(j-1)+l}{l+1}\right)^n+(1-\gamma)\frac{(n+j-1)!}{n!(j-1)!}\right\}}z^j,$$

$j \geq 2.$ (3.1.2.12)

Proof. It suffices to show that

$$\left|\frac{zf''(z)}{f'(z))}\right| \leq 1-\delta, \quad |z| < r_2.$$

Since

$$\left|\frac{zf''(z)}{f'(z)}\right| = \left|\frac{\sum_{j=2}^{\infty}j(j-1)a_jz^{j-1}}{1+\sum_{j=2}^{\infty}ja_jz^{j-1}}\right| \leq \frac{\sum_{j=2}^{\infty}j(j-1)a_j|z|^{j-1}}{1-\sum_{j=2}^{\infty}ja_j|z|^{j-1}},$$

to prove the theorem, we must show that

$$\frac{\sum_{j=2}^{\infty} j(j-1)a_j|z|^{j-1}}{1-\sum_{j=2}^{\infty} ja_j|z|^{j-1}} \le 1-\delta,$$

$$\sum_{j=2}^{\infty} j(j-\delta)a_j|z|^{j-1} \le 1-\delta,$$

using Theorem 3.7, we obtain

$$|z|^{j-1} \le \frac{(1-\delta)(1+\alpha(j-1))(j-1+\beta|d|)\left\{\gamma\left(\frac{1+\lambda(j-1)+l}{l+1}\right)^n + (1-\gamma)\frac{(n+j-1)!}{n!(j-1)!}\right\}}{2(j-\delta)\beta|d|},$$

or

$$|z| \le \left\{\frac{(1-\delta)(1+\alpha(j-1))(j-1+\beta|d|)\left\{\gamma\left(\frac{1+\lambda(j-1)+l}{l+1}\right)^n + (1-\gamma)\frac{(n+j-1)!}{n!(j-1)!}\right\}}{2(j-\delta)\beta|d|}\right\}^{\frac{1}{j-1}}.$$

Hence, the proof is complete. □

Theorem 3.16 (Alb Lupaş, 2015d). *Let the function* $f \in \mathcal{A}$ *be in the subclass* $\mathcal{RI}(d, \alpha, \beta)$. *Then* f *is univalent close-to-convex of order* δ, $0 \le \delta < 1$, *in* $|z| < r_3$, *where*

$$r_3 = \inf_j \left\{\frac{(1-\delta)(1+\alpha(j-1))(j-1+\beta|d|)\left\{\gamma\left(\frac{1+\lambda(j-1)+l}{l+1}\right)^n + (1-\gamma)\frac{(n+j-1)!}{n!(j-1)!}\right\}}{j\beta|d|}\right\}^{\frac{1}{j-1}}.$$

The result is sharp for the function $f(z)$ *given by* (3.1.2.12).

Proof. It suffices to show that

$$|f'(z)-1| \le 1-\delta, \quad |z| < r_3.$$

Then

$$|f'(z)-1| = \left|\sum_{j=2}^{\infty} ja_j z^{j-1}\right| \le \sum_{j=2}^{\infty} ja_j|z|^{j-1}.$$

Thus, $|f'(z)-1| \leq 1-\delta$ if $\sum_{j=2}^{\infty} \frac{ja_j}{1-\delta}|z|^{j-1} \leq 1$. Using Theorem 3.7, the above inequality holds true if

$$|z|^{j-1} \leq \frac{(1-\delta)(1+\alpha(j-1))(j-1+\beta|d|)\left\{\gamma\left(\frac{1+\lambda(j-1)+l}{l+1}\right)^n+(1-\gamma)\frac{(n+j-1)!}{n!(j-1)!}\right\}}{j\beta|d|}$$

or

$$|z| \leq \left\{\frac{(1-\delta)(1+\alpha(j-1))(j-1+\beta|d|)\left\{\gamma\left(\frac{1+\lambda(j-1)+l}{l+1}\right)^n+(1-\gamma)\frac{(n+j-1)!}{n!(j-1)!}\right\}}{j\beta|d|}\right\}^{\frac{1}{j-1}}.$$

Hence, the proof is complete. □

3.1.3 *On Certain Properties of a Special Subclass of Analytic Functions Involving Multiplier Transformation and Ruscheweyh Derivative*

The main object of the paper Alb Lupaş (2015f) is to study some properties of certain subclass of analytic functions in the open unit disc which is defined by the linear operator from Definition 2.7 $RI_{n,\lambda,l}^{\alpha}: \mathcal{A} \to \mathcal{A}$,

$$RI_{n,\lambda,l}^{\alpha} f(z) = (1-\alpha)R^n f(z) + \alpha I(n,\lambda,l) f(z), \quad z \in U,$$

where $R^n f(z)$ is the Ruscheweyh derivative, $I(n,\lambda,l)f(z)$ the multiplier transformation and $\mathcal{A}_n = \{f \in \mathcal{H}(U) : f(z) = z + a_{n+1}z^{n+1} + \cdots,\ z \in U\}$ is the class of normalized analytic functions with $\mathcal{A}_1 = \mathcal{A}$. These properties include a coefficient inequality, distortion theorem and extreme points of differential operator. We also discuss the boundedness properties associated with partial sums of functions in the subclass $\mathcal{TS}_{\lambda,l,\alpha}^{n}(\beta,\gamma)$.

Definition 3.4 (Alb Lupaş, 2015f). Let $0 < \beta \leq 1$, $\alpha, \lambda, l \geq 0$, $n \in \mathbb{N}$, $\gamma \in \mathbb{C}\backslash\{0\}$. Then the function $f \in \mathcal{A}$ is said to be in the subclass $\mathcal{S}_{\lambda,l,\alpha}^{n}(\beta,\gamma)$ if

$$\left|\frac{1}{\gamma}\left(\frac{z\left(RI_{n,\lambda,l}^{\alpha} f(z)\right)'}{RI_{n,\lambda,l}^{\alpha} f(z)}-1\right)\right| < \beta, \quad z \in U.$$

We define now the subclass $\mathcal{TS}_{\lambda,l,\alpha}^{n}(\beta,\gamma)$ by

$$\mathcal{TS}_{\lambda,l,\alpha}^{n}(\beta,\gamma) = \mathcal{S}_{\lambda,l,\alpha}^{n}(\beta,\gamma) \cap \mathcal{T}.$$

Coefficient Inequality

Theorem 3.17 (Alb Lupaş, 2015f). *Let the function $f \in \mathcal{T}$. Then $f(z)$ is in the subclass $\mathcal{TS}^n_{\lambda,l,\alpha}(\beta,\gamma)$ if and only if*

$$\sum_{j=2}^{\infty}(j-1+\beta|\gamma|)\left\{\alpha\left(\frac{1+\lambda(j-1)+l}{l+1}\right)^n+(1-\alpha)\frac{(n+j-1)!}{n!\,(j-1)!}\right\}a_j\le\beta|\gamma|, \tag{3.1.3.1}$$

where $\gamma \in \mathbb{C}-\{0\}$, $\lambda,\alpha,l\ge 0$, $n\in\mathbb{N}$, $z\in U$.

Proof. Let $f(z)\in\mathcal{TS}^n_{\lambda,l,\alpha}(\beta,\gamma)$. Then we have

$$\begin{aligned}
&\operatorname{Re}\left\{\frac{z(RI^{\alpha}_{n,\lambda,l}f(z))'}{RI^{\alpha}_{n,\lambda,l}f(z)}-1\right\}>-\beta|\gamma|\frac{z(RI^{\alpha}_{n,\lambda,l}f(z))'}{RI^{\alpha}_{n,\lambda,l}f(z)}\\
&\quad=\frac{z-\sum_{j=2}^{\infty}j\left\{\alpha\left(\frac{1+\lambda(j-1)+l}{l+1}\right)^n+(1-\alpha)\frac{(n+j-1)!}{n!(j-1)!}\right\}a_jz^j}{z-\sum_{j=2}^{\infty}\left\{\alpha\left(\frac{1+\lambda(j-1)+l}{l+1}\right)^n+(1-\alpha)\frac{(n+j-1)!}{n!(j-1)!}\right\}a_jz^j}\\
&\qquad\times\operatorname{Re}\left\{\frac{-\sum_{j=2}^{\infty}(j-1)\left\{\alpha\left(\frac{1+\lambda(j-1)+l}{l+1}\right)^n+(1-\alpha)\frac{(n+j-1)!}{n!(j-1)!}\right\}a_jz^j}{z-\sum_{j=2}^{\infty}\left\{\alpha\left(\frac{1+\lambda(j-1)+l}{l+1}\right)^n+(1-\alpha)\frac{(n+j-1)!}{n!(j-1)!}\right\}a_jz^j}\right\}\\
&\quad>-\beta\,|\gamma|.
\end{aligned}$$

By choosing the values of z on the real axis and letting $z\to 1^-$ through real values, the above inequality immediately yields the required condition (3.1.3.1).

Conversely, by applying the hypothesis (3.1.3.1) and letting $|z|=1$, we obtain

$$\begin{aligned}
&\left|\frac{z(RI^{\alpha}_{n,\lambda,l}f(z))'}{RI^{\alpha}_{n,\lambda,l}f(z)}-1\right|\\
&\quad=\left|\frac{\sum_{j=2}^{\infty}(j-1)\left\{\alpha\left(\frac{1+\lambda(j-1)+l}{l+1}\right)^n+(1-\alpha)\frac{(n+j-1)!}{n!(j-1)!}\right\}a_jz^j}{z-\sum_{j=2}^{\infty}\left\{\alpha\left(\frac{1+\lambda(j-1)+l}{l+1}\right)^n+(1-\alpha)\frac{(n+j-1)!}{n!(j-1)!}\right\}a_jz^j}\right|\\
&\quad\le\frac{\beta\,|\gamma|\left(1-\sum_{j=2}^{\infty}\left\{\alpha\left(\frac{1+\lambda(j-1)+l}{l+1}\right)^n+(1-\alpha)\frac{(n+j-1)!}{n!(j-1)!}\right\}a_j\right)}{1-\sum_{j=2}^{\infty}\left\{\alpha\left(\frac{1+\lambda(j-1)+l}{l+1}\right)^n+(1-\alpha)\frac{(n+j-1)!}{n!(j-1)!}\right\}a_j}\le\beta|\gamma|.
\end{aligned}$$

Hence, by maximum modulus theorem, we have $f \in \mathcal{TS}^n_{\lambda,l,\alpha}(\beta,\gamma)$, which evidently completes the proof of theorem.

Finally, the result is sharp with the extremal functions f_j be in the subclass $\mathcal{TS}^n_{\lambda,l,\alpha}(\beta,\gamma)$ given by

$$f_j(z) = z - \frac{\beta|\gamma|}{(j-1+\beta|\gamma|)\left\{\alpha\left(\frac{1+\lambda(j-1)+l}{l+1}\right)^n + (1-\alpha)\frac{(n+j-1)!}{n!(j-1)!}\right\}} z^j,$$

$$\text{for } j \geq 2. \tag{3.1.3.2}$$

□

Corollary 3.5 (Alb Lupaş, 2015f). *Let the function $f \in \mathcal{T}$ be in the subclass $\mathcal{TS}^n_{\lambda,\alpha}(\beta,\gamma)$. Then we have*

$$a_j \leq \frac{\beta|\gamma|}{(j-1+\beta|\gamma|)\left\{\alpha\left(\frac{1+\lambda(j-1)+l}{l+1}\right)^n + (1-\alpha)\frac{(n+j-1)!}{n!(j-1)!}\right\}}, \quad \text{for } j \geq 2. \tag{3.1.3.3}$$

The equality is attained for the functions f given by (3.1.3.2).

Growth and Distortion Theorems

Theorem 3.18 (Alb Lupaş, 2015f). *Let the function $f \in \mathcal{T}$ be in the subclass $\mathcal{TS}^n_{\lambda,l,\alpha}(\beta,\gamma)$. Then for $|z| = r$, we have*

$$r - \frac{\beta|\gamma|}{(1+\beta|\gamma|)\left[\alpha\left(\frac{1+\lambda+l}{l+1}\right)^n + (1-\alpha)(n+1)\right]} r^2 \leq |f(z)|$$
$$\leq r + \frac{\beta|\gamma|}{(1+\beta|\gamma|)\left[\alpha\left(\frac{1+\lambda+l}{l+1}\right)^n + (1-\alpha)(n+1)\right]} r^2,$$

with equality for

$$f(z) = z - \frac{\beta|\gamma|}{(1+\beta|\gamma|)\left[\alpha\left(\frac{1+\lambda+l}{l+1}\right)^n + (1-\alpha)(n+1)\right]} z^2.$$

Proof. In view of Theorem 3.17, we have

$$(1+\beta|\gamma|)\left[\alpha\left(\frac{1+\lambda+l}{l+1}\right)^n + (1-\alpha)(n+1)\right]\sum_{j=2}^{\infty} a_j$$
$$\leq \sum_{j=2}^{\infty}(j-1+\beta|\gamma|)\left[\alpha\left(\frac{1+\lambda+l}{l+1}\right)^n + (1-\alpha)(n+1)\right] a_j \leq \beta|\gamma|.$$

Hence,

$$|f(z)| \le r + \sum_{j=2}^{\infty} a_j r^j \le r + r^2 \sum_{j=2}^{\infty} a_j$$
$$\le r + \frac{\beta|\gamma|}{(1+\beta|\gamma|)\left[\alpha\left(\frac{1+\lambda+l}{l+1}\right)^n + (1-\alpha)(n+1)\right]} r^2$$

and

$$|f(z)| \ge r - \sum_{j=2}^{\infty} a_j r^j \ge r - r^2 \sum_{j=2}^{\infty} a_j$$
$$\ge r - \frac{\beta|\gamma|}{(1+\beta|\gamma|)\left[\alpha\left(\frac{1+\lambda+l}{l+1}\right)^n + (1-\alpha)(n+1)\right]} r^2.$$

This completes the proof. □

Theorem 3.19 (Alb Lupaş, 2015f). *Let the function $f \in \mathcal{T}$ be in the subclass $\mathcal{TS}^n_{\lambda,l,\alpha}(\beta,\gamma)$. Then for $|z| = r$, we have*

$$r - \frac{2\beta|\gamma|}{(1+\beta|\gamma|)\left[\alpha\left(\frac{1+\lambda+l}{l+1}\right)^n + (1-\alpha)(n+1)\right]} r^2 \le |f'(z)|$$
$$\le r + \frac{2\beta|\gamma|}{(1+\beta|\gamma|)\left[\alpha\left(\frac{1+\lambda+l}{l+1}\right)^n + (1-\alpha)(n+1)\right]} r^2,$$

with equality for

$$f(z) = z - \frac{\beta|\gamma|}{(1+\beta|\gamma|)\left[\alpha\left(\frac{1+\lambda+l}{l+1}\right)^n + (1-\alpha)(n+1)\right]} z^2.$$

Proof. We have

$$|f'(z)| \le r + \sum_{j=2}^{\infty} j a_j r^j \le r + 2r^2 \sum_{j=2}^{\infty} a_j$$
$$\le r + \frac{2\beta|\gamma|}{(1+\beta|\gamma|)\left[\alpha\left(\frac{1+\lambda+l}{l+1}\right)^n + (1-\alpha)(n+1)\right]} r^2$$

and

$$|f'(z)| \geq r - \sum_{j=2}^{\infty} j a_j r^j \geq r - 2r^2 \sum_{j=2}^{\infty} a_j$$

$$\geq r - \frac{2\beta|\gamma|}{(1+\beta|\gamma|)\left[\alpha\left(\frac{1+\lambda+l}{l+1}\right)^n + (1-\alpha)(n+1)\right]} r^2,$$

which completes the proof. □

Extreme Points

The extreme points of the subclass $\mathcal{TS}^n_{\lambda,l,\alpha}(\beta,\gamma)$ will be now determined.

Theorem 3.20 (Alb Lupaş, 2015f). *Let* $f_1(z) = z$ *and* $f_j(z) = z - \frac{\beta|\gamma|}{(j-1+\beta|\gamma|)\left\{\alpha\left(\frac{1+\lambda(j-1)+l}{l+1}\right)^n + (1-\alpha)\frac{(n+j-1)!}{n!(j-1)!}\right\}} z^j$, $j \geq 2$.

Assume that f *is analytic in* U. *Then* $f \in \mathcal{TS}^n_{\lambda,l,\alpha}(\beta,\gamma)$ *if and only if it can be expressed in the form*

$$f(z) = \sum_{j=1}^{\infty} \mu_j f_j(z),$$

where $\mu_j \geq 0$ *and* $\sum_{j=1}^{\infty} \mu_j = 1$.

Proof. Suppose that $f(z) = \sum_{j=1}^{\infty} \mu_j f_j(z)$ with $\mu_j \geq 0$ and $\sum_{j=1}^{\infty} \mu_j = 1$. Then

$$f(z) = \sum_{j=1}^{\infty} \mu_j f_j(z) = \mu_1 f_1(z) + \sum_{j=2}^{\infty} \mu_j f_j(z)$$

$$= \mu_1 z + \sum_{j=2}^{\infty} \mu_j \left(z - \frac{\beta|\gamma|}{(j-1+\beta|\gamma|)\left\{\alpha\left(\frac{1+\lambda(j-1)+l}{l+1}\right)^n + (1-\alpha)\frac{(n+j-1)!}{n!(j-1)!}\right\}} z^j \right)$$

$$= z - \sum_{j=2}^{\infty} \mu_j \frac{\beta|\gamma|}{(j-1+\beta|\gamma|)\left\{\alpha\left(\frac{1+\lambda(j-1)+l}{l+1}\right)^n + (1-\alpha)\frac{(n+j-1)!}{n!(j-1)!}\right\}} z^j.$$

Then

$$\sum_{j=2}^{\infty} \mu_j \frac{\beta|\gamma|}{(j-1+\beta|\gamma|)\left\{\alpha\left(\frac{1+\lambda(j-1)+l}{l+1}\right)^n + (1-\alpha)\frac{(n+j-1)!}{n!(j-1)!}\right\}} \cdot \frac{(j-1+\beta|\gamma|)\left\{\alpha\left(\frac{1+\lambda(j-1)+l}{l+1}\right)^n + (1-\alpha)\frac{(n+j-1)!}{n!(j-1)!}\right\}}{\beta|\gamma|}$$

$$= \sum_{j=2}^{\infty} \mu_j = \sum_{j=1}^{\infty} \mu_j - \mu_1 = 1 - \mu_1 \leq 1.$$

Thus, $f \in \mathcal{TS}_{\lambda,l,\alpha}^{n}(\beta,\gamma)$ by Theorem 3.17.

Conversely, suppose that $f \in \mathcal{TS}_{\lambda,l,\alpha}^{n}(\beta,\gamma)$. By using (3.1.3.3), we may set and

$$\mu_j = \frac{(j-1+\beta|\gamma|)\left\{\alpha\left(\frac{1+\lambda(j-1)+l}{l+1}\right)^n + (1-\alpha)\frac{(n+j-1)!}{n!(j-1)!}\right\}}{\beta|\gamma|} a_j$$

for $j \geq 2$ and $\mu_1 = 1 - \sum_{j=2}^{\infty} \mu_j$.

Then

$$f(z) = z - \sum_{j=1}^{\infty} a_j z^j$$

$$= z - \sum_{j=2}^{\infty} \mu_j \frac{\beta|\gamma|}{(j-1+\beta|\gamma|)\left\{\alpha\left(\frac{1+\lambda(j-1)+l}{l+1}\right)^n + (1-\alpha)\frac{(n+j-1)!}{n!(j-1)!}\right\}} z^j$$

$$= \mu_1 f_1(z) + \sum_{j=2}^{\infty} \mu_j f_j(z) = \sum_{j=1}^{\infty} \mu_j f_j(z),$$

with $\mu_j \geq 0$ and $\sum_{j=1}^{\infty} \mu_j = 1$, which completes the proof. □

Corollary 3.6 (Alb Lupaş, 2015f). *The extreme points of* $\mathcal{TS}_{\lambda,l,\alpha}^{n}(\beta,\gamma)$ *are the functions* $f_1(z) = z$ *and*

$$f_j(z) = \frac{\beta|\gamma|}{(j-1+\beta|\gamma|)\left\{\alpha\left(\frac{1+\lambda(j-1)+l}{l+1}\right)^n + (1-\alpha)\frac{(n+j-1)!}{n!(j-1)!}\right\}} z^j, \quad \text{for } j \geq 2.$$

Partial Sums

We investigate in the following the partial sums of functions in the subclass $\mathcal{TS}^n_{\lambda,l,\alpha}(\beta,\gamma)$. We shall obtain sharp lower bounds for the real part of its ratios. In what follows, we will use the well known result that for an analytic function ω in U,

$$\text{Re}\left(\frac{1+\omega(z)}{1-\omega(z)}\right) > 0, \quad z \in U,$$

if and only if the inequality $|\omega(z)| < 1$ is satisfied.

Theorem 3.21 (Alb Lupaş, 2015f). *Let* $f(z) = z - \sum_{j=2}^{\infty} a_j z^j \in \mathcal{TS}^n_{\lambda,l,\alpha}(\beta,\gamma)$ *and define its partial sums by* $f_1(z) = z$ *and* $f_m(z) = z - \sum_{j=2}^{m} a_j z^j$, $m \geq 2$. *Then*

$$\text{Re}\left(\frac{f(z)}{f_m(z)}\right) \geq 1 - \frac{1}{c_{m+1}}, \quad z \in U, \ m \in \mathbb{N}, \tag{3.1.3.4}$$

and

$$\text{Re}\left(\frac{f_m(z)}{f(z)}\right) \geq \frac{c_{m+1}}{1+c_{m+1}}, \quad z \in U, \ m \in \mathbb{N}, \tag{3.1.3.5}$$

where

$$c_j = \frac{(j+\beta|\gamma|)\left\{\alpha\left(\frac{1+\lambda(j-1)+l}{l+1}\right)^n + (1-\alpha)\frac{(n+j-1)!}{n!(j-1)!}\right\}}{\beta|\gamma|}. \tag{3.1.3.6}$$

This estimates in (3.1.3.4) *and* (3.1.3.5) *are sharp.*

Proof. To prove (3.1.3.4), it suffices to show that

$$c_{m+1}\left(\frac{f(z)}{f_m(z)} - \left(1 - \frac{1}{c_{m+1}}\right)\right) \prec \frac{1+z}{1-z}, \quad z \in U.$$

By the subordination property, we can write

$$\frac{1 - \sum_{j=2}^{m} a_j z^{j-1} - c_{m+1}\sum_{j=m+1}^{\infty} a_j z^{j-1}}{1 - \sum_{j=2}^{m} a_j z^{j-1}} = \frac{1+\omega(z)}{1-\omega(z)}$$

for certain ω analytic in U with $|\omega(z)| \leq 1$. Notice that $\omega(0) = 0$ and

$$|\omega(z)| \leq \frac{c_{m+1}\sum_{j=m+1}^{\infty} a_j}{2 - 2\sum_{j=2}^{m} a_j - c_{m+1}\sum_{j=m+1}^{\infty} a_j}.$$

Now $|\omega(z)| \leq 1$ if and only if

$$\sum_{j=2}^{m} a_j + c_{m+1} \sum_{j=m+1}^{\infty} a_j \leq \sum_{j=2}^{\infty} c_j a_j \leq 1.$$

The above inequality holds because c_j is a non-decreasing sequence. This completes the proof of (3.1.3.4).

Finally, it is observed that equality in (3.1.3.4) is attained for the function given by (3.1.3.3) when $z = re^{2\frac{\pi i}{m}}$ as $r \to 1^-$.

Similarly, we take

$$\begin{aligned}(1 + c_{m+1})&\left(\frac{f_m(z)}{f(z)} - \frac{c_{m+1}}{1 + c_{m+1}}\right)\\ &= \frac{1 - \sum_{j=2}^{m} a_j z^{j-1} + c_{m+1} \sum_{j=m+1}^{\infty} a_j z^{j-1}}{1 - \sum_{j=2}^{m} a_j z^{j-1}} = \frac{1 + \omega(z)}{1 - \omega(z)},\end{aligned}$$

where

$$|\omega(z)| \leq \frac{(1 + c_{m+1}) \sum_{j=m+1}^{\infty} a_j}{2 - 2\sum_{j=2}^{m} a_j + (1 - c_{m+1}) \sum_{j=m+1}^{\infty} a_j}.$$

Now $|\omega(z)| \leq 1$ if and only if

$$\sum_{j=2}^{m} a_j + c_{m+1} \sum_{j=m+1}^{\infty} a_j \leq \sum_{j=2}^{\infty} c_j a_j \leq 1.$$

This immediately leads to assertion (3.1.3.5) of theorem. This completes the proof. □

Using a similar method, we can prove the following theorem.

Theorem 3.22 (Alb Lupaş, 2015f). *If $f \in \mathcal{TS}_{\lambda,\alpha}^{n}(\beta, \gamma)$ and define the partial sums by $f_1(z) = z$ and $f_m(z) = z - \sum_{j=2}^{m} a_j z^j$. Then*

$$\operatorname{Re}\left(\frac{f'(z)}{f'_m(z)}\right) \geq 1 - \frac{m+1}{c_{m+1}}, \quad z \in U, \ m \in \mathbb{N},$$

and

$$\operatorname{Re}\left(\frac{f'_m(z)}{f(z)}\right) \geq \frac{c_{m+1}}{m + 1 + c_{m+1}},$$

where c_j is given by (3.1.3.6). The result is sharp for every m, with extremal functions given by (3.1.3.2).

3.1.4 *Characterization Properties Regarding New Subclasses of Analytic Functions by Making Use of Multiplier Transformation and Ruscheweyh Derivative*

In the paper Alb Lupaş (2015g), we introduced new subclasses containing the linear operator from Definition 2.7, $RI^{\alpha}_{n,\lambda,l} : \mathcal{A} \to \mathcal{A}$,

$$RI^{\alpha}_{n,\lambda,l}f(z) = (1-\alpha)R^n f(z) + \alpha I(n,\lambda,l) f(z), \quad z \in U,$$

where $R^n f(z)$ is the Ruscheweyh derivative, $I(n,\lambda,l) f(z)$ the multiplier transformation and $\mathcal{A}_n = \{f \in \mathcal{H}(U) : f(z) = z + a_{n+1}z^{n+1} + \cdots,\ z \in U\}$ is the class of normalized analytic functions with $\mathcal{A}_1 = \mathcal{A}$. Characterization and other properties of these subclasses are studied.

Definition 3.5 (Alb Lupaş, 2015g). Let $f \in \mathcal{A}$. Then $f(z)$ is in the subclass $\mathcal{S}^n_{\lambda,l,\alpha}(\mu)$ if and only if

$$\operatorname{Re}\left(\frac{z\left(RI^{\alpha}_{n,\lambda,l}f(z)\right)'}{RI^{\alpha}_{n,\lambda,l}f(z)}\right) > \mu, \quad 0 \le \mu < 1,\ z \in U.$$

Definition 3.6 (Alb Lupaş, 2015g). Let $f \in \mathcal{A}$. Then $f(z)$ is in the subclass $\mathcal{C}^n_{\lambda,l,\alpha}(\mu)$ if and only if

$$\operatorname{Re}\left(\frac{\left[z\left(RI^{\alpha}_{n,\lambda,l}f(z)\right)'\right]'}{\left(RI^{\alpha}_{n,\lambda,l}f(z)\right)'}\right) > \mu, \quad 0 \le \mu < 1,\ z \in U.$$

We study the characterization properties and distortion theorems for the function $f(z) \in \mathcal{A}$ to belong to the subclasses $\mathcal{S}^n_{\lambda,l,\alpha}(\mu)$ and $\mathcal{C}^n_{\lambda,l,\alpha}(\mu)$ by obtaining the coefficient bounds.

Theorem 3.23 (Alb Lupaş, 2015g). *Let $f \in \mathcal{A}$. If*

$$\sum_{j=2}^{\infty}(j-\mu)\left\{\alpha\left(\frac{1+\lambda(j-1)+l}{l+1}\right)^n + (1-\alpha)\frac{(n+j-1)!}{n!\,(j-1)!}\right\}|a_j| \le 1-\mu, \tag{3.1.4.1}$$

$0 \le \mu < 1$, then $f(z) \in \mathcal{S}^n_{\lambda,l,\alpha}(\mu)$. The result (3.1.4.1) is sharp.

Proof. Suppose that (3.1.4.1) holds. Since

$$1-\mu \geq \sum_{j=2}^{\infty}(j-\mu)\left\{\alpha\left(\frac{1+\lambda(j-1)+l}{l+1}\right)^{n}+(1-\alpha)\frac{(n+j-1)!}{n!\,(j-1)!}\right\}|a_j|$$

$$\geq \mu\sum_{j=2}^{\infty}\left\{\alpha\left(\frac{1+\lambda(j-1)+l}{l+1}\right)^{n}+(1-\alpha)\frac{(n+j-1)!}{n!\,(j-1)!}\right\}|a_j|$$

$$-\sum_{j=2}^{\infty}j\left\{\alpha\left(\frac{1+\lambda(j-1)+l}{l+1}\right)^{n}+(1-\alpha)\frac{(n+j-1)!}{n!\,(j-1)!}\right\}|a_j|$$

then this implies that

$$\frac{1+\sum_{j=2}^{\infty}j\left\{\alpha\left(\frac{1+\lambda(j-1)+l}{l+1}\right)^{n}+(1-\alpha)\frac{(n+j-1)!}{n!(j-1)!}\right\}|a_j|}{1+\sum_{j=2}^{\infty}\left\{\alpha\left(\frac{1+\lambda(j-1)+l}{l+1}\right)^{n}+(1-\alpha)\frac{(n+j-1)!}{n!(j-1)!}\right\}|a_j|}>\mu.$$

So, we deduce that

$$\operatorname{Re}\left(\frac{z\left(RI_{n,\lambda,l}^{\alpha}f(z)\right)'}{RI_{n,\lambda,l}^{\alpha}f(z)}\right)>\mu,\quad 0\leq\mu<1,\quad z\in U.$$

We have $f(z)\in\mathcal{S}_{\lambda,l,\alpha}^{n}(\mu)$, which evidently completes the proof.

The assertion (3.1.4.1) is sharp and the extremal function is given by

$$f(z)=z+\sum_{j=2}^{\infty}\frac{(1-\mu)}{(j-\mu)\left\{\alpha\left(\frac{1+\lambda(j-1)+l}{l+1}\right)^{n}+(1-\alpha)\frac{(n+j-1)!}{n!(j-1)!}\right\}}z^{j}.\qquad\square$$

Corollary 3.7 (Alb Lupaş, 2015g). *Let the hypotheses of Theorem* 3.23 *satisfy. Then*

$$|a_j|\leq\frac{1-\mu}{(j-\mu)\left\{\alpha\left(\frac{1+\lambda(j-1)+l}{l+1}\right)^{n}+(1-\alpha)\frac{(n+j-1)!}{n!(j-1)!}\right\}},\quad\forall\, j\geq 2.$$

Theorem 3.24 (Alb Lupaş, 2015g). *Let* $f\in\mathcal{A}$. *If*

$$\sum_{j=2}^{\infty}j(j-\mu)\left\{\alpha\left(\frac{1+\lambda(j-1)+l}{l+1}\right)^{n}+(1-\alpha)\frac{(n+j-1)!}{n!\,(j-1)!}\right\}|a_j|\leq 1-\mu, \tag{3.1.4.2}$$

$0\leq\mu<1$, *then* $f(z)\in\mathcal{C}_{\lambda,l,\alpha}^{n}(\mu)$. *The result* (3.1.4.2) *is sharp.*

Proof. Suppose that (3.1.4.2) holds. Since

$$\begin{aligned} 1-\mu &\geq \sum_{j=2}^{\infty} j(j-\mu)\left\{\alpha\left(\frac{1+\lambda(j-1)+l}{l+1}\right)^{n}+(1-\alpha)\frac{(n+j-1)!}{n!\,(j-1)!}\right\}|a_j| \\ &\geq \mu\sum_{j=2}^{\infty} j\left\{\alpha\left(\frac{1+\lambda(j-1)+l}{l+1}\right)^{n}+(1-\alpha)\frac{(n+j-1)!}{n!\,(j-1)!}\right\}|a_j| \\ &\quad -\sum_{j=2}^{\infty} j^2\left\{\alpha\left(\frac{1+\lambda(j-1)+l}{l+1}\right)^{n}+(1-\alpha)\frac{(n+j-1)!}{n!\,(j-1)!}\right\}|a_j| \end{aligned}$$

then this implies that

$$\frac{1+\sum_{j=2}^{\infty} j^2\left\{\alpha\left(\frac{1+\lambda(j-1)+l}{l+1}\right)^{n}+(1-\alpha)\frac{(n+j-1)!}{n!(j-1)!}\right\}|a_j|}{1+\sum_{j=2}^{\infty} j\left\{\alpha\left(\frac{1+\lambda(j-1)+l}{l+1}\right)^{n}+(1-\alpha)\frac{(n+j-1)!}{n!(j-1)!}\right\}|a_j|}>\mu.$$

So, we deduce that

$$\mathrm{Re}\left(\frac{\left[z\left(RI_{n,\lambda,l}^{\alpha}f(z)\right)'\right]'}{\left(RI_{n,\lambda,l}^{\alpha}f(z)\right)'}\right)>\mu, \quad 0\leq\mu<1, \;\; z\in U.$$

We have $f(z)\in\mathcal{C}_{\lambda,l,\alpha}^{n}(\mu)$, which evidently completes the proof.

The assertion (3.1.4.2) is sharp and the extremal function is given by

$$f(z)=z+\sum_{j=2}^{\infty}\frac{(1-\mu)}{j(j-\mu)\left\{\alpha\left(\frac{1+\lambda(j-1)+l}{l+1}\right)^{n}+(1-\alpha)\frac{(n+j-1)!}{n!(j-1)!}\right\}}z^{j}.$$

□

Corollary 3.8 (Alb Lupaş, 2015g). *Let the hypotheses of Theorem* 3.24 *be satisfied. Then*

$$|a_j|\leq\frac{1-\mu}{j(j-\mu)\left\{\alpha\left(\frac{1+\lambda(j-1)+l}{l+1}\right)^{n}+(1-\alpha)\frac{(n+j-1)!}{n!(j-1)!}\right\}}, \quad \forall\, j\geq 2.$$

Also, we have the following inclusion results:

Theorem 3.25 (Alb Lupaş, 2015g). *Let* $0\leq\mu_1\leq\mu_2<1$. *Then* $\mathcal{S}_{\lambda,l,\alpha}^{n}(\mu_1)\supseteq\mathcal{S}_{\lambda,l,\alpha}^{n}(\mu_2)$.

Proof. By Theorem 3.23. □

Theorem 3.26 (Alb Lupaş, 2015g). *Let* $0 \leq \mu_1 \leq \mu_2 < 1$. *Then* $\mathcal{C}^n_{\lambda,l,\alpha}(\mu_1) \supseteq \mathcal{C}^n_{\lambda,l,\alpha}(\mu_2)$.

Proof. By Theorem 3.24. □

We introduce the following distortion theorems.

Theorem 3.27 (Alb Lupaş, 2015g). *Let the function* $f \in \mathcal{A}$ *and*

$$\sum_{j=2}^{\infty}(j-\mu)\left\{\alpha\left(\frac{1+\lambda(j-1)+l}{l+1}\right)^n+(1-\alpha)\frac{(n+j-1)!}{n!\,(j-1)!}\right\}|a_j| \leq 1-\mu,$$

$0 \leq \mu < 1$.

Then for $z \in U$ *and* $0 \leq \mu < 1$,

$$\left|RI^{\alpha}_{n,\lambda,l}f(z)\right| \geq |z| - \frac{1-\mu}{2-\mu}|z|^2$$

and

$$\left|RI^{\alpha}_{n,\lambda,l}f(z)\right| \leq |z| + \frac{1-\mu}{2-\mu}|z|^2.$$

Proof. By using Theorem 3.23, one can verify that

$$(2-\mu)\sum_{j=2}^{\infty}\left\{\alpha\left(\frac{1+\lambda(j-1)+l}{l+1}\right)^n+(1-\alpha)\frac{(n+j-1)!}{n!\,(j-1)!}\right\}|a_j|$$

$$\leq \sum_{j=2}^{\infty}(j-\mu)\left\{\alpha\left(\frac{1+\lambda(j-1)+l}{l+1}\right)^n+(1-\alpha)\frac{(n+j-1)!}{n!\,(j-1)!}\right\}|a_j| \leq 1-\mu.$$

Hence,

$$\sum_{j=2}^{\infty}\left\{\alpha\left(\frac{1+\lambda(j-1)+l}{l+1}\right)^n+(1-\alpha)\frac{(n+j-1)!}{n!\,(j-1)!}\right\}|a_j| \leq \frac{1-\mu}{2-\mu}.$$

We obtain

$$\begin{aligned}
&\left|RI^{\alpha}_{n,\lambda,l}f(z)\right| \\
&= \left| z + \sum_{j=2}^{\infty} \left\{ \alpha \left(\frac{1+\lambda(j-1)+l}{l+1} \right)^n + (1-\alpha) \frac{(n+j-1)!}{n!\,(j-1)!} \right\} a_j z^j \right| \\
&\leq |z| + \sum_{j=2}^{\infty} \left\{ \alpha \left(\frac{1+\lambda(j-1)+l}{l+1} \right)^n + (1-\alpha) \frac{(n+j-1)!}{n!\,(j-1)!} \right\} |a_j|\,|z|^j \\
&\leq |z| + \sum_{j=2}^{\infty} \left\{ \alpha \left(\frac{1+\lambda(j-1)+l}{l+1} \right)^n + (1-\alpha) \frac{(n+j-1)!}{n!\,(j-1)!} \right\} |a_j|\,|z|^2 \\
&\leq |z| + \frac{1-\mu}{2-\mu}\,|z|^2.
\end{aligned}$$

The other assertion can be proved as follows

$$\begin{aligned}
&\left|RI^{\alpha}_{n,\lambda,l}f(z)\right| \\
&= \left| z + \sum_{j=2}^{\infty} \left\{ \alpha \left(\frac{1+\lambda(j-1)+l}{l+1} \right)^n + (1-\alpha) \frac{(n+j-1)!}{n!\,(j-1)!} \right\} a_j z^j \right| \\
&\geq |z| - \sum_{j=2}^{\infty} \left\{ \alpha \left(\frac{1+\lambda(j-1)+l}{l+1} \right)^n + (1-\alpha) \frac{(n+j-1)!}{n!\,(j-1)!} \right\} |a_j|\,|z|^j \\
&\geq |z| - \sum_{j=2}^{\infty} \left\{ \alpha \left(\frac{1+\lambda(j-1)+l}{l+1} \right)^n + (1-\alpha) \frac{(n+j-1)!}{n!\,(j-1)!} \right\} |a_j|\,|z|^2 \\
&\geq |z| - \frac{1-\mu}{2-\mu}\,|z|^2.
\end{aligned}$$

This completes the proof. □

Theorem 3.28 (Alb Lupaş, 2015g). *Let the function $f \in \mathcal{A}$ and*

$$\sum_{j=2}^{\infty} j(j-\mu) \left\{ \alpha \left(\frac{1+\lambda(j-1)+l}{l+1} \right)^n + (1-\alpha) \frac{(n+j-1)!}{n!\,(j-1)!} \right\} |a_j| \leq 1-\mu,$$

$0 \leq \mu < 1$.

Then for $z \in U$ and $0 \le \mu < 1$,

$$\left|RI^{\alpha}_{n,\lambda,l}f(z)\right| \ge |z| - \frac{1-\mu}{2(2-\mu)}|z|^2$$

and

$$\left|RI^{\alpha}_{n,\lambda,l}f(z)\right| \le |z| + \frac{1-\mu}{2(2-\mu)}|z|^2.$$

Proof. By using Theorem 3.24, one can verify that

$$\begin{aligned}
&2(2-\mu)\sum_{j=2}^{\infty}\left\{\alpha\left(\frac{1+\lambda(j-1)+l}{l+1}\right)^n + (1-\alpha)\frac{(n+j-1)!}{n!\,(j-1)!}\right\}|a_j| \\
&\le \sum_{j=2}^{\infty} j(j-\mu)\left\{\alpha\left(\frac{1+\lambda(j-1)+l}{l+1}\right)^n + (1-\alpha)\frac{(n+j-1)!}{n!\,(j-1)!}\right\}|a_j| \\
&\le 1-\mu.
\end{aligned}$$

Hence,

$$\sum_{j=2}^{\infty}\left\{\alpha\left(\frac{1+\lambda(j-1)+l}{l+1}\right)^n + (1-\alpha)\frac{(n+j-1)!}{n!\,(j-1)!}\right\}|a_j| \le \frac{1-\mu}{2(2-\mu)}.$$

We obtain

$$\begin{aligned}
&\left|RI^{\alpha}_{n,\lambda,l}f(z)\right| \\
&= \left|z + \sum_{j=2}^{\infty}\left\{\alpha\left(\frac{1+\lambda(j-1)+l}{l+1}\right)^n + (1-\alpha)\frac{(n+j-1)!}{n!\,(j-1)!}\right\}a_j z^j\right| \\
&\le |z| + \sum_{j=2}^{\infty}\left\{\alpha\left(\frac{1+\lambda(j-1)+l}{l+1}\right)^n + (1-\alpha)\frac{(n+j-1)!}{n!\,(j-1)!}\right\}|a_j|\,|z|^j \\
&\le |z| + \sum_{j=2}^{\infty}\left\{\alpha\left(\frac{1+\lambda(j-1)+l}{l+1}\right)^n + (1-\alpha)\frac{(n+j-1)!}{n!\,(j-1)!}\right\}|a_j|\,|z|^2 \\
&\le |z| + \frac{1-\mu}{2(2-\mu)}|z|^2.
\end{aligned}$$

The other assertion can be proved as follows

$$\begin{aligned}
&\left|RI^{\alpha}_{n,\lambda,l}f(z)\right| \\
&= \left|z+\sum_{j=2}^{\infty}\left\{\alpha\left(\frac{1+\lambda(j-1)+l}{l+1}\right)^{n}+(1-\alpha)\frac{(n+j-1)!}{n!\,(j-1)!}\right\}a_j z^j\right| \\
&\geq |z|-\sum_{j=2}^{\infty}\left\{\alpha\left(\frac{1+\lambda(j-1)+l}{l+1}\right)^{n}+(1-\alpha)\frac{(n+j-1)!}{n!\,(j-1)!}\right\}|a_j|\,|z|^j \\
&\geq |z|-\sum_{j=2}^{\infty}\left\{\alpha\left(\frac{1+\lambda(j-1)+l}{l+1}\right)^{n}+(1-\alpha)\frac{(n+j-1)!}{n!\,(j-1)!}\right\}|a_j|\,|z|^2 \\
&\geq |z|-\frac{1-\mu}{2(2-\mu)}|z|^2.
\end{aligned}$$

This completes the proof. □

Also, we have the following distortion results.

Theorem 3.29 (Alb Lupaş, 2015g). *Let the hypotheses of Theorem* 3.23 *be satisfied. Then*

$$|f(z)| \geq |z|-\frac{1-\mu}{(2-\mu)\left[\alpha\left(\frac{1+\lambda+l}{l+1}\right)^{n}+(1-\alpha)(n+1)\right]}|z|^2$$

and

$$|f(z)| \leq |z|+\frac{1-\mu}{(2-\mu)\left[\alpha\left(\frac{1+\lambda+l}{l+1}\right)^{n}+(1-\alpha)(n+1)\right]}|z|^2.$$

Proof. In virtue of Theorem 3.23, we have

$$\begin{aligned}
&(2-\mu)\left[\alpha\left(\frac{1+\lambda+l}{l+1}\right)^{n}+(1-\alpha)(n+1)\right]\sum_{j=2}^{\infty}|a_j| \\
&\leq \sum_{j=2}^{\infty}(j-\mu)\left\{\alpha\left(\frac{1+\lambda(j-1)+l}{l+1}\right)^{n}+(1-\alpha)\frac{(n+j-1)!}{n!\,(j-1)!}\right\}|a_j| \\
&\leq 1-\mu,
\end{aligned}$$

thus,

$$\sum_{j=2}^{\infty} |a_j| \leq \frac{1-\mu}{(2-\mu)\left[\alpha\left(\frac{1+\lambda+l}{l+1}\right)^n + (1-\alpha)(n+1)\right]}.$$

We obtain

$$|f(z)| = \left| z + \sum_{j=2}^{\infty} a_j z^j \right| \leq |z| + \sum_{j=2}^{\infty} |a_j| |z|^2$$
$$\leq |z| + \frac{1-\mu}{(2-\mu)\left[\alpha\left(\frac{1+\lambda+l}{l+1}\right)^n + (1-\alpha)(n+1)\right]} |z|^2.$$

The other assertion can be proved as follows:

$$|f(z)| \geq |z| - \sum_{j=2}^{\infty} |a_j| |z|^2$$
$$\geq |z| - \frac{1-\mu}{(2-\mu)\left[\alpha\left(\frac{1+\lambda+l}{l+1}\right)^n + (1-\alpha)(n+1)\right]} |z|^2.$$

This completes the proof. □

In the same way, we can get the following result.

Theorem 3.30 (Alb Lupaş, 2015g). *Let the hypotheses of Theorem* 3.24 *be satisfied. Then*

$$(j-\mu)\left\{\alpha\left[1+(j-1)\lambda\right]^n + (1-\alpha)\frac{(n+j-1)!}{n!\,(j-1)!}\right\} \geq 0$$

and $0 \leq \mu < 1$ *imply*

$$|f(z)| \geq |z| - \frac{1-\mu}{2(2-\mu)\left[\alpha\left(\frac{1+\lambda+l}{l+1}\right)^n + (1-\alpha)(n+1)\right]} |z|^2$$

and

$$|f(z)| \leq |z| + \frac{1-\mu}{2(2-\mu)\left[\alpha\left(\frac{1+\lambda+l}{l+1}\right)^n + (1-\alpha)(n+1)\right]} |z|^2.$$

Proof. In virtue of Theorem 3.24, we have

$$2(2-\mu)\left[\alpha\left(\frac{1+\lambda+l}{l+1}\right)^n+(1-\alpha)(n+1)\right]\sum_{j=2}^{\infty}|a_j|$$
$$\leq\sum_{j=2}^{\infty}j(j-\mu)\left\{\alpha\left(\frac{1+\lambda(j-1)+l}{l+1}\right)^n+(1-\alpha)\frac{(n+j-1)!}{n!(j-1)!}\right\}|a_j|$$
$$\leq 1-\mu.$$

Thus,

$$\sum_{j=2}^{\infty}|a_j|\leq\frac{1-\mu}{2(2-\mu)\left[\alpha\left(\frac{1+\lambda+l}{l+1}\right)^n+(1-\alpha)(n+1)\right]}.$$

We obtain

$$|f(z)|=\left|z+\sum_{j=2}^{\infty}a_jz^j\right|\leq|z|+\sum_{j=2}^{\infty}|a_j||z|^2$$
$$\leq|z|+\frac{1-\mu}{2(2-\mu)\left[\alpha\left(\frac{1+\lambda+l}{l+1}\right)^n+(1-\alpha)(n+1)\right]}|z|^2.$$

The other assertion can be proved as follows:

$$|f(z)|=\left|z+\sum_{j=2}^{\infty}a_jz^j\right|\geq|z|-\sum_{j=2}^{\infty}|a_j||z|^2$$
$$\geq|z|-\frac{1-\mu}{2(2-\mu)\left[\alpha\left(\frac{1+\lambda+l}{l+1}\right)^n+(1-\alpha)(n+1)\right]}|z|^2.$$

This completes the proof. □

3.1.5 *Fractional Integral of Confluent Hypergeometric Function Applied to Define a New Subclass of Analytic Functions*

The study on fractional integral of confluent hypergeometric function which provided interesting subordination and superordination results inspired the idea of using this operator for introducing a new subclass of analytic functions in Alb Lupaş and Oros (2022a).

Given the class of functions $\mathcal{A}_n = \{f \in \mathcal{H}(U) : f(z) = z + a_{n+1}z^{n+1} + \cdots,\ z \in U\}$, written simply $\mathcal{A}$ when $n = 1$, the newly introduced subclass involves functions $f \in \mathcal{A}$ having some special properties. The aim of the paper Alb Lupaş and Oros (2022a) is to present the outcome of the study done on the new subclass which include a coefficient inequality, a distortion theorem and extreme points of the subclass. Starlikeness and convexity properties of this subclass are also discussed and partial sums of functions from the subclass are evaluated in order to obtain subclass boundary properties.

The fractional integral of confluent hypergeometric function from Definition 2.20 can be written

$$D_z^{-\lambda}\phi\left(a, c; z\right) = \frac{\Gamma\left(c\right)}{\Gamma\left(a\right)} \sum_{k=0}^{\infty} \frac{\Gamma\left(a+k\right)}{\Gamma\left(c+k\right)\Gamma\left(\lambda+k+1\right)} z^{k+\lambda}, \tag{3.1.5.1}$$

after a simple calculation. Evidently, $D_z^{-\lambda}\phi\left(a, c; z\right) \in \mathcal{H}\left[0, \lambda\right]$.

Definition 3.7 (Alb Lupaş and Oros, 2022a). Denote by $D_z^{-\lambda}\Phi\left(a, c\right)$ the operator given by the convolution of the operator $D_z^{-\lambda}\phi\left(a, c; z\right)$ given by (3.1.5.1) and the analytic function $f \in \mathcal{A}$, $D_z^{-\lambda}\Phi\left(a, c\right) : \mathcal{A} \to \mathcal{A}$,

$$D_z^{-\lambda}\Phi\left(a, c\right)\left(z\right) = D_z^{-\lambda}\phi\left(a, c; z\right) * f\left(z\right).$$

Remark 3.2. For $f \in \mathcal{A}$, $f(z) = z + \sum_{k=2}^{\infty} a_k z^k$, we get

$$D_z^{-\lambda}\Phi\left(a, c\right)\left(z\right) = z + \frac{\Gamma\left(c\right)}{\Gamma\left(a\right)} \sum_{k=2}^{\infty} \frac{\Gamma\left(a+k\right)}{\Gamma\left(c+k\right)\Gamma\left(\lambda+k+1\right)} a_{k+\lambda} z^{k+\lambda}.$$

The new subclass of analytic functions is defined as:

Definition 3.8 (Alb Lupaş and Oros, 2022a). Denote by $\mathcal{D}\Phi(d, \alpha, \beta, a, c, \lambda)$ the subclass consisting of functions $f \in \mathcal{A}$ which satisfy the inequality:

$$\left| \frac{1}{d} \left(\frac{z(D_z^{-\lambda}\Phi\left(a, c\right)\left(z\right))' + \alpha z^2 (D_z^{-\lambda}\Phi\left(a, c\right)\left(z\right))''}{(1-\alpha)D_z^{-\lambda}\Phi\left(a, c\right)\left(z\right) + \alpha z (D_z^{-\lambda}\Phi\left(a, c\right)\left(z\right))'} - 1 \right) \right| < \beta, \tag{3.1.5.2}$$

where $z \in U$, $d \in \mathbb{C} - \{0\}$, $0 < \beta \leq 1$, $0 \leq \alpha \leq 1$, $a, c \in \mathbb{C}$, $c \neq 0, -1, -2, \ldots$, $\lambda > 0$.

Study Regarding Function Subclass Coefficients

Conditions which must be complied by functions in order to belong to subclass $\mathcal{D}\Phi(d,\alpha,\beta,a,c,\lambda)$ are first established.

Theorem 3.31 (Alb Lupaş and Oros, 2022a). *A function $f \in \mathcal{A}$ belongs to $\mathcal{D}\Phi(d,\alpha,\beta,a,c,\lambda)$ if and only if the next inequality holds:*

$$\sum_{k=2}^{\infty} \frac{\Gamma(a+k)}{\Gamma(c+k)\Gamma(\lambda+k+1)}(k+\lambda-1+\beta|d|)\left[1+\alpha(k+\lambda-1)\right]a_{k+\lambda} \leq \beta|d|\frac{\Gamma(a)}{\Gamma(c)}, \tag{3.1.5.3}$$

where $z \in U$, $d \in \mathbb{C}-\{0\}$, $0<\beta\leq 1$, $0\leq\alpha\leq 1$, $a,c\in\mathbb{C}$, $c\neq 0,-1,-2,\ldots$, $\lambda>0$.

Proof. Suppose that function $f \in \mathcal{A}$ and that inequality (3.1.5.3) holds. Then we obtain

$$\left|\frac{z(D_z^{-\lambda}\Phi(a,c)(z))' + \alpha z^2(D_z^{-\lambda}\Phi(a,c)(z))''}{(1-\alpha)D_z^{-\lambda}\Phi(a,c)(z) + \alpha z(D_z^{-\lambda}\Phi(a,c)(z))'} - 1\right|$$

$$= \left|\frac{\begin{array}{c} z+\frac{\Gamma(c)}{\Gamma(a)}\sum_{k=2}^{\infty}\frac{\Gamma(a+k)}{\Gamma(c+k)\Gamma(\lambda+k+1)}(k+\lambda)a_{k+\lambda}z^{k+\lambda} \\ +\alpha\frac{\Gamma(c)}{\Gamma(a)}\sum_{k=2}^{\infty}\frac{\Gamma(a+k)}{\Gamma(c+k)\Gamma(\lambda+k+1)}(k+\lambda)(k+\lambda-1)a_{k+\lambda}z^{k+\lambda}\end{array}}{\begin{array}{c}(1-\alpha)z+(1-\alpha)\frac{\Gamma(c)}{\Gamma(a)}\sum_{k=2}^{\infty}\frac{\Gamma(a+k)}{\Gamma(c+k)\Gamma(\lambda+k+1)}a_{k+\lambda}z^{k+\lambda} \\ +\alpha z+\alpha\frac{\Gamma(c)}{\Gamma(a)}\sum_{k=2}^{\infty}\frac{\Gamma(a+k)}{\Gamma(c+k)\Gamma(\lambda+k+1)}(k+\lambda)a_{k+\lambda}z^{k+\lambda}\end{array}} - 1\right|$$

$$= \left|\frac{\sum_{k=2}^{\infty}\frac{\Gamma(a+k)}{\Gamma(c+k)\Gamma(\lambda+k+1)}(k+\lambda-1)\left[1+\alpha(k+\lambda-1)\right]a_{k+\lambda}z^{k+\lambda}}{\frac{\Gamma(a)}{\Gamma(c)}z+\sum_{k=2}^{\infty}\frac{\Gamma(a+k)}{\Gamma(c+k)\Gamma(\lambda+k+1)}\left[1+\alpha(k+\lambda-1)\right]a_{k+\lambda}z^{k+\lambda}}\right|$$

$$\leq \frac{\sum_{k=2}^{\infty}\frac{\Gamma(a+k)}{\Gamma(c+k)\Gamma(\lambda+k+1)}(k+\lambda-1)\left[1+\alpha(k+\lambda-1)\right]a_{k+\lambda}|z|^{k+\lambda-1}}{\frac{\Gamma(a)}{\Gamma(c)}-\sum_{k=2}^{\infty}\frac{\Gamma(a+k)}{\Gamma(c+k)\Gamma(\lambda+k+1)}\left[1+\alpha(k+\lambda-1)\right]a_{k+\lambda}|z|^{k+\lambda-1}}$$

$$< \beta|d|.$$

Considering for z values on the real axis and putting $z\to 1^-$, we have

$$\sum_{k=2}^{\infty} \frac{\Gamma(a+k)}{\Gamma(c+k)\Gamma(\lambda+k+1)}(k+\lambda-1+\beta|d|)\left[1+\alpha(k+\lambda-1)\right]a_{k+\lambda} \leq \beta|d|\frac{\Gamma(a)}{\Gamma(c)}.$$

Conversely, supposing $f(z) \in \mathcal{D}\Phi(d, \alpha, \beta, a, c, \lambda)$, we obtain the following inequality

$$\operatorname{Re}\left\{\left|\frac{z(D_z^{-\lambda}\Phi(a,c)(z))' + \alpha z^2(D_z^{-\lambda}\Phi(a,c)(z))''}{(1-\alpha)D_z^{-\lambda}\Phi(a,c)(z) + \alpha z(D_z^{-\lambda}\Phi(a,c)(z))'} - 1\right|\right\} > -\beta|d|$$

$$\operatorname{Re}\left\{\frac{z+\frac{\Gamma(c)}{\Gamma(a)}\sum_{k=2}^{\infty}\frac{\Gamma(a+k)}{\Gamma(c+k)\Gamma(\lambda+k+1)}(k+\lambda)[1+\alpha(k+\lambda-1)]a_{k+\lambda}z^{k+\lambda}}{z+\frac{\Gamma(c)}{\Gamma(a)}\sum_{k=2}^{\infty}\frac{\Gamma(a+k)}{\Gamma(c+k)\Gamma(\lambda+k+1)}[1+\alpha(k+\lambda-1)]a_{k+\lambda}z^{k+\lambda}} - 1 + \beta|d|\right\} > 0$$

$$\operatorname{Re}\frac{\beta|d|z+\frac{\Gamma(c)}{\Gamma(a)}\sum_{k=2}^{\infty}\frac{\Gamma(a+k)}{\Gamma(c+k)\Gamma(\lambda+k+1)}(k+\lambda-1+\beta|d|)[1+\alpha(k+\lambda-1)]a_{k+\lambda}z^{k+\lambda}}{z+\frac{\Gamma(c)}{\Gamma(a)}\sum_{k=2}^{\infty}\frac{\Gamma(a+k)}{\Gamma(c+k)\Gamma(\lambda+k+1)}[1+\alpha(k+\lambda-1)]a_{k+\lambda}z^{k+\lambda}} > 0.$$

The inequality can be written in the following form considering $\operatorname{Re}(-e^{i\theta}) \geq -|e^{i\theta}| = -1$,

$$\frac{\beta|d|r - \frac{\Gamma(c)}{\Gamma(a)}\sum_{k=2}^{\infty}\frac{\Gamma(a+k)}{\Gamma(c+k)\Gamma(\lambda+k+1)}(k+\lambda-1+\beta|d|)[1+\alpha(k+\lambda-1)]a_{k+\lambda}r^{k+\lambda}}{r - \frac{\Gamma(c)}{\Gamma(a)}\sum_{k=2}^{\infty}\frac{\Gamma(a+k)}{\Gamma(c+k)\Gamma(\lambda+k+1)}[1+\alpha(k+\lambda-1)]a_{k+\lambda}r^{k+\lambda}} > 0.$$

Applying the mean value theorem for $r \to 1^-$, we get the inequality (3.1.5.3).

The proof of theorem is complete. □

Corollary 3.9 (Alb Lupaş and Oros, 2022a). *Considering $f \in \mathcal{D}\Phi(d, \alpha, \beta, a, c, \lambda)$, the following inequalities hold:*

$$a_{k+\lambda} \leq \frac{\beta|d|\Gamma(a)\Gamma(c+k)\Gamma(\lambda+k+1)}{\Gamma(c)\Gamma(a+k)(k+\lambda-1+\beta|d|)[1+\alpha(k+\lambda-1)]}, \quad k \geq 2.$$

Distortion Results for Subclass $\mathcal{D}\Phi(d, \alpha, \beta, a, c, \lambda)$

Theorem 3.32 (Alb Lupaş and Oros, 2022a). *The following inequalities are satisfied by any function $f \in \mathcal{D}\Phi(d, \alpha, \beta, a, c, \lambda)$ for $|z| = r < 1$:*

$$r - \frac{c(c+1)\beta|d|\Gamma(\lambda+3)}{a(a+1)(\lambda+1+\beta|d|)[1+\alpha(\lambda+1)]}r^2 \leq |f(z)|$$

$$\leq r + \frac{c(c+1)\beta|d|\Gamma(\lambda+3)}{a(a+1)(\lambda+1+\beta|d|)[1+\alpha(\lambda+1)]}r^2.$$

The equality is true for

$$f(z) = z + \frac{c(c+1)\beta|d|\Gamma(\lambda+3)}{a(a+1)(\lambda+1+\beta|d|)[1+\alpha(\lambda+1)]}z^2.$$

Proof. Given that $f \in \mathcal{D}\Phi(d, \alpha, \beta, a, c, \lambda)$, from the equation (3.1.5.3) and since

$$\frac{\Gamma(a+k)}{\Gamma(c+k)\Gamma(\lambda+k+1)}(k+\lambda-1+\beta|d|)[1+\alpha(k+\lambda-1)]$$

is increasing and positive for $j \geq 2$, then we have

$$\begin{aligned}
&\frac{a(a+1)(\lambda+1+\beta|d|)[1+\alpha(\lambda+1)]}{c(c+1)\beta|d|\Gamma(\lambda+3)}\sum_{k=2}^{\infty}a_{k+\lambda}\\
&\quad\leq\sum_{k=2}^{\infty}\frac{\Gamma(a+k)}{\Gamma(c+k)\Gamma(\lambda+k+1)}(k+\lambda-1+\beta|d|)[1+\alpha(k+\lambda-1)]a_{k+\lambda}\\
&\quad\leq\beta|d|\frac{\Gamma(a)}{\Gamma(c)},
\end{aligned}$$

which is equivalent to,

$$\sum_{k=2}^{\infty}a_{k+\lambda}\leq\frac{c(c+1)\beta|d|\Gamma(\lambda+3)}{a(a+1)(\lambda+1+\beta|d|)[1+\alpha(\lambda+1)]}. \tag{3.1.5.4}$$

We obtain using the properties of modulus function for

$$\begin{aligned}
f(z)&=z+\sum_{k=2}^{\infty}a_k z^k\\
r-r^2\sum_{k=2}^{\infty}a_k&\leq r-\sum_{k=2}^{\infty}a_k r^k\leq|z|-\sum_{k=2}^{\infty}a_k|z|^k\leq|f(z)|\\
&\leq|z|+\sum_{k=2}^{\infty}a_k|z|^k\leq r+\sum_{k=2}^{\infty}a_k r^k\leq r+r^2\sum_{k=2}^{\infty}a_k
\end{aligned}$$

and by (3.1.5.4), we get

$$\begin{aligned}
r-\frac{c(c+1)\beta|d|\Gamma(\lambda+3)}{a(a+1)(\lambda+1+\beta|d|)[1+\alpha(\lambda+1)]}r^2&\leq f(z)\\
\leq r+\frac{c(c+1)\beta|d|\Gamma(\lambda+3)}{a(a+1)(\lambda+1+\beta|d|)[1+\alpha(\lambda+1)]}r^2&.
\end{aligned}$$

The proof of theorem is complete. □

Theorem 3.33 (Alb Lupaş and Oros, 2022a). *The following inequalities are satisfied by the derivative of any function $f \in \mathcal{D}\Phi(d, \alpha, \beta, a, c, \lambda)$ for $|z| = r < 1$:*

$$-\frac{2c(c+1)\beta|d|\Gamma(\lambda+3)}{a(a+1)(\lambda+1+\beta|d|)[1+\alpha(\lambda+1)]}r \leq |f'(z)|$$
$$\leq \frac{2c(c+1)\beta|d|\Gamma(\lambda+3)}{a(a+1)(\lambda+1+\beta|d|)[1+\alpha(\lambda+1)]}r.$$

For

$$f(z) = z + \frac{c(c+1)\beta|d|\Gamma(\lambda+3)}{a(a+1)(\lambda+1+\beta|d|)[1+\alpha(\lambda+1)]}z^2$$

we have the equality.

Proof. The following result is obtained by applying the properties of the modulus function:

$$f'(z) = 1 + \sum_{k=2}^{\infty} k a_k z^{k-1}$$

$$1 - \sum_{k=2}^{\infty} k a_k |z| \leq 1 - \sum_{k=2}^{\infty} k a_k |z|^{k-1} \leq |f'(z)| \leq 1 + \sum_{k=2}^{\infty} k a_k |z|^{k-1}$$
$$\leq 1 + \sum_{k=2}^{\infty} k a_k |z|.$$

Now, using relation (3.1.5.4), we obtain:

$$1 - \frac{2c(c+1)\beta|d|\Gamma(\lambda+3)}{a(a+1)(\lambda+1+\beta|d|)[1+\alpha(\lambda+1)]}r \leq |f'(z)|$$
$$\leq 1 + \frac{2c(c+1)\beta|d|\Gamma(\lambda+3)}{a(a+1)(\lambda+1+\beta|d|)[1+\alpha(\lambda+1)]}r.$$

Hence, the proof is complete. □

Properties of Partial Sums of Functions from Subclass $\mathcal{D}\Phi(d, \alpha, \beta, a, c, \lambda)$

Theorem 3.34 (Alb Lupaş and Oros, 2022a). *For $i = 1, 2, \ldots, m$, let f_i be the function defined by*

$$f_i(z) = z + \sum_{k=2}^{\infty} a_{k,i} z^k, \quad (a_{k,i} \geq 0) \tag{3.1.5.5}$$

and

$$h(z)=\sum_{i=1}^{m}\mu_i f_i(z), \qquad \left(\mu_i\geq 0,\ \sum_{i=1}^{m}\mu_i=1\right).$$

If all the function $f_i\in\mathcal{D}\Phi(d,\alpha,\beta,a,c,\lambda)$, *then* $h\in\mathcal{D}\Phi(d,\alpha,\beta,a,c,\lambda)$.

Proof. Function h has the following form:

$$h(z)=\sum_{i=1}^{m}\mu_m z+\sum_{i=1}^{m}\sum_{k=2}^{\infty}\mu_i a_{k,i}z^k=z+\sum_{k=2}^{\infty}\sum_{i=1}^{m}\mu_i a_{k,i}z^k.$$

Applying Theorem 3.31 for the functions $f_i(z)$, we obtain

$$\sum_{k=2}^{\infty}\frac{\Gamma(a+k)}{\Gamma(c+k)\Gamma(\lambda+k+1)}(k+\lambda-1+\beta|d|)\left[1+\alpha(k+\lambda-1)\right]a_{k+\lambda,i}$$
$$\leq\beta|d|\frac{\Gamma(a)}{\Gamma(c)}.$$

Thus, it is enough to prove that:

$$\begin{aligned}&\frac{\Gamma(c)}{\Gamma(a)}\sum_{k=2}^{\infty}\frac{\Gamma(a+k)}{\Gamma(c+k)\Gamma(\lambda+k+1)}\\&\quad\times(k+\lambda-1+\beta|d|)\left[1+\alpha(k+\lambda-1)\right]\left(\sum_{i=1}^{m}\mu_i a_{k+\lambda,i}\right)\\&=\frac{\Gamma(c)}{\Gamma(a)}\sum_{i=1}^{m}\mu_i\sum_{k=2}^{\infty}\frac{\Gamma(a+k)}{\Gamma(c+k)\Gamma(\lambda+k+1)}\\&\quad\times(k+\lambda-1+\beta|d|)\left[1+\alpha(k+\lambda-1)\right]a_{k+\lambda,i}\\&\leq\sum_{i=1}^{m}\mu_i\beta|d|=\beta|d|,\end{aligned}$$

which completes the proof. □

Theorem 3.35 (Alb Lupaş and Oros, 2022a). *Let*

$$f_1(z)=z,$$

and

$$f_k(z) = z + \frac{\beta|d|\Gamma(a)\Gamma(c+k)\Gamma(\lambda+k+1)}{\Gamma(c)\Gamma(a+k)(k+\lambda-1+\beta|d|)[1+\alpha(k+\lambda-1)]} z^{k+\lambda}, \quad k \geq 2.$$

The function f belongs to the subclass $\mathcal{D}\Phi(d,\alpha,\beta,a,c,\lambda)$ if and only if it can be written in the following form:

$$f(z) = \mu_1 f_1(z) + \sum_{k=2}^{\infty} \mu_k f_k(z),$$

where $\mu_1 \geq 0$, $\mu_k \geq 0$, $k \geq 2$ and $\mu_1 + \sum_{k=2}^{\infty} \mu_k = 1$.

Proof. Suppose that f can be written as

$$\begin{aligned} f(z) &= \mu_1 f_1(z) + \sum_{k=2}^{\infty} \mu_k f_k(z) \\ &= z + \sum_{k=2}^{\infty} \frac{\beta|d|\Gamma(a)\Gamma(c+k)\Gamma(\lambda+k+1)}{\Gamma(c)\Gamma(a+k)(k+\lambda-1+\beta|d|)[1+\alpha(k+\lambda-1)]} \mu_k z^{k+\lambda}. \end{aligned}$$

Thus,

$$\begin{aligned} &\sum_{k=2}^{\infty} \frac{\Gamma(c)\Gamma(a+k)(k+\lambda-1+\beta|d|)[1+\alpha(k+\lambda-1)]}{\beta|d|\Gamma(a)\Gamma(c+k)\Gamma(\lambda+k+1)} \\ &\qquad \cdot \frac{\beta|d|\Gamma(a)\Gamma(c+k)\Gamma(\lambda+k+1)}{\Gamma(c)\Gamma(a+k)(k+\lambda-1+\beta|d|)[1+\alpha(k+\lambda-1)]} \mu_k \\ &\quad = \sum_{k=2}^{\infty} \mu_k = 1 - \mu_1 \leq 1. \end{aligned}$$

Hence, $f(z) \in \mathcal{D}\Phi(d,\alpha,\beta,a,c,\lambda)$.

Conversely, suppose that $f \in \mathcal{D}\Phi(d,\alpha,\beta,a,c,\lambda)$.

Considering

$$\mu_k = \frac{\Gamma(c)\Gamma(a+k)(k+\lambda-1+\beta|d|)[1+\alpha(k+\lambda-1)]}{\beta|d|\Gamma(a)\Gamma(c+k)\Gamma(\lambda+k+1)} a_k,$$

and

$$\mu_1 = 1 - \sum_{k=2}^{\infty} \mu_k,$$

we obtain that

$$f(z) = \mu_1 f_1(z) + \sum_{k=2}^{\infty} \mu_k f_k(z),$$

which complete the proof. □

Corollary 3.10 (Alb Lupaş and Oros, 2022a). *The extreme points of the subclass $\mathcal{D}\Phi(d, \alpha, \beta, a, c, \lambda)$ are the functions*

$$f_1(z) = z,$$

and

$$f_j(z) = z + \frac{\beta|d|\Gamma(a)\Gamma(c+k)\Gamma(\lambda+k+1)}{\Gamma(c)\Gamma(a+k)(k+\lambda-1+\beta|d|)[1+\alpha(k+\lambda-1)]} z^k, \quad k \geq 2.$$

Relations of Inclusion Involving Subclass $D\Phi(d, \alpha, \beta, a, c, \lambda)$ and Neighborhoods

We remind that the δ-neighborhood of a function $f(z) \in \mathcal{A}$ is

$$N_\delta(f) = \left\{ g \in \mathcal{A} : g(z) = z + \sum_{k=2}^{\infty} b_k z^k \text{ and } \sum_{k=2}^{\infty} k|a_k - b_k| \leq \delta \right\}. \quad (3.1.5.6)$$

A particular case when $e(z) = z$

$$N_\delta(e) = \left\{ g \in \mathcal{A} : g(z) = z + \sum_{k=2}^{\infty} b_k z^k \text{ and } \sum_{k=2}^{\infty} k|b_k| \leq \delta \right\}. \quad (3.1.5.7)$$

A function $f \in \mathcal{A}$ belongs to the subclass $\mathcal{D}\Phi^{\zeta}(d, \alpha, \beta, a, c, \lambda)$ if there exists a function $h \in \mathcal{D}\Phi(d, \alpha, \beta, a, c, \lambda)$ such that

$$\left| \frac{f(z)}{h(z)} - 1 \right| < 1 - \zeta, \quad z \in U, \ 0 \leq \zeta < 1. \quad (3.1.5.8)$$

Theorem 3.36 (Alb Lupaş and Oros, 2022a). *If*

$$\frac{\Gamma(a+k)}{\Gamma(c+k)\Gamma(\lambda+k+1)}[1+\alpha(\lambda+k-1)] \geq \frac{\Gamma(a+2)}{\Gamma(c+2)\Gamma(\lambda+3)}[1+\alpha(\lambda+1)], \quad k \geq 2$$

and

$$\delta = \frac{2\beta|d|c\,(c+1)\,\Gamma\,(\lambda+3)}{a\,(a+1)\,(\lambda+1+\beta|d|)\,[1+\alpha\,(\lambda+1)]},$$

then

$$\mathcal{D}\Phi(d,\alpha,\beta,a,c,\lambda) \subset N_\delta(e).$$

Proof. Consider $f \in \mathcal{D}\Phi(d,\alpha,\beta,a,c,\lambda)$. Using relation (3.1.5.3) of Theorem 3.31 and the condition

$$\frac{\Gamma\,(a+k)}{\Gamma\,(c+k)\,\Gamma\,(\lambda+k+1)}\,[1+\alpha\,(\lambda+k-1)]$$
$$\geq \frac{\Gamma\,(a+2)}{\Gamma\,(c+2)\,\Gamma\,(\lambda+3)}\,[1+\alpha\,(\lambda+1)] \quad \text{for } k \geq 2,$$

we get

$$\frac{\Gamma\,(a+2)}{\Gamma\,(c+2)\,\Gamma\,(\lambda+3)}\,(\lambda+1+\beta|d|)\,[1+\alpha\,(\lambda+1)]\sum_{k=2}^{\infty} a_{k+\lambda}$$
$$\leq \sum_{k=2}^{\infty} \frac{\Gamma\,(a+k)}{\Gamma\,(c+k)\,\Gamma\,(\lambda+k+1)}\,(k+\lambda-1+\beta|d|)\,[1+\alpha\,(k+\lambda-1)]\,a_{k+\lambda}$$
$$\leq \beta|d|\frac{\Gamma\,(a)}{\Gamma\,(c)}$$

which implies

$$\sum_{k=2}^{\infty} a_{k+\lambda} \leq \frac{\beta|d|c\,(c+1)\,\Gamma\,(\lambda+3)}{a\,(a+1)\,(\lambda+1+\beta|d|)\,[1+\alpha\,(\lambda+1)]}. \tag{3.1.5.9}$$

Using relation (3.1.5.3) of Theorem 3.31 and relation (3.1.5.9), we obtain

$$\frac{\beta|d|c\,(c+1)\,\Gamma\,(\lambda+3)}{a\,(a+1)\,(\lambda+1+\beta|d|)\,[1+\alpha\,(\lambda+1)]}\sum_{k=2}^{\infty} a_{k+\lambda} \leq \beta|d|,$$

$$\frac{2\beta|d|c\,(c+1)\,\Gamma\,(\lambda+3)}{a\,(a+1)\,(\lambda+1+\beta|d|)\,[1+\alpha\,(\lambda+1)]}\sum_{k=2}^{\infty} a_{k+\lambda} \leq 2\beta|d|$$

$$\sum_{k=2}^{\infty} k a_{k+\lambda} \leq \frac{2\beta|d|c\,(c+1)\,\Gamma\,(\lambda+3)}{a\,(a+1)\,(\lambda+1+\beta|d|)\,[1+\alpha\,(\lambda+1)]} = \delta,$$

applying (3.1.5.6), we get $f \in N_\delta(e)$.

The proof of the theorem is complete. □

Theorem 3.37 (Alb Lupaş and Oros, 2022a). *If* $h \in \mathcal{D}\Phi(d, \alpha, \beta, a, c, \lambda)$ *and*

$$\zeta = 1 - \frac{\delta}{2} \frac{a(a+1)(\lambda+1+\beta|d|)[1+\alpha(\lambda+1)]}{a(a+1)(\lambda+1+\beta|d|)[1+\alpha(\lambda+1)] - \beta|d|c(c+1)\Gamma(\lambda+3)}, \tag{3.1.5.10}$$

then

$$N_\delta(h) \subset \mathcal{D}\Phi^\zeta(d, \alpha, \beta, a, c, \lambda).$$

Proof. Assume that $f \in N_\delta(h)$, using (3.1.5.6), we find

$$\sum_{k=2}^{\infty} k|a_k - b_k| \le \delta,$$

which give the coefficient inequality

$$\sum_{k=2}^{\infty} |a_k - b_k| \le \frac{\delta}{2}. \tag{3.1.5.11}$$

Applying relation (3.1.5.9), taking account that $h \in \mathcal{D}\Phi(d, \alpha, \beta, a, c, \lambda)$, we obtain

$$\sum_{k=2}^{\infty} b_k \le \frac{\beta|d|c(c+1)\Gamma(\lambda+3)}{a(a+1)(\lambda+1+\beta|d|)[1+\alpha(\lambda+1)]}. \tag{3.1.5.12}$$

Using (3.1.5.11) and (3.1.5.12), we get

$$\begin{aligned}
\left|\frac{f(z)}{h(z)} - 1\right| &\le \frac{\sum_{k=2}^{\infty} |a_k - b_k|}{1 - \sum_{k=2}^{\infty} b_k} \le \frac{\delta}{2\left(1 - \frac{\beta|d|c(c+1)\Gamma(\lambda+3)}{a(a+1)(\lambda+1+\beta|d|)[1+\alpha(\lambda+1)]}\right)} \\
&\le \frac{\delta}{2} \frac{a(a+1)(\lambda+1+\beta|d|)[1+\alpha(\lambda+1)]}{a(a+1)(\lambda+1+\beta|d|)[1+\alpha(\lambda+1)] - \beta|d|c(c+1)\Gamma(\lambda+3)} \\
&= 1 - \zeta.
\end{aligned}$$

Using condition (3.1.5.8), we obtain that $f \in \mathcal{D}\Phi^\zeta(d, \alpha, \beta, a, c, \lambda)$, where ζ is given by (3.1.5.10). □

Starlikeness and Convexity Properties

Theorem 3.38 (Alb Lupaş and Oros, 2022a). *Consider the function* $f \in \mathcal{A}$ *from the subclass* $\mathcal{D}\Phi(d, \alpha, \beta, a, c, \lambda)$. *The function* f *is univalent* *starlike of order* δ, $0 \leq \delta < 1$, *in* $|z| < r_1$, *where*

$$r_1 = \inf_k \left\{ \frac{(1-\delta)\Gamma(a+k)\Gamma(c)(k+\lambda-1+\beta|d|)[1+\alpha(k+\lambda-1)]}{\Gamma(a)\Gamma(c+k)\Gamma(\lambda+k+1)\beta|d|(2-\delta)} \right\}^{\frac{1}{k-1}}.$$

When function f *has the form given by*

$$f_k(z) = z + \frac{\Gamma(a)\Gamma(c+k)\Gamma(\lambda+k+1)\beta|d|}{\Gamma(a+k)\Gamma(c)(k+\lambda-1+\beta|d|)[1+\alpha(k+\lambda-1)]} z^k, \quad k \geq 2,$$

the result is sharp.

Proof. We have to show that

$$\left| \frac{zf'(z)}{f(z)} - 1 \right| \leq 1 - \delta \quad \text{if } |z| < r_1.$$

Taking account that

$$\left| \frac{zf'(z)}{f(z)} - 1 \right| = \left| \frac{\sum_{k=2}^{\infty}(k-1)a_k z^{k-1}}{1+\sum_{k=2}^{\infty} a_k z^{k-1}} \right| \leq \frac{\sum_{k=2}^{\infty}(k-1)a_k |z|^{k-1}}{1-\sum_{k=2}^{\infty} a_k |z|^{k-1}},$$

to prove the theorem, we must show that

$$\frac{\sum_{k=2}^{\infty}(k-1)a_k |z|^{k-1}}{1-\sum_{k=2}^{\infty} a_k |z|^{k-1}} \leq 1 - \delta,$$

which is equivalent to

$$\sum_{k=2}^{\infty} (k-\delta) a_k |z|^{k-1} \leq 1 - \delta,$$

and applying Theorem 3.31, we have

$$|z| \leq \left\{ \frac{(1-\delta)\Gamma(a+k)\Gamma(c)(k+\lambda-1+\beta|d|)[1+\alpha(k+\lambda-1)]}{\Gamma(a)\Gamma(c+k)\Gamma(\lambda+k+1)\beta|d|(2-\delta)} \right\}^{\frac{1}{k-1}}.$$

Hence, the proof is complete. □

Theorem 3.39 (Alb Lupaş and Oros, 2022a). *Consider the function* $f \in \mathcal{A}$ *from the subclass* $\mathcal{D}\Phi(d, \alpha, \beta, a, c, \lambda)$. *The function* f *is univalent convex of order* δ, $0 \leq \delta \leq 1$, *in* $|z| < r_2$, *where*

$$r_2 = \inf_k \left\{ \frac{(1-\delta)\Gamma(a+k)\Gamma(c)(k+\lambda-1+\beta|d|)[1+\alpha(k+\lambda-1)]}{2(k-\delta)\Gamma(a)\Gamma(c+k)\Gamma(\lambda+k+1)\beta|d|(2-\delta)} \right\}^{\frac{1}{k-1}}.$$

When function f has the form given by

$$f_k(z) = z + \frac{\Gamma(a)\Gamma(c+k)\Gamma(\lambda+k+1)\beta|d|}{\Gamma(a+k)\Gamma(c)(k+\lambda-1+\beta|d|)[1+\alpha(k+\lambda-1)]} z^k, \quad k \geq 2, \tag{3.1.5.13}$$

the result is sharp.

Proof. It is sufficient to prove that

$$\left|\frac{zf''(z)}{f'(z))}\right| \leq 1-\delta \quad \text{if } |z| < r_2.$$

Taking account that

$$\left|\frac{zf''(z)}{f'(z)}\right| = \left|\frac{\sum_{k=2}^{\infty} k(k-1)a_k z^{k-1}}{1+\sum_{k=2}^{\infty} k a_k z^{k-1}}\right| \leq \frac{\sum_{k=2}^{\infty} k(k-1)a_k |z|^{k-1}}{1-\sum_{k=2}^{\infty} k a_k |z|^{k-1}},$$

to prove the theorem, we must show that

$$\frac{\sum_{k=2}^{\infty} k(k-1)a_k |z|^{k-1}}{1-\sum_{k=2}^{\infty} k a_k |z|^{k-1}} \leq 1-\delta,$$

$$\sum_{k=2}^{\infty} k(k-\delta)a_k |z|^{k-1} \leq 1-\delta,$$

and applying Theorem 3.31, we get

$$|z|^{k-1} \leq \frac{(1-\delta)\Gamma(a+k)\Gamma(c)(k+\lambda-1+\beta|d|)[1+\alpha(k+\lambda-1)]}{2(k-\delta)\Gamma(a)\Gamma(c+k)\Gamma(\lambda+k+1)\beta|d|},$$

or

$$|z| \leq \left\{\frac{(1-\delta)\Gamma(a+k)\Gamma(c)(k+\lambda-1+\beta|d|)[1+\alpha(k+\lambda-1)]}{2(k-\delta)\Gamma(a)\Gamma(c+k)\Gamma(\lambda+k+1)\beta|d|}\right\}^{\frac{1}{k-1}}.$$

Hence, the proof is complete. □

Theorem 3.40 (Alb Lupaş and Oros, 2022a). *Consider the function $f \in \mathcal{A}$ from the subclass $\mathcal{D}\Phi(d,\alpha,\beta,a,c,\lambda)$. The function f is univalent close-to-convex of order δ, $0 \leq \delta < 1$, in $|z| < r_3$, where*

$$r_3 = \inf_k \left\{\frac{(1-\delta)\Gamma(a+k)\Gamma(c)(k+\lambda-1+\beta|d|)[1+\alpha(k+\lambda-1)]}{k\Gamma(a)\Gamma(c+k)\Gamma(\lambda+k+1)\beta|d|}\right\}^{\frac{1}{k-1}}.$$

The result is sharp for the function f given by (3.1.5.13).

Proof. It is sufficient to prove that

$$|f'(z) - 1| \leq 1 - \delta \quad \text{if } |z| < r_3.$$

We obtain

$$|f'(z) - 1| = \left| \sum_{k=2}^{\infty} k a_k z^{k-1} \right| \leq \sum_{k=2}^{\infty} k a_k |z|^{k-1},$$

thus $|f'(z) - 1| \leq 1 - \delta$ if $\sum_{k=2}^{\infty} \frac{k a_k}{1-\delta} |z|^{k-1} \leq 1$. Using Theorem 3.31, the inequality holds when

$$|z|^{k-1} \leq \frac{(1-\delta)\Gamma(a+k)\Gamma(c)(k+\lambda-1+\beta|d|)[1+\alpha(k+\lambda-1)]}{k\Gamma(a)\Gamma(c+k)\Gamma(\lambda+k+1)\beta|d|}$$

or

$$|z| \leq \left\{ \frac{(1-\delta)\Gamma(a+k)\Gamma(c)(k+\lambda-1+\beta|d|)[1+\alpha(k+\lambda-1)]}{k\Gamma(a)\Gamma(c+k)\Gamma(\lambda+k+1)\beta|d|} \right\}^{\frac{1}{k-1}}.$$

Hence, the proof is complete. □

3.1.6 *The Study of a Subclass of Analytic Functions Defined by the Fractional Integral of Confluent Hypergeometric Function*

The study done for obtaining the original results of the paper Alb Lupaş and Oros (2022b) involves the fractional integral of confluent hypergeometric function and presents its new applications for introducing a certain subclass of analytic functions $\mathcal{D}\Phi(\mu, \lambda, \beta, \gamma, a, c)$. Conditions for functions to belong to this subclass are determined and the subclass is investigated considering aspects regarding coefficient bounds as well as partial sums of these functions. Distortion properties of the functions belonging to the subclass are proved and radii estimates are established for starlikeness and convexity properties of the subclass.

A new approach was taken when the convolution product between fractional integral of confluent hypergeometric function and an analytic function resulted in an interesting operator in the paper Alb Lupaş and Oros (2022a).

Denote by $D_z^{-\lambda}\Phi(a,c)$ the operator given by the Hadamard product (the convolution product) of the fractional integral of confluent hypergeometric function and the analytic function $f \in \mathcal{A}$, $D_z^{-\lambda}\Phi(a,c) : \mathcal{A} \to \mathcal{A}$,

$$D_z^{-\lambda}\Phi(a,c)(z) = D_z^{-\lambda}\phi(a,c;z) * f(z).$$

If $f \in \mathcal{A}$, $f(z) = z + \sum_{k=2}^{\infty} a_k z^k$, then

$$D_z^{-\lambda}\Phi(a,c)(z) = z + \frac{\Gamma(c)}{\Gamma(a)} \sum_{k=2}^{\infty} \frac{\Gamma(a+k)}{\Gamma(c+k)\Gamma(\lambda+k+1)} a_{k+\lambda} z^{k+\lambda}.$$

A new subclass of functions is introduced as follows:

Definition 3.9 (Alb Lupaş and Oros, 2022b). For $\mu \geq 0$, $\lambda \in \mathbb{N}$, $\gamma \in \mathbb{C} - \{0\}$, $0 < \beta \leq 1$, $a, c \in \mathbb{C}$, $c \neq 0, -1, -2, \ldots$, let $\mathcal{D}\Phi(\mu, \lambda, \beta, \gamma, a, c)$ be the subclass of $\mathcal{A}$ consisting of functions that satisfying the inequality

$$\left| \frac{\lambda(1-\mu)\frac{D_z^{-\lambda}\Phi(a,c)(z)}{z} + \mu\left(D_z^{-\lambda}\Phi(a,c)(z)\right)'}{\lambda(1-\mu)\frac{D_z^{-\lambda}\Phi(a,c)(z)}{z} + \mu\left(D_z^{-\lambda}\Phi(a,c)(z)\right)' - \gamma} \right| < \beta. \tag{3.1.6.1}$$

Coefficient Estimates and Partial Sums

The study on subclass $\mathcal{D}\Phi(\mu, \lambda, \beta, \gamma, a, c)$ begins with giving necessary and sufficient conditions for membership of functions $f \in \mathcal{A}$ to the subclass.

Theorem 3.41 (Alb Lupaş and Oros, 2022b). *A function $f \in \mathcal{A}$ belongs to the subclass $\mathcal{D}\Phi(\mu, \lambda, \beta, \gamma, a, c)$ if and only if it satisfies:*

$$\sum_{k=2}^{\infty} \frac{\Gamma(a+k)}{\Gamma(c+k)\Gamma(\lambda+k+1)} (\lambda + k\mu) a_{k+\lambda} < \frac{\beta|\gamma|\Gamma(a)}{(\beta+1)\Gamma(c)} - \frac{|\lambda+\mu-\lambda\mu|\Gamma(a)}{\Gamma(c)}. \tag{3.1.6.2}$$

The result is sharp for the function F defined by

$$F(z) = z + \frac{(\beta|\gamma| - (\beta+1)|\lambda+\mu-\lambda\mu|)\Gamma(a)\Gamma(c+k)\Gamma(\lambda+k+1)}{(\beta+1)(\lambda+k\mu)\Gamma(c)\Gamma(a+k)} z^{k+\lambda}, \tag{3.1.6.3}$$

$k \geq 2$.

Proof. Suppose that function f satisfies (3.1.6.2). Then for $|z| < 1$, we have

$$\left|\lambda(1-\mu)\frac{D_z^{-\lambda}\Phi(a,c)(z)}{z} + \mu\left(D_z^{-\lambda}\Phi(a,c)(z)\right)'\right|$$

$$-\beta\left|\lambda(1-\mu)\frac{D_z^{-\lambda}\Phi(a,c)(z)}{z} + \mu\left(D_z^{-\lambda}\Phi(a,c)(z)\right)' - \gamma\right|$$

$$= \left|(\lambda+\mu-\lambda\mu) + \frac{\Gamma(c)}{\Gamma(a)}\sum_{k=2}^{\infty}\frac{\Gamma(a+k)}{\Gamma(c+k)\Gamma(\lambda+k+1)}(\lambda+k\mu)a_{k+\lambda}z^{k+\lambda-1}\right|$$

$$-\beta\left|(\lambda+\mu-\lambda\mu) + \frac{\Gamma(c)}{\Gamma(a)}\sum_{k=2}^{\infty}\frac{\Gamma(a+k)}{\Gamma(c+k)\Gamma(\lambda+k+1)}\right.$$

$$\left.\times(\lambda+k\mu)a_{k+\lambda}z^{k+\lambda-1} - \gamma\right|$$

$$\leq |\lambda+\mu-\lambda\mu|$$

$$+\left|\frac{\Gamma(c)}{\Gamma(a)}\sum_{k=2}^{\infty}\frac{\Gamma(a+k)}{\Gamma(c+k)\Gamma(\lambda+k+1)}(\lambda+k\mu)a_{k+\lambda}z^{k+\lambda-1}\right|$$

$$-\beta|\gamma| + \beta|\lambda+\mu-\lambda\mu|$$

$$+\beta\left|\frac{\Gamma(c)}{\Gamma(a)}\sum_{k=2}^{\infty}\frac{\Gamma(a+k)}{\Gamma(c+k)\Gamma(\lambda+k+1)}(\lambda+k\mu)a_{k+\lambda}z^{k+\lambda-1}\right|$$

$$< (\beta+1)|\lambda+\mu-\lambda\mu| - \beta|\gamma|$$

$$+\frac{(\beta+1)\Gamma(c)}{\Gamma(a)}\sum_{k=2}^{\infty}\frac{\Gamma(a+k)}{\Gamma(c+k)\Gamma(\lambda+k+1)}(\lambda+k\mu)a_{k+\lambda}z^{k+\lambda-1} < 0.$$

Hence, by using the maximum modulus theorem and (3.1.6.1), $f \in \mathcal{D}\Phi(\mu,\lambda,\beta,\gamma,a,c)$.

Conversely, assume that

$$\left|\frac{\lambda(1-\mu)\frac{D_z^{-\lambda}\Phi(a,c)(z)}{z} + \mu\left(D_z^{-\lambda}\Phi(a,c)(z)\right)'}{\lambda(1-\mu)\frac{D_z^{-\lambda}\Phi(a,c)(z)}{z} + \mu\left(D_z^{-\lambda}\Phi(a,c)(z)\right)' - \gamma}\right|$$

$$= \left|\frac{(\lambda+\mu-\lambda\mu) + \frac{\Gamma(c)}{\Gamma(a)}\sum_{k=2}^{\infty}\frac{\Gamma(a+k)}{\Gamma(c+k)\Gamma(\lambda+k+1)}(\lambda+k\mu)a_{k+\lambda}z^{k+\lambda-1}}{(\lambda+\mu-\lambda\mu) + \frac{\Gamma(c)}{\Gamma(a)}\sum_{k=2}^{\infty}\frac{\Gamma(a+k)}{\Gamma(c+k)\Gamma(\lambda+k+1)}(\lambda+k\mu)a_{k+\lambda}z^{k+\lambda-1} - \gamma}\right| < \beta,$$

$z \in U$.

Since $\mathrm{Re}(z) \leq |z|$ for all $z \in U$, we have

$$\mathrm{Re}\left\{\frac{(\lambda+\mu-\lambda\mu)+\frac{\Gamma(c)}{\Gamma(a)}\sum_{k=2}^{\infty}\frac{\Gamma(a+k)}{\Gamma(c+k)\Gamma(\lambda+k+1)}(\lambda+k\mu)\,a_{k+\lambda}z^{k+\lambda-1}}{(\lambda+\mu-\lambda\mu)+\frac{\Gamma(c)}{\Gamma(a)}\sum_{k=2}^{\infty}\frac{\Gamma(a+k)}{\Gamma(c+k)\Gamma(\lambda+k+1)}(\lambda+k\mu)\,a_{k+\lambda}z^{k+\lambda-1}-\gamma}\right\}<\beta. \tag{3.1.6.4}$$

By choosing choose values of z on the real axis so that $\lambda(1-\mu)\frac{D_z^{-\lambda}\Phi(a,c)(z)}{z}+\mu(D_z^{-\lambda}\Phi(a,c)(z))'$ is real and letting $z\to 1$ through real values, we obtain the desired inequality (3.1.6.2). □

Corollary 3.11 (Alb Lupaş and Oros, 2022b). *If a function f belongs to subclass $\mathcal{D}\Phi(\mu,\lambda,\beta,\gamma,a,c)$, then the following inequality holds:*

$$a_{k+\lambda}\leq\frac{(\beta\,|\gamma|-(\beta+1)\,|\lambda+\mu-\lambda\mu|)\,\Gamma\,(a)\,\Gamma\,(c+k)\,\Gamma\,(\lambda+k+1)}{(\beta+1)\,(\lambda+k\mu)\,\Gamma\,(c)\,\Gamma\,(a+k)},\quad k\geq 2, \tag{3.1.6.5}$$

with equality only for functions defined by (3.1.6.3).

Theorem 3.42 (Alb Lupaş and Oros, 2022b). *Let $f_1(z)=z$ and*

$$f_k(z)=z-\frac{(\beta\,|\gamma|-(\beta+1)\,|\lambda+\mu-\lambda\mu|)\,\Gamma\,(a)\,\Gamma\,(c+k)\,\Gamma\,(\lambda+k+1)}{(\beta+1)\,(\lambda+k\mu)\,\Gamma\,(c)\,\Gamma\,(a+k)}z^{k+\lambda}, \tag{3.1.6.6}$$

$k\geq 2$, *for* $\mu\geq 0$, $\lambda\in\mathbb{N}$, $\gamma\in\mathbb{C}-\{0\}$, $0<\beta\leq 1$, $a,c\in\mathbb{C}$, $c\neq 0,-1,-2,\ldots$.

Then f is in the subclass $\mathcal{D}\Phi(\mu,\lambda,\beta,\gamma,a,c)$ if and only if it can be expressed in the form

$$f(z)=\sum_{k=1}^{\infty}\omega_k f_k(z), \tag{3.1.6.7}$$

where $\omega_k\geq 0$ and $\sum_{k=1}^{\infty}\omega_k=1$.

Proof. Suppose f can be written as in (3.1.6.7). Then

$$f(z)=z-\sum_{k=2}^{\infty}\omega_k\frac{(\beta\,|\gamma|-(\beta+1)\,|\lambda+\mu-\lambda\mu|)\,\Gamma\,(a)\,\Gamma\,(c+k)\,\Gamma\,(\lambda+k+1)}{(\beta+1)\,(\lambda+k\mu)\,\Gamma\,(c)\,\Gamma\,(a+k)}\times z^{k+\lambda}.$$

Now,

$$\sum_{k=2}^{\infty}\frac{(\beta+1)(\lambda+k\mu)\Gamma(c)\Gamma(a+k)}{(\beta|\gamma|-(\beta+1)|\lambda+\mu-\lambda\mu|)\Gamma(a)\Gamma(c+k)\Gamma(\lambda+k+1)}\omega_k$$
$$\times\frac{(\beta|\gamma|-(\beta+1)|\lambda+\mu-\lambda\mu|)\Gamma(a)\Gamma(c+k)\Gamma(\lambda+k+1)}{(\beta+1)(\lambda+k\mu)\Gamma(c)\Gamma(a+k)}$$
$$=\sum_{k=2}^{\infty}\omega_k=1-\omega_1\leq 1.$$

Thus, $f\in\mathcal{D}\Phi(\mu,\lambda,\beta,\gamma,a,c)$.

Conversely, let $f\in\mathcal{D}\Phi(\mu,\lambda,\beta,\gamma,a,c)$. Then by using (3.1.6.5), setting

$$\omega_k=\frac{(\beta|\gamma|-(\beta+1)|\lambda+\mu-\lambda\mu|)\Gamma(a)\Gamma(c+k)\Gamma(\lambda+k+1)}{(\beta+1)(\lambda+k\mu)\Gamma(c)\Gamma(a+k)}a_{k+\lambda},\quad k\geq 2$$

and $\omega_1=1-\sum_{k=2}^{\infty}\omega_k$, we have $f(z)=\sum_{k=1}^{\infty}\omega_k f_k(z)$. And this completes the proof of theorem. □

Distortion Bounds

Next we obtain distortion bounds for the subclass $\mathcal{D}\Phi(\mu,\lambda,\beta,\gamma,a,c)$.

Theorem 3.43 (Alb Lupaş and Oros, 2022b). *If $f\in\mathcal{D}\Phi(\mu,\lambda,\beta,\gamma,a,c)$, then*

$$r-\frac{c(c+1)(\beta|\gamma|-(\beta+1)|\lambda+\mu-\lambda\mu|)\Gamma(\lambda+3)}{a(a+1)(\beta+1)(\lambda+2\mu)}r^2\leq|f(z)|$$
$$\leq r+\frac{c(c+1)(\beta|\gamma|-(\beta+1)|\lambda+\mu-\lambda\mu|)\Gamma(\lambda+3)}{a(a+1)(\beta+1)(\lambda+2\mu)}r^2 \tag{3.1.6.8}$$

holds if the sequence $\{\sigma_k(\mu,\lambda,a,c)\}_{k=2}^{\infty}$ is non-decreasing, and

$$1-2\frac{c(c+1)(\beta|\gamma|-(\beta+1)|\lambda+\mu-\lambda\mu|)\Gamma(\lambda+3)}{a(a+1)(\beta+1)(\lambda+2\mu)}r^2\leq|f'(z)|$$
$$\leq 1+2\frac{c(c+1)(\beta|\gamma|-(\beta+1)|\lambda+\mu-\lambda\mu|)\Gamma(\lambda+3)}{a(a+1)(\beta+1)(\lambda+2\mu)}r^2 \tag{3.1.6.9}$$

holds if the sequence $\{\frac{\sigma_k(\mu,\lambda,a,c)}{k}\}_{k=2}^{\infty}$ *is non-decreasing, where*

$$\sigma_k(\mu,\lambda,a,c) = \frac{(\lambda+k\mu)\,\Gamma\,(a+k)}{\Gamma\,(c+k)\,\Gamma\,(\lambda+k+1)}.$$

The bounds in (3.1.6.8) *and* (3.1.6.9) *are sharp, for* f *given by*

$$f(z) = z + \frac{c\,(c+1)\,(\beta\,|\gamma| - (\beta+1)\,|\lambda+\mu-\lambda\mu|)\,\Gamma\,(\lambda+3)}{a\,(a+1)\,(\beta+1)\,(\lambda+2\mu)}r^2, \quad z = \pm r. \tag{3.1.6.10}$$

Proof. In view of Theorem 3.41, we have

$$\sum_{k=2}^{\infty} a_{k+\lambda} \le \frac{c\,(c+1)\,(\beta\,|\gamma| - (\beta+1)\,|\lambda+\mu-\lambda\mu|)\,\Gamma\,(\lambda+3)}{a\,(a+1)\,(\beta+1)\,(\lambda+2\mu)}. \tag{3.1.6.11}$$

We obtain

$$|z| - |z|^2 \sum_{k=2}^{\infty} a_{k+\lambda} \le |f(z)| \le |z| + |z|^2 \sum_{k=2}^{\infty} a_{k+\lambda}.$$

Thus,

$$\begin{aligned} r - \frac{c\,(c+1)\,(\beta\,|\gamma| - (\beta+1)\,|\lambda+\mu-\lambda\mu|)\,\Gamma\,(\lambda+3)}{a\,(a+1)\,(\beta+1)\,(\lambda+2\mu)}r^2 &\le |f(z)| \\ \le r + \frac{c\,(c+1)\,(\beta\,|\gamma| - (\beta+1)\,|\lambda+\mu-\lambda\mu|)\,\Gamma\,(\lambda+3)}{a\,(a+1)\,(\beta+1)\,(\lambda+2\mu)}r^2. \end{aligned} \tag{3.1.6.12}$$

Hence, (3.1.6.8) follows from (3.1.6.12). Further,

$$\sum_{k=2}^{\infty} (k+\lambda)\,a_{k+\lambda} \le \frac{c\,(c+1)\,(\beta\,|\gamma| - (\beta+1)\,|\lambda+\mu-\lambda\mu|)\,\Gamma\,(\lambda+3)}{a\,(a+1)\,(\beta+1)\,(\lambda+2\mu)}.$$

Hence, (3.1.6.9) follows from

$$1 - r^2 \sum_{k=2}^{\infty} (k+\lambda)\,a_{k+\lambda} \le |f'(z)| \le 1 + r^2 \sum_{k=2}^{\infty} (k+\lambda)\,a_{k+\lambda}.$$

□

Radius of Starlikeness and Convexity

The radii of close-to-convexity, starlikeness and convexity for the subclass $\mathcal{D}\Phi(\mu,\lambda,\beta,\gamma,a,c)$ are given in the following.

Theorem 3.44 (Alb Lupaş and Oros, 2022b). *Let the function $f\in\mathcal{A}$ belongs to the subclass $\mathcal{D}\Phi(\mu,\lambda,\beta,\gamma,a,c)$, Then f is close-to-convex of order δ, $0\le\delta<1$ in the disc $|z|<r$, where*

$$r := \inf_{k\geq 2}\sqrt{\frac{(1-\delta)\,(\beta+1)\,(\lambda+k\mu)\,\Gamma\,(c)\,\Gamma\,(a+k)}{(\beta\,|\gamma|-(\beta+1)|\lambda+\mu-\lambda\mu|)(k+\lambda)\,\Gamma(a)\,\Gamma(c+k)\,\Gamma(\lambda+k+1)}}. \tag{3.1.6.13}$$

The result is sharp, with extremal function f given by (3.1.6.3).

Proof. For given $f\in\mathcal{A}$, we must show that

$$|f'(z)-1|<1-\delta. \tag{3.1.6.14}$$

By a simple calculation we have

$$|f'(z)-1|\leq\sum_{k=2}^{\infty}(k+\lambda)\,a_{k+\lambda}\,|z|^2.$$

The last expressions is less than $1-\delta$ if

$$\sum_{k=2}^{\infty}\frac{(k+\lambda)}{1-\delta}a_{k+\lambda}\,|z|^2<1.$$

Using the fact that $f\in\mathcal{D}\Phi(\mu,\lambda,\beta,\gamma,a,c)$ if and only if

$$\sum_{k=2}^{\infty}\frac{(\beta+1)\,(\lambda+k\mu)\,\Gamma\,(c)\,\Gamma\,(a+k)}{(\beta\,|\gamma|-(\beta+1)\,|\lambda+\mu-\lambda\mu|)\,\Gamma\,(a)\,\Gamma\,(c+k)\,\Gamma\,(\lambda+k+1)}a_{k+\lambda}<1,$$

(3.1.6.13) holds true if

$$\frac{k+\lambda}{1-\delta}\,|z|^2\leq\sum_{k=2}^{\infty}\frac{(\beta+1)\,(\lambda+k\mu)\,\Gamma\,(c)\,\Gamma\,(a+k)}{(\beta\,|\gamma|-(\beta+1)\,|\lambda+\mu-\lambda\mu|)\,\Gamma\,(a)\,\Gamma\,(c+k)\,\Gamma\,(\lambda+k+1)}.$$

Or, equivalently,

$$|z|^2 \leq \sum_{k=2}^{\infty} \frac{(1-\delta)(\beta+1)(\lambda+k\mu)\Gamma(c)\Gamma(a+k)}{(\beta|\gamma|-(\beta+1)|\lambda+\mu-\lambda\mu|)(k+\lambda)\Gamma(a)\Gamma(c+k)\Gamma(\lambda+k+1)},$$

which completes the proof. □

Theorem 3.45 (Alb Lupaş and Oros, 2022b). *Let* $f \in \mathcal{D}\Phi(\mu,\lambda,\beta,\gamma,a,c)$. *Then*

1. f *is starlike of order* δ, $0 \leq \delta < 1$, *in the disc* $|z| < r_1$ *where,*

$$r_1 = \inf_{k\geq 2} \frac{(1-\delta)(\beta+1)(\lambda+k\mu)\Gamma(c)\Gamma(a+k)}{(\beta|\gamma|-(\beta+1)|\lambda+\mu-\lambda\mu|)(k+\delta-2)\Gamma(a)\Gamma(c+k)\Gamma(\lambda+k+1)}.$$

2. f *is convex of order* δ, $0 \leq \delta < 1$, *in the disc* $|z| < r_2$ *where,*

$$r_2 = \inf_{k\geq 2} \frac{(1-\delta)(\beta+1)(\lambda+k\mu)\Gamma(c)\Gamma(a+k)}{k(k-1)(\beta|\gamma|-(\beta+1)|\lambda+\mu-\lambda\mu|)\Gamma(a)\Gamma(c+k)\Gamma(\lambda+k+1)}.$$

Each of these results is sharp for the extremal function f *given by* (3.1.6.3).

Proof. 1. For $0 \leq \delta < 1$, we need to show that

$$\left|\frac{zf'(z)}{f(z)} - 1\right| < 1-\delta. \tag{3.1.6.15}$$

We have

$$\left|\frac{zf'(z)}{f(z)} - 1\right| \leq \left|\frac{\sum_{k=2}^{\infty}(k-1)a_k|z|}{1+\sum_{k=2}^{\infty}a_k|z|}\right|.$$

The last expression is less than $1-\delta$ if

$$\sum_{k=2}^{\infty} \frac{(k+\delta-2)}{1-\delta} a_k |z| < 1.$$

Using the fact that $f \in \mathcal{D}\Phi(\mu,\lambda,\beta,\gamma,a,c)$ if and only if

$$\sum_{k=2}^{\infty} \frac{(\beta+1)(\lambda+k\mu)\Gamma(c)\Gamma(a+k)}{(\beta|\gamma|-(\beta+1)|\lambda+\mu-\lambda\mu|)\Gamma(a)\Gamma(c+k)\Gamma(\lambda+k+1)} a_{k+\lambda} < 1.$$

(3.1.6.15) holds true if

$$\frac{k+\delta-2}{1-\delta}|z| < \frac{(\beta+1)(\lambda+k\mu)\Gamma(c)\Gamma(a+k)}{(\beta|\gamma|-(\beta+1)|\lambda+\mu-\lambda\mu|)\Gamma(a)\Gamma(c+k)\Gamma(\lambda+k+1)}.$$

Or, equivalently,

$$|z| < \frac{(1-\delta)(\beta+1)(\lambda+k\mu)\Gamma(c)\Gamma(a+k)}{(\beta|\gamma|-(\beta+1)|\lambda+\mu-\lambda\mu|)(k+\delta-2)\Gamma(a)\Gamma(c+k)\Gamma(\lambda+k+1)},$$

which yields the starlikeness of the family.

2. Using the fact that f is convex if and only zf' is starlike, we can prove (2) with a similar way of the proof of (1). The function f is convex if and only if

$$|zf''(z)| < 1-\delta. \tag{3.1.6.16}$$

We have

$$|zf''(z)| \le \left|\sum_{k=2}^{\infty} k(k-1)a_k|z|\right| < 1-\delta,$$

$$\sum_{k=2}^{\infty} \frac{k(k-1)}{1-\delta} a_k |z| < 1.$$

Using the fact that $f \in \mathcal{D}\Phi(\mu,\lambda,\beta,\gamma,a,c)$ if and only if

$$\sum_{k=2}^{\infty} \frac{(\beta+1)(\lambda+k\mu)\Gamma(c)\Gamma(a+k)}{(\beta|\gamma|-(\beta+1)|\lambda+\mu-\lambda\mu|)\Gamma(a)\Gamma(c+k)\Gamma(\lambda+k+1)} a_{k+\lambda} < 1.$$

(3.1.6.16) holds true if

$$\frac{k(k-1)}{1-\delta}|z| < \frac{(\beta+1)(\lambda+k\mu)\Gamma(c)\Gamma(a+k)}{(\beta|\gamma|-(\beta+1)|\lambda+\mu-\lambda\mu|)\Gamma(a)\Gamma(c+k)\Gamma(\lambda+k+1)},$$

or, equivalently,

$$|z| < \frac{(1-\delta)(\beta+1)(\lambda+k\mu)\Gamma(c)\Gamma(a+k)}{k(k-1)(\beta|\gamma|-(\beta+1)|\lambda+\mu-\lambda\mu|)\Gamma(a)\Gamma(c+k)\Gamma(\lambda+k+1)},$$

which yields the convexity of the family. □

3.1.7 *Properties of a Subclass of Analytic Functions Defined by Riemann–Liouville Fractional Integral Applied to Convolution Product of Multiplier Transformation and Ruscheweyh Derivative*

The contribution of fractional calculus in the development of different areas of research is well known. The paper Alb Lupaş and Acu (2023) presents investigations involving fractional calculus in the study of analytic functions. Riemann–Liouville fractional integral is known for its extensive applications in geometric function theory. New contributions were previously obtained by applying the Riemann–Liouville fractional integral to the convolution product of multiplier transformation and Ruscheweyh derivative. For the study presented in the paper Alb Lupaş and Acu (2023), the resulting operator is used following the line of research which concerns the study of certain new subclasses of analytic functions using fractional operators. Riemann–Liouville fractional integral of the convolution product of multiplier transformation and Ruscheweyh derivative is applied here for introducing a new subclass of analytic functions. Investigations regarding this newly introduced subclass concern the usual aspects considered by researchers in geometric function theory targeting the conditions that a function must meet to be part of this subclass and the properties that characterize the functions that fulfill these conditions. Theorems and corollaries regarding neighborhoods and their inclusion relation involving the newly defined subclass are stated, closure and distortion theorems are proved and coefficient estimates are obtained involving the functions belonging to this subclass. Geometrical properties such as radii of convexity, starlikeness and close-to-convexity are also obtained for this new subclass of functions.

The Riemann–Liouville fractional integral applied to the convolution product of multiplier transformation and Ruscheweyh derivative is defined in Definition 2.18 by

$$\begin{aligned} D_z^{-\lambda} IR_{\alpha,l}^{m,n} f(z) &= \frac{1}{\Gamma(\lambda)} \int_0^z \frac{IR_{\alpha,l}^{m,n} f(t)}{(z-t)^{1-\lambda}} dt = \frac{1}{\Gamma(\lambda)} \int_0^z \frac{t}{(z-t)^{1-\lambda}} dt \\ &\quad + \frac{1}{\Gamma(\lambda)\Gamma(n+1)} \sum_{k=2}^{\infty} \frac{\Gamma(n+k)}{\Gamma(k)} \left(\frac{1+\alpha(k-1)+l}{l+1} \right)^m a_k^2 \\ &\quad \times \int_0^z \frac{t^k}{(z-t)^{1-\lambda}} dt, \end{aligned}$$

which can be written making a simple calculation as follows:

$$\begin{aligned} D_z^{-\lambda} IR_{\alpha,l}^{m,n} f(z) &= \frac{1}{\Gamma(\lambda+2)} z^{\lambda+1} + \frac{1}{\Gamma(n+1)} \\ &\quad \times \sum_{k=2}^{\infty} \frac{k\Gamma(n+k)}{\Gamma(k+\lambda+1)} \left(\frac{1+\alpha(k-1)+l}{l+1} \right)^m a_k^2 z^{k+\lambda}, \end{aligned}$$

considering the function $f(z) = z + \sum_{k=2}^{\infty} a_k z^k \in \mathcal{A}$. We observe that $D_z^{-\lambda} IR_{\alpha,l}^{m,n} f(z) \in \mathcal{H}(\lambda+1, 1)$.

Inspired by the results obtained in Alb Lupaş and Oros (2022a), we define a new subclass of analytic functions using the operator introduced in Alb Lupaş (2022d).

Definition 3.10 (Alb Lupaş and Acu, 2023). The subclass $\mathcal{IR}^{m,n}_{\lambda,\alpha,l}(d,\beta,\gamma)$ consists the function $f \in \mathcal{A}$ which satisfies the following relation:

$$\left|\frac{1}{d}\left(\frac{z(D_z^{-\lambda}IR^{m,n}_{\alpha,l}f(z))' + \gamma z^2(D_z^{-\lambda}IR^{m,n}_{\alpha,l}f(z))''}{(1-\gamma)D_z^{-\lambda}IR^{m,n}_{\alpha,l}f(z) + \gamma z(D_z^{-\lambda}IR^{m,n}_{\alpha,l}f(z))'} - 1\right)\right| < \beta, \quad (3.1.7.1)$$

with $m,n \in \mathbb{N}$, $\lambda > 0$, $\alpha, l \geq 0$, $d \in \mathbb{C} - \{0\}$, $0 < \beta \leq 1$, $0 \leq \gamma \leq 1$, $z \in U$.

Properties Related to Coefficient Inequality

Theorem 3.46 (Alb Lupaş and Acu, 2023). *The function $f \in \mathcal{A}$ is contained in the subclass $\mathcal{IR}^{m,n}_{\lambda,\alpha,l}(d,\beta,\gamma)$ if and only if*

$$\sum_{k=2}^{\infty} k\left\{\gamma k^2 + \left[1+\gamma\left(2\lambda \quad 2+\beta|d|\right)\right]k + \left[1+\gamma(\lambda-1)\right](\lambda-1+\beta|d|)\right\}$$
$$\times\left(\frac{1+\alpha(k-1)+l}{l+1}\right)^m \frac{\Gamma(n+k)}{\Gamma(k+\lambda+1)}a_k^2 \leq (\gamma\lambda+1)(\beta|d|-\lambda)\frac{\Gamma(n+1)}{\Gamma(\lambda+2)}, \quad (3.1.7.2)$$

with $m,n \in \mathbb{N}$, $\lambda > 0$, $\alpha, l \geq 0$, $d \in \mathbb{C} - \{0\}$, $0 < \beta \leq 1$, $0 \leq \gamma \leq 1$, $z \in U$.

Proof. Let $f \in \mathcal{A}$. Considering the inequality (3.1.7.2) and making an easy calculus, we obtain

$$\left|\frac{z(D_z^{-\lambda}IR^{m,n}_{\alpha,l}f(z))' + \gamma z^2(D_z^{-\lambda}IR^{m,n}_{\alpha,l}f(z))''}{(1-\gamma)D_z^{-\lambda}IR^{m,n}_{\alpha,l}f(z) + \gamma z(D_z^{-\lambda}IR^{m,n}_{\alpha,l}(z))'} - 1\right|$$

$$= \left|\frac{\frac{\lambda(\gamma\lambda+1)\Gamma(n+1)}{\Gamma(\lambda+2)}z^{\lambda+1} + \sum_{k=2}^{\infty} k\{\gamma k^2 + [1+2\gamma(\lambda-1)]k + (\lambda-1)[1+\gamma(\lambda-1)]\} \times \frac{\Gamma(n+k)}{\Gamma(k+\lambda+1)}\left(\frac{1+\alpha(k-1)+l}{l+1}\right)^m a_k^2 z^{k+\lambda}}{\frac{(\gamma\lambda+1)\Gamma(n+1)}{\Gamma(\lambda+2)}z^{\lambda+1} + \sum_{k=2}^{\infty} k[1+\gamma(k+\lambda-1)]\frac{\Gamma(n+k)}{\Gamma(k+\lambda+1)} \times \left(\frac{1+\alpha(k-1)+l}{l+1}\right)^m a_k^2 z^{k+\lambda}}\right| \leq$$

and applying properties of modulus function, we get:

$$\frac{\frac{\lambda(\alpha\lambda+1)\Gamma(n+1)}{\Gamma(\lambda+2)} + \sum_{k=2}^{\infty} k\left\{\gamma k^2 + [1+2\gamma(\lambda-1)]k + (\lambda-1)[1+\gamma(\lambda-1)]\right\} \times \left(\frac{1+\alpha(k-1)+l}{l+1}\right)^m \frac{\Gamma(n+k)}{\Gamma(k+\lambda+1)}a_k^2\left|z^{k-1}\right|}{\frac{(\alpha\lambda+1)\Gamma(n+1)}{\Gamma(\lambda+2)} - \sum_{k=2}^{\infty} k[1+\gamma(k+\lambda-1)]\left(\frac{1+\alpha(k-1)+l}{l+1}\right)^m \frac{\Gamma(n+k)}{\Gamma(k+\lambda+1)}a_k^2\left|z^{k-1}\right|}.$$

Considering on real axis values of z and taking $z \to 1^-$, we get using the relation (3.1.7.2)

$$\sum_{k=2}^{\infty} k \left\{\gamma k^2 + [1+\gamma(2\lambda+\beta|d|-2)]k + [1+\gamma(\lambda-1)](\lambda+\beta|d|-1)\right\}$$

$$\times \left(\frac{1+\alpha(k-1)+l}{l+1}\right)^m \frac{\Gamma(n+k)}{\Gamma(k+\lambda+1)} a_k^2 \le (\gamma\lambda+1)(\beta|d|-\lambda)\frac{\Gamma(n+1)}{\Gamma(\lambda+2)},$$

that

$$\left|\frac{z(D_z^{-\lambda} IR_{\alpha,l}^{m,n} f(z))' + \gamma z^2 (D_z^{-\lambda} IR_{\alpha,l}^{m,n} f(z))''}{(1-\gamma) D_z^{-\lambda} IR_{\alpha,l}^{m,n} f(z) + \gamma z (D_z^{-\lambda} IR_{\alpha,l}^{m,n}(z))'} - 1\right| \le \beta|d|,$$

equivalently with $f \in \mathcal{IR}_{\lambda,\alpha,l}^{m,n}(d,\beta,\gamma)$.

Conversely, suppose that $f \in \mathcal{IR}_{\lambda,\alpha,l}^{m,n}(d,\beta,\gamma)$. In this condition we obtain the following equivalently inequalities

$$\operatorname{Re}\left\{\frac{z(D_z^{-\lambda} IR_{\alpha,l}^{m,n} f(z))' + \gamma z^2 (D_z^{-\lambda} IR_{\alpha,l}^{m,n} f(z))''}{(1-\gamma) D_z^{-\lambda} IR_{\alpha,l}^{m,n} f(z) + \gamma z (D_z^{-\lambda} IR_{\alpha,l}^{m,n} f(z))'} - 1\right\} > -\beta|d|,$$

$$\operatorname{Re}\left\{\frac{\begin{array}{c}\frac{\lambda(\gamma\lambda+1)\Gamma(n+1)}{\Gamma(\lambda+2)} z^{\lambda+1} + \sum_{k=2}^{\infty} k\left\{\gamma k^2 + [2\gamma(\lambda-1)+1]k + (\lambda-1)[\gamma(\lambda-1)+1]\right\} \\ \times \frac{\Gamma(n+k)}{\Gamma(k+\lambda+1)} \left(\frac{1+\alpha(k-1)+l}{l+1}\right)^m a_k^2 z^{k+\lambda}\end{array}}{\begin{array}{c}\frac{(\gamma\lambda+1)\Gamma(n+1)}{\Gamma(\lambda+2)} z^{\lambda+1} + \sum_{k=2}^{\infty} k[1+\gamma(k+\lambda-1)] \frac{\Gamma(n+k)}{\Gamma(k+\lambda+1)} \\ \times \left(\frac{1+\alpha(k-1)+l}{l+1}\right)^m a_k^2 z^{k+\lambda}\end{array}} - 1 + \beta|d|\right\} > 0,$$

$$\operatorname{Re}\left\{\frac{\frac{(\gamma\lambda+1)(\beta|d|-\lambda)\Gamma(n+1)}{\Gamma(\lambda+2)} z^{\lambda+1}}{\frac{(\gamma\lambda+1)\Gamma(n+1)}{\Gamma(\lambda+2)} z^{\lambda+1} + \sum_{k=2}^{\infty} j[1+\gamma(k+\lambda-1)] \frac{\Gamma(n+k)}{\Gamma(k+\lambda+1)} a_k^2 z^{k+\lambda}}\right.$$

$$\left. + \frac{\begin{array}{c}\sum_{k=2}^{\infty} k\left\{\gamma k^2 + [\gamma(2\lambda+\beta|d|-2)+1]k + [\gamma(\lambda-1)+1](\lambda+\beta|d|-1)\right\} \\ \times \frac{\Gamma(n+k)}{\Gamma(k+\lambda+1)} \left(\frac{1+\alpha(k-1)+l}{l+1}\right)^m a_k^2 z^{k+\lambda}\end{array}}{\frac{(\gamma\lambda+1)\Gamma(n+1)}{\Gamma(\lambda+2)} z^{\lambda+1} + \sum_{k=2}^{\infty} j[1+\gamma(k+\lambda-1)] \frac{\Gamma(n+k)}{\Gamma(k+\lambda+1)} a_k^2 z^{k+\lambda}}\right\} > 0.$$

Because $\mathrm{Re}(e^{i\theta}) = r$ and $\mathrm{Re}(-e^{i\theta}) \geq -|e^{i\theta}| = -1$, the inequality becomes

$$\frac{\frac{(\gamma\lambda+1)(\beta|d|-\lambda)\Gamma(n+1)}{\Gamma(\lambda+2)}r^{\lambda+1}}{\frac{(\gamma\lambda+1)\Gamma(n+1)}{\Gamma(\lambda+2)}r^{\lambda+1} - \sum_{k=2}^{\infty} j\left[1+\gamma\left(k+\lambda-1\right)\right]\frac{\Gamma(n+k)}{\Gamma(k+\lambda+1)}\left(\frac{1+\alpha(k-1)+l}{l+1}\right)^m a_k^2 r^{k+\lambda}}$$

$$-\frac{\sum_{k=2}^{\infty} k\left\{\gamma k^2 + \left[1+\left(2\lambda+\beta\left|d\right|-2\right)\right]k + \left[1+\gamma\left(\lambda-1\right)\right]\left(\lambda+\beta\left|d\right|-1\right)\right\} \times \frac{\Gamma(n+k)}{\Gamma(k+\lambda+1)}\left(\frac{1+\alpha(k-1)+l}{l+1}\right)^m a_k^2 r^{k+\lambda}}{\frac{(\gamma\lambda+1)\Gamma(n+1)}{\Gamma(\lambda+2)}r^{\lambda+1} - \sum_{k=2}^{\infty} j\left[1+\gamma\left(k+\lambda-1\right)\right]\frac{\Gamma(n+k)}{\Gamma(k+\lambda+1)} \times \left(\frac{1+\alpha(k-1)+l}{l+1}\right)^m a_k^2 r^{k+\lambda}} > 0.$$

Letting $r \to 1^-$ and by using the mean value theorem, we get the inequality (3.1.7.2), which finalized the proof of theorem. □

Corollary 3.12 (Alb Lupaş and Acu, 2023). *We get for the function $f \in \mathcal{IR}_{\lambda,\alpha,l}^{m,n}(d,\beta,\gamma)$, the following coefficient inequality:*

$$a_k \leq \sqrt{\frac{(\beta|d|-\lambda)(\gamma\lambda+1)\frac{\Gamma(n+1)}{\Gamma(\lambda+2)}}{k\{\gamma k^2+[1+\gamma(2\lambda+\beta|d|-2)]k+[1+\gamma(\lambda-1)](\lambda+\beta|d|-1)\}\frac{\Gamma(n+k)}{\Gamma(k+\lambda+1)}\left(\frac{1+\alpha(k-1)+l}{l+1}\right)^m}},$$

$k \geq 2$.

Properties Related to Distortion

Theorem 3.47 (Alb Lupaş and Acu, 2023). *The function $f \in \mathcal{IR}_{\lambda,\alpha,l}^{m,n}(d,\beta,\gamma)$ has the property*

$$r - \sqrt{\frac{\left(\gamma\lambda+1\right)\left(\beta\left|d\right|-\lambda\right)\left(\lambda+2\right)}{2\left(n+1\right)\left(\frac{1+\alpha+l}{l+1}\right)^m\left\{\beta\left|d\right|+\left(\lambda+1\right)\left[1+\gamma\left(\lambda+\beta\left|d\right|+1\right)\right]\right\}}r^2} \leq |f(z)|$$

$$\leq r + \sqrt{\frac{\left(\gamma\lambda+1\right)\left(\beta\left|d\right|-\lambda\right)\left(\lambda+2\right)}{2\left(n+1\right)\left(\frac{1+\alpha+l}{l+1}\right)^m\left\{\beta\left|d\right|+\left(\lambda+1\right)\left[1+\gamma\left(\lambda+\beta\left|d\right|+1\right)\right]\right\}}r^2},$$

with $|z| = r < 1$.

The equality holds for

$$f(z) = z + \sqrt{\frac{(\lambda+2)(\beta|d|-\lambda)(\gamma\lambda+1)}{2(n+1)\left(\frac{1+\alpha+l}{l+1}\right)^m \{\beta|d| + (\lambda+1)[1+\gamma(\lambda+1+\beta|d|)]\}}}z^2,$$

with $z \in U$.

Proof. Considering $f \in \mathcal{IR}_{m,n}(\lambda, \alpha, l, d, \beta, \gamma)$, from the relation (3.1.7.2) and taking account that the sequence

$$k\left\{\gamma k^2 + [1+\gamma(2\lambda+\beta|d|-2)]k + [1+\gamma(\lambda-1)](\lambda+\beta|d|-1)\right\}$$
$$\times\left(\frac{1+\alpha(k-1)+l}{l+1}\right)^m \frac{\Gamma(n+k)}{\Gamma(k+\lambda+1)}$$

is increasing and positive for $k \geq 2$, in these conditions we get

$$\sqrt{2(n+1)\left(\frac{1+\alpha+l}{l+1}\right)^m \{\beta|d| + (\lambda+1)[1+\gamma(\lambda+\beta|d|+1)]\}}\sum_{k=2}^{\infty} a_k$$
$$\leq \sum_{k=2}^{\infty}\sqrt{k\{\gamma k^2 + [1+\gamma(2\lambda+\beta|d|-2)]k + [1+\gamma(\lambda-1)](\lambda+\beta|d|-1)\}}$$
$$\cdot\sqrt{\left(\frac{1+\alpha(k-1)+l}{l+1}\right)^m \frac{\Gamma(n+k)}{\Gamma(k+\lambda+1)}}a_k$$
$$\leq \sqrt{(\beta|d|-\lambda)(\gamma\lambda+1)\frac{\Gamma(n+1)}{\Gamma(\lambda+2)}},$$

equivalently with

$$\sum_{k=2}^{\infty} a_k \leq \sqrt{\frac{(\lambda+2)(\beta|d|-\lambda)(\gamma\lambda+1)}{2(n+1)\left(\frac{1+\alpha+l}{l+1}\right)^m \{\beta|d| + (\lambda+1)[1+\gamma(\lambda+\beta|d|+1)]\}}}. \tag{3.1.7.3}$$

Using the properties of modulus function for

$$f(z) = z + \sum_{k=2}^{\infty} a_k z^k,$$

we obtain

$$r - r^2\sum_{k=2}^{\infty} a_k \leq r - \sum_{k=2}^{\infty} a_k r^k \leq |z| - \sum_{k=2}^{\infty} a_k |z|^k \leq |f(z)|$$
$$\leq |z| + \sum_{k=2}^{\infty} a_k |z|^k \leq r + \sum_{k=2}^{\infty} a_k r^k \leq r + r^2\sum_{k=2}^{\infty} a_k,$$

and applying relation (3.1.7.3) we get

$$r - \sqrt{\frac{(\lambda+2)(\beta|d|-\lambda)(\gamma\lambda+1)}{2(n+1)\left(\frac{1+\alpha+l}{l+1}\right)^m\{\beta|d|+(\lambda+1)[1+\gamma(\lambda+\beta|d|+1)]\}}}r^2 \leq |f(z)|$$
$$\leq r + \sqrt{\frac{(\lambda+2)(\beta|d|-\lambda)(\gamma\lambda+1)}{2(n+1)\left(\frac{1+\alpha+l}{l+1}\right)^m\{\beta|d|+(\lambda+1)[\gamma(\lambda+\beta|d|+1)+1]\}}}r^2,$$

which complete the proof of theorem. □

Theorem 3.48 (Alb Lupaş and Acu, 2023). *The function* $f \in \mathcal{IR}_{\lambda,\alpha,l}^{m,n}(d,\beta,\gamma)$ *has the property*

$$-\sqrt{\frac{2(\lambda+2)(\beta|d|-\lambda)(\gamma\lambda+1)}{(n+1)\left(\frac{1+\alpha+l}{l+1}\right)^m\{\beta|d|+(\lambda+1)[1+\beta|d|+\gamma(\lambda+1)]\}}}r \leq |f'(z)|$$
$$\leq \sqrt{\frac{2(\lambda+2)(\beta|d|-\lambda)(\gamma\lambda+1)}{(n+1)\left(\frac{1+\alpha+l}{l+1}\right)^m\{\beta|d|+(\lambda+1)[1+\beta|d|+\gamma(\lambda+1)]\}}}r,$$

with $|z| = r < 1$.

The equality holds for

$$f(z) = z + \sqrt{\frac{(\lambda+2)(\beta|d|-\lambda)(\gamma\lambda+1)}{2(n+1)\left(\frac{1+\alpha+l}{l+1}\right)^m\{\beta|d|+(\lambda+1)[1+\beta|d|+\gamma(\lambda+1)]\}}}z^2,$$

$z \in U$.

Proof. The following result is obtained by applying the properties of the modulus function for

$$f'(z) = 1 + \sum_{k=2}^{\infty} k a_k z^{k-1},$$

getting

$$1-\sum_{k=2}^{\infty}ka_k|z|\leq 1-\sum_{k=2}^{\infty}ka_k|z|^{k-1}\leq|f'(z)|\leq 1+\sum_{k=2}^{\infty}ka_k|z|^{k-1}$$
$$\leq 1+\sum_{k=2}^{\infty}ka_k|z|.$$

Now, using relation (3.1.7.3), we obtain:

$$1-\sqrt{\frac{2(\lambda+2)(\beta|d|-\lambda)(\gamma\lambda+1)}{(n+1)\left(\frac{1+\alpha+l}{l+1}\right)^m\{\beta|d|+(\lambda+1)[1+\gamma(\lambda+\beta|d|+1)]\}}}r\leq|f'(z)|$$
$$\leq 1+\sqrt{\frac{2(\lambda+2)(\beta|d|-\lambda)(\gamma\lambda+1)}{(n+1)\left(\frac{1+\alpha+l}{l+1}\right)^m\{\beta|d|+(\lambda+1)[1+\gamma(\lambda+\beta|d|+1)]\}}}r,$$

which finalized the proof of theorem. □

Properties Related to Closure

Theorem 3.49 (Alb Lupaş and Acu, 2023). *Consider the functions* f_p*,* $p=1,2,\ldots,q$*, given by*

$$f_p(z)=z+\sum_{k=2}^{\infty}a_{k,p}z^k,\quad a_{k,p}\geq 0,\ \ z\in U,\tag{3.1.7.4}$$

from the subclass $\mathcal{IR}_{\lambda,\alpha,l}^{m,n}(d,\beta,\gamma)$*.*

The function

$$h(z)=\sum_{p=1}^{q}\mu_p f_p(z),\quad z\in U,\ \ \mu_p\geq 0,$$

is in the subclass $\mathcal{IR}_{\lambda,\alpha,l}^{m,n}(d,\beta,\gamma)$*, when*

$$\sum_{p=1}^{q}\mu_p=1.$$

Proof. For the function f, we can write

$$h(z)=\sum_{p=1}^{q}\mu_p z+\sum_{p=1}^{q}\sum_{k=2}^{\infty}\mu_p a_{k,p}z^k=z+\sum_{k=2}^{\infty}\sum_{p=1}^{q}\mu_p a_{k,p}z^k,$$

and since the functions f_p, $p = 1, 2, \ldots, q$, belong to the subclass $\mathcal{IR}^{m,n}_{\lambda,\alpha,l}(d, \beta, \gamma)$, applying Corollary 3.12, we obtain

$$\sum_{k=2}^{\infty} \sqrt{k\left\{\gamma k^2 + \left[1+\gamma\left(2\lambda+\beta\left|d\right|-2\right)\right]k + \left[1+\gamma\left(\lambda-1\right)\right]\left(\lambda+\beta\left|d\right|-1\right)\right\}}$$

$$\cdot\sqrt{\frac{\Gamma\left(n+k\right)}{\Gamma\left(k+\lambda+1\right)}\left(\frac{1+\alpha\left(k-1\right)+l}{l+1}\right)^m} a_k$$

$$\leq \sqrt{\left(\beta\left|d\right|-\lambda\right)\left(\gamma\lambda+1\right)\frac{\Gamma\left(n+1\right)}{\Gamma\left(\lambda+2\right)}}.$$

In this condition, it is enough to demonstrate that

$$\sum_{k=2}^{\infty} \sqrt{k\left\{\gamma k^2 + \left[1+\gamma\left(2\lambda+\beta\left|d\right|-2\right)\right]k + \left[1+\gamma\left(\lambda-1\right)\right]\left(\lambda+\beta\left|d\right|-1\right)\right\}}$$

$$\cdot\sqrt{\frac{\Gamma\left(n+k\right)}{\Gamma\left(k+\lambda+1\right)}\left(\frac{1+\alpha\left(k-1\right)+l}{l+1}\right)^m}\left(\sum_{p=1}^{q}\mu_p a_{k,p}\right)$$

$$= \sum_{p=1}^{q}\mu_p \sum_{k=2}^{\infty} \sqrt{k\left\{\gamma k^2 + \left[1+\gamma\left(2\lambda+\beta\left|d\right|-2\right)\right]k + \left[1+\gamma\left(\lambda-1\right)\right]\left(\lambda+\beta\left|d\right|-1\right)\right\}}$$

$$\cdot\sqrt{\frac{\Gamma\left(n+k\right)}{\Gamma\left(k+\lambda+1\right)}\left(\frac{1+\alpha\left(k-1\right)+l}{l+1}\right)^m} a_{k,p}$$

$$\leq \sum_{p=1}^{q}\mu_p\sqrt{\left(\beta\left|d\right|-\lambda\right)\left(\gamma\lambda+1\right)\frac{\Gamma\left(n+1\right)}{\Gamma\left(\lambda+2\right)}} = \sqrt{\left(\beta\left|d\right|-\lambda\right)\left(\gamma\lambda+1\right)\frac{\Gamma\left(n+1\right)}{\Gamma\left(\lambda+2\right)}}.$$

Hence, the proof is finalized. □

Corollary 3.13 (Alb Lupaş and Acu, 2023). *Considering the functions f_p, $p = 1, 2$, given by (3.1.7.4) from the subclass $\mathcal{IR}^{m,n}_{\lambda,\alpha,l}(d, \beta, \gamma)$, the function*

$$h(z) = (1-\xi)f_1(z) + \xi f_2(z), \quad 0 \leq \xi \leq 1, \;\; z \in U,$$

belongs to the subclass $\mathcal{IR}^{m,n}_{\lambda,\alpha,l}(d, \beta, \gamma)$.

Theorem 3.50 (Alb Lupaş and Acu, 2023). *Let*

$$f_1(z) = z,$$

and

$$f_k(z) = z + \sqrt{\frac{(\beta|d|-\lambda)(\gamma\lambda+1)\frac{\Gamma(n+1)}{\Gamma(\lambda+2)}}{k\{\gamma k^2+[1+\gamma(2\lambda+\beta|d|-2)]k+[1+\gamma(\lambda-1)](\lambda+\beta|d|-1)\}\frac{\Gamma(n+k)}{\Gamma(k+\lambda+1)}\left(\frac{1+\alpha(k-1)+l}{l+1}\right)^m}}\, z^k,$$

$k \geq 2, \quad z \in U.$

The function f is contained in the subclass $\mathcal{IR}^{m,n}_{\lambda,\alpha,l}(d,\beta,\gamma)$ if and only if it has the following form

$$f(z) = \mu_1 f_1(z) + \sum_{k=2}^{\infty} \mu_k f_k(z), \quad z \in U,$$

with $\mu_k \geq 0$, $k \geq 1$ and $\sum_{k=1}^{\infty} \mu_k = 1$.

Proof. Considering the function

$$f(z) = \mu_1 f_1(z) + \sum_{k=2}^{\infty} \mu_k f_k(z)$$

$$= z + \sum_{k=2}^{\infty} \sqrt{\frac{(\beta|d|-\lambda)(\gamma\lambda+1)\frac{\Gamma(n+1)}{\Gamma(\lambda+2)}}{k\{\gamma k^2+[1+\gamma(2\lambda+\beta|d|-2)]k+[1+\gamma(\lambda-1)](\lambda+\beta|d|-1)\} \times \frac{\Gamma(n+k)}{\Gamma(k+\lambda+1)}\left(\frac{1+\alpha(k-1)+l}{l+1}\right)^m}}\,\mu_k z^k,$$

we obtain

$$\sum_{k=2}^{\infty} \sqrt{\frac{k\left\{\gamma k^2+[1+\gamma(2\lambda+\beta|d|-2)]k+[1+\gamma(\lambda-1)](\lambda+\beta|d|-1)\right\}\frac{\Gamma(n+k)}{\Gamma(k+\lambda+1)}\left(\frac{1+\alpha(k-1)+l}{l+1}\right)^m}{(\beta|d|-\lambda)(\gamma\lambda+1)\frac{\Gamma(n+1)}{\Gamma(\lambda+2)}}}$$
$$\cdot \sqrt{\frac{(\beta|d|-\lambda)(\gamma\lambda+1)\frac{\Gamma(n+1)}{\Gamma(\lambda+2)}}{k\{\gamma k^2+[1+\gamma(2\lambda+\beta|d|-2)]k+[1+\gamma(\lambda-1)](\lambda+\beta|d|-1)\}\frac{\Gamma(n+k)}{\Gamma(k+\lambda+1)}\left(\frac{1+\alpha(k-1)+l}{l+1}\right)^m}}\,\mu_k$$
$$= \sum_{k=2}^{\infty} \mu_k = 1 - \mu_1 \leq 1.$$

Hence, $f \in \mathcal{IR}^{m,n}_{\lambda,\alpha,l}(d,\beta,\gamma)$.

Conversely, we assume that $f \in \mathcal{IR}^{m,n}_{\lambda,\alpha,l}(d,\beta,\gamma)$.

Putting

$$\mu_k = \sqrt{\frac{k\left\{\gamma k^2+[1+\gamma(2\lambda+\beta|d|-2)]k+[1+\gamma(\lambda-1)](\lambda+\beta|d|-1)\right\}\frac{\Gamma(n+k)}{\Gamma(k+\lambda+1)}\left(\frac{1+\alpha(k-1)+l}{l+1}\right)^m}{(\beta|d|-\lambda)(\gamma\lambda+1)\frac{\Gamma(n+1)}{\Gamma(\lambda+2)}}}\,a_k,$$

since

$$\mu_1 = 1 - \sum_{k=2}^{\infty} \mu_k.$$

Thus,

$$f(z) = \mu_1 f_1(z) + \sum_{k=2}^{\infty} \mu_k f_k(z).$$

Hence, the proof is finalized. □

Corollary 3.14 (Alb Lupaş and Acu, 2023). *The functions*

$$f_1(z) = z,$$

and

$$f_k(z) = z + \sqrt{\frac{(\gamma\lambda+1)(\beta|d|-\lambda)\frac{\Gamma(n+1)}{\Gamma(\lambda+2)}}{k\{\gamma k^2+[1+\gamma(2\lambda-2+\beta|d|)]k+[1+\gamma(\lambda-1)](\lambda-1+\beta|d|)\}\left(\frac{1+\alpha(k-1)+l}{l+1}\right)^m \frac{\Gamma(n+k)}{\Gamma(k+\lambda+1)}}}\, z^k,$$

$k \geq 2$, $z \in U$, are the extreme points for the subclass $\mathcal{IR}_{\lambda,\alpha,l}^{m,n}(d,\beta,\gamma)$.

Properties Related to Inclusion and Neighborhood

The δ-neighborhood of a function $f \in \mathcal{A}$ is

$$N_\delta(f) = \left\{ g \in \mathcal{A} : g(z) = z + \sum_{k=2}^{\infty} b_k z^k \text{ and } \sum_{k=2}^{\infty} k|a_k - b_k| \leq \delta \right\}, \quad (3.1.7.5)$$

and for a particular function $e(z) = z$, we get

$$N_\delta(e) = \left\{ g \in \mathcal{A} : g(z) = z + \sum_{k=2}^{\infty} b_k z^k \text{ and } \sum_{k=2}^{\infty} k|b_k| \leq \delta \right\}. \quad (3.1.7.6)$$

A function $f \in \mathcal{A}$ is contained in the subclass $\mathcal{IR}_{\lambda,\alpha,l}^{m,n}(d,\beta,\gamma,\zeta)$ if there exists a function $h \in \mathcal{IR}_{\lambda,\alpha,l}^{m,n}(d,\beta,\gamma)$ such that

$$\left| \frac{f(z)}{h(z)} - 1 \right| < 1 - \zeta, \quad 0 \leq \zeta < 1, \ z \in U. \quad (3.1.7.7)$$

Theorem 3.51 (Alb Lupaş and Acu, 2023). *If*

$$\delta = \sqrt{\frac{2(\lambda+2)(\beta|d|-\lambda)(\gamma\lambda+1)}{(n+1)\{\beta|d|+(\lambda+1)[1+\gamma(\lambda+\beta|d|+1)]\}\left(\frac{1+\alpha+l}{l+1}\right)^m}},$$

then

$$\mathcal{IR}^{m,n}_{\lambda,\alpha,l}(d,\beta,\gamma) \subset N_\delta(e).$$

Proof. Consider $f \in \mathcal{IR}^{m,n}_{\lambda,\alpha,l}(d,\beta,\gamma)$. By applying Corollary 3.12 taking account that

$$\begin{aligned} &k\left\{\gamma k^2+[1+\gamma(2\lambda+\beta|d|-2)]k+[1+\gamma(\lambda-1)](\lambda+\beta|d|-1)\right\} \\ &\quad\cdot\frac{\Gamma(n+k)}{\Gamma(k+\lambda+1)}\left(\frac{1+\alpha(k-1)+l}{l+1}\right)^m \\ &\geq \{\beta|d|+(\lambda+1)[1+\gamma(\lambda+\beta|d|+1)]\}\frac{\Gamma(n+2)}{4\Gamma(\lambda+3)}\left(\frac{1+\alpha+l}{l+1}\right)^m, \end{aligned}$$

for $k \geq 2$, we obtain

$$\begin{aligned} &\sqrt{\{\beta|d|+(\lambda+1)[1+\gamma(\lambda+\beta|d|+1)]\}\frac{\Gamma(n+2)}{4\Gamma(\lambda+3)}\left(\frac{1+\alpha+l}{l+1}\right)^m}\sum_{k=2}^{\infty}a_k \\ &\leq \sum_{k=2}^{\infty}\sqrt{k\{\gamma k^2+[1+\gamma(2\lambda+\beta|d|-2)]k+[1+\gamma(\lambda-1)](\lambda+\beta|d|-1)\}} \\ &\quad\cdot\sqrt{\frac{\Gamma(n+k)}{\Gamma(k+\lambda+1)}\left(\frac{1+\alpha(k-1)+l}{l+1}\right)^m}a_k \\ &\leq \sqrt{(\beta|d|-\lambda)(\gamma\lambda+1)\frac{\Gamma(n+1)}{\Gamma(\lambda+2)}}, \end{aligned}$$

therefore

$$\sum_{k=2}^{\infty}a_k \leq \sqrt{\frac{(\lambda+2)(\beta|d|-\lambda)(\gamma\lambda+1)}{2(n+1)\{\beta|d|+(\lambda+1)[1+\gamma(\lambda+\beta|d|+1)]\}\left(\frac{1+\alpha+l}{l+1}\right)^m}}. \tag{3.1.7.8}$$

Applying Corollary 3.12 and relation (3.1.7.8), we get

$$\sum_{k=2}^{\infty} ka_k \leq \sqrt{\frac{2(\lambda+2)(\beta|d|-\lambda)(\gamma\lambda+1)}{(n+1)\{\beta|d|+(\lambda+1)[1+\gamma(\lambda+\beta|d|+1)]\}\left(\frac{1+\alpha+l}{l+1}\right)^m}} = \delta,$$

by virtue of (3.1.7.5), we obtain $f \in N_\delta(e)$, which finalized the proof of the theorem. □

Theorem 3.52 (Alb Lupaş and Acu, 2023). *If* $h \in \mathcal{IR}^{m,n}_{\lambda,\alpha,l}(d,\beta,\gamma)$ *and*

$$\zeta = 1 + \frac{\delta}{2}\sqrt{\frac{(\lambda+2)(\beta|d|-\lambda)(\gamma\lambda+1)}{2(n+1)\{\beta|d|+(\lambda+1)[1+\gamma(\lambda+\beta|d|+1)]\}\left(\frac{1+\alpha+l}{l+1}\right)^m}}, \tag{3.1.7.9}$$

then

$$N_\delta(h) \subset \mathcal{IR}^{m,n}_{\lambda,\alpha,l}(d,\beta,\gamma,\zeta).$$

Proof. Considering $f \in N_\delta(h)$, from relation (3.1.7.5), we deduce that

$$\sum_{k=2}^{\infty} k|a_k - b_k| \leq \delta,$$

which give the coefficient inequality

$$\sum_{k=2}^{\infty} |a_k - b_k| \leq \frac{\delta}{2}. \tag{3.1.7.10}$$

Using relation (3.1.7.8) and taking account that $h \in \mathcal{IR}^{m,n}_{\lambda,\alpha,l}(d,\beta,\gamma)$, we get

$$\sum_{k=2}^{\infty} b_k \leq \sqrt{\frac{(\lambda+2)(\beta|d|-\lambda)(\gamma\lambda+1)}{2(n+1)\{\beta|d|+(\lambda+1)[1+\gamma(\lambda+\beta|d|+1)]\}\left(\frac{1+\alpha+l}{l+1}\right)^m}}. \tag{3.1.7.11}$$

Using (3.1.7.10) and (3.1.7.11), we have

$$\left|\frac{f(z)}{h(z)} - 1\right| \leq \frac{\sum_{k=2}^{\infty}|a_k - b_k|}{1 - \sum_{k=2}^{\infty} b_k}$$

$$\leq \frac{\delta}{2\left(1 - \sqrt{\frac{(\lambda+2)(\beta|d|-\lambda)(\gamma\lambda+1)}{2(n+1)\{\beta|d|+(\lambda+1)[1+\gamma(\lambda+\beta|d|+1)]\}\left(\frac{1+\alpha+l}{l+1}\right)^m}}\right)} = 1 - \zeta,$$

thus by condition (3.1.7.7), $f \in \mathcal{IR}^{m,n}_{\lambda,\alpha,l}(d,\beta,\gamma,\zeta)$, and ζ is given by (3.1.7.9). □

Properties Related to Radii of Convexity, Starlikeness and Close-to-Convexity

Theorem 3.53 (Alb Lupaş and Acu, 2023). *The function $f \in \mathcal{IR}^{m,n}_{\lambda,\alpha,l}(d,\beta,\gamma)$ is analytic starlike of order δ, with $|z| < r_1$, for $0 \le \delta < 1$ and*

$$r_1 = \inf_k \left\{ \frac{k\left\{\gamma k^2 + [1+\gamma(2\lambda+\beta|d|-2)]k + [1+\gamma(\lambda-1)](\lambda+\beta|d|-1)\right\} \times (1-\delta)^2 \frac{\Gamma(n+k)}{\Gamma(k+\lambda+1)}\left(\frac{1+\alpha(k-1)+l}{l+1}\right)^m}{(\beta|d|-\lambda)(\gamma\lambda+1)\frac{\Gamma(n+1)}{\Gamma(\lambda+2)}(k-\delta)^2} \right\}^{\frac{1}{2(k-1)}}.$$

The function

$$f_k(z) = z + \sqrt{\frac{(\beta|d|-\lambda)(\gamma\lambda+1)\frac{\Gamma(n+1)}{\Gamma(\lambda+2)}}{k\left\{\gamma k^2 + [1+\gamma(2\lambda+\beta|d|-2)]k + [1+\gamma(\lambda-1)](\lambda+\beta|d|-1)\right\} \times \frac{\Gamma(n+k)}{\Gamma(k+\lambda+1)}\left(\frac{1+\alpha(k-1)+l}{l+1}\right)^m}}\, z^k,$$

$k \ge 2$, gives the sharp result.

Proof. We have to show that

$$\left|\frac{zf'(z)}{f(z)} - 1\right| \le 1-\delta, \quad \text{for } |z| < r_1.$$

Taking account that

$$\left|\frac{zf'(z)}{f(z)} - 1\right| = \left|\frac{\sum_{k=2}^{\infty}(k-1)a_k z^{k-1}}{1+\sum_{k=2}^{\infty} a_k z^{k-1}}\right| \le \frac{\sum_{k=2}^{\infty}(k-1)a_k |z|^{k-1}}{1-\sum_{k=2}^{\infty} a_k |z|^{k-1}},$$

we will demonstrate that

$$\frac{\sum_{k=2}^{\infty}(k-1)a_k |z|^{k-1}}{1-\sum_{k=2}^{\infty} a_k |z|^{k-1}} \le 1-\delta,$$

equivalently to

$$\sum_{k=2}^{\infty}(k-\delta)a_k|z|^{k-1} \le 1-\delta.$$

Using Theorem 3.46, we get

$$|z| \leq \left\{ \frac{\begin{array}{c} k\left\{\gamma k^2 + \left[1+\gamma\left(2\lambda+\beta\left|d\right|-2\right)\right]k + \left[1+\gamma\left(\lambda-1\right)\right]\left(\lambda+\beta\left|d\right|-1\right)\right\} \\ \times\left(1-\delta\right)^2 \frac{\Gamma(n+k)}{\Gamma(k+\lambda+1)}\left(\frac{1+\alpha(k-1)+l}{l+1}\right)^m \end{array}}{\left(\beta\left|d\right|-\lambda\right)\left(\gamma\lambda+1\right)\frac{\Gamma(n+1)}{\Gamma(\lambda+2)}\left(k-\delta\right)^2} \right\}^{\frac{1}{2(k-1)}},$$

and the proof is finalized. □

Theorem 3.54 (Alb Lupaş and Acu, 2023). *The function $f \in \mathcal{IR}^{m,n}_{\lambda,\alpha,l}(d,\beta,\gamma)$ is analytic convex of order δ, with $|z| < r_2$, for $0 \leq \delta \leq 1$ and*

$$r_2 = \inf_k \left\{ \frac{\begin{array}{c} k\left\{\gamma k^2 + \left[1+\gamma\left(2\lambda+\beta\left|d\right|-2\right)\right]k + \left[1+\gamma\left(\lambda-1\right)\right]\left(\lambda+\beta\left|d\right|-1\right)\right\} \\ \times\left(1-\delta\right)^2 \frac{\Gamma(n+k)}{\Gamma(k+\lambda+1)}\left(\frac{1+\alpha(k-1)+l}{l+1}\right)^m \end{array}}{\left(\beta\left|d\right|-\lambda\right)\left(\gamma\lambda+1\right)\frac{\Gamma(n+1)}{\Gamma(\lambda+2)}\left(k-\delta\right)^2} \right\}^{\frac{1}{2(k-1)}}.$$

The function

$$f_k(z) = z + \sqrt{\frac{(\gamma\lambda+1)(\beta|d|-\lambda)\frac{\Gamma(n+1)}{\Gamma(\lambda+2)}}{k\{\gamma k^2+[1+\gamma(2\lambda-2+\beta|d|)]k+[1+\gamma(\lambda-1)](\lambda-1+\beta|d|)\}\left(\frac{1+\alpha(k-1)+l}{l+1}\right)^m\frac{\Gamma(n+k)}{\Gamma(k+\lambda+1)}}}\, z^k, \tag{3.1.7.12}$$

$k \geq 2$, gives the sharp result.

Proof. It is enough to demonstrate that

$$\left|\frac{zf''(z)}{f'(z))}\right| \leq 1-\delta, \quad |z| < r_2.$$

Taking account that

$$\left|\frac{zf''(z)}{f'(z)}\right| = \left|\frac{\sum_{k=2}^{\infty} k(k-1)a_k z^{k-1}}{1+\sum_{k=2}^{\infty} k a_k z^{k-1}}\right| \leq \frac{\sum_{k=2}^{\infty} k(k-1)a_k |z|^{k-1}}{1-\sum_{k=2}^{\infty} k a_k |z|^{k-1}},$$

we have to show that

$$\frac{\sum_{k=2}^{\infty} k(k-1)a_k |z|^{k-1}}{1-\sum_{k=2}^{\infty} k a_k |z|^{k-1}} \leq 1-\delta,$$

$$\sum_{k=2}^{\infty} k(k-\delta)a_k |z|^{k-1} \leq 1-\delta,$$

and using Theorem 3.46, we have the following result:

$$|z|^{k-1} \leq \frac{(1-\delta)}{k(k-\delta)} \times \sqrt{\frac{k\left\{\gamma k^2+[1+\gamma(2\lambda+\beta|d|-2)]k+[1+\gamma(\lambda-1)](\lambda+\beta|d|-1)\right\}\frac{\Gamma(n+k)}{\Gamma(k+\lambda+1)}\left(\frac{1+\alpha(k-1)+l}{l+1}\right)^m}{(\beta|d|-\lambda)(\gamma\lambda+1)\frac{\Gamma(n+1)}{\Gamma(\lambda+2)}}},$$

written as

$$|z| \leq \left\{\frac{k\left\{\gamma k^2+[1+\gamma(2\lambda+\beta|d|-2)]k+[1+\gamma(\lambda-1)](\lambda+\beta|d|-1)\right\} \times (1-\delta)^2\frac{\Gamma(n+k)}{\Gamma(k+\lambda+1)}\left(\frac{1+\alpha(k-1)+l}{l+1}\right)^m}{(\beta|d|-\lambda)(\gamma\lambda+1)\frac{\Gamma(n+1)}{\Gamma(\lambda+2)}(k-\delta)^2}\right\}^{\frac{1}{2(k-1)}},$$

and the proof is finalized. □

Theorem 3.55 (Alb Lupaş and Acu, 2023). *The function* $f \in \mathcal{IR}^{m,n}_{\lambda,\alpha,l}(d,\beta,\gamma)$ *is analytic close-to-convex of order* δ, *with* $|z| < r_3$, *for* $0 \leq \delta < 1$ *and*

$$r_3 = \inf_k \left\{\frac{\frac{\left\{\gamma k^2+[1+\gamma(2\lambda+\beta|d|-2)]k+[1+\gamma(\lambda-1)](\lambda+\beta|d|-1)\right\}}{k^2}(1-\delta)^2 \times \frac{\Gamma(n+k)}{\Gamma(k+\lambda+1)}\left(\frac{1+\alpha(k-1)+l}{l+1}\right)^m}{(\beta|d|-\lambda)(\gamma\lambda+1)\frac{\Gamma(n+1)}{\Gamma(\lambda+2)}}\right\}^{\frac{1}{2(k-1)}}.$$

For the function f *defined by* (3.1.7.12), *the result is sharp.*

Proof. It is enough to demonstrate that

$$|f'(z)-1| \leq 1-\delta, \quad \text{where } |z| < r_3.$$

Then

$$|f'(z)-1| = \left|\sum_{k=2}^{\infty} k a_k z^{k-1}\right| \leq \sum_{k=2}^{\infty} k a_k |z|^{k-1},$$

and we can write $|f'(z)-1| \leq 1-\delta$, if we have $\sum_{k=2}^{\infty} \frac{k a_k}{1-\delta}|z|^{k-1} \leq 1$. Using Theorem 3.46, the above inequality gives the result:

$$|z|^{k-1} \leq \frac{(1-\delta)}{k} \times \sqrt{\frac{k\left\{\gamma k^2+[1+\gamma(2\lambda+\beta|d|-2)]k+[1+\gamma(\lambda-1)](\lambda+\beta|d|-1)\right\}\frac{\Gamma(n+k)}{\Gamma(k+\lambda+1)}\left(\frac{1+\alpha(k-1)+l}{l+1}\right)^m}{(\beta|d|-\lambda)(\gamma\lambda+1)\frac{\Gamma(n+1)}{\Gamma(\lambda+2)}}},$$

written as

$$|z| \leq \left\{ \frac{\frac{\{\gamma k^2+[1+\gamma(2\lambda+\beta|d|-2)]k+[1+\gamma(\lambda-1)](\lambda+\beta|d|-1)\}}{k^2}(1-\delta)^2 \times \frac{\Gamma(n+k)}{\Gamma(k+\lambda+1)}\left(\frac{1+\alpha(k-1)+l}{l+1}\right)^m}{(\beta\,|d|-\lambda)\,(\gamma\lambda+1)\,\frac{\Gamma(n+1)}{\Gamma(\lambda+2)}} \right\}^{\frac{1}{2(k-1)}},$$

which finalized the proof. □

3.1.8 A Survey of Fractional Integral Applied to the Convolution Product of Multiplier Transformation and Ruscheweyh Derivative for Introducing Subclasses of Analytic Functions

Fractional integral is prolific in interesting outcomes when associated to different operators. For the study presented in the paper Alb Lupaş (2022d), fractional integral is associated with the convolution product of multiplier transformation and Ruscheweyh derivative. Using the operator obtained as a result of this association and inspired by previously published results obtained with similarly introduced operators, the subclass of analytic functions $\mathcal{IR}(\mu, \lambda, \beta, \gamma, \alpha, l, m, n)$ is defined and investigated concerning various characteristics such as distortion bounds, extreme points and radii of close-to-convexity, starlikeness and convexity for functions belonging to this subclass.

The fractional integral associated with the convolution product of multiplier transformation and Ruscheweyh derivative is introduced in Definition 2.18 by

$$D_z^{-\lambda} IR_{\alpha,l}^{m,n} f(z) = \frac{1}{\Gamma(\lambda)} \int_0^z \frac{IR_{\alpha,l}^{m,n} f(t)}{(z-t)^{1-\lambda}} dt = \frac{1}{\Gamma(\lambda)} \int_0^z \frac{t}{(z-t)^{1-\lambda}} dt$$
$$+ \frac{1}{\Gamma(\lambda)\Gamma(n+1)} \sum_{k=2}^{\infty} \left(\frac{1+\alpha(k-1)+l}{l+1}\right)^m \frac{\Gamma(n+k)}{\Gamma(k)} a_k^2$$
$$\times \int_0^z \frac{t^k}{(z-t)^{1-\lambda}} dt,$$

which has the following form, after a simple calculation,

$$D_z^{-\lambda} IR_{\alpha,l}^{m,n} f(z) = \frac{1}{\Gamma(\lambda+2)} z^{\lambda+1} + \frac{1}{\Gamma(n+1)} \sum_{k=2}^{\infty} \left(\frac{1+\alpha(k-1)+l}{l+1} \right)^m \times \frac{k\Gamma(n+k)}{\Gamma(k+\lambda+1)} a_k^2 z^{k+\lambda},$$

for the function $f(z) = z + \sum_{k=2}^{\infty} a_k z^k \in \mathcal{A}$. We note that $D_z^{-\lambda} IR_{\alpha,l}^{m,n} f(z) \in \mathcal{A}(\lambda+1,1)$.

A new subclass of analytic functions is defined using the operator defined above.

Definition 3.11 (Alb Lupaş, 2022d). For $\mu, \alpha, l \geq 0$, $\lambda, m, n \in \mathbb{N}$, $\gamma \in \mathbb{C} - \{0\}$ and $\frac{\lambda+\mu}{\lambda+\mu+|\gamma|\Gamma(\lambda+2)} < \beta \leq 1$, let $\mathcal{IR}(\mu, \lambda, \beta, \gamma, \alpha, l, m, n)$ be the subclass of $\mathcal{A}$ consisting of functions that satisfying the inequality

$$\left| \frac{\lambda(1-\mu) \frac{D_z^{-\lambda} IR_{\alpha,l}^{m,n} f(z)}{z} + \mu \left(D_z^{-\lambda} IR_{\alpha,l}^{m,n} f(z) \right)'}{\lambda(1-\mu) \frac{D_z^{-\lambda} IR_{\alpha,l}^{m,n} f(z)}{z} + \mu \left(D_z^{-\lambda} IR_{\alpha,l}^{m,n} f(z) \right)' - \gamma} \right| < \beta \qquad (3.1.8.1)$$

Coefficient Bounds

Coefficient bounds and extreme points for functions in $\mathcal{IR}(\mu, \lambda, \beta, \gamma, \alpha, l, m, n)$ are obtained.

Theorem 3.56 (Alb Lupaş, 2022d). *The function $f \in \mathcal{A}$ belongs to the subclass $\mathcal{IR}(\mu, \lambda, \beta, \gamma, \alpha, l, m, n)$ if and only if*

$$\sum_{k=2}^{\infty} \frac{(\lambda+\mu k) k \left(\frac{1+\alpha(k-1)+l}{l+1} \right)^m \Gamma(n+k)}{\Gamma(k+\lambda+1)} a_k^2 < \frac{\beta|\gamma|\Gamma(n+1)}{\beta+1} - \frac{(\lambda+\mu)\Gamma(n+1)}{\Gamma(\lambda+2)}. \qquad (3.1.8.2)$$

The result is sharp for the function

$$F(z) = z + \sqrt{\frac{\left(\frac{\beta|\gamma|}{\beta+1} - \frac{(\lambda+\mu)}{\Gamma(\lambda+2)} \right) \Gamma(n+1)\Gamma(k+\lambda+1)}{(\lambda+\mu k) k \left(\frac{1+\alpha(k-1)+l}{l+1} \right)^m \Gamma(n+k)}} z^k, \quad k \geq 2. \qquad (3.1.8.3)$$

Proof. Assume that function $f \in \mathcal{A}$ and that inequality (3.1.8.2) holds. Then we obtain

$$\left|\frac{\lambda(1-\mu)\frac{D_z^{-\lambda}IR_{\alpha,l}^{m,n}f(z)}{z}+\mu\left(D_z^{-\lambda}IR_{\alpha,l}^{m,n}f(z)\right)'}{\lambda(1-\mu)\frac{D_z^{-\lambda}IR_{\alpha,l}^{m,n}f(z)}{z}+\mu\left(D_z^{-\lambda}IR_{\alpha,l}^{m,n}f(z)\right)'-\gamma}\right|$$

$$=\left|\frac{\frac{\lambda+\mu}{\Gamma(\lambda+2)}z^{\lambda}+\frac{1}{\Gamma(n+1)}\sum_{k=2}^{\infty}\frac{(\lambda+\mu k)k\left(\frac{1+\alpha(k-1)+l}{l+1}\right)^m\Gamma(n+k)}{\Gamma(k+\lambda+1)}a_k^2z^{k+\lambda-1}}{\frac{\lambda+\mu}{\Gamma(\lambda+2)}z^{\lambda}+\frac{1}{\Gamma(n+1)}\sum_{k=2}^{\infty}\frac{(\lambda+\mu k)k\left(\frac{1+\alpha(k-1)+l}{l+1}\right)^m\Gamma(n+k)}{\Gamma(k+\lambda+1)}a_k^2z^{k+\lambda-1}-\gamma}\right|$$

$$=\frac{\left|\frac{\lambda+\mu}{\Gamma(\lambda+2)}z^{\lambda}+\frac{1}{\Gamma(n+1)}\sum_{k=2}^{\infty}\frac{(\lambda+\mu k)k\left(\frac{1+\alpha(k-1)+l}{l+1}\right)^m\Gamma(n+k)}{\Gamma(k+\lambda+1)}a_k^2z^{k+\lambda-1}\right|}{\left|\frac{\lambda+\mu}{\Gamma(\lambda+2)}z^{\lambda}+\frac{1}{\Gamma(n+1)}\sum_{k=2}^{\infty}\frac{(\lambda+\mu k)k\left(\frac{1+\alpha(k-1)+l}{l+1}\right)^m\Gamma(n+k)}{\Gamma(k+\lambda+1)}a_k^2z^{k+\lambda-1}-\gamma\right|}$$

$$<\frac{\left|\frac{\lambda+\mu}{\Gamma(\lambda+2)}z^{\lambda}\right|+\left|\frac{1}{\Gamma(n+1)}\sum_{k=2}^{\infty}\frac{(\lambda+\mu k)k\left(\frac{1+\alpha(k-1)+l}{l+1}\right)^m\Gamma(n+k)}{\Gamma(k+\lambda+1)}a_k^2z^{k+\lambda-1}\right|}{|\gamma|-\left|\frac{\lambda+\mu}{\Gamma(\lambda+2)}z^{\lambda}\right|-\left|\frac{1}{\Gamma(n+1)}\sum_{k=2}^{\infty}\frac{(\lambda+\mu k)k\left(\frac{1+\alpha(k-1)+l}{l+1}\right)^m\Gamma(n+k)}{\Gamma(k+\lambda+1)}a_k^2z^{k+\lambda-1}\right|}$$

$$=\frac{\frac{\lambda+\mu}{\Gamma(\lambda+2)}\left|z^{\lambda}\right|+\frac{1}{\Gamma(n+1)}\sum_{k=2}^{\infty}\frac{(\lambda+\mu k)k\left(\frac{1+\alpha(k-1)+l}{l+1}\right)^m\Gamma(n+k)}{\Gamma(k+\lambda+1)}a_k^2\left|z^{k+\lambda-1}\right|}{|\gamma|-\frac{\lambda+\mu}{\Gamma(\lambda+2)}\left|z^{\lambda}\right|-\frac{1}{\Gamma(n+1)}\sum_{k=2}^{\infty}\frac{(\lambda+\mu k)k\left(\frac{1+\alpha(k-1)+l}{l+1}\right)^m\Gamma(n+k)}{\Gamma(k+\lambda+1)}a_k^2\left|z^{k+\lambda-1}\right|}$$

$$<\beta,\quad z\in U.$$

Choosing values of z on real axis and considering $z\to 1^-$, we have

$$\frac{\lambda+\mu}{\Gamma(\lambda+2)}+\frac{1}{\Gamma(n+1)}\sum_{k=2}^{\infty}\frac{(\lambda+\mu k)k\left(\frac{1+\alpha(k-1)+l}{l+1}\right)^m\Gamma(n+k)}{\Gamma(k+\lambda+1)}a_k^2$$

$$<\beta|\gamma|-\beta\frac{\lambda+\mu}{\Gamma(\lambda+2)}-\beta\frac{1}{\Gamma(n+1)}$$

$$\times\sum_{k=2}^{\infty}\frac{(\lambda+\mu k)k\left(\frac{1+\alpha(k-1)+l}{l+1}\right)^m\Gamma(n+k)}{\Gamma(k+\lambda+1)}a_k^2,$$

equivalently with

$$\sum_{k=2}^{\infty} \frac{(\lambda+\mu k)\, k \left(\frac{1+\alpha(k-1)+l}{l+1}\right)^{m} \Gamma(n+k)}{\Gamma(k+\lambda+1)} a_k^2 < \frac{\beta|\gamma|\Gamma(n+1)}{\beta+1} - \frac{(\lambda+\mu)\Gamma(n+1)}{\Gamma(\lambda+2)},$$

obtaining that $f \in \mathcal{IR}(\mu,\lambda,\beta,\gamma,\alpha,l,m,n)$.

Conversely, suppose that $f \in \mathcal{IR}(\mu,\lambda,\beta,\gamma,\alpha,l,m,n)$, then we get the following inequality

$$\operatorname{Re}\left\{\left|\frac{\lambda(1-\mu)\frac{D_z^{-\lambda} IR_{\alpha,l}^{m,n} f(z)}{z} + \mu\left(D_z^{-\lambda} IR_{\alpha,l}^{m,n} f(z)\right)'}{\lambda(1-\mu)\frac{D_z^{-\lambda} IR_{\alpha,l}^{m,n} f(z)}{z} + \mu\left(D_z^{-\lambda} IR_{\alpha,l}^{m,n} f(z)\right)' - \gamma}\right|\right\} > -\beta$$

$$\operatorname{Re}\left\{\frac{\frac{\lambda+\mu}{\Gamma(\lambda+2)} z^{\lambda} + \frac{1}{\Gamma(n+1)}\sum_{k=2}^{\infty}\frac{(\lambda+\mu k)k\left(\frac{1+\alpha(k-1)+l}{l+1}\right)^{m}\Gamma(n+k)}{\Gamma(k+\lambda+1)} a_k^2 z^{k+\lambda-1}}{\frac{\lambda+\mu}{\Gamma(\lambda+2)} z^{\lambda} + \frac{1}{\Gamma(n+1)}\sum_{k=2}^{\infty}\frac{(\lambda+\mu k)k\left(\frac{1+\alpha(k-1)+l}{l+1}\right)^{m}\Gamma(n+k)}{\Gamma(k+\lambda+1)} a_k^2 z^{k+\lambda-1} - \gamma} + \beta\right\} > 0$$

$$\operatorname{Re}\left\{\frac{(1+\beta)\frac{\lambda+\mu}{\Gamma(\lambda+2)} z^{\lambda} + \frac{1+\beta}{\Gamma(n+1)}\sum_{k=2}^{\infty}\frac{(\lambda+\mu k)k\left(\frac{1+\alpha(k-1)+l}{l+1}\right)^{m}\Gamma(n+k)}{\Gamma(k+\lambda+1)} a_k^2 z^{k+\lambda-1} - \beta\gamma}{\frac{\lambda+\mu}{\Gamma(\lambda+2)} z^{\lambda} + \frac{1}{\Gamma(n+1)}\sum_{k=2}^{\infty}\frac{(\lambda+\mu k)k\left(\frac{1+\alpha(k-1)+l}{l+1}\right)^{m}\Gamma(n+k)}{\Gamma(k+\lambda+1)} a_k^2 z^{k+\lambda-1} - \gamma}\right\} > 0.$$

Taking account that $\operatorname{Re}(-e^{i\theta}) \geq -|e^{i\theta}| = -1$, the above inequality reduces to

$$\frac{-(1+\beta)\frac{\lambda+\mu}{\Gamma(\lambda+2)} r^{\lambda} - \frac{1+\beta}{\Gamma(n+1)}\sum_{k=2}^{\infty}\frac{(\lambda+\mu k)k\left(\frac{1+\alpha(k-1)+l}{l+1}\right)^{m}\Gamma(n+k)}{\Gamma(k+\lambda+1)} a_k^2 r^{k+\lambda-1} + \beta|\gamma|}{\frac{\lambda+\mu}{\Gamma(\lambda+2)} r^{\lambda} + \frac{1}{\Gamma(n+1)}\sum_{k=2}^{\infty}\frac{(\lambda+\mu k)k\left(\frac{1+\alpha(k-1)+l}{l+1}\right)^{m}\Gamma(n+k)}{\Gamma(k+\lambda+1)} a_k^2 r^{k+\lambda-1} - \gamma} > 0.$$

Letting $r \to 1^-$ and applying the mean value theorem, we have desired inequality (3.1.8.2).

This completes the proof of theorem □

Corollary 3.15 (Alb Lupaş, 2022d). *If* $f \in \mathcal{IR}(\mu,\lambda,\beta,\gamma,\alpha,l,m,n)$, *then*

$$a_k \leq \sqrt{\frac{\left(\frac{\beta|\gamma|}{\beta+1} - \frac{(\lambda+\mu)}{\Gamma(\lambda+2)}\right)\Gamma(n+1)\Gamma(k+\lambda+1)}{(\lambda+\mu k)\, k\left(\frac{1+\alpha(k-1)+l}{l+1}\right)^{m}\Gamma(n+k)}}, \quad k \geq 2, \tag{3.1.8.4}$$

with equality only for functions defined by (3.1.8.3).

Theorem 3.57 (Alb Lupaş, 2022d). *Consider* $f_1(z) = z$ *and*

$$f_k(z) = z - \sqrt{\frac{\left(\frac{\beta|\gamma|}{\beta+1} - \frac{(\lambda+\mu)}{\Gamma(\lambda+2)}\right)\Gamma(n+1)\Gamma(k+\lambda+1)}{(\lambda+\mu k)\, k \left(\frac{1+\alpha(k-1)+l}{l+1}\right)^m \Gamma(n+k)}} z^k, \quad k \geq 2, \tag{3.1.8.5}$$

for $\mu, \alpha, l \geq 0$, $\lambda, m, n \in \mathbb{N}$, $\gamma \in \mathbb{C} - \{0\}$ *and* $0 < \beta \leq 1$.

Then $f \in \mathcal{IR}(\mu, \lambda, \beta, \gamma, \alpha, l, m, n)$ *if and only if it can be written in the form*

$$f(z) = \sum_{k=1}^{\infty} \omega_k f_k(z), \tag{3.1.8.6}$$

where $\omega_k \geq 0$ *and* $\sum_{k=1}^{\infty} \omega_k = 1$.

Proof. Assume f can be written as in (3.1.8.6). Then

$$f(z) = z - \sum_{k=2}^{\infty} \omega_k \sqrt{\frac{\left(\frac{\beta|\gamma|}{\beta+1} - \frac{(\lambda+\mu)}{\Gamma(\lambda+2)}\right)\Gamma(n+1)\Gamma(k+\lambda+1)}{(\lambda+\mu k)\, k \left(\frac{1+\alpha(k-1)+l}{l+1}\right)^m \Gamma(n+k)}} z^k.$$

Now,

$$\sum_{k=2}^{\infty} \sqrt{\frac{(\lambda+\mu k)\, k \left(\frac{1+\alpha(k-1)+l}{l+1}\right)^m \Gamma(n+k)}{\left(\frac{\beta|\gamma|}{\beta+1} - \frac{(\lambda+\mu)}{\Gamma(\lambda+2)}\right)\Gamma(n+1)\Gamma(k+\lambda+1)}} \omega_k$$

$$\times \sqrt{\frac{\left(\frac{\beta|\gamma|}{\beta+1} - \frac{(\lambda+\mu)}{\Gamma(\lambda+2)}\right)\Gamma(n+1)\Gamma(k+\lambda+1)}{(\lambda+\mu k)\, k \left(\frac{1+\alpha(k-1)+l}{l+1}\right)^m \Gamma(n+k)}} = \sum_{k=2}^{\infty} \omega_k = 1 - \omega_1 \leq 1.$$

Thus, $f \in \mathcal{IR}(\mu, \lambda, \beta, \gamma, \alpha, l, m, n)$.

Conversely, let $f \in \mathcal{IR}(\mu, \lambda, \beta, \gamma, \alpha, l, m, n)$. Then by using (3.1.8.4), setting

$$\omega_k = \sqrt{\frac{\left(\frac{\beta|\gamma|}{\beta+1} - \frac{(\lambda+\mu)}{\Gamma(\lambda+2)}\right)\Gamma(n+1)\Gamma(k+\lambda+1)}{(\lambda+\mu k)\, k \left(\frac{1+\alpha(k-1)+l}{l+1}\right)^m \Gamma(n+k)}} a_k, \quad k \geq 2$$

and $\omega_1 = 1 - \sum_{k=2}^{\infty} \omega_k$, we obtain $f(z) = \sum_{k=1}^{\infty} \omega_k f_k(z)$, completing the proof of theorem. □

Distortion Bounds

Next distortion bounds for the subclass $\mathcal{IR}(\mu,\lambda,\beta,\gamma,\alpha,l,m,n)$ are obtained.

Theorem 3.58 (Alb Lupaş, 2022d). *For $f\in\mathcal{IR}(\mu,\lambda,\beta,\gamma,\alpha,l,m,n)$, inequality*

$$r-\sqrt{\frac{\left(\frac{\beta|\gamma|}{\beta+1}-\frac{(\lambda+\mu)}{\Gamma(\lambda+2)}\right)\Gamma(\lambda+3)}{2(n+1)(\lambda+2\mu)\left(\frac{1+\alpha+l}{l+1}\right)^m}}r^2\le|f(z)|$$

$$\le r+\sqrt{\frac{\left(\frac{\beta|\gamma|}{\beta+1}-\frac{(\lambda+\mu)}{\Gamma(\lambda+2)}\right)\Gamma(\lambda+3)}{2(n+1)(\lambda+2\mu)\left(\frac{1+\alpha+l}{l+1}\right)^m}}r^2 \tag{3.1.8.7}$$

holds if the sequence $\{\sigma_k(\mu,\lambda,\alpha,l,m,n)\}_{j=2}^{\infty}$ is non-decreasing, and

$$1-2\sqrt{\frac{\left(\frac{\beta|\gamma|}{\beta+1}-\frac{(\lambda+\mu)}{\Gamma(\lambda+2)}\right)\Gamma(\lambda+3)}{2(n+1)(\lambda+2\mu l)\left(\frac{1+\alpha+l}{l+1}\right)^m}}r^2\le|f'(z)|$$

$$\le 1+2\sqrt{\frac{\left(\frac{\beta|\gamma|}{\beta+1}-\frac{(\lambda+\mu)}{\Gamma(\lambda+2)}\right)\Gamma(\lambda+3)}{2(n+1)(\lambda+2\mu)\left(\frac{1+\alpha+l}{l+1}\right)^m}}r^2 \tag{3.1.8.8}$$

holds if the sequence $\{\frac{\sigma_k(\mu,\lambda,\alpha,l,m,n)}{k}\}_{k=2}^{\infty}$ is non-decreasing, where

$$\sigma_k(\mu,\lambda,\alpha,l,m,n)=\sqrt{\frac{(\lambda+\mu k)k\left(\frac{1+\alpha(k-1)+l}{l+1}\right)^m\Gamma(n+k)}{\Gamma(k+\lambda+1)}}.$$

The bounds in (3.1.8.7) *and* (3.1.8.8) *are sharp, for f given by*

$$f(z)=z+\sqrt{\frac{\left(\frac{\beta|\gamma|}{\beta+1}-\frac{(\lambda+\mu)}{\Gamma(\lambda+2)}\right)\Gamma(\lambda+3)}{2(n+1)(\lambda+\mu)\left(\frac{1+\alpha+l}{l+1}\right)^m}}z^2,\quad z=\pm r. \tag{3.1.8.9}$$

Proof. Using Theorem 3.56, we obtain

$$\sum_{k=2}^{\infty}a_k\le\sqrt{\frac{\left(\frac{\beta|\gamma|}{\beta+1}-\frac{(\lambda+\mu)}{\Gamma(\lambda+2)}\right)\Gamma(\lambda+3)}{2(n+1)(\lambda+2\mu)\left(\frac{1+\alpha+l}{l+1}\right)^m}}. \tag{3.1.8.10}$$

We have

$$|z| - |z|^2 \sum_{k=2}^{\infty} a_2 \leq |f(z)| \leq |z| + |z|^2 \sum_{k=2}^{\infty} a_2.$$

Thus,

$$r - \sqrt{\frac{\left(\frac{\beta|\gamma|}{\beta+1} - \frac{(\lambda+\mu)}{\Gamma(\lambda+2)}\right)\Gamma(\lambda+3)}{2(n+1)(\lambda+2\mu)\left(\frac{1+\alpha+l}{l+1}\right)^m}r^2} \leq |f(z)|$$

$$\leq r + \sqrt{\frac{\left(\frac{\beta|\gamma|}{\beta+1} - \frac{(\lambda+\mu)}{\Gamma(\lambda+2)}\right)\Gamma(\lambda+3)}{2(n+1)(\lambda+2\mu)\left(\frac{1+\alpha+l}{l+1}\right)^m}r^2}. \qquad (3.1.8.11)$$

Hence, (3.1.8.7) follows from (3.1.8.11). Further,

$$\sum_{k=2}^{\infty} k a_k \leq \sqrt{\frac{\left(\frac{\beta|\gamma|}{\beta+1} - \frac{(\lambda+\mu)}{\Gamma(\lambda+2)}\right)\Gamma(\lambda+3)}{2(n+1)(\lambda+2\mu)\left(\frac{1+\alpha+l}{l+1}\right)^m}}.$$

Hence, (3.1.8.8) follows from

$$1 - r\sum_{k=2}^{\infty} k a_k \leq |f'(z)| \leq 1 + r\sum_{k=2}^{\infty} k a_k. \qquad \square$$

Radius of Starlikeness and Convexity

We give the radii of close-to-convexity, starlikeness and convexity for the subclass $\mathcal{IR}(\mu, \lambda, \beta, \gamma, \alpha, l, m, n)$.

Theorem 3.59 (Alb Lupaş, 2022d). *The function $f \in \mathcal{IR}(\mu, \lambda, \beta, \gamma, \alpha, l, m, n)$ is close-to-convex of order δ, $0 \leq \delta < 1$ in the disc $|z| < r$, where*

$$r := \inf_{k \geq 2} \sqrt{\frac{(n+1)(1-\delta)^2(\lambda+\mu k)\left(\frac{1+\alpha(k-1)+l}{l+1}\right)^m}{\left(\frac{\beta|\gamma|}{\beta+1} - \frac{(\lambda+\mu)}{\Gamma(\lambda+2)}\right)k\Gamma(k+\lambda+1)}}. \qquad (3.1.8.12)$$

The result is sharp, with extremal function f given by (3.1.8.3).

Proof. For function $f \in \mathcal{A}$, we have to show that

$$|f'(z) - 1| < 1 - \delta, \quad \text{for } |z| < r. \qquad (3.1.8.13)$$

By a simple calculation, we obtain

$$\left|f'(z)-1\right| \leq \sum_{k=2}^{\infty} k a_k \left|z\right|,$$

which is less than $1-\delta$ if

$$\sum_{k=2}^{\infty} \frac{k}{1-\delta} a_k \left|z\right| < 1.$$

Function $f \in \mathcal{IR}(\mu,\lambda,\beta,\gamma,\alpha,l,m,n)$ if and only if

$$\frac{1}{\Gamma(n+1)} \sum_{k=2}^{\infty} \frac{(\lambda+\mu k)\, k \left(\frac{1+\alpha(k-1)+l}{l+1}\right)^m \Gamma(n+k)}{\left(\frac{\beta|\gamma|}{\beta+1} - \frac{(\lambda+\mu)}{\Gamma(\lambda+2)}\right) \Gamma(k+\lambda+1)} a_k^2 < 1,$$

relation (3.1.8.13) is true if

$$\frac{k}{1-\delta} \left|z\right| \leq \sum_{k=2}^{\infty} \sqrt{\frac{(\lambda+\mu k)\, k \left(\frac{1+\alpha(k-1)+l}{l+1}\right)^m \Gamma(n+k)}{\left(\frac{\beta|\gamma|}{\beta+1} - \frac{(\lambda+\mu)}{\Gamma(\lambda+2)}\right) \Gamma(n+1)\,\Gamma(k+\lambda+1)}},$$

or, equivalently,

$$\left|z\right| \leq \sum_{k=2}^{\infty} \sqrt{\frac{(1-\delta)^2 (\lambda+\mu k) \left(\frac{1+\alpha(k-1)+l}{l+1}\right)^m \Gamma(n+k)}{\left(\frac{\beta|\gamma|}{\beta+1} - \frac{(\lambda+\mu)}{\Gamma(\lambda+2)}\right) k\Gamma(n+1)\,\Gamma(k+\lambda+1)}},$$

which completes the proof. □

Theorem 3.60 (Alb Lupaş, 2022d). *Consider $f \in \mathcal{IR}(\mu,\lambda,\beta,\gamma,\alpha,l,m,n)$. Then*

1. *f is starlike of order δ, $0 \leq \delta < 1$, in the disc $|z| < r_1$ where,*

$$r_1 = \inf_{k\geq 2} \sqrt{\frac{(1-\delta)^2 (\lambda+\mu k)\, k \left(\frac{1+\alpha(k-1)+l}{l+1}\right)^m \Gamma(n+k)}{\left(\frac{\beta|\gamma|}{\beta+1} - \frac{(\lambda+\mu)}{\Gamma(\lambda+2)}\right) (k+\delta-2)^2\, \Gamma(n+1)\,\Gamma(k+\lambda+1)}}.$$

2. *f is convex of order δ, $0 \leq \delta < 1$, in the disc $|z| < r_2$ where,*

$$r_2 = \inf_{k\geq 2} \sqrt{\frac{(1-\delta)^2 (\lambda+\mu k) \left(\frac{1+\alpha(k-1)+l}{l+1}\right)^m \Gamma(n+k)}{\left(\frac{\beta|\gamma|}{\beta+1} - \frac{(\lambda+\mu)}{\Gamma(\lambda+2)}\right) k\,(k-1)^2\, \Gamma(n+1)\,\Gamma(k+\lambda+1)}}.$$

Each of these results is sharp for the extremal function f given by (3.1.8.3).

Proof. 1. For $0 \leq \delta < 1$, we have to prove that

$$\left| \frac{zf'(z)}{f(z)} - 1 \right| < 1 - \delta, \quad \text{for } |z| < r_1. \tag{3.1.8.14}$$

We obtain

$$\left| \frac{zf'(z)}{f(z)} - 1 \right| \leq \left| \frac{\sum_{k=2}^{\infty}(k-1)a_k |z|}{1 + \sum_{k=2}^{\infty} a_k |z|} \right|,$$

which is less than $1 - \delta$ if

$$\sum_{k=2}^{\infty} \frac{(k+\delta-2)}{1-\delta} a_k |z| < 1.$$

Function $f \in \mathcal{IR}(\mu, \lambda, \beta, \gamma, \alpha, l, m, n)$ if and only if

$$\frac{1}{\Gamma(n+1)} \sum_{k=2}^{\infty} \frac{(\lambda + \mu k) k \left(\frac{1+\alpha(k-1)+l}{l+1}\right)^m \Gamma(n+k)}{\left(\frac{\beta|\gamma|}{\beta+1} - \frac{(\lambda+\mu)}{\Gamma(\lambda+2)}\right) \Gamma(k+\lambda+1)} a_k^2 < 1.$$

Relation (3.1.8.14) holds if

$$\frac{k+\delta-2}{1-\delta} |z| < \sqrt{\frac{(\lambda + \mu k) k \left(\frac{1+\alpha(k-1)+l}{l+1}\right)^m \Gamma(n+k)}{\left(\frac{\beta|\gamma|}{\beta+1} - \frac{(\lambda+\mu)}{\Gamma(\lambda+2)}\right) \Gamma(n+1) \Gamma(k+\lambda+1)}},$$

equivalently,

$$|z| < \sqrt{\frac{(1-\delta)^2 (\lambda + \mu k) k \left(\frac{1+\alpha(k-1)+l}{l+1}\right)^m \Gamma(n+k)}{\left(\frac{\beta|\gamma|}{\beta+1} - \frac{(\lambda+\mu)}{\Gamma(\lambda+2)}\right) (k+\delta-2)^2 \Gamma(n+1) \Gamma(k+\lambda+1)}},$$

which yields the starlikeness of the family.

2. The function f is convex if and only the function zf' is starlike, therefore it is enough to prove (2) with a similar way of the proof of (1). So, the function f is convex if and only if

$$|zf''(z)| < 1 - \delta. \tag{3.1.8.15}$$

We obtain

$$|zf''(z)| \leq \left| \sum_{k=3}^{\infty} k(k-1)a_k |z| \right| < 1 - \delta,$$

equivalently

$$\sum_{k=2}^{\infty} \frac{k(k-1)}{1-\delta} a_k |z| < 1.$$

Function $f \in \mathcal{IR}(\mu, \lambda, \beta, \gamma, \alpha, l, m, n)$ if and only if

$$\frac{1}{\Gamma(n+1)} \sum_{k=2}^{\infty} \frac{(\lambda + \mu k) k \left(\frac{1+\alpha(k-1)+l}{l+1}\right)^m \Gamma(n+k)}{\left(\frac{\beta|\gamma|}{\beta+1} - \frac{(\lambda+\mu)}{\Gamma(\lambda+2)}\right) \Gamma(k+\lambda+1)} a_k^2 < 1$$

and relation (3.1.8.15) is true if

$$\frac{k(k-1)}{1-\delta} |z| < \sqrt{\frac{(\lambda + \mu k) k \left(\frac{1+\alpha(k-1)+l}{l+1}\right)^m \Gamma(n+j)}{\left(\frac{\beta|\gamma|}{\beta+1} - \frac{(\lambda+\mu)}{\Gamma(\lambda+2)}\right) \Gamma(n+1) \Gamma(j+\lambda+1)}} |z|,$$

equivalently with

$$|z| < \sqrt{\frac{(1-\delta)^2 (\lambda + \mu k) \left(\frac{1+\alpha(k-1)+l}{l+1}\right)^m \Gamma(n+k)}{\left(\frac{\beta|\gamma|}{\beta+1} - \frac{(\lambda+\mu)}{\Gamma(\lambda+2)}\right) k(k-1)^2 \Gamma(n+1) \Gamma(k+\lambda+1)}},$$

which yields the convexity of the family. □

3.2 Applications of Differential Operators in the Theory of Differential Subordinations and Superordinations

3.2.1 *Some Differential Subordinations Using Ruscheweyh Derivative and a Multiplier Transformation*

In the paper Alb Lupaş (2010d), we defined a new operator using the multiplier transformation and Ruscheweyh derivative. Denote by $RI_{n,\lambda,l}^{\alpha}$, the operator from Definition 2.7 given by $RI_{n,\lambda,l}^{\alpha} : \mathcal{A} \to \mathcal{A}$,

$$RI_{n,\lambda,l}^{\alpha} f(z) = (1-\alpha) R^n f(z) + \alpha I(n, \lambda, l) f(z),$$

for $z \in U$, where $R^n f(z)$ denote the Ruscheweyh derivative, $I(n, \lambda, l) f(z)$ is the multiplier transformation and $\mathcal{A}_n = \{f \in \mathcal{H}(U) : f(z) = z + a_{n+1} z^{n+1} + \cdots,\ z \in U\}$ is the class of normalized analytic functions with $\mathcal{A}_1 = \mathcal{A}$.

A certain subclass, denoted by $\mathcal{RI}_n(\delta, \lambda, l, \alpha)$, of analytic functions in the open unit disc is introduced by means of the new operator. By making use of the concept of differential subordination we will derive various properties and characteristics of the subclass $\mathcal{RI}_n(\delta, \lambda, l, \alpha)$. Also, several differential subordinations are established regarding the operator $RI^{\alpha}_{n,\lambda,l}$ in the papers Alb Lupaş (2010d, 2015b).

If $f \in \mathcal{A}$, $f(z) = z + \sum_{j=2}^{\infty} a_j z^j$, then

$$RI^{\alpha}_{n,\lambda,l}f(z) = z + \sum_{j=2}^{\infty} \left\{ \alpha \left(\frac{1 + \lambda(j-1) + l}{l+1} \right)^n + (1-\alpha) C^n_{n+j-1} \right\} a_j z^j,$$

for $z \in U$.

Definition 3.12 (Alb Lupaş, 2010d). Let $\delta \in [0,1)$, $\alpha, \lambda, l \geq 0$ and $n \in \mathbb{N}$. A function $f \in \mathcal{A}$ is said to be in the subclass $\mathcal{RI}_n(\delta, \lambda, l, \alpha)$ if it satisfies the inequality

$$\mathrm{Re}\left(RI^{\alpha}_{n,\lambda,l}f(z)\right)' > \delta, \quad \text{for } z \in U.$$

Theorem 3.61 (Alb Lupaş, 2010d). *The set* $\mathcal{RI}_n(\delta, \lambda, l, \alpha)$ *is convex.*

Proof. Let the functions

$$f_j(z) = z + \sum_{j=2}^{\infty} a_{jk} z^j, \quad \text{for } k = 1, 2, \ z \in U,$$

be in the subclass $\mathcal{RI}_n(\delta, \lambda, l, \alpha)$. It is sufficient to show that the function

$$h(z) = \eta_1 f_1(z) + \eta_2 f_2(z)$$

is in the subclass $\mathcal{RI}_n(\delta, \lambda, l, \alpha)$, with η_1 and η_2 nonnegative such that $\eta_1 + \eta_2 = 1$.

Since $h(z) = z + \sum_{j=2}^{\infty} (\eta_1 a_{j1} + \eta_2 a_{j2}) z^j$, for $z \in U$, then

$$RI^{\alpha}_{n,\lambda,l}h(z) = z + \sum_{j=2}^{\infty} \left\{ \alpha \left(\frac{1 + \lambda(j-1) + l}{l+1} \right)^n + (1-\alpha) C^n_{n+j-1} \right\} (\eta_1 a_{j1} + \eta_2 a_{j2}) z^j,$$

for $z \in U$. (3.2.1.1)

Differentiating (3.2.1.1), we obtain

$$\begin{aligned}&\left(RI^{\alpha}_{n,\lambda,l}h(z)\right)'\\&\quad=1+\sum_{j=2}^{\infty}\left\{\alpha\left(\frac{1+\lambda(j-1)+l}{l+1}\right)^{n}+(1-\alpha)C^{n}_{n+j-1}\right\}\\&\quad\times(\eta_1 a_{j1}+\eta_2 a_{j2})jz^{j-1},\quad \text{for } z\in U.\end{aligned}$$

Hence,

$$\begin{aligned}&\operatorname{Re}\left(RI^{\alpha}_{n,\lambda,l}h(z)\right)'\\&=1+\operatorname{Re}\left(\eta_1\sum_{j=2}^{\infty}j\left\{\alpha\left(\frac{1+\lambda(j-1)+l}{l+1}\right)^{n}+(1-\alpha)C^{n}_{n+j-1}\right\}a_{j1}z^{j-1}\right)\\&+\operatorname{Re}\left(\eta_2\sum_{j=2}^{\infty}j\left\{\alpha\left(\frac{1+\lambda(j-1)+l}{l+1}\right)^{n}+(1-\alpha)C^{n}_{n+j-1}\right\}a_{j2}z^{j-1}\right).\end{aligned}\tag{3.2.1.2}$$

Taking into account that $f_1, f_2\in\mathcal{RI}_n(\delta,\lambda,l,\alpha)$ we deduce

$$\begin{aligned}&\operatorname{Re}\left(\eta_k\sum_{j=2}^{\infty}j\left\{\alpha\left(\frac{1+\lambda(j-1)+l}{l+1}\right)^{n}+(1-\alpha)C^{n}_{n+j-1}\right\}a_{jk}z^{j-1}\right)\\&\quad>\eta_k(\delta-1),\quad \text{for } k=1,2.\end{aligned}\tag{3.2.1.3}$$

Using (3.2.1.3), we get from (3.2.1.2)

$$\operatorname{Re}\left(RI^{\alpha}_{n,\lambda,l}h(z)\right)'>1+\eta_1(\delta-1)+\eta_2(\delta-1)=\delta,\quad \text{for } z\in U,$$

which is equivalent that $\mathcal{RI}_n(\delta,\lambda,l,\alpha)$ is convex. □

Theorem 3.62 (Alb Lupaş, 2010d). *Let g be a convex function in U and let $h(z)=g(z)+\frac{1}{c+2}zg'(z)$, where $z\in U$, $c>0$.*

If $f\in\mathcal{RI}_n(\delta,\lambda,l,\alpha)$ and $F(z)=I_c(f)(z)=\frac{c+2}{z^{c+1}}\int_0^z t^c f(t)\,dt$, for $z\in U$, then

$$\left(RI^{\alpha}_{n,\lambda,l}f(z)\right)'\prec h(z),\quad \text{for } z\in U,\tag{3.2.1.4}$$

implies

$$\left(RI_{n,\lambda,l}^{\alpha}F(z)\right)' \prec g(z), \quad \text{for } z \in U,$$

and this result is sharp.

Proof. We obtain that

$$z^{c+1}F(z) = (c+2)\int_0^z t^c f(t)\,dt. \tag{3.2.1.5}$$

Differentiating (3.2.1.5), with respect to z, we have $(c+1)F(z)+zF'(z) = (c+2)f(z)$ and

$$(c+1)RI_{n,\lambda,l}^{\alpha}F(z) + z\left(RI_{n,\lambda,l}^{\alpha}F(z)\right)' = (c+2)RI_{n,\lambda,l}^{\alpha}f(z), \quad \text{for } z \in U. \tag{3.2.1.6}$$

Differentiating (3.2.1.6), we have

$$\left(RI_{n,\lambda,l}^{\alpha}F(z)\right)' + \frac{1}{c+2}z\left(RI_{n,\lambda,l}^{\alpha}F(z)\right)'' = \left(RI_{n,\lambda,l}^{\alpha}f(z)\right)', \quad \text{for } z \in U. \tag{3.2.1.7}$$

Using (3.2.1.7), the differential subordination (3.2.1.4) becomes

$$\left(RI_{n,\lambda,l}^{\alpha}F(z)\right)' + \frac{1}{c+2}z\left(RI_{n,\lambda,l}^{\alpha}F(z)\right)'' \prec g(z) + \frac{1}{c+2}zg'(z). \tag{3.2.1.8}$$

If we denote

$$p(z) = \left(RI_{n,\lambda,l}^{\alpha}F(z)\right)', \quad \text{for } z \in U, \tag{3.2.1.9}$$

then $p \in \mathcal{H}[1,1]$.

Replacing (3.2.1.9) in (3.2.1.8), we obtain

$$p(z) + \frac{1}{c+2}zp'(z) \prec g(z) + \frac{1}{c+2}zg'(z), \quad \text{for } z \in U.$$

Using Lemma 1.2, we have

$$p(z) \prec g(z), \quad \text{for } z \in U, \quad \text{i.e. } \left(RI_{n,\lambda,l}^{\alpha}F(z)\right)' \prec g(z), \quad \text{for } z \in U,$$

and g is the best dominant. □

Example 3.1 (Alb Lupaş, 2010d). If $f \in RD_1\left(1,1,0,\frac{1}{2}\right)$, then $f'(z) + zf''(z) \prec \frac{3-2z}{3(1-z)^2}$ implies $F'(z) + zF''(z) \prec \frac{1}{1-z}$, where $F(z) = \frac{3}{z^2}\int_0^z tf(t)\,dt$.

Theorem 3.63 (Alb Lupaş, 2010d). *Let* $h(z) = \frac{1+(2\delta-1)z}{1+z}$, *where* $\delta \in [0,1)$ *and* $c > 0$.

If $\alpha, \lambda, l \geq 0$, $n \in \mathbb{N}$ *and* $I_c(f)(z) = \frac{c+2}{z^{c+1}} \int_0^z t^c f(t)\, dt$, *for* $z \in U$, *then*

$$I_c\left[\mathcal{RI}_n(\delta, \lambda, l, \alpha)\right] \subset \mathcal{RI}_n(\delta^*, \lambda, l, \alpha), \tag{3.2.1.10}$$

where $\delta^* = 2\delta - 1 + (c+2)(2-2\delta) \int_0^1 \frac{t^{c+1}}{t+1} dt$.

Proof. The function h is convex and using the same steps as in the proof of Theorem 3.62, we get from the hypothesis of Theorem 3.63 that

$$p(z) + \frac{1}{c+2} z p'(z) \prec h(z),$$

where $p(z)$ is defined in (3.2.1.9).

Using Lemma 1.1, we deduce that

$$p(z) \prec g(z) \prec h(z), \quad \text{i.e. } \left(RI^{\alpha}_{n,\lambda,l} F(z)\right)' \prec g(z) \prec h(z),$$

where

$$\begin{aligned} g(z) &= \frac{c+2}{z^{c+2}} \int_0^z t^{c+1} \frac{1+(2\delta-1)t}{1+t} dt \\ &= 2\delta - 1 + \frac{(c+2)(2-2\delta)}{z^{c+2}} \int_0^z \frac{t^{c+1}}{t+1} dt. \end{aligned}$$

Since g is convex and $g(U)$ is symmetric with respect to the real axis, we deduce

$$\begin{aligned} \operatorname{Re}\left(RI^{\alpha}_{n,\lambda,l} F(z)\right)' &\geq \min_{|z|=1} \operatorname{Re} g(z) = \operatorname{Re} g(1) \\ &= \delta^* = 2\delta - 1 + (c+2)(2-2\delta) \int_0^1 \frac{t^{c+1}}{t+1} dt. \end{aligned} \tag{3.2.1.11}$$

From (3.2.1.11), we deduce inclusion (3.2.1.10). □

Theorem 3.64 (Alb Lupaş, 2010d). *Let* g *be a convex function,* $g(0) = 1$ *and let* h *be the function* $h(z) = g(z) + zg'(z)$, *for* $z \in U$.

If $\alpha, \lambda, l \geq 0$, $n \in \mathbb{N}$, $f \in \mathcal{A}$ *and satisfies the differential subordination*

$$\left(RI^{\alpha}_{n,\lambda,l} f(z)\right)' \prec h(z), \quad \text{for } z \in U, \tag{3.2.1.12}$$

then

$$\frac{RI^{\alpha}_{n,\lambda,l}f(z)}{z} \prec g(z), \quad \text{for } z \in U,$$

and this result is sharp.

Proof. By using the properties of operator $RI^{\alpha}_{n,\lambda,l}$, we have

$$RI^{\alpha}_{n,\lambda,l}f(z) = z + \sum_{j=2}^{\infty}\left\{\alpha\left(\frac{1+\lambda(j-1)+l}{l+1}\right)^{n} + (1-\alpha)C^{n}_{n+j-1}\right\}a_j z^j,$$

for $z \in U$.

Consider

$$\begin{aligned}
p(z) &= \frac{RI^{\alpha}_{n,\lambda,l}f(z)}{z} \\
&= \frac{z + \sum_{j=2}^{\infty}\left\{\alpha\left(\frac{1+\lambda(j-1)+l}{l+1}\right)^{n} + (1-\alpha)C^{n}_{n+j-1}\right\}a_j z^j}{z} \\
&= 1 + p_1 z + p_2 z^2 + \cdots, \quad \text{for } z \in U.
\end{aligned}$$

We deduce that $p \in \mathcal{H}[1,1]$.

Let $RI^{\alpha}_{n,\lambda,l}f(z) = zp(z)$, for $z \in U$. Differentiating, we obtain $(RI^{\alpha}_{n,\lambda,l}f(z))' = p(z) + zp'(z)$, for $z \in U$.

Then, (3.2.1.12) becomes

$$p(z) + zp'(z) \prec h(z) = g(z) + zg'(z), \quad \text{for } z \in U.$$

By using Lemma 1.2, we have

$$p(z) \prec g(z), \quad \text{for } z \in U, \quad \text{i.e. } \frac{RI^{\alpha}_{n,\lambda,l}f(z)}{z} \prec g(z), \quad \text{for } z \in U. \qquad \square$$

Theorem 3.65 (Alb Lupaş, 2010d). *Let h be an holomorphic function which satisfies the inequality* $\operatorname{Re}\left(1 + \frac{zh''(z)}{h'(z)}\right) > -\frac{1}{2}$, *for $z \in U$, and* $h(0) = 1$.

If $\alpha, \lambda, l \geq 0$, $n \in \mathbb{N}$, $f \in \mathcal{A}$ and satisfies the differential subordination

$$\left(RI^{\alpha}_{n,\lambda,l}f(z)\right)' \prec h(z), \quad \text{for } z \in U, \tag{3.2.1.13}$$

then

$$\frac{RI^{\alpha}_{n,\lambda,l}f(z)}{z} \prec q(z), \quad \text{for } z \in U,$$

where $q(z) = \frac{1}{z}\int_0^z h(t)dt$. *The function* q *is convex and it is the best dominant.*

Proof. Let

$$\begin{aligned} p(z) &= \frac{RI^{\alpha}_{n,\lambda,l}f(z)}{z} \\ &= \frac{z + \sum_{j=2}^{\infty}\left\{\alpha\left(\frac{1+\lambda(j-1)+l}{l+1}\right)^n + (1-\alpha)\,C^n_{n+j-1}\right\}a_j z^j}{z} \\ &= 1 + \sum_{j=2}^{\infty}\left\{\alpha\left(\frac{1+\lambda(j-1)+l}{l+1}\right)^n + (1-\alpha)\,C^n_{n+j-1}\right\}a_j z^{j-1} \\ &= 1 + \sum_{j=2}^{\infty} p_j z^{j-1}, \end{aligned}$$

for $z \in U$, $p \in \mathcal{H}[1,1]$.

Differentiating, we obtain $\left(RI^{\alpha}_{n,\lambda,l}f(z)\right)' = p(z) + zp'(z)$, for $z \in U$ and (3.2.1.13) becomes

$$p(z) + zp'(z) \prec h(z), \quad \text{for } z \in U.$$

Using Lemma 1.1, we have

$$p(z) \prec q(z), \quad z \in U, \quad \text{i.e. } \frac{RI^{\alpha}_{n,\lambda,l}f(z)}{z} \prec q(z) = \frac{1}{z}\int_0^z h(t)dt, \quad z \in U,$$

and q is the best dominant. □

Theorem 3.66 (Alb Lupaş, 2010d). *Let* g *be a convex function such that* $g(0) = 1$ *and let* h *be the function* $h(z) = g(z) + zg'(z)$, *for* $z \in U$.

If $\alpha, \lambda, l \geq 0$, $n \in \mathbb{N}$, $f \in \mathcal{A}$ *and the differential subordination*

$$\left(\frac{zRI^{\alpha}_{n+1,\lambda,l}f(z)}{RI^{\alpha}_{n,\lambda,l}f(z)}\right)' \prec h(z), \quad \text{for } z \in U \tag{3.2.1.14}$$

holds, then

$$\frac{RI^{\alpha}_{n+1,\lambda,l}f(z)}{RI^{\alpha}_{n,\lambda,l}f(z)} \prec g(z), \quad \text{for } z \in U,$$

and this result is sharp.

Proof. For $f \in \mathcal{A}$, $f(z) = z + \sum_{j=2}^{\infty} a_j z^j$, we have

$$RI^{\alpha}_{n,\lambda,l}f(z) = z + \sum_{j=2}^{\infty} \left\{ \alpha \left(\frac{1+\lambda(j-1)+l}{l+1} \right)^n + (1-\alpha) C^n_{n+j-1} \right\} a_j z^j,$$

for $z \in U$.

Consider

$$p(z) = \frac{RI^{\alpha}_{n+1,\lambda,l}f(z)}{RI^{\alpha}_{n,\lambda,l}f(z)} = \frac{z + \sum_{j=2}^{\infty} \left\{ \alpha \left(\frac{1+\lambda(j-1)+l}{l+1} \right)^{n+1} + (1-\alpha) C^{n+1}_{n+j} \right\} a_j z^j}{z + \sum_{j=2}^{\infty} \left\{ \alpha \left(\frac{1+\lambda(j-1)+l}{l+1} \right)^{n} + (1-\alpha) C^{n}_{n+j-1} \right\} a_j z^j}.$$

We have $p'(z) = \frac{\left(RI^{\alpha}_{n+1,\lambda,l}f(z)\right)'}{RI^{\alpha}_{n,\lambda,l}f(z)} - p(z) \cdot \frac{\left(RI^{\alpha}_{n,\lambda,l}f(z)\right)'}{RI^{\alpha}_{n,\lambda,l}f(z)}$ and we obtain $p(z) + z \cdot p'(z) = \left(\frac{zRI^{\alpha}_{n+1,\lambda,l}f(z)}{RI^{\alpha}_{n,\lambda,l}f(z)} \right)'$.

Relation (3.2.1.14) becomes

$$p(z) + zp'(z) \prec h(z) = g(z) + zg'(z), \quad \text{for } z \in U.$$

By using Lemma 1.2, we have

$$p(z) \prec g(z), \quad \text{for } z \in U, \quad \text{i.e. } \frac{RI^{\alpha}_{n+1,\lambda,l}f(z)}{RI^{\alpha}_{n,\lambda,l}f(z)} \prec g(z), \quad \text{for } z \in U. \qquad \square$$

Theorem 3.67 (Alb Lupaş, 2010d). *Let g be a convex function such that $g(0) = 0$ and let h be the function $h(z) = g(z) + zg'(z)$, for $z \in U$.*

If $\alpha, \lambda, l \geq 0$, $n \in \mathbb{N}$, $f \in \mathcal{A}$ and the differential subordination

$$(n+1) RI^{\alpha}_{n+1,\lambda,l}f(z) - (n-1) RI^{\alpha}_{n,\lambda,l}f(z) - \alpha \left(n+1 - \frac{l+1}{\lambda} \right) [I(n+1,\lambda,l) f(z) - I(n,\lambda,l) f(z)] \prec h(z),$$

for $z \in U$ (3.2.1.15)

holds, then

$$RI^{\alpha}_{n,\lambda,l}f(z) \prec g(z), \quad \text{for } z \in U.$$

This result is sharp.

Proof. Let

$$\begin{aligned} p(z) &= RI^{\alpha}_{n,\lambda,\alpha}f\,(z) = (1-\alpha)R^n f(z) + \alpha I\,(n,\lambda,l)\,f(z) \\ &= z + \sum_{j=2}^{\infty}\left\{\alpha\left(\frac{1+\lambda\,(j-1)+l}{l+1}\right)^n + (1-\alpha)\,C^n_{n+j-1}\right\}a_j z^j \\ &= p_1 z + p_2 z^2 + \cdots . \end{aligned} \tag{3.2.1.16}$$

We deduce that $p \in \mathcal{H}[0,1]$.

By using the properties of operators $RI^{\alpha}_{n,\lambda,l}$, R^n and $I\,(n,\lambda,l)$, after a short calculation, we obtain $p\,(z) + zp'\,(z) = (n+1)\,RI^{\alpha}_{n+1,\lambda,l}f\,(z) - (n-1)\,RI^{\alpha}_{n,\lambda,l}f\,(z) - \alpha\left(n+1-\frac{l+1}{\lambda}\right)[I\,(n+1,\lambda,l)\,f\,(z) - I\,(n,\lambda,l)\,f\,(z)]$.

Using the notation in (3.2.1.16), the differential subordination becomes

$$p(z) + zp'(z) \prec h(z) = g(z) + zg'(z).$$

By using Lemma 1.2, we have

$$p(z) \prec g(z), \quad \text{for } z \in U, \quad \text{i.e. } RI^{\alpha}_{n,\lambda,l}f(z) \prec g(z), \quad \text{for } z \in U,$$

and this result is sharp. □

Theorem 3.68 (Alb Lupaş, 2010d). *Let $h(z) = \frac{1+(2\beta-1)z}{1+z}$ be a convex function in U, where $0 \le \beta < 1$.*

If $\alpha, \lambda, l \ge 0$, $n \in \mathbb{N}$, $f \in \mathcal{A}$ and satisfies the differential subordination

$$\begin{aligned} &(n+1)\,RI^{\alpha}_{n+1,\lambda,l}f\,(z) - (n-1)\,RI^{\alpha}_{n,\lambda,l}f\,(z) \\ &\quad - \alpha\left(n+1-\frac{l+1}{\lambda}\right)[I\,(n+1,\lambda,l)\,f\,(z) - I\,(n,\lambda,l)\,f\,(z)] \prec h(z), \\ &\text{for } z \in U, \end{aligned} \tag{3.2.1.17}$$

then

$$RI^{\alpha}_{n,\lambda,l}f(z) \prec q(z), \quad \text{for } z \in U,$$

where q is given by $q(z) = 2\beta - 1 + 2(1-\beta)\frac{\ln(1+z)}{z}$, for $z \in U$. The function q is convex and it is the best dominant.

Proof. Following the same steps as in the proof of Theorem 3.2.1.3 and considering $p(z) = RI^{\alpha}_{n,\lambda,l}f(z)$, the differential subordination (3.2.1.17) becomes

$$p(z) + zp'(z) \prec h(z) = \frac{1 + (2\beta - 1)z}{1+z}, \quad \text{for } z \in U.$$

By using Lemma 1.1 for $\gamma = 1$ and $n = 1$, we have $p(z) \prec q(z)$, i.e.

$$\begin{aligned} RI^{\alpha}_{n,\lambda,l}f(z) \prec q(z) &= \frac{1}{z}\int_0^z h(t)dt = \frac{1}{z}\int_0^z \frac{1+(2\beta-1)t}{1+t}dt \\ &= 2\beta - 1 + 2(1-\beta)\frac{1}{z}\ln(z+1), \quad \text{for } z \in U. \end{aligned}$$ □

Theorem 3.69 (Alb Lupaş, 2010d). *Let h be an holomorphic function which satisfies the inequality* $\operatorname{Re}\left[1 + \frac{zh''(z)}{h'(z)}\right] > -\frac{1}{2}$, *for $z \in U$, and* $h(0) = 0$.

If $\alpha, \lambda, l \geq 0$, $n \in \mathbb{N}$, $f \in \mathcal{A}$ and satisfies the differential subordination

$$\begin{aligned} &(n+1)\, RI^{\alpha}_{n+1,\lambda,l}f(z) - (n-1)\, RI^{\alpha}_{n,\lambda,l}f(z) \\ &\quad - \alpha\left(n+1-\frac{l+1}{\lambda}\right)\left[I(n+1,\lambda,l)f(z) - I(n,\lambda,l)f(z)\right] \prec h(z), \\ &\quad \text{for } z \in U, \end{aligned} \tag{3.2.1.18}$$

then

$$RI^{\alpha}_{n,\lambda,l}f(z) \prec q(z), \quad \text{for } z \in U,$$

where q is given by $q(z) = \frac{1}{z}\int_0^z h(t)dt$. The function q is convex and it is the best dominant.

Proof. Using the properties of operator $RI^{\alpha}_{n,\lambda,l}$ and considering $p(z) = RI^{\alpha}_{n,\lambda,l}f(z)$, we obtain

$$\begin{aligned} p(z) + zp'(z) &= (n+1)\, RI^{\alpha}_{n+1,\lambda,l}f(z) - (n-1)\, RI^{\alpha}_{n,\lambda,l}f(z) \\ &\quad - \alpha\left(n+1-\frac{l+1}{\lambda}\right)\left[I(n+1,\lambda,l)f(z) - I(n,\lambda,l)f(z)\right], \\ &\quad \text{for } z \in U. \end{aligned}$$

Then (3.2.1.18) becomes

$$p(z) + zp'(z) \prec h(z), \quad \text{for } z \in U.$$

Since $p \in \mathcal{H}[0,1]$, using Lemma 1.1, we deduce

$$p(z) \prec q(z), \quad z \in U, \quad \text{i.e. } RI^{\alpha}_{n,\lambda,l}f(z) \prec q(z) = \frac{1}{z}\int_0^z h(t)dt, \quad z \in U,$$

and q is the best dominant. □

Theorem 3.70 (Alb Lupaş, 2010d). *Let g be a convex function such that $g(0) = 1$ and let h be the function $h(z) = g(z) + zg'(z)$, for $z \in U$.*
If $\alpha, \lambda, l \geq 0$, $n \in \mathbb{N}$, $f \in \mathcal{A}$ and the differential subordination

$$\begin{aligned}
&\frac{(n+1)(n+2)}{z} RI^{\alpha}_{n+2,\lambda,l}f(z) - \frac{(n+1)(2n+1)}{z} RI^{\alpha}_{n+1,\lambda,l}f(z) \\
&\quad + \frac{n^2}{z} RI^{\alpha}_{n,\lambda,l}f(z) + \frac{\alpha}{z}\left[\frac{(l+1)^2}{\lambda^2} - (n+1)(n+2)\right] I(n+2,\lambda,l) f(z) \\
&\quad - \frac{\alpha}{z}\left[\frac{2(l+1-\lambda)(\bar{l}+1)}{\lambda^2} - (n+1)(2n+1)\right] I(n+1,\lambda,l) f(z) \\
&\quad + \frac{\alpha}{z}\left[\frac{(l+1-\lambda)^2}{\lambda^2} - n^2\right] I(n,\lambda,l) f(z) \prec h(z), \quad \text{for } z \in U
\end{aligned} \tag{3.2.1.19}$$

holds, then

$$[RI^{\alpha}_{n,\lambda,l}f(z)]' \prec g(z), \quad \text{for } z \in U.$$

This result is sharp.

Proof. Let

$$\begin{aligned}
p(z) &= \left(RI^{\alpha}_{n,\lambda,l}f(z)\right)' = (1-\alpha)\left(R^n f(z)\right)' + \alpha\left(I(n,\lambda,l) f(z)\right)' \\
&= 1 + \sum_{j=2}^{\infty}\left\{\alpha\left(\frac{1+\lambda(j-1)+l}{l+1}\right)^n + (1-\alpha) C^n_{n+j-1}\right\} ja_j z^{j-1} \\
&= 1 + p_1 z + p_2 z^2 + \cdots .
\end{aligned} \tag{3.2.1.20}$$

We deduce that $p \in \mathcal{H}[1,1]$.

By using the properties of operators $RI_{n,\lambda,l}^{\alpha}$, R^n and $I(n,\lambda,l)$, after a short calculation, we obtain

$$\begin{aligned} p(z)+zp'(z) = {} & \frac{(n+1)(n+2)}{z} RI_{n+2,\lambda,l}^{\alpha} f(z) \\ & - \frac{(n+1)(2n+1)}{z} RI_{n+1,\lambda,l}^{\alpha} f(z) + \frac{n^2}{z} RI_{n,\lambda,l}^{\alpha} f(z) \\ & + \frac{\alpha}{z}\left[\frac{(l+1)^2}{\lambda^2} - (n+1)(n+2)\right] I(n+2,\lambda,l) f(z) \\ & - \frac{\alpha}{z}\left[\frac{2(l+1-\lambda)(\bar{l}+1)}{\lambda^2} - (n+1)(2n+1)\right] \\ & \times I(n+1,\lambda,l) f(z) \\ & + \frac{\alpha}{z}\left[\frac{(l+1-\lambda)^2}{\lambda^2} - n^2\right] I(n,\lambda,l) f(z). \end{aligned}$$

Using the notation in (3.2.1.20), the differential subordination becomes

$$p(z) + zp'(z) \prec h(z) = g(z) + zg'(z).$$

By using Lemma 1.2, we have

$$p(z) \prec g(z), \quad z \in U, \quad \text{i.e. } \left(RI_{n,\lambda,l}^{\alpha} f(z)\right)' \prec g(z), \quad z \in U,$$

and this result is sharp. □

Example 3.2 (Alb Lupaş, 2010d). If $n = 1$, $\alpha = 1$, $\lambda = 1$, $l = 0$, $f \in \mathcal{A}$, we deduce that $f'(z) + 3zf''(z) + z^2 f'''(z) \prec g(z) + zg'(z)$, which yields that $f'(z) + zf''(z) \prec g(z)$, for $z \in U$.

Theorem 3.71 (Alb Lupaş, 2010d). *Let* $h(z) = \frac{1+(2\beta-1)z}{1+z}$ *be a convex function in* U, *where* $0 \le \beta < 1$.

If $\alpha, \lambda, l \geq 0$, $n \in \mathbb{N}$, $f \in \mathcal{A}$ *and satisfies the differential subordination*

$$\begin{aligned}
&\frac{(n+1)(n+2)}{z} RI^{\alpha}_{n+2,\lambda,l} f(z) - \frac{(n+1)(2n+1)}{z} RI^{\alpha}_{n+1,\lambda,l} f(z) \\
&+ \frac{n^2}{z} RI^{\alpha}_{n,\lambda,l} f(z) + \frac{\alpha}{z}\left[\frac{(l+1)^2}{\lambda^2} - (n+1)(n+2)\right] I(n+2,\lambda,l) f(z) \\
&- \frac{\alpha}{z}\left[\frac{2(l+1-\lambda)(\bar{l}+1)}{\lambda^2} - (n+1)(2n+1)\right] I(n+1,\lambda,l) f(z) \\
&+ \frac{\alpha}{z}\left[\frac{(l+1-\lambda)^2}{\lambda^2} - n^2\right] I(n,\lambda,l) f(z) \prec h(z), \quad \text{for } z \in U,
\end{aligned} \tag{3.2.1.21}$$

then

$$\left(RI^{\alpha}_{n,\lambda,l} f(z)\right)' \prec q(z), \quad \text{for } z \in U,$$

where q *is given by* $q(z) = 2\beta - 1 + 2(1-\beta)\frac{\ln(1+z)}{z}$, *for* $z \in U$. *The function* q *is convex and it is the best dominant.*

Proof. Following the same steps as in the proof of Theorem 3.70 and considering $p(z) = (RI^{\alpha}_{n,\lambda,l} f(z))'$, the differential subordination (3.2.1.21) becomes

$$p(z) + zp'(z) \prec h(z) = \frac{1 + (2\beta - 1)z}{1+z}, \quad \text{for } z \in U.$$

By using Lemma 1.1 for $\gamma = 1$ and $n = 1$, we have $p(z) \prec q(z)$, i.e.

$$\begin{aligned}
\left(RI^{\alpha}_{n,\lambda,l} f(z)\right)' \prec q(z) &= \frac{1}{z}\int_0^z h(t)dt = \frac{1}{z}\int_0^z \frac{1 + (2\beta-1)t}{1+t} dt \\
&= 2\beta - 1 + 2(1-\beta)\frac{1}{z}\ln(z+1), \quad \text{for } z \in U. \quad \square
\end{aligned}$$

Theorem 3.72 (Alb Lupaş, 2010d). *Let* h *be an holomorphic function which satisfies the inequality* $\operatorname{Re}\left[1 + \frac{zh''(z)}{h'(z)}\right] > -\frac{1}{2}$, *for* $z \in U$, *and* $h(0) = 1$.

If $\alpha, \lambda, l \geq 0$, $n \in \mathbb{N}$, $f \in \mathcal{A}$ and satisfies the differential subordination

$$\frac{(n+1)(n+2)}{z} RI^{\alpha}_{n+2,\lambda,l} f(z) - \frac{(n+1)(2n+1)}{z} RI^{\alpha}_{n+1,\lambda,l} f(z)$$
$$+ \frac{n^2}{z} RI^{\alpha}_{n,\lambda,l} f(z) + \frac{\alpha}{z}\left[\frac{(l+1)^2}{\lambda^2} - (n+1)(n+2)\right] I(n+2,\lambda,l) f(z)$$
$$- \frac{\alpha}{z}\left[\frac{2(l+1-\lambda)(\bar{l}+1)}{\lambda^2} - (n+1)(2n+1)\right] I(n+1,\lambda,l) f(z)$$
$$+ \frac{\alpha}{z}\left[\frac{(l+1-\lambda)^2}{\lambda^2} - n^2\right] I(n,\lambda,l) f(z) \prec h(z), \quad \text{for } z \in U, \tag{3.2.1.22}$$

then

$$\left(RI^{\alpha}_{n,\lambda,l} f(z)\right)' \prec q(z), \quad \text{for } z \in U,$$

where q is given by $q(z) = \frac{1}{z}\int_0^z h(t)dt$. The function q is convex and it is the best dominant.

Proof. Using the properties of operator $RI^{\alpha}_{n,\lambda,l}$ and considering $p(z) = \left(RI^{\alpha}_{n,\lambda,l} f(z)\right)'$, we obtain

$$\begin{aligned} p(z) + zp'(z) = {} & \frac{(n+1)(n+2)}{z} RI^{\alpha}_{n+2,\lambda,l} f(z) \\ & - \frac{(n+1)(2n+1)}{z} RI^{\alpha}_{n+1,\lambda,l} f(z) + \frac{n^2}{z} RI^{\alpha}_{n,\lambda,l} f(z) \\ & + \frac{\alpha}{z}\left[\frac{(l+1)^2}{\lambda^2} - (n+1)(n+2)\right] I(n+2,\lambda,l) f(z) \\ & - \frac{\alpha}{z}\left[\frac{2(l+1-\lambda)(\bar{l}+1)}{\lambda^2} - (n+1)(2n+1)\right] \\ & \times I(n+1,\lambda,l) f(z) \\ & + \frac{\alpha}{z}\left[\frac{(l+1-\lambda)^2}{\lambda^2} - n^2\right] I(n,\lambda,l) f(z), \quad \text{for } z \in U. \end{aligned}$$

Then, (3.2.1.22) becomes

$$p(z) + zp'(z) \prec h(z), \quad \text{for } z \in U.$$

Since $p \in \mathcal{H}[1,1]$, using Lemma 1.1, we deduce

$$p(z) \prec q(z), \quad z \in U, \quad \text{i.e.} \ \left(RI^{\alpha}_{n,\lambda,l}f(z)\right)' \prec q(z) = \frac{1}{z}\int_0^z h(t)dt, \quad z \in U,$$

and q is the best dominant. □

Theorem 3.73 (Alb Lupaş, 2015b). *Let g be a convex function, $g(0) = 1$ and let h be the function $h(z) = g(z) + \frac{z}{\delta}g'(z)$, $z \in U$. If $\alpha, \lambda, l, \delta \geq 0$, $n \in \mathbb{N}$, $f \in \mathcal{A}$ and satisfies the differential subordination*

$$\left(\frac{RI^{\alpha}_{n,\lambda,l}f(z)}{z}\right)^{\delta-1}\left(RI^{\alpha}_{n,\lambda,l}f(z)\right)' \prec h(z), \quad z \in U, \tag{3.2.1.23}$$

then

$$\left(\frac{RI^{\alpha}_{n,\lambda,l}f(z)}{z}\right)^{\delta} \prec g(z), \quad z \in U,$$

and this result is sharp.

Proof. Consider

$$\begin{aligned} p(z) &= \left(\frac{RI^{\alpha}_{n,\lambda,l}f(z)}{z}\right)^{\delta} \\ &= \left(\frac{z + \sum_{j=2}^{\infty}\left\{\alpha\left(\frac{1+\lambda(j-1)+l}{l+1}\right)^n + (1-\alpha)\frac{(n+j-1)!}{n!(j-1)!}\right\}a_j z^j}{z}\right)^{\delta} \\ &= 1 + p_1 z + p_2 z^2 + \cdots, \quad z \in U. \end{aligned}$$

We deduce that $p \in \mathcal{H}[1,1]$.

Differentiating, we obtain $\left(\frac{RI^{\alpha}_{n,\lambda,l}f(z)}{z}\right)^{\delta-1}\left(RI^{\alpha}_{n,\lambda,l}f(z)\right)' = p(z) + \frac{1}{\delta}zp'(z)$, $z \in U$.

Then (3.2.1.23) becomes $p(z) + \frac{1}{\delta}zp'(z) \prec h(z) = g(z) + \frac{z}{\delta}g'(z)$, $z \in U$.

By using Lemma 1.2, we have

$$p(z) \prec g(z), \quad z \in U, \quad \text{i.e.} \ \left(\frac{RI^{\alpha}_{n,\lambda,l}f(z)}{z}\right)^{\delta} \prec g(z), \quad z \in U.$$ □

Theorem 3.74 (Alb Lupaş, 2015b). *Let h be an holomorphic function which satisfies the inequality* $\operatorname{Re}\left(1+\frac{zh''(z)}{h'(z)}\right)>-\frac{1}{2}$, $z\in U$, *and* $h(0)=1$. *If* $\alpha,\lambda,l,\delta\geq 0$, $n\in\mathbb{N}$, $f\in\mathcal{A}$ *and satisfies the differential subordination*

$$\left(\frac{RI_{n,\lambda,l}^{\alpha}f(z)}{z}\right)^{\delta-1}\left(RI_{n,\lambda,l}^{\alpha}f(z)\right)'\prec h(z),\quad z\in U, \tag{3.2.1.24}$$

then

$$\left(\frac{RI_{n,\lambda,l}^{\alpha}f(z)}{z}\right)^{\delta}\prec q(z),\quad z\in U,$$

where $q(z)=\frac{\delta}{z^{\delta}}\int_0^z h(t)t^{\delta-1}dt$.

Proof. Let

$$\begin{aligned}
p(z)&=\left(\frac{RI_{n,\lambda,l}^{\alpha}f(z)}{z}\right)^{\delta}\\
&=\left(\frac{z+\sum_{j=2}^{\infty}\left\{\alpha\left(\frac{1+\lambda(j-1)+l}{l+1}\right)^n+(1-\alpha)\frac{(n+j-1)!}{n!(j-1)!}\right\}a_jz^j}{z}\right)^{\delta}\\
&=\left(1+\sum_{j=2}^{\infty}\left\{\alpha\left(\frac{1+\lambda(j-1)+l}{l+1}\right)^n+(1-\alpha)\frac{(n+j-1)!}{n!\,(j-1)!}\right\}a_jz^{j-1}\right)^{\delta}\\
&=1+\sum_{j=2}^{\infty}p_jz^{j-1},
\end{aligned}$$

for $z\in U$, $p\in\mathcal{H}[1,1]$.

Differentiating, we obtain $\left(\frac{RI_{n,\lambda,l}^{\alpha}f(z)}{z}\right)^{\delta-1}\left(RI_{n,\lambda,l}^{\alpha}f(z)\right)'=p(z)+\frac{1}{\delta}zp'(z)$, $z\in U$, and (3.2.1.24) becomes $p(z)+\frac{1}{\delta}zp'(z)\prec h(z)$, $z\in U$.

Using Lemma 1.1, we have

$$\begin{aligned}
&p(z)\prec q(z),\quad z\in U,\\
&\text{i.e. }\left(\frac{RI_{n,\lambda,l}^{\alpha}f(z)}{z}\right)^{\delta}\prec q(z)=\frac{\delta}{z^{\delta}}\int_0^z h(t)t^{\delta-1}dt,\quad z\in U,
\end{aligned}$$

and q is the best dominant. □

Corollary 3.16 (Alb Lupaş, 2015b). *Let* $h(z) = \frac{1+(2\beta-1)z}{1+z}$ *be a convex function in* U*, where* $0 \leq \beta < 1$*. If* $\alpha, \delta, l, \lambda \geq 0$*,* $n \in \mathbb{N}$*,* $f \in \mathcal{A}$ *and satisfies the differential subordination*

$$\left(\frac{RI^{\alpha}_{n,\lambda,l}f(z)}{z}\right)^{\delta-1} \left(RI^{\alpha}_{n,\lambda,l}f(z)\right)' \prec h(z), \quad z \in U, \tag{3.2.1.25}$$

then

$$\left(\frac{RI^{\alpha}_{n,\lambda,l}f(z)}{z}\right)^{\delta} \prec q(z), \quad z \in U,$$

where q *is given by* $q(z) = (2\beta-1) + \frac{2(1-\beta)\delta}{z^{\delta}} \int_0^z \frac{t^{\delta-1}}{1+t} dt$, $z \in U$*. The function* q *is convex and it is the best dominant.*

Proof. Following the same steps as in the proof of Theorem 3.74 and considering $p(z) = \left(\frac{RI^{\alpha}_{n,\lambda,l}f(z)}{z}\right)^{\delta}$, the differential subordination (3.2.1.25) becomes $p(z) + \frac{z}{\delta}p'(z) \prec h(z) = \frac{1+(2\beta-1)z}{1+z}$, $z \in U$.

By using Lemma 1.1 for $\gamma = \delta$, we have $p(z) \prec q(z)$, i.e.

$$\begin{aligned}\left(\frac{RI^{\alpha}_{n,\lambda,l}f(z)}{z}\right)^{\delta} \prec q(z) &= \frac{\delta}{z^{\delta}} \int_0^z h(t)t^{\delta-1}dt \\ &= \frac{\delta}{z^{\delta}} \int_0^z t^{\delta-1} \frac{1+(2\beta-1)t}{1+t} dt \\ &= \frac{\delta}{z^{\delta}} \int_0^z \left[(2\beta-1)\, t^{\delta-1} + 2\,(1-\beta) \frac{t^{\delta-1}}{1+t}\right] dt \\ &= (2\beta-1) + \frac{2\,(1-\beta)\,\delta}{z^{\delta}} \int_0^z \frac{t^{\delta-1}}{1+t} dt, \quad z \in U.\end{aligned}$$

□

Theorem 3.75 (Alb Lupaş, 2015b). *Let* g *be a convex function such that* $g(0) = 1$ *and let* h *be the function* $h(z) = g(z) + \frac{z}{\delta}g'(z)$, $z \in U$*. If* $\alpha, \lambda, l, \delta \geq 0$*,* $n \in \mathbb{N}$*,* $f \in \mathcal{A}$ *and the differential subordination*

$$\begin{aligned} &z\frac{\delta+1}{\delta} \frac{RI^{\alpha}_{n,\lambda,l}f(z)}{\left(RI^{\alpha}_{n+1,\lambda,l}f(z)\right)^2} \\ &+ \frac{z^2}{\delta} \frac{RI^{\alpha}_{n,\lambda,l}f(z)}{\left(RI^{\alpha}_{n+1,\lambda,l}f(z)\right)^2} \left[\frac{\left(RI^{\alpha}_{n,\lambda,l}f(z)\right)'}{RI^{\alpha}_{n,\lambda,l}f(z)} - 2\frac{\left(RI^{\alpha}_{n+1,\lambda,l}f(z)\right)'}{RI^{\alpha}_{n+1,\lambda,l}f(z)}\right] \prec h(z), \end{aligned} \tag{3.2.1.26}$$

$z \in U$, holds, then

$$z\frac{RI_{n,\lambda,l}^{\alpha}f(z)}{\left(RI_{n+1,\lambda,l}^{\alpha}f(z)\right)^{2}} \prec g(z), \quad z \in U,$$

and this result is sharp.

Proof. Consider $p(z) = z\frac{RI_{n,\lambda,l}^{\alpha}f(z)}{\left(RI_{n+1,\lambda,l}^{\alpha}f(z)\right)^{2}}$ and we obtain

$$\begin{aligned} p(z) + \frac{z}{\delta}p'(z) &= z\frac{\delta+1}{\delta}\frac{RI_{n,\lambda,l}^{\alpha}f(z)}{\left(RI_{n+1,\lambda,l}^{\alpha}f(z)\right)^{2}} \\ &\quad + \frac{z^2}{\delta}\frac{RI_{n,\lambda,l}^{\alpha}f(z)}{\left(RI_{n+1,\lambda,l}^{\alpha}f(z)\right)^{2}} \\ &\quad \times \left[\frac{\left(RI_{n,\lambda,l}^{\alpha}f(z)\right)'}{RI_{n,\lambda,l}^{\alpha}f(z)} - 2\frac{\left(RI_{n+1,\lambda,l}^{\alpha}f(z)\right)'}{RI_{n+1,\lambda,l}^{\alpha}f(z)}\right]. \end{aligned}$$

Relation (3.2.1.26) becomes $p(z)+\frac{z}{\delta}p'(z) \prec h(z) = g(z)+\frac{z}{\delta}g'(z)$, $z \in U$. By using Lemma 1.2, we have

$$p(z) \prec g(z), \quad z \in U, \quad \text{i.e. } z\frac{RI_{n,\lambda,l}^{\alpha}f(z)}{\left(RI_{n+1,\lambda,l}^{\alpha}f(z)\right)^{2}} \prec g(z), \quad z \in U.$$

□

Theorem 3.76 (Alb Lupaş, 2015b). *Let h be an holomorphic function which satisfies the inequality $\operatorname{Re}\left(1+\frac{zh''(z)}{h'(z)}\right) > -\frac{1}{2}$, $z \in U$, and $h(0) = 1$. If $\alpha, \lambda, l, \delta \geq 0$, $n \in \mathbb{N}$, $f \in \mathcal{A}$ and satisfies the differential subordination*

$$\begin{aligned} &z\frac{\delta+1}{\delta}\frac{RI_{n,\lambda,l}^{\alpha}f(z)}{\left(RI_{n+1,\lambda,l}^{\alpha}f(z)\right)^{2}} \\ &\quad + \frac{z^2}{\delta}\frac{RI_{n,\lambda,l}^{\alpha}f(z)}{\left(RI_{n+1,\lambda,l}^{\alpha}f(z)\right)^{2}}\left[\frac{\left(RI_{n,\lambda,l}^{\alpha}f(z)\right)'}{RI_{n,\lambda,l}^{\alpha}f(z)} - 2\frac{\left(RI_{n+1,\lambda,l}^{\alpha}f(z)\right)'}{RI_{n+1,\lambda,l}^{\alpha}f(z)}\right] \prec h(z), \end{aligned} \tag{3.2.1.27}$$

$z \in U$, then

$$z\frac{RI^{\alpha}_{n,\lambda,l}f(z)}{\left(RI^{\alpha}_{n+1,\lambda,l}f(z)\right)^2} \prec q(z), \quad z \in U,$$

where $q(z) = \frac{\delta}{z^\delta}\int_0^z h(t)t^{\delta-1}dt$.

Proof. Let $p(z) = z\frac{RI^{\alpha}_{n,\lambda,l}f(z)}{\left(RI^{\alpha}_{n+1,\lambda,l}f(z)\right)^2}$, $z \in U$, $p \in \mathcal{H}[1,1]$.

Differentiating, we obtain

$$p(z) + \frac{z}{\delta}p'(z) = z\frac{\delta+1}{\delta}\frac{RI^{\alpha}_{n,\lambda,l}f(z)}{\left(RI^{\alpha}_{n+1,\lambda,l}f(z)\right)^2} + \frac{z^2}{\delta}\frac{RI^{\alpha}_{n,\lambda,l}f(z)}{\left(RI^{\alpha}_{n+1,\lambda,l}f(z)\right)^2}$$
$$\times\left[\frac{\left(RI^{\alpha}_{n,\lambda,l}f(z)\right)'}{RI^{\alpha}_{n,\lambda,l}f(z)} - 2\frac{\left(RI^{\alpha}_{n+1,\lambda,l}f(z)\right)'}{RI^{\alpha}_{n+1,\lambda,l}f(z)}\right],$$

$z \in U$, and (3.2.1.27) becomes $p(z) + \frac{z}{\delta}p'(z) \prec h(z)$, $z \in U$.

Using Lemma 1.1, we have

$$p(z) \prec q(z), \quad \text{i.e. } z\frac{RI^{\alpha}_{n,\lambda,l}f(z)}{\left(RI^{\alpha}_{n+1,\lambda,l}f(z)\right)^2} \prec q(z) = \frac{\delta}{z^\delta}\int_0^z h(t)t^{\delta-1}dt, \quad z \in U,$$

and q is the best dominant. □

Theorem 3.77 (Alb Lupaş, 2015b). *Let g be a convex function such that $g(0) = 0$ and let h be the function $h(z) = g(z) + \frac{z}{\delta}g'(z)$, $z \in U$. If $\alpha, \lambda, l, \delta \geq 0$, $n \in \mathbb{N}$, $f \in \mathcal{A}$ and the differential subordination*

$$z^2\frac{\delta+2}{\delta}\frac{\left(RI^{\alpha}_{n,\lambda,l}f(z)\right)'}{RI^{\alpha}_{n,\lambda,l}f(z)}$$
$$+\frac{z^3}{\delta}\left[\frac{\left(RI^{\alpha}_{n,\lambda,l}f(z)\right)''}{RI^{\alpha}_{n,\lambda,l}f(z)} - \left(\frac{\left(RI^{\alpha}_{n,\lambda,l}f(z)\right)'}{RI^{\alpha}_{n,\lambda,l}f(z)}\right)^2\right] \prec h(z), \quad (3.2.1.28)$$

$z \in U$ holds, then

$$z^2\frac{\left(RI^{\alpha}_{n,\lambda,l}f(z)\right)'}{RI^{\alpha}_{n,\lambda,l}f(z)} \prec g(z), \quad z \in U.$$

This result is sharp.

Proof. Let $p(z) = z^2 \frac{\left(RI^{\alpha}_{n,\lambda,l}f(z)\right)'}{RI^{\alpha}_{n,\lambda,l}f(z)}$. We deduce that $p \in \mathcal{H}[0,1]$.

Differentiating, we obtain

$$p(z) + \frac{z}{\delta}p'(z) = z^2 \frac{\delta+2}{\delta} \frac{\left(RI^{\alpha}_{n,\lambda,l}f(z)\right)'}{RI^{\alpha}_{n,\lambda,l}f(z)} + \frac{z^3}{\delta}\left[\frac{\left(RI^{\alpha}_{n,\lambda,l}f(z)\right)''}{RI^{\alpha}_{n,\lambda,l}f(z)} - \left(\frac{\left(RI^{\alpha}_{n,\lambda,l}f(z)\right)'}{RI^{\alpha}_{n,\lambda,l}f(z)}\right)^2\right], \quad z \in U.$$

Using the notation in (3.2.1.28), the differential subordination becomes $p(z) + \frac{1}{\delta}zp'(z) \prec h(z) = g(z) + \frac{z}{\delta}g'(z)$.

By using Lemma 1.2, we have

$$p(z) \prec g(z), \quad z \in U, \quad \text{i.e. } z^2 \frac{\left(RI^{\alpha}_{n,\lambda,l}f(z)\right)'}{RI^{\alpha}_{n,\lambda,l}f(z)} \prec g(z), \quad z \in U,$$

and this result is sharp. □

Theorem 3.78 (Alb Lupaş, 2015b). *Let h be an holomorphic function which satisfies the inequality* $\operatorname{Re}\left(1 + \frac{zh''(z)}{h'(z)}\right) > -\frac{1}{2}$, $z \in U$, *and* $h(0) = 0$. *If* $\alpha, \lambda, l, \delta \geq 0$, $n \in \mathbb{N}$, $f \in \mathcal{A}$ *and satisfies the differential subordination*

$$z^2 \frac{\delta+2}{\delta} \frac{\left(RI^{\alpha}_{n,\lambda,l}f(z)\right)'}{RI^{\alpha}_{n,\lambda,l}f(z)} + \frac{z^3}{\delta}\left[\frac{\left(RI^{\alpha}_{n,\lambda,l}f(z)\right)''}{RI^{\alpha}_{n,\lambda,l}f(z)} - \left(\frac{\left(RI^{\alpha}_{n,\lambda,l}f(z)\right)'}{RI^{\alpha}_{n,\lambda,l}f(z)}\right)^2\right] \prec h(z), \quad (3.2.1.29)$$

$z \in U$, *then*

$$z^2 \frac{\left(RI^{\alpha}_{n,\lambda,l}f(z)\right)'}{RI^{\alpha}_{n,\lambda,l}f(z)} \prec q(z), \quad z \in U,$$

where $q(z) = \frac{\delta}{z^{\delta}} \int_0^z h(t)t^{\delta-1}dt$.

Proof. Let $p(z) = z^2 \frac{\left(RI^{\alpha}_{n,\lambda,l}f(z)\right)'}{RI^{\alpha}_{n,\lambda,l}f(z)}$, $z \in U$, $p \in \mathcal{H}[0,1]$.

Differentiating, we obtain

$$p(z) + \frac{z}{\delta}p'(z) = z^2 \frac{\delta+2}{\delta} \frac{\left(RI^{\alpha}_{n,\lambda,l}f(z)\right)'}{RI^{\alpha}_{n,\lambda,l}f(z)} + \frac{z^3}{\delta}\left[\frac{\left(RI^{\alpha}_{n,\lambda,l}f(z)\right)''}{RI^{\alpha}_{n,\lambda,l}f(z)} - \left(\frac{\left(RI^{\alpha}_{n,\lambda,l}f(z)\right)'}{RI^{\alpha}_{n,\lambda,l}f(z)}\right)^2\right],$$

$z \in U$, and (3.2.1.29) becomes $p(z) + \frac{1}{\delta}zp'(z) \prec h(z)$, $z \in U$.

Using Lemma 1.1, we have

$$p(z) \prec q(z), \quad \text{i.e. } z^2 \frac{\left(RI^{\alpha}_{n,\lambda,l}f(z)\right)'}{RI^{\alpha}_{n,\lambda,l}f(z)} \prec q(z) = \frac{\delta}{z^{\delta}} \int_0^z h(t)t^{\delta-1}dt, \quad z \in U,$$

and q is the best dominant. □

Theorem 3.79 (Alb Lupaş, 2015b). *Let g be a convex function such that $g(0) = 1$ and let h be the function $h(z) = g(z) + zg'(z)$, $z \in U$. If $\alpha, \lambda, l \geq 0$, $n \in \mathbb{N}$, $f \in \mathcal{A}$ and the differential subordination*

$$1 - \frac{RI^{\alpha}_{n,\lambda,l}f(z) \cdot \left(RI^{\alpha}_{n,\lambda,l}f(z)\right)''}{\left[\left(RI^{\alpha}_{n,\lambda,l}f(z)\right)'\right]^2} \prec h(z), \quad z \in U \tag{3.2.1.30}$$

holds, then

$$\frac{RI^{\alpha}_{n,\lambda,l}f(z)}{z\left(RI^{\alpha}_{n,\lambda,l}f(z)\right)'} \prec g(z), \quad z \in U.$$

This result is sharp.

Proof. Let $p(z) = \frac{RI^{\alpha}_{n,\lambda,l}f(z)}{z\left(RI^{\alpha}_{n,\lambda,l}f(z)\right)'}$. We deduce that $p \in \mathcal{H}[1,1]$.

Differentiating, we obtain $1 - \frac{RI^{\alpha}_{n,\lambda,l}f(z)\cdot\left(RI^{\alpha}_{n,\lambda,l}f(z)\right)''}{\left[\left(RI^{\alpha}_{n,\lambda,l}f(z)\right)'\right]^2} = p(z) + zp'(z)$, $z \in U$.

Using the notation in (3.2.1.30), the differential subordination becomes $p(z) + zp'(z) \prec h(z) = g(z) + zg'(z)$.

By using Lemma 1.2, we have

$$p(z) \prec g(z), \quad z \in U, \quad \text{i.e.} \ \frac{RI^{\alpha}_{n,\lambda,l}f(z)}{z\left(RI^{\alpha}_{n,\lambda,l}f(z)\right)'} \prec g(z), \quad z \in U,$$

and this result is sharp. □

Theorem 3.80 (Alb Lupaş, 2015b). *Let h be an holomorphic function which satisfies the inequality* $\operatorname{Re}\left(1+\frac{zh''(z)}{h'(z)}\right) > -\frac{1}{2}$, *$z \in U$, and $h(0)=1$. If $\alpha, \lambda, l \geq 0$, $n \in \mathbb{N}$, $f \in \mathcal{A}$ and satisfies the differential subordination*

$$1 - \frac{RI^{\alpha}_{n,\lambda,l}f(z) \cdot \left(RI^{\alpha}_{n,\lambda,l}f(z)\right)''}{\left[\left(RI^{\alpha}_{n,\lambda,l}f(z)\right)'\right]^2} \prec h(z), \quad z \in U, \qquad (3.2.1.31)$$

then

$$\frac{RI^{\alpha}_{n,\lambda,l}f(z)}{z\left(RI^{\alpha}_{n,\lambda,l}f(z)\right)'} \prec q(z), \quad z \in U,$$

where $q(z) = \frac{1}{z}\int_0^z h(t)dt$.

Proof. Let $p(z) = \frac{RI^{\alpha}_{n,\lambda,l}f(z)}{z\left(RI^{\alpha}_{n,\lambda,l}f(z)\right)'}$, $z \in U$, $p \in \mathcal{H}[1,1]$.

Differentiating, we obtain $1 - \frac{RI^{\alpha}_{n,\lambda,l}f(z)\cdot\left(RI^{\alpha}_{n,\lambda,l}f(z)\right)''}{\left[\left(RI^{\alpha}_{n,\lambda,l}f(z)\right)'\right]^2} = p(z) + zp'(z)$, $z \in U$, and (3.2.1.31) becomes $p(z) + zp'(z) \prec h(z)$, $z \in U$.

Using Lemma 1.1, we have

$$p(z) \prec q(z), \ z \in U, \ \text{i.e.} \ \frac{RI^{\alpha}_{n,\lambda,l}f(z)}{z\left(RI^{\alpha}_{n,\lambda,l}f(z)\right)'} \prec q(z) = \frac{1}{z}\int_0^z h(t)dt, \ z \in U,$$

and q is the best dominant. □

Corollary 3.17 (Alb Lupaş, 2015b). *Let* $h(z) = \frac{1+(2\beta-1)z}{1+z}$ *be a convex function in U, where $0 \leq \beta < 1$. If $\alpha, \lambda, l \geq 0$, $n \in \mathbb{N}$, $f \in \mathcal{A}$ and satisfies the differential subordination*

$$1 - \frac{RI^{\alpha}_{n,\lambda,l}f(z) \cdot \left(RI^{\alpha}_{n,\lambda,l}f(z)\right)''}{\left[\left(RI^{\alpha}_{n,\lambda,l}f(z)\right)'\right]^2} \prec h(z), \quad z \in U, \qquad (3.2.1.32)$$

then

$$\frac{RI^{\alpha}_{n,\lambda,l}f\left(z\right)}{z\left(RI^{\alpha}_{n,\lambda,l}f\left(z\right)\right)'} \prec q(z), \quad z\in U,$$

where q is given by $q(z)=(2\beta-1)+2\left(1-\beta\right)\frac{\ln(1+z)}{z}$, $z\in U$. The function q is convex and it is the best dominant.

Proof. Following the same steps as in the proof of Theorem 3.80 and considering $p(z)=\frac{RI^{\alpha}_{n,\lambda,l}f(z)}{z\left(RI^{\alpha}_{n,\lambda,l}f(z)\right)'}$, the differential subordination (3.2.1.32) becomes $p(z)+zp'(z)\prec h(z)=\frac{1+(2\beta-1)z}{1+z}$, $z\in U$.

By using Lemma 1.1 for $\gamma=1$, we have $p(z)\prec q(z)$, i.e.

$$\begin{aligned}\frac{RI^{\alpha}_{n,\lambda,l}f\left(z\right)}{z\left(RI^{\alpha}_{n,\lambda,l}f\left(z\right)\right)'} \prec q(z) &= \frac{1}{z}\int_0^z h(t)dt=\frac{1}{z}\int_0^z \frac{1+(2\beta-1)t}{1+t}dt\\ &=\frac{1}{z}\int_0^z\left[(2\beta-1)+\frac{2\left(1-\beta\right)}{1+t}\right]dt\\ &=(2\beta-1)+2\left(1-\beta\right)\frac{\ln\left(1+z\right)}{z}, \quad z\in U. \quad \square\end{aligned}$$

Example 3.3 (Alb Lupaş, 2015b). Let $h\left(z\right)=\frac{1-z}{1+z}$ a convex function in U with $h\left(0\right)=1$ and $\operatorname{Re}\left(\frac{zh''(z)}{h'(z)}+1\right)>-\frac{1}{2}$.

Let $f\left(z\right)=z+z^2$, $z\in U$. For $n=1$, $m=1$, l=2, $\lambda=1$, $\alpha=\frac{1}{2}$, we obtain $RI^{\frac{1}{2}}_{1,1,2}f\left(z\right)=\frac{1}{2}R^1f\left(z\right)+\frac{1}{2}I\left(1,1,2\right)f\left(z\right)=\frac{1}{3}f\left(z\right)+\frac{2}{3}zf'\left(z\right)=z+\frac{5}{3}z^2$, $z\in U$.

Then $\left(RI^{\frac{1}{2}}_{1,1,2}f\left(z\right)\right)'=1+\frac{10}{3}z$ and $\left(RI^{\frac{1}{2}}_{1,1,2}f\left(z\right)\right)''=\frac{10}{3}$, $\frac{RI^{\frac{1}{2}}_{1,1,2}f(z)}{z\left(RI^{\frac{1}{2}}_{1,1,2}f(z)\right)'}=\frac{z+\frac{5}{3}z^2}{z\left(1+\frac{10}{3}z\right)}=\frac{3+5z}{3+10z}$,

$$1-\frac{RI^{\frac{1}{2}}_{1,1,2}f\left(z\right)\cdot\left(RI^{\frac{1}{2}}_{1,1,2}f\left(z\right)\right)''}{\left[\left(RI^{\frac{1}{2}}_{1,1,2}f\left(z\right)\right)'\right]^2}=1-\frac{\left(z+\frac{5}{3}z^2\right)\cdot\frac{10}{3}}{\left(1+\frac{10}{3}z\right)^2}=\frac{50z^2+30z+9}{\left(3+10z\right)^2}.$$

We have $q\left(z\right)=\frac{1}{z}\int_0^z\frac{1-t}{1+t}dt=-1+\frac{2\ln(1+z)}{z}$.

Using Theorem 3.80, we obtain $\frac{50z^2+30z+9}{(3+10z)^2}\prec\frac{1-z}{1+z}$, $z\in U$, induce $\frac{3+5z}{3+10z}\prec-1+\frac{2\ln(1+z)}{z}$, $z\in U$.

Theorem 3.81 (Alb Lupaş, 2015b). *Let g be a convex function such that $g(0) = 1$ and let h be the function $h(z) = g(z) + zg'(z)$, $z \in U$. If $\alpha, \lambda, l \geq 0$, $n \in \mathbb{N}$, $f \in \mathcal{A}$ and the differential subordination*

$$\left[\left(RI^{\alpha}_{n,\lambda,l}f(z)\right)'\right]^2 + RI^{\alpha}_{n,\lambda,l}f(z) \cdot \left(RI^{\alpha}_{n,\lambda,l}f(z)\right)'' \prec h(z), \quad z \in U \tag{3.2.1.33}$$

holds, then

$$\frac{RI^{\alpha}_{n,\lambda,l}f(z) \cdot \left(RI^{\alpha}_{n,\lambda,l}f(z)\right)'}{z} \prec g(z), \quad z \in U.$$

This result is sharp.

Proof. Let $p(z) = \frac{RI^{\alpha}_{n,\lambda,l}f(z) \cdot \left(RI^{\alpha}_{n,\lambda,l}f(z)\right)'}{z}$. We deduce that $p \in \mathcal{H}[1,1]$.

Differentiating, we obtain

$$\left[\left(RI^{\alpha}_{n,\lambda,l}f(z)\right)'\right]^2 + RI^{\alpha}_{n,\lambda,l}f(z) \cdot \left(RI^{\alpha}_{n,\lambda,l}f(z)\right)'' = p(z) + zp'(z), \quad z \in U.$$

Using the notation in (3.2.1.33), the differential subordination becomes $p(z) + zp'(z) \prec h(z) = g(z) + zg'(z)$.

By using Lemma 1.2, we have

$$p(z) \prec g(z), \quad \text{i.e.} \quad \frac{RI^{\alpha}_{n,\lambda,l}f(z) \cdot \left(RI^{\alpha}_{n,\lambda,l}f(z)\right)'}{z} \prec g(z), \quad z \in U,$$

and this result is sharp. □

Theorem 3.82. *Let h be an holomorphic function which satisfies the inequality $\operatorname{Re}\left(1 + \frac{zh''(z)}{h'(z)}\right) > -\frac{1}{2}$, $z \in U$, and $h(0) = 1$. If $\alpha, \lambda, l \geq 0$, $n \in \mathbb{N}$, $f \in \mathcal{A}$ and satisfies the differential subordination*

$$\left[\left(RI^{\alpha}_{n,\lambda,l}f(z)\right)'\right]^2 + RI^{\alpha}_{n,\lambda,l}f(z) \cdot \left(RI^{\alpha}_{n,\lambda,l}f(z)\right)'' \prec h(z), \quad z \in U, \tag{3.2.1.34}$$

then

$$\frac{RI^{\alpha}_{n,\lambda,l}f(z) \cdot \left(RI^{\alpha}_{n,\lambda,l}f(z)\right)'}{z} \prec q(z), \quad z \in U,$$

where $q(z) = \frac{1}{z}\int_0^z h(t)dt$.

Proof. Let $p(z) = \frac{RI^{\alpha}_{n,\lambda,l}f(z)\cdot\left(RI^{\alpha}_{n,\lambda,l}f(z)\right)'}{z}$, $z \in U$, $p \in \mathcal{H}[1,1]$.

Differentiating, we obtain

$$\left[\left(RI^{\alpha}_{n,\lambda,l}f(z)\right)'\right]^2 + RI^{\alpha}_{n,\lambda,l}f(z)\cdot\left(RI^{\alpha}_{n,\lambda,l}f(z)\right)'' = p(z) + zp'(z), \quad z \in U,$$

and (3.2.1.34) becomes $p(z) + zp'(z) \prec h(z)$, $z \in U$.

Using Lemma 1.1, we have

$$p(z) \prec q(z), \quad \text{i.e.} \quad \frac{RI^{\alpha}_{n,\lambda,l}f(z)\cdot\left(RI^{\alpha}_{n,\lambda,l}f(z)\right)'}{z} \prec q(z) = \frac{1}{z}\int_0^z h(t)dt,$$

$z \in U$,

and q is the best dominant. □

Corollary 3.18 (Alb Lupaş, 2015b). *Let* $h(z) = \frac{1+(2\beta-1)z}{1+z}$ *be a convex function in* U, *where* $0 \leq \beta < 1$. *If* $\alpha, \lambda, l \geq 0$, $n \in \mathbb{N}$, $f \in \mathcal{A}$ *and satisfies the differential subordination*

$$\left[\left(RI^{\alpha}_{n,\lambda,l}f(z)\right)'\right]^2 + RI^{\alpha}_{n,\lambda,l}f(z)\cdot\left(RI^{\alpha}_{n,\lambda,l}f(z)\right)'' \prec h(z), \quad z \in U, \tag{3.2.1.35}$$

then

$$\frac{RI^{\alpha}_{n,\lambda,l}f(z)\cdot\left(RI^{\alpha}_{n,\lambda,l}f(z)\right)'}{z} \prec q(z), \quad z \in U,$$

where q *is given by* $q(z) = (2\beta-1) + 2(1-\beta)\frac{\ln(1+z)}{z}$, $z \in U$. *The function* q *is convex and it is the best dominant.*

Proof. Following the same steps as in the proof of Theorem 3.82 and considering $p(z) = \frac{RI^{\alpha}_{n,\lambda,l}f(z)\cdot\left(RI^{\alpha}_{n,\lambda,l}f(z)\right)'}{z}$, the differential subordination (3.2.1.35) becomes $p(z) + zp'(z) \prec h(z) = \frac{1+(2\beta-1)z}{1+z}$, $z \in U$.

By using Lemma 1.1 for $\gamma = 1$, we have $p(z) \prec q(z)$, i.e.

$$\begin{aligned}\frac{RI^{\alpha}_{n,\lambda,l}f(z)\cdot\left(RI^{\alpha}_{n,\lambda,l}f(z)\right)'}{z} \prec q(z) &= \frac{1}{z}\int_0^z h(t)dt = \frac{1}{z}\int_0^z \frac{1+(2\beta-1)t}{1+t}dt \\ &= \frac{1}{z}\int_0^z \left[(2\beta-1) + \frac{2(1-\beta)}{1+t}\right]dt \\ &= (2\beta-1) + 2(1-\beta)\frac{\ln(1+z)}{z}, \quad z \in U.\end{aligned}$$

□

Example 3.4 (Alb Lupaş, 2015b). Let $h(z) = \frac{1-z}{1+z}$ a convex function in U with $h(0) = 1$ and $\mathrm{Re}\left(\frac{zh''(z)}{h'(z)} + 1\right) > -\frac{1}{2}$.

Let $f(z) = z + z^2$, $z \in U$. For $n = 1$, $m = 1$, $l = 2$, $\lambda = 1$, $\alpha = \frac{1}{2}$, we obtain $RI^{\frac{1}{2}}_{1,1,2}f(z) = \frac{1}{2}R^1 f(z) + \frac{1}{2}I(1,1,2) f(z) = \frac{1}{3}f(z) + \frac{2}{3}zf'(z) = z + \frac{5}{3}z^2$, $z \in U$.

Then $\left(RI^{\frac{1}{2}}_{1,1,2}f(z)\right)' = 1 + \frac{10}{3}z$, $\left(RI^{\frac{1}{2}}_{1,1,2}f(z)\right)'' = \frac{10}{3}$, $\frac{RI^{\frac{1}{2}}_{1,1,2}f(z)\cdot\left(RI^{\frac{1}{2}}_{1,1,2}f(z)\right)'}{z} = \frac{\left(z+\frac{5}{3}z^2\right)\left(1+\frac{10}{3}z\right)}{z} = \frac{50}{9}z^2 + 5z + 1$,

$$\left[\left(RI^{\frac{1}{2}}_{1,1,2}f(z)\right)'\right]^2 + RI^{\frac{1}{2}}_{1,1,2}f(z)\cdot\left(RI^{\frac{1}{2}}_{1,1,2}f(z)\right)''$$
$$= \left(1 + \frac{10}{3}z\right)^2 + \left(z + \frac{5}{3}z^2\right)\cdot\frac{10}{3} = \frac{50}{3}z^2 + 10z + 1.$$

We have $q(z) = \frac{1}{z}\int_0^z \frac{1-t}{1+t}dt - -1 + \frac{2\ln(1+z)}{z}$.

Using Theorem 3.82, we obtain $\frac{50}{3}z^2 + 10z + 1 \prec \frac{1-z}{1+z}$, $z \in U$, induce $\frac{50}{9}z^2 + 5z + 1 \prec -1 + \frac{2\ln(1+z)}{z}$, $z \in U$.

Theorem 3.83 (Alb Lupaş, 2015b). *Let g be a convex function such that $g(0) = 1$ and let h be the function $h(z) = g(z) + \frac{z}{1-\delta}g'(z)$, $z \in U$. If $\alpha, \lambda, l \geq 0$, $\delta \in (0,1)$, $n \in \mathbb{N}$, $f \in \mathcal{A}$ and the differential subordination*

$$\left(\frac{z}{RI^{\alpha}_{n,\lambda,l}f(z)}\right)^{\delta}\frac{RI^{\alpha}_{n+1,\lambda,l}f(z)}{1-\delta}$$
$$\times\left(\frac{\left(RI^{\alpha}_{n+1,\lambda,l}f(z)\right)'}{RI^{\alpha}_{n+1,\lambda,l}f(z)} - \delta\frac{\left(RI^{\alpha}_{n,\lambda,l}f(z)\right)'}{RI^{\alpha}_{n,\lambda,l}f(z)}\right) \prec h(z), \tag{3.2.1.36}$$

$z \in U$ holds, then

$$\frac{RI^{\alpha}_{n+1,\lambda,l}f(z)}{z}\cdot\left(\frac{z}{RI^{\alpha}_{n,\lambda,l}f(z)}\right)^{\delta} \prec g(z), \quad z \in U.$$

This result is sharp.

Proof. Let $p(z) = \frac{RI^{\alpha}_{n+1,\lambda,l}f(z)}{z} \cdot \left(\frac{z}{RI^{\alpha}_{n,\lambda,l}f(z)}\right)^{\delta}$. We deduce that $p \in \mathcal{H}[1,1]$.

Differentiating, we obtain

$$\left(\frac{z}{RI^{\alpha}_{n,\lambda,l}f(z)}\right)^{\delta} \frac{RI^{\alpha}_{n+1,\lambda,l}f(z)}{1-\delta} \left(\frac{\left(RI^{\alpha}_{n+1,\lambda,l}f(z)\right)'}{RI^{\alpha}_{n+1,\lambda,l}f(z)} - \delta \frac{\left(RI^{\alpha}_{n,\lambda,l}f(z)\right)'}{RI^{\alpha}_{n,\lambda,l}f(z)}\right)$$

$$= p(z) + \frac{1}{1-\delta} zp'(z), \quad z \in U.$$

Using the notation in (3.2.1.36), the differential subordination becomes $p(z) + \frac{1}{1-\delta} zp'(z) \prec h(z) = g(z) + \frac{z}{1-\delta} g'(z)$.

By using Lemma 1.2, we have

$$p(z) \prec g(z), \quad \text{i.e.} \ \frac{RI^{\alpha}_{n+1,\lambda,l}f(z)}{z} \cdot \left(\frac{z}{RI^{\alpha}_{n,\lambda,l}f(z)}\right)^{\delta} \prec g(z), \quad z \in U,$$

and this result is sharp. □

Theorem 3.84 (Alb Lupaş, 2015b). *Let h be an holomorphic function which satisfies the inequality* $\mathrm{Re}\left(1 + \frac{zh''(z)}{h'(z)}\right) > -\frac{1}{2}$, $z \in U$, *and* $h(0) = 1$. *If* $\alpha, \lambda, l \geq 0$, $\delta \in (0,1)$, $n \in \mathbb{N}$, $f \in \mathcal{A}$ *and satisfies the differential subordination*

$$\left(\frac{z}{RI^{\alpha}_{n,\lambda,l}f(z)}\right)^{\delta} \frac{RI^{\alpha}_{n+1,\lambda,l}f(z)}{1-\delta} \times \left(\frac{\left(RI^{\alpha}_{n+1,\lambda,l}f(z)\right)'}{RI^{\alpha}_{n+1,\lambda,l}f(z)} - \delta \frac{\left(RI^{\alpha}_{n,\lambda,l}f(z)\right)'}{RI^{\alpha}_{n,\lambda,l}f(z)}\right) \prec h(z), \tag{3.2.1.37}$$

$z \in U$, *then*

$$\frac{RI^{\alpha}_{n+1,\lambda,l}f(z)}{z} \cdot \left(\frac{z}{RI^{\alpha}_{n,\lambda,l}f(z)}\right)^{\delta} \prec q(z), \quad z \in U,$$

where $q(z) = \frac{1-\delta}{z^{1-\delta}} \int_0^z h(t) t^{-\delta} dt$.

Proof. Let $p(z) = \frac{RI^{\alpha}_{n+1,\lambda,l}f(z)}{z} \cdot \left(\frac{z}{RI^{\alpha}_{n,\lambda,l}f(z)}\right)^{\delta}$, $z \in U$, $p \in \mathcal{H}[1,1]$.

Differentiating, we obtain

$$\left(\frac{z}{RI^{\alpha}_{n,\lambda,l}f(z)}\right)^{\delta} \frac{RI^{\alpha}_{n+1,\lambda,l}f(z)}{1-\delta} \left(\frac{\left(RI^{\alpha}_{n+1,\lambda,l}f(z)\right)'}{RI^{\alpha}_{n+1,\lambda,l}f(z)} - \delta \frac{\left(RI^{\alpha}_{n,\lambda,l}f(z)\right)'}{RI^{\alpha}_{n,\lambda,l}f(z)}\right)$$

$$= p(z) + \frac{1}{1-\delta} zp'(z), \quad z \in U,$$

and (3.2.1.37) becomes $p(z) + \frac{1}{1-\delta} zp'(z) \prec h(z)$, $z \in U$.

Using Lemma 1.1, we have

$$p(z) \prec q(z), \quad z \in U, \quad \text{i.e.}$$

$$\frac{RI^{\alpha}_{n+1,\lambda,l}f(z)}{z} \cdot \left(\frac{z}{RI^{\alpha}_{n,\lambda,l}f(z)}\right)^{\delta} \prec q(z) = \frac{1-\delta}{z^{1-\delta}} \int_0^z h(t)t^{-\delta}dt, \quad z \in U,$$

and q is the best dominant. □

3.2.2 *Subordination Results for a Fractional Integral Applied to the Convolution Product of Sălăgean Operator and Ruscheweyh Derivative*

In the paper Alb Lupaş (2022b), we established several differential subordinations regarding the operator $D_z^{-\lambda}SR^{m,n}$ defined using the fractional integral of the differential operator $SR^{m,n}$ obtained as a convolution product of Sălăgean operator S^m and Ruscheweyh derivative R^n. By means of the newly obtained operator, a new subclass of analytic functions denoted by $\mathcal{SR}_{m,n,\lambda}(\delta)$ is introduced and various properties and characteristics of this subclass are derived making use of the concept of differential subordination.

Let $\lambda > 0$ and $m, n \in \mathbb{N}$. The fractional integral associated with the linear differential operator $SR^{m,n}f$ is defined in Definition 2.16 by

$$D_z^{-\lambda}SR^{m,n}f(z) = \frac{1}{\Gamma(\lambda)} \int_0^z \frac{SR^{m,n}f(t)}{(z-t)^{1-\lambda}}dt$$

$$= \frac{1}{\Gamma(\lambda)} \left(\int_0^z \frac{t}{(z-t)^{1-\lambda}}dt + \sum_{j=l+1}^{\infty} \left(j^m \frac{\Gamma(n+j)}{\Gamma(n+1)\Gamma(j)}\right) a_j^2 \int_0^z \frac{t^j}{(z-t)^{1-\lambda}}dt\right),$$

which has the following form, after a simple calculation,

$$D_z^{-\lambda} SR^{m,n} f(z) = \frac{1}{\Gamma(\lambda+2)} z^{\lambda+1} + \sum_{j=l+1}^{\infty} \frac{j^{m+1}\Gamma(n+j)}{\Gamma(n+1)\Gamma(j+\lambda+1)} a_j^2 z^{j+\lambda},$$

for a function $f(z) = z + \sum_{j=l+1}^{\infty} a_j z^j \in \mathcal{A}_l$. We note that $D_z^{-\lambda} SR^{m,n} f(z) \in \mathcal{A}(\lambda+1, l)$.

After a simple calculation we obtain the relation

$$z\left(D_z^{-\lambda} SR^{m,n} f(z)\right)' = \lambda D_z^{-\lambda} SR^{m,n} f(z) + D_z^{-\lambda} SR^{m+1,n} f(z), \quad z \in U. \tag{3.2.2.1}$$

Firstly, we define and study a subclass of analytic functions using the differential subordinations regarding to the differential operator $D_z^{-\lambda} SR^{m,n} : \mathcal{A}_l \to \mathcal{A}_l$.

Definition 3.13 (Alb Lupaş, 2022b). Let $\delta \in [0,1)$ and $m, n \in \mathbb{N}, \lambda > 0$. A function $f \in \mathcal{A}_l$ is said to be in the subclass $\mathcal{SR}_{m,n,\lambda}(\delta)$ if it satisfies the inequality

$$\operatorname{Re}\left(D_z^{-\lambda} SR^{m,n} f(z)\right)' > \delta, \quad z \in U.$$

Theorem 3.85 (Alb Lupaş, 2022b). *Let g be a convex function in U and let $h(z) = g(z) + \frac{1}{c+2} z g'(z)$, $z \in U$, where $c > 0$. If $f \in \mathcal{SR}_{m,n,\lambda}(\delta)$ and $F(z) = I_c(f)(z) = \frac{c+2}{z^{c+1}} \int_0^z t^c f(t)\, dt$, $z \in U$, then*

$$\left(D_z^{-\lambda} SR^{m,n} f(z)\right)' \prec h(z), \quad z \in U, \tag{3.2.2.2}$$

implies

$$\left(D_z^{-\lambda} SR^{m,n} F(z)\right)' \prec g(z), \quad z \in U,$$

and this result is sharp.

Proof. We have $z^{c+1} F(z) = (c+2) \int_0^z t^c f(t)\, dt$. Differentiating, with respect to z, we obtain

$$(c+1) F(z) + z F'(z) = (c+2) f(z)$$

and

$$(c+1) D_z^{-\lambda} SR^{m,n} F(z) + z\left(D_z^{-\lambda} SR^{m,n} F(z)\right)' = (c+2) D_z^{-\lambda} SR^{m,n} f(z), \tag{3.2.2.3}$$

$$z \in U.$$

Differentiating (3.2.2.3), we have

$$\left(D_z^{-\lambda}SR^{m,n}F(z)\right)' + \frac{1}{c+2}z\left(D_z^{-\lambda}SR^{m,n}F(z)\right)'' = \left(D_z^{-\lambda}SR^{m,n}f(z)\right)', \tag{3.2.2.4}$$

$z \in U$.

Using (3.2.2.4), the differential subordination (3.2.2.2) becomes

$$\left(D_z^{-\lambda}SR^{m,n}F(z)\right)' + \frac{1}{c+2}z\left(D_z^{-\lambda}SR^{m,n}F(z)\right)'' \prec g(z) + \frac{1}{c+2}zg'(z). \tag{3.2.2.5}$$

If we denote

$$p(z) = \left(D_z^{-\lambda}SR^{m,n}F(z)\right)', \tag{3.2.2.6}$$

then $p \in \mathcal{H}[\lambda+1, l]$.

Replacing (3.2.2.6) in (3.2.2.5), we obtain

$$p(z) + \frac{1}{c+2}zp'(z) \prec g(z) + \frac{1}{c+2}zg'(z), \quad z \in U.$$

Using Lemma 1.2, we have

$$p(z) \prec g(z) \quad \text{i.e.} \quad \left(D_z^{-\lambda}SR^{m,n}F(z)\right)' \prec g(z), \quad z \in U,$$

and g is the best dominant. □

Theorem 3.86 (Alb Lupaş, 2022b). *Let* $h(z) = \frac{1+(2\delta-1)z}{1+z}$, $\delta \in [0,1)$ *and* $c > 0$. *If* $m, n \in \mathbb{N}$, $\lambda > 0$, *and* I_c *is as in Theorem* 3.85, *then*

$$I_c\left[\mathcal{SR}_{m,n,\lambda}(\delta)\right] \subset \mathcal{SR}_{m,n,\lambda}(\delta^*), \tag{3.2.2.7}$$

where $\delta^* = 2\delta - 1 + \frac{(c+2)(2-2\delta)}{l}\beta\left(\frac{c+2}{l} - 2\right)$ *and* $\beta(x) = \int_0^1 \frac{t^{x+1}}{t+1}dt$.

Proof. The function h is convex and using the same steps as in the proof of Theorem 3.85, we get from the hypothesis of Theorem 3.86 that

$$p(z) + \frac{1}{c+2}zp'(z) \prec h(z),$$

where $p(z)$ is defined in (3.2.2.6).

Using Lemma 1.1, we deduce that

$$p(z) \prec g(z) \prec h(z),$$

that is

$$\left(D_z^{-\lambda}SR^{m,n}F(z)\right)' \prec g(z) \prec h(z),$$

where

$$\begin{aligned} g(z) &= \frac{c+2}{lz^{\frac{c+2}{l}}} \int_0^z t^{\frac{c+2}{l}-1} \frac{1+(2\delta-1)t}{1+t} dt \\ &= (2\delta-1) + \frac{(c+2)(2-2\delta)}{lz^{\frac{c+2}{l}}} \int_0^z \frac{t^{\frac{c+2}{l}-1}}{1+t} dt. \end{aligned}$$

Since g is convex and $g(U)$ is symmetric with respect to the real axis, we deduce

$$\begin{aligned} \operatorname{Re}\left(D_z^{-\lambda}SR^{m,n}F(z)\right)' &\geq \min_{|z|=1} \operatorname{Re}\, g(z) = \operatorname{Re}\, g(1) = \delta^* \\ &= 2\delta - 1 + \frac{(c+2)(2-2\delta)}{l} \beta\left(\frac{c+2}{l} - 2\right). \end{aligned} \tag{3.2.2.8}$$

From (3.2.2.8), we deduce inclusion (3.2.2.7). □

Theorem 3.87 (Alb Lupaş, 2022b). *Let g be a convex function, $g(0)=0$, and let h be the function $h(z) = g(z) + zg'(z)$, $z \in U$. If $m, n \in \mathbb{N}$, $f \in \mathcal{A}_l$ and verifies the differential subordination*

$$\left(D_z^{-\lambda}SR^{m,n}f(z)\right)' \prec h(z), \quad z \in U, \tag{3.2.2.9}$$

then

$$\frac{D_z^{-\lambda}SR^{m,n}f(z)}{z} \prec g(z), \quad z \in U,$$

and this result is sharp.

Proof. For $f \in \mathcal{A}_l$, $f(z) = z + \sum_{j=l+1}^{\infty} a_j z^j$, we have

$$D_z^{-\lambda}SR^{m,n}f(z) = \frac{1}{\Gamma(\lambda+2)} z^{\lambda+1} + \sum_{j=l+1}^{\infty} \frac{j^{m+1}\Gamma(n+j)}{\Gamma(n+1)\Gamma(j+\lambda+1)} a_j^2 z^{j+\lambda},$$

$$z \in U.$$

Consider

$$p(z) = \frac{D_z^{-\lambda} SR^{m,n} f(z)}{z}$$

$$= \frac{\frac{1}{\Gamma(\lambda+2)} z^{\lambda+1} + \sum_{j=l+1}^{\infty} \frac{j^{m+1}\Gamma(n+j)}{\Gamma(n+1)\Gamma(j+\lambda+1)} a_j^2 z^{j+\lambda}}{z}$$

$$= \frac{1}{\Gamma(\lambda+2)} z^{\lambda} + \sum_{j=l+1}^{\infty} \frac{j^{m+1}\Gamma(n+j)}{\Gamma(n+1)\Gamma(j+\lambda+1)} a_j^2 z^{j+\lambda-1}, \quad p \in \mathcal{H}[0, \lambda].$$

We have

$$p(z) + zp'(z) = \left(D_z^{-\lambda} SR^{m,n} f(z)\right)', \quad z \in U.$$

Then

$$\left(D_z^{-\lambda} SR^{m,n} f(z)\right)' \prec h(z), \quad z \in U,$$

becomes

$$p(z) + zp'(z) \prec h(z) = g(z) + zg'(z), \quad z \in U.$$

By using Lemma 1.2, we obtain

$$p(z) \prec g(z), \quad z \in U, \quad \text{i.e. } \frac{D_z^{-\lambda} SR^{m,n} f(z)}{z} \prec g(z), \quad z \in U. \qquad \square$$

Theorem 3.88 (Alb Lupaş, 2022b). *Let* $h \in \mathcal{H}(U)$, *with* $h(0) = 0$, *which verifies the inequality* $\operatorname{Re}\left(1 + \frac{zh''(z)}{h'(z)}\right) > -\frac{1}{2}$, $z \in U$. *If* $m, n \in \mathbb{N}$, $\lambda > 0$, $f \in \mathcal{A}_l$ *and verifies the differential subordination*

$$\left(D_z^{-\lambda} SR^{m,n} f(z)\right)' \prec h(z), \quad z \in U, \tag{3.2.2.10}$$

then

$$\frac{D_z^{-\lambda} SR^{m,n} f(z)}{z} \prec q(z), \quad z \in U,$$

where $q(z) = \frac{1}{lz^{\frac{1}{l}}} \int_0^z h(t) t^{\frac{1}{l}-1} dt$. *The function* q *is convex and it is the best dominant.*

Proof. Let

$$p(z) = \frac{D_z^{-\lambda} SR^{m,n} f(z)}{z}$$
$$= \frac{1}{\Gamma(\lambda+2)} z^{\lambda} + \sum_{j=l+1}^{\infty} \frac{j^{m+1}\Gamma(n+j)}{\Gamma(n+1)\Gamma(j+\lambda+1)} a_j^2 z^{j+\lambda-1},$$
$$z \in U,\ p \in \mathcal{H}[0,\lambda].$$

Differentiating, we obtain

$$\left(D_z^{-\lambda} SR^{m,n} f(z)\right)' = p(z) + zp'(z), \quad z \in U,$$

and (3.2.2.10) becomes

$$p(z) + zp'(z) \prec h(z), \quad z \in U.$$

Using Lemma 1.1, we have

$$p(z) \prec q(z) = \frac{1}{lz^{\frac{1}{l}}} \int_0^z h(t) t^{\frac{1}{l}-1} dt, \quad z \in U,$$

i.e.

$$\frac{D_z^{-\lambda} SR^{m,n} f(z)}{z} \prec q(z) = \frac{1}{lz^{\frac{1}{l}}} \int_0^z h(t) t^{\frac{1}{l}-1} dt, \quad z \in U,$$

and q is the best dominant. □

Corollary 3.19 (Alb Lupaş, 2022b). *Let $h(z) = \frac{(2\beta-1)z}{1+z}$ a convex function in U, $0 \le \beta < 1$. If $m, n \in \mathbb{N}$, $\lambda > 0$, $f \in \mathcal{A}_l$ and verifies the differential subordination*

$$\left(D_z^{-\lambda} SR^{m,n} f(z)\right)' \prec h(z), \quad z \in U, \tag{3.2.2.11}$$

then

$$\frac{D_z^{-\lambda} SR^{m,n}(z)}{z} \prec q(z), \quad z \in U,$$

where q is given by $q(z) = \frac{2\beta-1}{lz^{\frac{1}{l}}} \int_0^z \frac{t^{\frac{1}{l}}}{1+t} dt$, $z \in U$. The function q is convex and it is the best dominant.

Proof. Following the same steps as in the proof of Theorem 3.88 and considering $p(z) = \frac{D_z^{-\lambda} SR^{m,n} f(z)}{z}$, the differential subordination (3.2.2.11) becomes

$$p(z) + zp'(z) \prec h(z) = \frac{(2\beta - 1)z}{1+z}, \quad z \in U.$$

By using Lemma 1.1 for $\gamma = 1$, we have $p(z) \prec q(z)$, i.e.

$$\begin{aligned}\frac{D_z^{-\lambda} SR^{m,n} f(z)}{z} \prec q(z) &= \frac{1}{lz^{\frac{1}{l}}} \int_0^z h(t)\, t^{\frac{1}{l}-1} dt \\ &= \frac{1}{lz^{\frac{1}{l}}} \int_0^z t^{\frac{1}{l}-1} \frac{(2\beta-1)\,t}{1+t} dt = \frac{2\beta - 1}{lz^{\frac{1}{l}}} \int_0^z \frac{t^{\frac{1}{l}}}{1+t} dt, \\ & z \in U. \end{aligned}$$

□

Theorem 3.89 (Alb Lupaş, 2022b). *Let g be a convex function such that $g(0) = 0$ and let h be the function $h(z) = g(z) + \frac{l}{1-\lambda} zg'(z)$, $z \in U$, $m, n \in \mathbb{N}$, $\lambda > 0$. If $f \in \mathcal{A}_l$ and the differential subordination*

$$\frac{\lambda}{1-\lambda} \frac{D_z^{-\lambda} SR^{m+1,n} f(z)}{z} + \frac{1}{1-\lambda} \frac{D_z^{-\lambda} SR^{m+2,n} f(z)}{z} \prec h(z), \quad z \in U, \tag{3.2.2.12}$$

holds, then

$$\left(D_z^{-\lambda} SR^{m,n} f(z)\right)' \prec g(z), \quad z \in U,$$

and this result is sharp.

Proof. With notation

$$\begin{aligned} p(z) &= \left(D_z^{-\lambda} SR^{m,n} f(z)\right)' \\ &= \frac{1}{\Gamma(\lambda+1)} z^{\lambda} + \sum_{j=l+1}^{\infty} \frac{j^{m+1}\Gamma(n+j)}{\Gamma(n+1)\Gamma(j+\lambda)} a_j^2 z^{j+\lambda-1} \end{aligned}$$

and $p(0) = 0$, we obtain for $f(z) = z + \sum_{j=l+1}^{\infty} a_j z^j$, taking account relation (3.2.2.1),

$$p(z) + zp'(z) = \lambda p(z) + \frac{\lambda D_z^{-\lambda} SR^{m+1,n} f(z)}{z} + \frac{D_z^{-\lambda} SR^{m+2,n} f(z)}{z}.$$

We have

$$p(z) + \frac{1}{1-\lambda} zp'(z) \prec h(z) = g(z) + \frac{l}{1-\lambda} zg'(z), \quad z \in U.$$

By using Lemma 1.2, we obtain

$$p(z) \prec g(z), \quad z \in U, \quad \text{i.e.} \ \left(D_z^{-\lambda} SR^{m,n} f(z)\right)' \prec g(z), \quad z \in U,$$

and this result is sharp. □

Theorem 3.90 (Alb Lupaş, 2022b). *Let $h \in \mathcal{H}(U)$ with $h(0) = 0$, which verifies the inequality* $\operatorname{Re}\left[1 + \frac{zh''(z)}{h'(z)}\right] > -\frac{1}{2}$, $z \in U$. *If $m, n \in \mathbb{N}$, $\lambda > 0$, $f \in \mathcal{A}_l$ and satisfies the differential subordination*

$$\frac{\lambda}{1-\lambda} \frac{D_z^{-\lambda} SR^{m+1,n} f(z)}{z} + \frac{1}{1-\lambda} \frac{D_z^{-\lambda} SR^{m+2,n} f(z)}{z} \prec h(z), \quad z \in U, \tag{3.2.2.13}$$

then

$$\left(D_z^{-\lambda} SR^{m,n} f(z)\right)' \prec q(z), \quad z \in U,$$

where q is given by $q(z) = \frac{1-\lambda}{l z^{\frac{1-\lambda}{l}}} \int_0^z h(t) t^{\frac{1-\lambda}{l} - 1} dt$. The function q is convex and it is the best dominant.

Proof. Using the properties of operator $D_z^{-\lambda} SR^{m,n}$ and considering $p(z) = \left(D_z^{-\lambda} SR^{m,n} f(z)\right)'$, we obtain

$$\frac{\lambda}{1-\lambda} \frac{D_z^{-\lambda} SR^{m+1,n} f(z)}{z} + \frac{1}{1-\lambda} \frac{D_z^{-\lambda} SR^{m+2,n} f(z)}{z} = p(z) + \frac{1}{1-\lambda} z p'(z), \quad z \in U.$$

Then (3.2.2.13) becomes

$$p(z) + \frac{1}{1-\lambda} z p'(z) \prec h(z), \quad z \in U.$$

Since $p \in \mathcal{H}[0, \lambda]$, using Lemma 1.1 for $\gamma = 1 - \lambda$, we deduce

$$p(z) \prec q(z), \quad z \in U,$$

where

$$q(z) = \frac{1-\lambda}{l z^{\frac{1-\lambda}{l}}} \int_0^z h(t) t^{\frac{1-\lambda}{l} - 1} dt, \quad z \in U,$$

i.e.

$$\left(D_z^{-\lambda} SR^{m,n} f(z)\right)' \prec q(z) = \frac{1-\lambda}{l z^{\frac{1-\lambda}{l}}} \int_0^z h(t) t^{\frac{1-\lambda}{l} - 1} dt, \quad z \in U,$$

and q is the best dominant. □

Corollary 3.20 (Alb Lupaş, 2022b). *Let* $h(z) = \frac{(2\beta-1)z}{1+z}$ *be a convex function in* U, $0 \leq \beta < 1$. *If* $m, n \in \mathbb{N}$, $\lambda > 0$, $f \in \mathcal{A}_l$ *and verifies the differential subordination*

$$\frac{\lambda}{1-\lambda}\frac{D_z^{-\lambda}SR^{m+1,n}f(z)}{z} + \frac{1}{1-\lambda}\frac{D_z^{-\lambda}SR^{m+2,n}f(z)}{z} \prec h(z), \quad z \in U, \tag{3.2.2.14}$$

then

$$\left(D_z^{-\lambda}SR^{m,n}f(z)\right)' \prec q(z), \quad z \in U,$$

where q *is given by* $q(z) = \frac{(1-\lambda)(2\beta-1)}{lz^{\frac{1-\lambda}{l}}}\int_0^z \frac{t^{\frac{1-\lambda}{l}}}{1+t}dt$, $z \in U$. *The function* q *is convex and it is the best dominant.*

Proof. Following the same steps as in the proof of Theorem 3.89 and considering $p(z) = \left(D_z^{-\lambda}SR^{m,n}f(z)\right)'$, the differential subordination (3.2.2.14) becomes

$$p(z) + \frac{1}{1-\lambda}zp'(z) \prec h(z) = \frac{(2\beta-1)z}{1+z}, \quad z \in U.$$

By using Lemma 1.1 for $\gamma = 1-\lambda$, we have $p(z) \prec q(z)$, i.e.

$$\begin{aligned}\left(D_z^{-\lambda}SR^{m,n}f(z)\right)' \prec q(z) &= \frac{1-\lambda}{lz^{\frac{1-\lambda}{l}}}\int_0^z h(t)t^{\frac{1-\lambda}{l}-1}dt \\ &= \frac{1-\lambda}{lz^{\frac{1-\lambda}{l}}}\int_0^z t^{\frac{1-\lambda}{l}-1}\frac{(2\beta-1)t}{1+t}dt \\ &= \frac{(1-\lambda)(2\beta-1)}{lz^{\frac{1-\lambda}{l}}}\int_0^z \frac{t^{\frac{1-\lambda}{l}}}{1+t}dt, \\ & z \in U.\end{aligned}$$

□

3.2.3 *On Special Differential Subordinations Using Fractional Integral of Sălăgean and Ruscheweyh Operators*

In the paper Alb Lupaş and Oros (2021a), a new operator denoted by $D_z^{-\lambda}L_\alpha^n$ is defined using the fractional integral of Sălăgean and Ruscheweyh operators. By means of the newly obtained operator, the subclass $S_n(\delta, \alpha, \lambda)$ of analytic functions in the unit disc is introduced and various properties and characteristics of this subclass are derived by applying techniques specific to the differential subordination concept. By studying the operator $D_z^{-\lambda}L_\alpha^n$, some interesting differential subordinations are also given.

Let $\alpha \geq 0$, $\lambda > 0$ and $n \in \mathbb{N}$. The fractional integral of the operator $L_\alpha^n f$ is introduced in Definition 2.15 by

$$\begin{aligned} D_z^{-\lambda}L_\alpha^n f(z) &= \frac{1}{\Gamma(\lambda)}\int_0^z \frac{L_\alpha^n f(t)}{(z-t)^{1-\lambda}}dt \\ &= \frac{1}{\Gamma(\lambda)}\int_0^z \frac{t}{(z-t)^{1-\lambda}}dt \\ &\quad + \sum_{j=2}^{\infty}\left(\alpha j^n + (1-\alpha)\frac{\Gamma(n+j)}{\Gamma(n+1)\Gamma(j)}\right)a_j\int_0^z \frac{t^j}{(z-t)^{1-\lambda}}dt. \end{aligned}$$

The following form can be obtained easily:

$$\begin{aligned} D_z^{-\lambda}L_\alpha^n f(z) &= \frac{1}{\Gamma(\lambda+2)}z^{\lambda+1} \\ &\quad + \sum_{j=2}^{\infty}\left[\frac{\alpha j^m\Gamma(j+1)}{\Gamma(j+\lambda+1)} + \frac{(1-\alpha)j\Gamma(m+j)}{\Gamma(m+1)\Gamma(j+\lambda+1)}\right]a_j z^{j+\lambda}, \end{aligned}$$

considering $f(z) = z + \sum_{j=2}^{\infty} a_j z^j \in \mathcal{A}$. We note that $D_z^{-\lambda}L_\alpha^n f(z) \in \mathcal{H}[0, \lambda+1]$.

Using this operator, the subclass of analytic functions studied in Alb Lupaş and Oros (2021a) is defined as follows:

Definition 3.14 (Alb Lupaş and Oros, 2021a). The subclass $S_n(\delta, \alpha, \lambda)$ consists of functions $f \in \mathcal{A}$ which satisfy the inequality

$$\operatorname{Re}\left(D_z^{-\lambda}L_\alpha^n f(z)\right)' > \delta, \quad z \in U, \tag{3.2.3.1}$$

when $\delta \in [0,1)$, $\alpha \geq 0$, $\lambda > 0$ and n is a natural number.

We start the study of the subclass by proving that the set of functions contained in it is convex.

Theorem 3.91 (Alb Lupaş and Oros, 2021a). *The set $S_n(\delta, \alpha, \lambda)$ is convex.*

Proof. Consider the functions

$$f_j(z) = z + \sum_{j=2}^{\infty} a_{jk}z^j, \quad k = 1, 2, \ z \in U,$$

from the subclass $S_n(\delta, \alpha, \lambda)$. We have to show that the function

$$h(z) = \eta_1 f_1(z) + \eta_2 f_2(z)$$

belongs to the subclass $S_n(\delta, \alpha, \lambda)$ with $\eta_1, \eta_2 \geq 0$, $\eta_1 + \eta_2 = 1$.

We have $h(z) = z + \sum_{j=2}^{\infty} (\eta_1 a_{j1} + \eta_2 a_{j2}) z^j$, $z \in U$, so

$$\begin{aligned} D_z^{-\lambda} L_\alpha^n h(z) &= \frac{1}{\Gamma(\lambda+2)} z^{\lambda+1} \\ &\quad + \sum_{j=2}^{\infty} \left[\frac{\alpha j^m \Gamma(j+1)}{\Gamma(j+\lambda+1)} + \frac{(1-\alpha) j \Gamma(m+j)}{\Gamma(m+1)\Gamma(j+\lambda+1)} \right] \\ &\quad \times (\eta_1 a_{j1} + \eta_2 a_{j2}) z^{j+\lambda}, \quad z \in U, \end{aligned}$$

and differentiating this relation we get $\left(D_z^{-\lambda} L_\alpha^n h(z)\right)' = \frac{1}{\Gamma(\lambda+2)} z^\lambda + \sum_{j=2}^{\infty} \left[\frac{\alpha j^m \Gamma(j+1)}{\Gamma(j+\lambda+1)} + \frac{(1-\alpha) j \Gamma(m+j)}{\Gamma(m+1)\Gamma(j+\lambda+1)} \right] (\eta_1 a_{j1} + \eta_2 a_{j2}) (j+\lambda) z^{j+\lambda-1}$, $z \in U$, and

$$\begin{aligned} \operatorname{Re}\left(D_z^{-\lambda} L_\alpha^n h(z)\right)' &= \operatorname{Re}\left(\frac{1}{\Gamma(\lambda+1)} z^\lambda\right) \\ &\quad + \operatorname{Re}\left(\eta_1 \sum_{j=2}^{\infty} (j+\lambda) \left[\frac{\alpha j^m \Gamma(j+1)}{\Gamma(j+\lambda+1)} + \frac{(1-\alpha) j \Gamma(m+j)}{\Gamma(m+1)\Gamma(j+\lambda+1)} \right] a_{j1} z^{j+\lambda-1}\right) \\ &\quad + \operatorname{Re}\left(\eta_2 \sum_{j=2}^{\infty} (j+\lambda) \left[\frac{\alpha j^m \Gamma(j+1)}{\Gamma(j+\lambda+1)} + \frac{(1-\alpha) j \Gamma(m+j)}{\Gamma(m+1)\Gamma(j+\lambda+1)} \right] a_{j2} z^{j+\lambda-1}\right). \end{aligned}$$

Since $f_1, f_2 \in S_n(\delta, \alpha, \lambda)$, we have

$$\begin{aligned} &\operatorname{Re}\left(\eta_k \sum_{j=2}^{\infty} j \left[\alpha j^n + (1-\alpha) C_{n+j-1}^n\right] a_{jk} z^{j-1}\right) \\ &\quad > \eta_k \left(\delta - \operatorname{Re}\left(\frac{1}{\Gamma(\lambda+1)} z^\lambda\right)\right), \end{aligned}$$

$k = 1, 2$, which implies

$$\begin{aligned} \operatorname{Re}\left(D_z^{-\lambda} L_a^n h(z)\right)' &> \operatorname{Re}\left(\frac{1}{\Gamma(\lambda+1)} z^\lambda\right) + \eta_1 \left(\delta - \operatorname{Re}\left(\frac{1}{\Gamma(\lambda+1)} z^\lambda\right)\right) \\ &\quad + \eta_2 \left(\delta - \operatorname{Re}\left(\frac{1}{\Gamma(\lambda+1)} z^\lambda\right)\right) = \delta, \quad z \in U, \end{aligned}$$

equivalently with $S_n(\delta, \alpha, \lambda)$ is convex. □

Next, several interesting differential subordinations are proved involving the operator $D_z^{-\lambda} L_a^n f$.

Theorem 3.92 (Alb Lupaş and Oros, 2021a). *Taking g a convex function in U and defining $h(z) = g(z) + \frac{1}{c+2} z g'(z)$, with $c > 0$, $z \in U$, if $f \in S_n(\delta, \alpha, \lambda)$ and $F(z) = I_c(f)(z) = \frac{c+2}{z^{c+1}} \int_0^z t^c f(t)\, dt$, $z \in U$, the differential subordination*

$$\left(D_z^{-\lambda} L_a^n f(z)\right)' \prec h(z), \quad z \in U, \tag{3.2.3.2}$$

implies

$$\left(D_z^{-\lambda} L_\alpha^n F(z)\right)' \prec g(z), \quad z \in U,$$

and this result is sharp.

Proof. From the definition of F, we have $z^{c+1} F(z) = (c+2) \int_0^z t^c f(t)\, dt$ and differentiating it, we obtain $(c+1) F(z) + z F'(z) = (c+2) f(z)$. By applying the operator $D_z^{-\lambda}$, that is equivalent to

$$(c+1) D_z^{-\lambda} L_\alpha^n F(z) + z \left(D_z^{-\lambda} L_\alpha^n F(z)\right)' = (c+2) D_z^{-\lambda} L_\alpha^n f(z), \quad z \in U.$$

Differentiating the last relation, the following equality is obtained

$$\left(D_z^{-\lambda} L_\alpha^n F(z)\right)' + \frac{1}{c+2} z \left(D_z^{-\lambda} L_\alpha^n F(z)\right)'' = \left(D_z^{-\lambda} L_\alpha^n f(z)\right)', \quad z \in U,$$

which transforms relation (3.2.3.2) into

$$\left(D_z^{-\lambda} L_\alpha^n F(z)\right)' + \frac{1}{c+2} z \left(D_z^{-\lambda} L_\alpha^n F(z)\right)'' \prec g(z) + \frac{1}{c+2} z g'(z).$$

By using the notation

$$p(z) = \left(D_z^{-\lambda} L_\alpha^n F(z)\right)', \quad z \in U,$$

we get that $p \in \mathcal{H}[0, \lambda]$ and previous differential subordination becomes

$$p(z) + \frac{1}{c+2} z p'(z) \prec g(z) + \frac{1}{c+2} z g'(z), \quad z \in U.$$

Applying Lemma 1.2, we obtain

$$p(z) \prec g(z), \quad z \in U, \quad \text{i.e. } \left(D_z^{-\lambda} L_\alpha^n F(z)\right)' \prec g(z), \quad z \in U,$$

and g is the best dominant. □

An interesting inclusion result is proved for the subclass $S_n(\delta, \alpha, \lambda)$ in the next theorem.

Theorem 3.93 (Alb Lupaş and Oros, 2021a). *Let* $h(z) = \frac{1+(2\delta-1)z}{1+z}$, $\delta \in [0,1)$ *and* $c > 0$. *If* $\alpha \geq 0$, $\lambda > 0$, $n \in \mathbb{N}$ *and* $I_c(f)(z) = \frac{c+2}{z^{c+1}} \int_0^z t^c f(t)\, dt$, $z \in U$, *then*

$$I_c[S_n(\delta, \alpha, \lambda)] \subset S_n(\delta^*, \alpha, \lambda), \tag{3.2.3.3}$$

where $\delta^* = 2\delta - 1 + (c+2)(2-2\delta) \int_0^1 \frac{t^{c+1}}{t+1} dt$.

Proof. Using the same reasoning as in the proof of Theorem 3.92, since h is convex, we obtain that

$$p(z) + \frac{1}{c+2} zp'(z) \prec h(z),$$

with $p(z) = \left(D_z^{-\lambda} L_\alpha^n F(z)\right)'$.

Applying Lemma 1.1, we get

$$p(z) \prec g(z) \prec h(z), \quad \text{i.e. } \left(D_z^{-\lambda} L_\alpha^n F(z)\right)' \prec g(z) \prec h(z),$$

where

$$\begin{aligned} g(z) &= \frac{c+2}{z^{c+2}} \int_0^z t^{c+1} \frac{1+(2\delta-1)t}{1+t} dt \\ &= 2\delta - 1 + \frac{(c+2)(2-2\delta)}{z^{c+2}} \int_0^z \frac{t^{c+1}}{t+1} dt. \end{aligned}$$

From convexity of g and using that $g(U)$ is symmetric with respect to the real axis, we obtain

$$\begin{aligned} \operatorname{Re}\left(D_z^{-\lambda} L_\alpha^n F(z)\right)' &\geq \min_{|z|=1} \operatorname{Re} g(z) = \operatorname{Re} g(1) \\ &= \delta^* = 2\delta - 1 + (c+2)(2-2\delta) \int_0^1 \frac{t^{c+1}}{t+1} dt. \end{aligned}$$ □

Theorem 3.94 (Alb Lupaş and Oros, 2021a). *Let* $h(z) = g(z) + zg'(z)$, $z \in U$, *when* g *is a convex function in* U *with* $g(0) = 0$. *If a function* $f \in \mathcal{A}$ *satisfies*

$$\left(D_z^{-\lambda} L_\alpha^n f(z)\right)' \prec h(z), \quad z \in U, \tag{3.2.3.4}$$

then

$$\frac{D_z^{-\lambda}L_\alpha^n f(z)}{z} \prec g(z), \quad z \in U,$$

for $\alpha \geq 0$, $\lambda > 0$, $n \in \mathbb{N}$ and this result is sharp.

Proof. Denote by $p(z) = \frac{D_z^{-\lambda}L_\alpha^n f(z)}{z} \in \mathcal{H}[0,\lambda]$ and we obtain $zp(z) = D_z^{-\lambda}L_\alpha^n f(z)$, $z \in U$. After differentiating it, we get $p(z) + zp'(z) = \left(D_z^{-\lambda}L_\alpha^n f(z)\right)'$, $z \in U$, and relation (3.2.3.4) becomes

$$p(z) + zp'(z) \prec h(z) = g(z) + zg'(z), \quad z \in U.$$

Applying Lemma 1.2, we obtain

$$p(z) \prec g(z), \quad z \in U, \quad \text{i.e.} \ \frac{D_z^{-\lambda}L_\alpha^n f(z)}{z} \prec g(z), \quad z \in U. \qquad \square$$

Theorem 3.95 (Alb Lupaş and Oros, 2021a). *Consider h be a convex function of order $-\frac{1}{2}$ with $h(0) = 0$. If a function $f \in \mathcal{A}$ satisfies*

$$\left(D_z^{-\lambda}L_\alpha^n f(z)\right)' \prec h(z), \quad z \in U, \tag{3.2.3.5}$$

then

$$\frac{D_z^{-\lambda}L_\alpha^n f(z)}{z} \prec q(z), \quad z \in U,$$

for $\alpha \geq 0$, $\lambda > 0$, $n \in \mathbb{N}$ and $q(z) = \frac{1}{z}\int_0^z h(t)dt$ is convex and is the best dominant.

Proof. Let $p(z) = \frac{D_z^{-\lambda}L_\alpha^n f(z)}{z} \in \mathcal{H}[0,\lambda]$, $z \in U$. Differentiating it, we get $\left(D_z^{-\lambda}L_\alpha^n f(z)\right)' = p(z) + zp'(z)$, $z \in U$ and relation (3.2.3.5) can be written

$$p(z) + zp'(z) \prec h(z), \quad z \in U.$$

Applying Lemma 1.1, we obtain

$$p(z) \prec q(z), \quad z \in U, \quad \text{i.e.} \ \frac{D_z^{-\lambda}L_\alpha^n f(z)}{z} \prec q(z) = \frac{1}{z}\int_0^z h(t)dt, \quad z \in U,$$

and q is the best dominant. $\square$

Corollary 3.21 (Alb Lupaş and Oros, 2021a). *Let $h(z) = \frac{2(\beta-1)z}{1+z}$ a convex function in U, where $0 \leq \beta < 1$. If $f \in \mathcal{A}$ and satisfies the differential subordination*

$$\left(D_z^{-\lambda} L_\alpha^n f(z)\right)' \prec h(z), \quad z \in U, \tag{3.2.3.6}$$

then

$$\frac{D_z^{-\lambda} L_\alpha^n f(z)}{z} \prec q(z), \quad z \in U,$$

where $q(z) = 2(\beta-1) + 2(1-\beta)\frac{\ln(1+z)}{z}$, $z \in U$. The function q is convex and it is the best dominant.

Proof. Using the same reasoning as in the proof of Theorem 3.95 for $p(z) = \frac{D_z^{-\lambda} L_\alpha^n f(z)}{z}$, the differential subordination (3.2.3.6) becomes

$$p(z) + zp'(z) \prec h(z) = \frac{2(\beta-1)z}{1+z}, \quad z \in U.$$

Applying Lemma 1.1, we get $p(z) \prec q(z)$, i.e.

$$\begin{aligned}\frac{D_z^{-\lambda} L_\alpha^n f(z)}{z} \prec q(z) &= \frac{1}{z}\int_0^z h(t)dt \\ &= \frac{1}{z}\int_0^z \frac{2(\beta-1)t}{1+t}dt = 2(\beta-1) + 2(1-\beta)\frac{1}{z}\ln(z+1),\end{aligned}$$

$z \in U$. □

Example 3.5 (Alb Lupaş and Oros, 2021a). Let $h(z) = \frac{-2z}{1+z}$ with $h(0) = 0$, $h'(z) = \frac{-2}{(1+z)^2}$ and $h''(z) = \frac{4}{(1+z)^3}$.

Since $\operatorname{Re}\left(\frac{zh''(z)}{h'(z)} + 1\right) = \operatorname{Re}\left(\frac{1-z}{1+z}\right) = \operatorname{Re}\left(\frac{1-\rho\cos\theta - i\rho\sin\theta}{1+\rho\cos\theta + i\rho\sin\theta}\right) = \frac{1-\rho^2}{1+2\rho\cos\theta+\rho^2} > 0 > -\frac{1}{2}$, the function h is convex in U.

Let $f(z) = z + z^2$, $z \in U$. For $\alpha = 2$, $n = 1$, we have $L_2^1 f(z) = zf'(z) = z + 2z^2$ and $D_z^{-\lambda} L_2^1 f(z) = \frac{1}{\Gamma(\lambda)}\int_0^z \frac{L_2^1 f(t)}{(z-t)^{1-\lambda}}dt = \frac{1}{\Gamma(\lambda)}\int_0^z \frac{t+2t^2}{(z-t)^{1-\lambda}}dt$.

After a simple computation, we get $D_z^{-\lambda} L_2^1 f(z) = \frac{1}{\Gamma(\lambda+2)}z^{\lambda+1} + \frac{4}{\Gamma(\lambda+3)}z^{\lambda+2}$ and $\left(D_z^{-\lambda} L_2^1 f(z)\right)' = \frac{1}{\Gamma(\lambda+1)}z^{\lambda} + \frac{4}{\Gamma(\lambda+2)}z^{\lambda+1}$.

Also, we obtain $q(z) = \frac{1}{z}\int_0^z \frac{-2t}{1+t}dt = -2 + \frac{2\ln(1+z)}{z}$.

Using Theorem 3.95, we obtain

$$\frac{1}{\Gamma(\lambda+1)}z^{\lambda}+\frac{4}{\Gamma(\lambda+2)}z^{\lambda+1}\prec\frac{2z}{1+z},\quad z\in U,$$

induces

$$\frac{1}{\Gamma(\lambda+2)}z^{\lambda+1}+\frac{4}{\Gamma(\lambda+3)}z^{\lambda+2}\prec-2+\frac{2\ln(1+z)}{z},\quad z\in U.$$

Theorem 3.96 (Alb Lupaş and Oros, 2021a). *Let $h(z)=g(z)+zg'(z)$, $z\in U$, when g is a convex function in U with $g(0)=0$. If a function $f\in\mathcal{A}$ satisfies*

$$\left(\frac{zD_z^{-\lambda}L_\alpha^{n+1}f(z)}{D_z^{-\lambda}L_\alpha^n f(z)}\right)'\prec h(z),\quad z\in U,\tag{3.2.3.7}$$

then

$$\frac{D_z^{-\lambda}L_\alpha^{n+1}f(z)}{D_z^{-\lambda}L_\alpha^n f(z)}\prec g(z),\quad z\in U,$$

for $\alpha\ge0$, $\lambda>0$, $n\in\mathbb{N}$ and this result is sharp.

Proof. Let $p(z)=\frac{D_z^{-\lambda}L_\alpha^{n+1}f(z)}{D_z^{-\lambda}L_\alpha^n f(z)}$ and we get $p(z)+zp'(z)=\left(\frac{zL_\alpha^{n+1}f(z)}{L_\alpha^n f(z)}\right)'$. With this notation, relation (3.2.3.7) can be written as

$$p(z)+zp'(z)\prec h(z)=g(z)+zg'(z),\quad z\in U,$$

and applying Lemma 1.2, we obtain

$$p(z)\prec g(z),\quad z\in U,\quad \text{i.e. } \frac{D_z^{-\lambda}L_\alpha^{n+1}f(z)}{D_z^{-\lambda}L_\alpha^n f(z)}\prec g(z),\quad z\in U. \qquad \square$$

Theorem 3.97 (Alb Lupaş and Oros, 2021a). *Consider a convex function g with $g(0)=0$ and $h(z)=g(z)+\lambda zg'(z)$, $z\in U$, $\delta,\lambda>0$, $\alpha\ge0$. If $f\in\mathcal{A}$ and satisfies the differential subordination*

$$\left(\frac{D_z^{-\lambda}L_\alpha^n f(z)}{z}\right)^{\delta-1}\left(D_z^{-\lambda}L_\alpha^n f(z)\right)'\prec h(z),\quad z\in U,\tag{3.2.3.8}$$

then

$$\left(\frac{D_z^{-\lambda}L_\alpha^n f(z)}{z}\right)^{\delta}\prec g(z),\quad z\in U,$$

and this result is sharp.

Proof. Let $p(z) = \left(\frac{D_z^{-\lambda} L_\alpha^n f(z)}{z}\right)^\delta \in \mathcal{H}[0, \lambda\delta]$, $z \in U$. Differentiating it, we have

$$zp'(z) = \delta \left(\frac{D_z^{-\lambda} L_\alpha^n f(z)}{z}\right)^{\delta-1} \left(D_z^{-\lambda} L_\alpha^n f(z)\right)' - \delta \left(\frac{D_z^{-\lambda} L_\alpha^n f(z)}{z}\right)^\delta$$
$$= \delta \left(\frac{D_z^{-\lambda} L_\alpha^n f(z)}{z}\right)^{\delta-1} \left(D_z^{-\lambda} L_\alpha^n f(z)\right)' - \delta p(z),$$

so $p(z) + \frac{1}{\delta} zp'(z) = \left(\frac{D_z^{-\lambda} L_\alpha^n f(z)}{z}\right)^{\delta-1} \left(D_z^{-\lambda} L_\alpha^n f(z)\right)'$, $z \in U$.

Relation (3.2.3.8) can be written

$$p(z) + \frac{1}{\delta} zp'(z) \prec h(z) = g(z) + \lambda z g'(z), \quad z \in U.$$

Applying Lemma 1.2 for $\alpha = \frac{1}{\delta}$ and $n = \lambda\delta$, we have

$$p(z) \prec g(z), \quad z \in U, \quad \text{i.e.} \left(\frac{D_z^{-\lambda} L_\alpha^n f(z)}{z}\right)^\delta \prec g(z), \quad z \in U. \qquad \square$$

Theorem 3.98 (Alb Lupaş and Oros, 2021a). *Consider h the convex function of order $-\frac{1}{2}$ with $h(0) = 0$. If a function $f \in \mathcal{A}$ satisfies*

$$\left(\frac{D_z^{-\lambda} L_\alpha^n f(z)}{z}\right)^{\delta-1} \left(D_z^{-\lambda} L_\alpha^n f(z)\right)' \prec h(z), \quad z \in U, \tag{3.2.3.9}$$

then

$$\left(\frac{D_z^{-\lambda} L_\alpha^n f(z)}{z}\right)^\delta \prec q(z), \quad z \in U,$$

for $\alpha \geq 0$, $\delta, \lambda > 0$, $n \in \mathbb{N}$ and $q(z) = \frac{1}{z}\int_0^z h(t)dt$ is convex and is the best dominant.

Proof. Consider $p(z) = \left(\frac{D_z^{-\lambda} L_\alpha^n f(z)}{z}\right)^\delta \in \mathcal{H}[0, \lambda\delta]$, $z \in U$. After a short computation, we get

$$p(z) + \frac{1}{\delta} zp'(z) = \left(\frac{D_z^{-\lambda} L_\alpha^n (z)}{z}\right)^{\delta-1} \left(D_z^{-\lambda} L_\alpha^n f(z)\right)', \quad z \in U,$$

and relation (3.2.3.9) becomes

$$p(z) + \frac{1}{\delta} zp'(z) \prec h(z), \quad z \in U.$$

Applying Lemma 1.1, we get

$$p(z) \prec q(z), \quad \text{i.e.} \left(\frac{D_z^{-\lambda} L_\alpha^n f(z)}{z} \right)^\delta \prec q(z) = \frac{\delta}{z^\delta} \int_0^z h(t) t^{\delta - 1} dt, \quad z \in U,$$

and q is the best dominant. □

Theorem 3.99 (Alb Lupaş and Oros, 2021a). *Let $h(z) = g(z) + zg'(z)$, $z \in U$, where g is a convex function in U with $g(0) = \frac{1}{\lambda+1}$. If a function $f \in \mathcal{A}$ satisfies*

$$1 - \frac{D_z^{-\lambda} L_\alpha^n f(z) \left(D_z^{-\lambda} L_\alpha^n f(z) \right)''}{\left[\left(D_z^{-\lambda} L_\alpha^n f(z) \right)' \right]^2} \prec h(z), \quad z \in U, \tag{3.2.3.10}$$

then

$$\frac{D_z^{-\lambda} L_\alpha^n f(z)}{z \left(D_z^{-\lambda} L_\alpha^n f(z) \right)'} \prec g(z), \quad z \in U,$$

for $\alpha \geq 0$, $\lambda > 0$, $n \in \mathbb{N}$ and this result is sharp.

Proof. Consider $p(z) = \frac{D_z^{-\lambda} L_\alpha^n f(z)}{z \left(D_z^{-\lambda} L_\alpha^n f(z) \right)'}$ and differentiating it we get

$$p(z) + zp'(z) = 1 - \frac{D_z^{-\lambda} L_\alpha^n f(z) \left(D_z^{-\lambda} L_\alpha^n f(z) \right)''}{\left[\left(D_z^{-\lambda} L_\alpha^n f(z) \right)' \right]^2}, \quad z \in U.$$

Using this notation, the differential subordination becomes

$$p(z) + zp'(z) \prec h(z) = g(z) + zg'(z),$$

and applying Lemma 1.2, we get

$$p(z) \prec g(z), \quad z \in U, \quad \text{i.e.} \frac{D_z^{-\lambda} L_\alpha^n f(z)}{z \left(D_z^{-\lambda} L_\alpha^n f(z) \right)'} \prec g(z), \quad z \in U,$$

and this result is sharp. □

3.2.4 Certain Special Differential Superordinations Using Multiplier Transformation and Ruscheweyh Derivative

In the paper Alb Lupaş (2011e), we established several differential superordinations regarding the new operator $RI^{\alpha}_{m,\lambda,l}$ defined in Alb Lupaş (2010d) by using the multiplier transformation $I(m,\lambda,l)f(z)$ and Ruscheweyh derivative $R^m f(z)$, as in Definition 2.7, $RI^{\alpha}_{m,\lambda,l}: \mathcal{A}_n \to \mathcal{A}_n$,

$$RI^{\alpha}_{m,\lambda,l}f(z) = (1-\alpha)R^m f(z) + \alpha I(m,\lambda,l)f(z), \quad z \in U,$$

where $m,n \in \mathbb{N}$, $\lambda,\alpha,l \geq 0$ and $f \in \mathcal{A}_n = \{f \in \mathcal{H}(U) : f(z) = z + \sum_{j=n+1}^{\infty} a_j z^j,\ z \in U\}$. A number of interesting consequences of some of these superordination results are discussed.

If $f \in \mathcal{A}_n$, $f(z) = z + \sum_{j=n+1}^{\infty} a_j z^j$, then

$$RI^{\alpha}_{m,\lambda,l}f(z) = z + \sum_{j=n+1}^{\infty}\left\{\alpha\left(\frac{1+\lambda(j-1)+l}{l+1}\right)^m + (1-\alpha)C^m_{m+j-1}\right\}a_j z^j,$$

for $z \in U$.

Theorem 3.100 (Alb Lupaş, 2011e). *Let h be a convex function in U with $h(0)=1$. Let $m \in \mathbb{N}$, $\lambda,\alpha,l \geq 0$, $f \in \mathcal{A}_n$, $F(z) = I_c(f)(z) = \frac{c+2}{z^{c+1}}\int_0^z t^c f(t)\,dt$, $z \in U$, $\operatorname{Re} c > -2$, and suppose that $\left(RI^{\alpha}_{m,\lambda,l}f(z)\right)'$ is univalent in U, $\left(RI^{\alpha}_{m,\lambda,l}F(z)\right)' \in \mathcal{H}[1,n] \cap Q$ and*

$$h(z) \prec \left(RI^{\alpha}_{m,\lambda,l}f(z)\right)', \quad z \in U. \tag{3.2.4.1}$$

Then

$$q(z) \prec \left(RI^{\alpha}_{m,\lambda,l}F(z)\right)', \quad z \in U,$$

where $q(z) = \frac{c+2}{nz^{\frac{c+2}{n}}}\int_0^z h(t)t^{\frac{c+2}{n}-1}dt$. The function q is convex and it is the best subordinant.

Proof. We have $z^{c+1}F(z) = (c+2)\int_0^z t^c f(t)\,dt$ and differentiating it, with respect to z, we obtain $(c+1)F(z) + zF'(z) = (c+2)f(z)$ and $(c+1)RI^{\alpha}_{m,\lambda,l}F(z) + z\left(RI^{\alpha}_{m,\lambda,l}F(z)\right)' = (c+2)RI^{\alpha}_{m,\lambda,l}f(z)$, $z \in U$.

Differentiating the last relation, we have

$$\left(RI_{m,\lambda,l}^{\alpha}F(z)\right)' + \frac{1}{c+2}z\left(RI_{m,\lambda,l}^{\alpha}F(z)\right)'' = \left(RI_{m,\lambda,l}^{\alpha}f(z)\right)', \quad z \in U. \tag{3.2.4.2}$$

Using (3.2.4.2), the differential superordination (3.2.4.1) becomes

$$h(z) \prec \left(RI_{m,\lambda,l}^{\alpha}F(z)\right)' + \frac{1}{c+2}z\left(RI_{m,\lambda,l}^{\alpha}F(z)\right)''. \tag{3.2.4.3}$$

Denote

$$p(z) = \left(RI_{m,\lambda,l}^{\alpha}F(z)\right)', \quad z \in U. \tag{3.2.4.4}$$

Replacing (3.2.4.4) in (3.2.4.3), we obtain $h(z) \prec p(z) + \frac{1}{c+2}zp'(z)$, $z \in U$. Using Lemma 1.4 for $\gamma = c+2$, we have $q(z) \prec p(z)$, $z \in U$, i.e. $q(z) \prec \left(RI_{m,\lambda,l}^{\alpha}F(z)\right)'$, $z \in U$, where $q(z) = \frac{c+2}{nz^{\frac{c+2}{n}}}\int_0^z h(t)t^{\frac{c+2}{n}-1}dt$. The function q is convex and it is the best subordinant. □

Theorem 3.101 (Alb Lupaş, 2011e). *Let q be a convex function in U and let $h(z) = q(z) + \frac{1}{c+2}zq'(z)$, where $z \in U$, $\operatorname{Re} c > -2$. Let $m \in \mathbb{N}$, $\lambda, \alpha, l \geq 0$, $f \in \mathcal{A}_n$, $F(z) = I_c(f)(z) = \frac{c+2}{z^{c+1}}\int_0^z t^c f(t)\,dt$, $z \in U$ and suppose that $\left(RI_{m,\lambda,l}^{\alpha}f(z)\right)'$ is univalent in U, $\left(RI_{m,\lambda,l}^{\alpha}F(z)\right)' \in \mathcal{H}[1,n] \cap Q$ and*

$$h(z) \prec \left(RI_{m,\lambda,l}^{\alpha}f(z)\right)', \quad z \in U, \tag{3.2.4.5}$$

then

$$q(z) \prec \left(RI_{m,\lambda,l}^{\alpha}F(z)\right)', \quad z \in U,$$

where $q(z) = \frac{c+2}{nz^{\frac{c+2}{n}}}\int_0^z h(t)t^{\frac{c+2}{n}-1}dt$. The function q is the best subordinant.

Proof. Following the same steps as in the proof of Theorem 3.100 and considering $p(z) = \left(RI_{m,\lambda,l}^{\alpha}F(z)\right)'$, $z \in U$, the differential superordination (3.2.4.5) becomes $h(z) = q(z) + \frac{1}{c+2}zq'(z) \prec p(z) + \frac{1}{c+2}zp'(z)$, $z \in U$. Using Lemma 1.5 for $\gamma = c+2$, we have $q(z) \prec p(z)$, $z \in U$, i.e. $q(z) \prec \left(RI_{m,\lambda,l}^{\alpha}F(z)\right)'$, $z \in U$, where $q(z) = \frac{c+2}{nz^{\frac{c+2}{n}}}\int_0^z h(t)t^{\frac{c+2}{n}-1}dt$. The function q is the best subordinant. □

Theorem 3.102 (Alb Lupaş, 2011e). *Let* $h(z) = \frac{1+(2\beta-1)z}{1+z}$, *where* $\beta \in [0,1)$. *Let* $m \in \mathbb{N}$, $\lambda, \alpha, l \geq 0$, $f \in \mathcal{A}_n$, $F(z) = I_c(f)(z) = \frac{c+2}{z^{c+1}} \int_0^z t^c f(t)\, dt$, $z \in U$, Re $c > -2$, *and suppose that* $\left(RI^{\alpha}_{m,\lambda,l} f(z)\right)'$ *is univalent in* U, $\left(RI^{\alpha}_{m,\lambda,l} F(z)\right)' \in \mathcal{H}[1,n] \cap Q$ *and*

$$h(z) \prec \left(RI^{\alpha}_{m,\lambda,l} f(z)\right)', \quad z \in U, \tag{3.2.4.6}$$

then

$$q(z) \prec \left(RI^{\alpha}_{m,\lambda,l} F(z)\right)', \quad z \in U,$$

where q *is given by* $q(z) = 2\beta - 1 + \frac{(c+2)(2-2\beta)}{nz^{\frac{c+2}{n}}} \int_0^z \frac{t^{\frac{c+2}{n}-1}}{t+1} dt$, $z \in U$. *The function* q *is convex and it is the best subordinant.*

Proof. Following the same steps as in the proof of Theorem 3.100 and considering $p(z) = \left(RI^{\alpha}_{m,\lambda,l} F(z)\right)'$, the differential superordination (3.2.4.6) becomes $h(z) = \frac{1+(2\beta-1)z}{1+z} \prec p(z) + \frac{1}{c+2} zp'(z)$, $z \in U$. By using Lemma 1.4 for $\gamma = c+2$, we have $q(z) \prec p(z)$, i.e. $q(z) = \frac{c+2}{nz^{\frac{c+2}{n}}} \int_0^z h(t) t^{\frac{c+2}{n}-1} dt = \frac{c+2}{nz^{\frac{c+2}{n}}} \int_0^z \frac{1+(2\beta-1)t}{1+t} t^{\frac{c+2}{n}-1} dt = 2\beta - 1 + \frac{(c+2)(2-2\beta)}{nz^{\frac{c+2}{n}}} \int_0^z \frac{t^{\frac{c+2}{n}-1}}{t+1} dt \prec \left(RI^{\alpha}_{m,\lambda,l} F(z)\right)'$, $z \in U$. The function q is convex and it is the best subordinant. □

Theorem 3.103 (Alb Lupaş, 2011e). *Let* h *be a convex function,* $h(0) = 1$. *Let* $m \in \mathbb{N}$, $\lambda, \alpha, l \geq 0$, $f \in \mathcal{A}_n$ *and suppose that* $\left(RI^{\alpha}_{m,\lambda,l} f(z)\right)'$ *is univalent and* $\frac{RI^{\alpha}_{m,\lambda,l} f(z)}{z} \in \mathcal{H}[1,n] \cap Q$. *If*

$$h(z) \prec \left(RI^{\alpha}_{m,\lambda,l} f(z)\right)', \quad z \in U, \tag{3.2.4.7}$$

then

$$q(z) \prec \frac{RI^{\alpha}_{m,\lambda,l} f(z)}{z}, \quad z \in U,$$

where $q(z) = \frac{1}{nz^{\frac{1}{n}}} \int_0^z h(t) t^{\frac{1}{n}-1} dt$. *The function* q *is convex and it is the best subordinant.*

Proof. By using the properties of operator $RI_{m,\lambda,l}^{\alpha}$, we have

$$RI_{m,\lambda,l}^{\alpha}f(z) = z + \sum_{j=n+1}^{\infty} \left\{ \alpha \left(\frac{1+\lambda(j-1)+l}{l+1} \right)^m + (1-\alpha)\, C_{m+j-1}^{m} \right\} a_j z^j,$$

$z \in U$.

Consider

$$\begin{aligned} p(z) &= \frac{RI_{m,\lambda,l}^{\alpha}f(z)}{z} = \frac{z + \sum_{j=n+1}^{\infty} \left\{ \alpha \left(\frac{1+\lambda(j-1)+l}{l+1} \right)^m + (1-\alpha)\, C_{m+j-1}^{m} \right\} a_j z^j}{z} \\ &= 1 + p_n z^n + p_{n+1} z^{n+1} + \cdots, \quad z \in U. \end{aligned}$$

We deduce that $p \in \mathcal{H}[1,n]$.

Let $RI_{m,\lambda,l}^{\alpha}f(z) = zp(z)$, $z \in U$. Differentiating, we obtain $\left(RI_{m,\lambda,l}^{\alpha}f(z)\right)' = p(z) + zp'(z)$, $z \in U$.

Then (3.2.4.7) becomes $h(z) \prec p(z)+zp'(z)$, $z \in U$. By using Lemma 1.4 for $\gamma = 1$, we have $q(z) \prec p(z)$, $z \in U$, i.e. $q(z) \prec \frac{RI_{m,\lambda,l}^{\alpha}f(z)}{z}$, $z \in U$, where $q(z) = \frac{1}{nz^{\frac{1}{n}}} \int_0^z h(t) t^{\frac{1}{n}-1} dt$. The function q is convex and it is the best subordinant. □

Theorem 3.104 (Alb Lupaş, 2011e). *Let q be convex in U and let h be defined by $h(z) = q(z) + zq'(z)$. If $m \in \mathbb{N}$, $\lambda, \alpha, l \geq 0$, $f \in \mathcal{A}_n$, suppose that $\left(RI_{m,\lambda,l}^{\alpha}f(z)\right)'$ is univalent and $\frac{RI_{m,\lambda,l}^{\alpha}f(z)}{z} \in \mathcal{H}[1,n] \cap Q$ and satisfies the differential superordination*

$$h(z) = q(z) + zq'(z) \prec \left(RI_{m,\lambda,l}^{\alpha}f(z)\right)', \quad z \in U, \tag{3.2.4.8}$$

then

$$q(z) \prec \frac{RI_{m,\lambda,l}^{\alpha}f(z)}{z}, \quad z \in U,$$

where $q(z) = \frac{1}{nz^{\frac{1}{n}}} \int_0^z h(t) t^{\frac{1}{n}-1} dt$. The function q is the best subordinant.

Proof. Following the same steps as in the proof of Theorem 3.103 and considering $p(z) = \frac{RI_{m,\lambda,l}^{\alpha}f(z)}{z}$, the differential superordination (3.2.4.8) becomes $q(z) + zq'(z) \prec p(z) + zp'(z)$, $z \in U$. Using Lemma 1.5 for $\gamma = 1$, we have $q(z) \prec p(z)$, $z \in U$, i.e. $q(z) = \frac{1}{nz^{\frac{1}{n}}} \int_0^z h(t) t^{\frac{1}{n}-1} dt \prec \frac{RI_{m,\lambda,l}^{\alpha}f(z)}{z}$, $z \in U$, and q is the best subordinant. □

Theorem 3.105 (Alb Lupaş, 2011e). *Let* $h(z) = \frac{1+(2\beta-1)z}{1+z}$ *be a convex function in* U, *where* $0 \leq \beta < 1$. *Let* $m \in \mathbb{N}$, $\lambda, \alpha, l \geq 0$, $f \in \mathcal{A}_n$ *and suppose that* $\left(RI^{\alpha}_{m,\lambda,l}f(z)\right)'$ *is univalent and* $\frac{RI^{\alpha}_{m,\lambda,l}f(z)}{z} \in \mathcal{H}[1,n] \cap Q$. *If*

$$h(z) \prec \left(RI^{\alpha}_{m,\lambda,l}f(z)\right)', \quad z \in U, \tag{3.2.4.9}$$

then

$$q(z) \prec \frac{RI^{\alpha}_{m,\lambda,l}f(z)}{z}, \quad z \in U,$$

where q *is given by* $q(z) = 2\beta - 1 + \frac{2-2\beta}{nz^{\frac{1}{n}}} \int_0^z \frac{t^{\frac{1}{n}-1}}{t+1} dt$, $z \in U$. *The function* q *is convex and it is the best subordinant.*

Proof. Following the same steps as in the proof of Theorem 3.103 and considering $p(z) = \frac{RI^{\alpha}_{m,\lambda,l}f(z)}{z}$, the differential superordination (3.2.4.9) becomes $h(z) = \frac{1+(2\beta-1)z}{1+z} \prec p(z) + zp'(z)$, $z \in U$. By using Lemma 1.4 for $\gamma = 1$, we have $q(z) \prec p(z)$, i.e. $q(z) = \frac{1}{nz^{\frac{1}{n}}} \int_0^z h(t) t^{\frac{1}{n}-1} dt = \frac{1}{nz^{\frac{1}{n}}} \int_0^z \frac{1+(2\beta-1)t}{1+t} t^{\frac{1}{n}-1} dt = 2\beta - 1 + \frac{2-2\beta}{nz^{\frac{1}{n}}} \int_0^z \frac{t^{\frac{1}{n}-1}}{t+1} dt \prec \frac{RI^{\alpha}_{m,\lambda,l}f(z)}{z}$, $z \in U$. The function q is convex and it is the best subordinant. □

Theorem 3.106 (Alb Lupaş, 2011e). *Let* h *be a convex function,* $h(0) = 1$. *Let* $m \in \mathbb{N}$, $\lambda, \alpha, l \geq 0$, $f \in \mathcal{A}_n$ *and suppose that* $\left(\frac{zRI^{\alpha}_{m+1,\lambda,l}f(z)}{RI^{\alpha}_{m,\lambda,l}f(z)}\right)'$ *is univalent and* $\frac{RI^{\alpha}_{m+1,\lambda,l}f(z)}{RI^{\alpha}_{m,\lambda,l}f(z)} \in \mathcal{H}[1,n] \cap Q$. *If*

$$h(z) \prec \left(\frac{zRI^{\alpha}_{m+1,\lambda,l}f(z)}{RI^{\alpha}_{m,\lambda,l}f(z)}\right)', \quad z \in U, \tag{3.2.4.10}$$

then

$$q(z) \prec \frac{RI^{\alpha}_{m+1,\lambda,l}f(z)}{RI^{\alpha}_{m,\lambda,l}f(z)}, \quad z \in U,$$

where $q(z) = \frac{1}{nz^{\frac{1}{n}}} \int_0^z h(t) t^{\frac{1}{n}-1} dt$. *The function* q *is convex and it is the best subordinant.*

Proof. For $f \in \mathcal{A}_n$, $f(z) = z + \sum_{j=n+1}^{\infty} a_j z^j$, we have

$$RI^{\alpha}_{m,\lambda,l}f(z) = z + \sum_{j=n+1}^{\infty} \left\{\alpha\left(\frac{1+\lambda(j-1)+l}{l+1}\right)^m + (1-\alpha) C^m_{m+j-1}\right\} a_j z^j,$$

$$z \in U.$$

Consider

$$p(z)=\frac{RI^{\alpha}_{m+1,\lambda,l}f(z)}{RI^{\alpha}_{m,\lambda,l}f\left(z\right)}$$
$$=\frac{z+\sum_{j=n+1}^{\infty}\left\{\alpha\left(\frac{1+\lambda(j-1)+l}{l+1}\right)^{m+1}+\left(1-\alpha\right)C^{m+1}_{m+j}\right\}a_jz^j}{z+\sum_{j=n+1}^{\infty}\left\{\alpha\left(\frac{1+\lambda(j-1)+l}{l+1}\right)^{m}+\left(1-\alpha\right)C^{m}_{m+j-1}\right\}a_jz^j}.$$

We have $p'\left(z\right)=\frac{\left(RI^{\alpha}_{m+1,\lambda,l}f(z)\right)'}{RI^{\alpha}_{m,\lambda,l}f(z)}-p\left(z\right)\cdot\frac{\left(RI^{\alpha}_{m,\lambda,l}f(z)\right)'}{RI^{\alpha}_{m,\lambda,l}f(z)}$ and we obtain $p\left(z\right)+z\cdot p'\left(z\right)=\left(\frac{zRI^{\alpha}_{m+1,\lambda,l}f(z)}{RI^{\alpha}_{m,\lambda,l}f(z)}\right)'$.

Relation (3.2.4.10) becomes $h(z)\prec p(z)+zp'(z)$, $z\in U$. By using Lemma 1.4 for $\gamma=1$, we have $q(z)\prec p(z)$, $z\in U$, i.e. $q(z)\prec\frac{RI^{\alpha}_{m+1,\lambda,l}f(z)}{RI^{\alpha}_{m,\lambda,l}f(z)}$, $z\in U$, where $q(z)=\frac{1}{nz^{\frac{1}{n}}}\int_0^z h(t)t^{\frac{1}{n}-1}dt$. The function q is convex and it is the best subordinant. □

Theorem 3.107 (Alb Lupaş, 2011e). *Let q be a convex function and h be defined by $h\left(z\right)=q\left(z\right)+zq'\left(z\right)$. Let $\lambda,\alpha,l\geq 0$, $m\in\mathbb{N}$, $f\in\mathcal{A}_n$ and suppose that $\left(\frac{zRI^{\alpha}_{m+1,\lambda,l}f(z)}{RI^{\alpha}_{m,\lambda,l}f(z)}\right)'$ is univalent and $\frac{RI^{\alpha}_{m+1,\lambda,l}f(z)}{RI^{\alpha}_{m,\lambda,l}f(z)}\in\mathcal{H}\left[1,n\right]\cap Q$. If*

$$h(z)=q\left(z\right)+zq'\left(z\right)\prec\left(\frac{zRI^{\alpha}_{m+1,\lambda,l}f(z)}{RI^{\alpha}_{m,\lambda,l}f\left(z\right)}\right)',\quad z\in U,\qquad(3.2.4.11)$$

then

$$q(z)\prec\frac{RI^{\alpha}_{m+1,\lambda,l}f(z)}{RI^{\alpha}_{m,\lambda,l}f\left(z\right)},\quad z\in U,$$

where $q(z)=\frac{1}{nz^{\frac{1}{n}}}\int_0^z h(t)t^{\frac{1}{n}-1}dt$. The function q is the best subordinant.

Proof. Following the same steps as in the proof of Theorem 3.106 and considering $p(z)=\frac{RI^{\alpha}_{m+1,\lambda,l}f(z)}{RI^{\alpha}_{m,\lambda,l}f(z)}$, the differential superordination (3.2.4.11) becomes $h\left(z\right)=q(z)+zq'(z)\prec p(z)+zp'(z)$, $z\in U$. By using Lemma 1.5 for $\gamma=1$, we have $q(z)\prec p(z)$, $z\in U$, i.e. $q(z)=\frac{1}{nz^{\frac{1}{n}}}\int_0^z h(t)t^{\frac{1}{n}-1}dt\prec\frac{RI^{\alpha}_{m+1,\lambda,l}f(z)}{RI^{\alpha}_{m,\lambda,l}f(z)}$, $z\in U$, and q is the best subordinant. □

Theorem 3.108 (Alb Lupaş, 2011e). *Let $h(z) = \frac{1+(2\beta-1)z}{1+z}$ be a convex function in U, where $0 \leq \beta < 1$. Let $\lambda, \alpha, l \geq 0$, $m \in \mathbb{N}$, $f \in \mathcal{A}_n$ and suppose that $\left(\frac{zRI^{\alpha}_{m+1,\lambda,l}f(z)}{RI^{\alpha}_{m,\lambda,l}f(z)}\right)'$ is univalent and $\frac{RI^{\alpha}_{m+1,\lambda,l}f(z)}{RI^{\alpha}_{m,\lambda,l}f(z)} \in \mathcal{H}[1,n] \cap Q$. If*

$$h(z) \prec \left(\frac{zRI^{\alpha}_{m+1,\lambda,l}f(z)}{RI^{\alpha}_{m,\lambda,l}f(z)}\right)', \quad z \in U, \tag{3.2.4.12}$$

then

$$q(z) \prec \frac{RI^{\alpha}_{m+1,\lambda,l}f(z)}{RI^{\alpha}_{m,\lambda,l}f(z)}, \quad z \in U,$$

where q is given by $q(z) = 2\beta - 1 + \frac{2-2\beta}{nz^{\frac{1}{n}}}\int_0^z \frac{t^{\frac{1}{n}-1}}{t+1}dt$, $z \in U$. The function q is convex and it is the best subordinant.

Proof. Following the same steps as in the proof of Theorem 3.106 and considering $p(z) = \frac{RI^{\alpha}_{m+1,\lambda,l}f(z)}{RI^{\alpha}_{m,\lambda,l}f(z)}$, the differential superordination (3.2.4.12) becomes $h(z) = \frac{1+(2\beta-1)z}{1+z} \prec p(z) + zp'(z)$, $z \in U$. By using Lemma 1.4 for $\gamma = 1$, we have $q(z) \prec p(z)$, i.e. $q(z) = \frac{1}{nz^{\frac{1}{n}}}\int_0^z h(t)t^{\frac{1}{n}-1} = \frac{1}{nz^{\frac{1}{n}}}\int_0^z \frac{1+(2\beta-1)t}{1+t}t^{\frac{1}{n}-1}dt = 2\beta - 1 + \frac{2-2\beta}{nz^{\frac{1}{n}}}\int_0^z \frac{t^{\frac{1}{n}-1}}{t+1}dt \prec \frac{RI^{\alpha}_{m+1,\lambda,l}f(z)}{RI^{\alpha}_{m,\lambda,l}f(z)}$, $z \in U$. The function q is convex and it is the best subordinant. □

Theorem 3.109 (Alb Lupaş, 2011e). *Let $h \in \mathcal{H}(U)$ be a convex function in U with $h(0) = 1$ and let $\lambda, \alpha, l \geq 0$, $m \in \mathbb{N}$, $f \in \mathcal{A}_n$,*

$$\frac{(m+1)(m+2)}{z}RI^{\alpha}_{m+2,\lambda,l}f(z) - \frac{(m+1)(2m+1)}{z}RI^{\alpha}_{m+1,\lambda,l}f(z)$$
$$+\frac{m^2}{z}RI^{\alpha}_{m,\lambda,l}f(z) + \frac{\alpha}{z}\left[\frac{(l+1)^2}{\lambda^2} - (m+1)(m+2)\right]I(m+2,\lambda,l)f(z)$$
$$-\frac{\alpha}{z}\left[\frac{2(l+1-\lambda)(\bar{l}+1)}{\lambda^2} - (m+1)(2m+1)\right]I(m+1,\lambda,l)f(z)$$
$$+\frac{\alpha}{z}\left[\frac{(l+1-\lambda)^2}{\lambda^2} - m^2\right]I(m,\lambda,l)f(z)$$

is univalent and $[RD^m_{\lambda,\alpha}f(z)]' \in \mathcal{H}[1,n] \cap Q$. *If*

$$h(z) \prec \frac{(m+1)(m+2)}{z} RI^\alpha_{m+2,\lambda,l} f(z) - \frac{(m+1)(2m+1)}{z} RI^\alpha_{m+1,\lambda,l} f(z)$$

$$+ \frac{m^2}{z} RI^\alpha_{m,\lambda,l} f(z) + \frac{\alpha}{z}\left[\frac{(l+1)^2}{\lambda^2} - (m+1)(m+2)\right]$$

$$\times I(m+2,\lambda,l) f(z)$$

$$- \frac{\alpha}{z}\left[\frac{2(l+1-\lambda)(\bar{l}+1)}{\lambda^2} - (m+1)(2m+1)\right] I(m+1,\lambda,l) f(z)$$

$$+ \frac{\alpha}{z}\left[\frac{(l+1-\lambda)^2}{\lambda^2} - m^2\right] I(m,\lambda,l) f(z), \quad z \in U, \tag{3.2.4.13}$$

holds, then

$$q(z) \prec [RI^\alpha_{m,\lambda,l} f(z)]', \quad z \in U,$$

where $q(z) = \frac{1}{nz^{\frac{1}{n}}} \int_0^z h(t) t^{\frac{1}{n}-1}$. *The function* q *is convex and it is the best subordinant.*

Proof. For $f \in \mathcal{A}_n$, $f(z) = z + \sum_{j=n+1}^{\infty} a_j z^j$, we have

$$RI^\alpha_{m,\lambda,l} f(z) = z + \sum_{j=n+1}^{\infty} \left\{\alpha\left(\frac{1+\lambda(j-1)+l}{l+1}\right)^m + (1-\alpha) C^m_{m+j-1}\right\} a_j z^j,$$

$$z \in U.$$

Let

$$\begin{aligned} p(z) &= \left(RI^\alpha_{m,\lambda,l} f(z)\right)' \\ &= 1 + \sum_{j=n+1}^{\infty} \left\{\alpha\left(\frac{1+\lambda(j-1)+l}{l+1}\right)^m + (1-\alpha) C^m_{m+j-1}\right\} j a_j z^{j-1} \\ &= 1 + p_n z^n + p_{n+1} z^{n+1} + \cdots. \end{aligned} \tag{3.2.4.14}$$

By using the properties of operators $RI_{m,\lambda,l}^{\alpha}$, R^m and $I(m,\lambda,l)$, after a short calculation, we obtain

$$\begin{aligned}
p(z)+zp'(z) = {} & \frac{(m+1)(m+2)}{z} RI_{m+2,\lambda,l}^{\alpha} f(z) \\
& - \frac{(m+1)(2m+1)}{z} RI_{m+1,\lambda,l}^{\alpha} f(z) + \frac{m^2}{z} RI_{m,\lambda,l}^{\alpha} f(z) \\
& + \frac{\alpha}{z}\left[\frac{(l+1)^2}{\lambda^2} - (m+1)(m+2)\right] I(m+2,\lambda,l) f(z) \\
& - \frac{\alpha}{z}\left[\frac{2(l+1-\lambda)\left(\bar{l}+1\right)}{\lambda^2} - (m+1)(2m+1)\right] \\
& \times I(m+1,\lambda,l) f(z) \\
& + \frac{\alpha}{z}\left[\frac{(l+1-\lambda)^2}{\lambda^2} - m^2\right] I(m,\lambda,l) f(z).
\end{aligned}$$

Using the notation in (3.2.4.14), the differential superordination becomes $h(z) \prec p(z) + zp'(z)$. By using Lemma 1.4 for $\gamma = 1$, we have $q(z) \prec p(z)$, $z \in U$, i.e. $q(z) \prec \left(RI_{m,\lambda,l}^{\alpha} f(z)\right)'$, $z \in U$, where $q(z) = \frac{1}{nz^{\frac{1}{n}}} \int_0^z h(t) t^{\frac{1}{n}-1}$. The function q is convex and it is the best subordinant. □

Theorem 3.110 (Alb Lupaş, 2011e). *Let q be a convex function in U and $h(z) = q(z) + zq'(z)$. Let $\lambda, \alpha, l \geq 0$, $m \in \mathbb{N}$, $f \in \mathcal{A}_n$, suppose that*

$$\begin{aligned}
& \frac{(m+1)(m+2)}{z} RI_{m+2,\lambda,l}^{\alpha} f(z) - \frac{(m+1)(2m+1)}{z} RI_{m+1,\lambda,l}^{\alpha} f(z) \\
& + \frac{m^2}{z} RI_{m,\lambda,l}^{\alpha} f(z) + \frac{\alpha}{z}\left[\frac{(l+1)^2}{\lambda^2} - (m+1)(m+2)\right] I(m+2,\lambda,l) f(z) \\
& - \frac{\alpha}{z}\left[\frac{2(l+1-\lambda)\left(\bar{l}+1\right)}{\lambda^2} - (m+1)(2m+1)\right] I(m+1,\lambda,l) f(z) \\
& + \frac{\alpha}{z}\left[\frac{(l+1-\lambda)^2}{\lambda^2} - m^2\right] I(m,\lambda,l) f(z)
\end{aligned}$$

is univalent in U and $[RI^{\alpha}_{m,\lambda,l}f(z)]' \in \mathcal{H}[1,n] \cap Q$. *If*

$$\begin{aligned}
h(z) &= q(z) + zq'(z) \\
&\prec \frac{(m+1)(m+2)}{z} RI^{\alpha}_{m+2,\lambda,l}f(z) - \frac{(m+1)(2m+1)}{z} RI^{\alpha}_{m+1,\lambda,l}f(z) \\
&\quad + \frac{m^2}{z} RI^{\alpha}_{m,\lambda,l}f(z) + \frac{\alpha}{z}\left[\frac{(l+1)^2}{\lambda^2} - (m+1)(m+2)\right] \\
&\quad \times I(m+2,\lambda,l)f(z) \\
&\quad - \frac{\alpha}{z}\left[\frac{2(l+1-\lambda)(\bar{l}+1)}{\lambda^2} - (m+1)(2m+1)\right] I(m+1,\lambda,l)f(z) \\
&\quad + \frac{\alpha}{z}\left[\frac{(l+1-\lambda)^2}{\lambda^2} - m^2\right] I(m,\lambda,l)f(z), \quad z \in U, \qquad (3.2.4.15)
\end{aligned}$$

then

$$q(z) \prec \left(RI^{\alpha}_{m,\lambda,l}f(z)\right)', \quad z \in U,$$

where $q(z) = \frac{1}{nz^{\frac{1}{n}}} \int_0^z h(t)t^{\frac{1}{n}-1}$. *The function q is the best subordinant.*

Proof. Following the same steps as in the proof of Theorem 3.109 and considering $p(z) = \left(RI^{\alpha}_{m,\lambda,l}f(z)\right)'$, the differential superordination (3.2.4.15) becomes $h(z) = q(z)+zq'(z) \prec p(z)+zp'(z)$, $z \in U$. By using Lemma 1.5 for $\gamma = 1$, we have $q(z) \prec p(z)$, i.e. $q(z) = \frac{1}{nz^{\frac{1}{n}}} \int_0^z h(t)t^{\frac{1}{n}-1} \prec \left(RI^{\alpha}_{m,\lambda,l}f(z)\right)'$, $z \in U$. The function q is the best subordinant. □

Theorem 3.111 (Alb Lupaş, 2011e). *Let* $h(z) = \frac{1+(2\beta-1)z}{1+z}$ *be a convex function in U, where* $0 \le \beta < 1$. *Let* $\lambda, \alpha, l \ge 0$, $m \in \mathbb{N}$, $f \in \mathcal{A}_n$, *suppose that*

$$\begin{aligned}
&\frac{(m+1)(m+2)}{z} RI^{\alpha}_{m+2,\lambda,l}f(z) - \frac{(m+1)(2m+1)}{z} RI^{\alpha}_{m+1,\lambda,l}f(z) \\
&\quad + \frac{m^2}{z} RI^{\alpha}_{m,\lambda,l}f(z) + \frac{\alpha}{z}\left[\frac{(l+1)^2}{\lambda^2} - (m+1)(m+2)\right] I(m+2,\lambda,l)f(z) \\
&\quad - \frac{\alpha}{z}\left[\frac{2(l+1-\lambda)(\bar{l}+1)}{\lambda^2} - (m+1)(2m+1)\right] I(m+1,\lambda,l)f(z) \\
&\quad + \frac{\alpha}{z}\left[\frac{(l+1-\lambda)^2}{\lambda^2} - m^2\right] I(m,\lambda,l)f(z)
\end{aligned}$$

is univalent in U and $[RI^{\alpha}_{m,\lambda,l}f(z)]' \in \mathcal{H}[1,n] \cap Q$. If

$$h(z) \prec \frac{(m+1)(m+2)}{z} RI^{\alpha}_{m+2,\lambda,l}f(z) - \frac{(m+1)(2m+1)}{z} RI^{\alpha}_{m+1,\lambda,l}f(z)$$
$$+ \frac{m^2}{z} RI^{\alpha}_{m,\lambda,l}f(z) + \frac{\alpha}{z}\left[\frac{(l+1)^2}{\lambda^2} - (m+1)(m+2)\right]$$
$$\times I(m+2,\lambda,l)f(z)$$
$$- \frac{\alpha}{z}\left[\frac{2(l+1-\lambda)(\bar{l}+1)}{\lambda^2} - (m+1)(2m+1)\right] I(m+1,\lambda,l)f(z)$$
$$+ \frac{\alpha}{z}\left[\frac{(l+1-\lambda)^2}{\lambda^2} - m^2\right] I(m,\lambda,l)f(z), \quad z \in U, \tag{3.2.4.16}$$

then

$$q(z) \prec \left(RI^{\alpha}_{m,\lambda,l}f(z)\right)', \quad z \in U,$$

where q is given by $q(z) = 2\beta - 1 + \frac{2-2\beta}{nz^{\frac{1}{n}}}\int_0^z \frac{t^{\frac{1}{n}-1}}{t+1}dt$, $z \in U$. The function q is convex and it is the best subordinant.

Proof. Following the same steps as in the proof of Theorem 3.109 and considering $p(z) = \left(RI^{\alpha}_{m,\lambda,l}f(z)\right)'$, the differential superordination (3.2.4.16) becomes $h(z) = \frac{1+(2\beta-1)z}{1+z} \prec p(z) + zp'(z)$, $z \in U$. By using Lemma 1.4 for $\gamma = 1$, we have $q(z) \prec p(z)$, i.e. $q(z) = \frac{1}{nz^{\frac{1}{n}}}\int_0^z h(t)t^{\frac{1}{n}-1} = \frac{1}{nz^{\frac{1}{n}}}\int_0^z \frac{1+(2\beta-1)t}{1+t}t^{\frac{1}{n}-1}dt = 2\beta - 1 + \frac{2-2\beta}{nz^{\frac{1}{n}}}\int_0^z \frac{t^{\frac{1}{n}-1}}{t+1}dt \prec \left(RI^{\alpha}_{m,\lambda,l}f(z)\right)'$, $z \in U$. The function q is convex and it is the best subordinant. □

3.2.5 *About Some Differential Sandwich Theorems Involving a Multiplier Transformation and Ruscheweyh Derivative*

In the paper Alb Lupaş (2018b), we studied a new operator $IR^{m,n}_{\lambda,l}$ defined in Alb Lupaş (2010c) as the Hadamard product of the multiplier transformation $I(m,\lambda,l)$ and Ruscheweyh derivative R^n, given in Definition 2.10 by $IR^{m,n}_{\lambda,l} : \mathcal{A} \to \mathcal{A}$,

$$IR^{m,n}_{\lambda,l}f(z) = (I(m,\lambda,l) * R^n)f(z)$$

and $\mathcal{A}_n = \{f \in \mathcal{H}(U) : f(z) = z + a_{n+1}z^{n+1} + \cdots,\ z \in U\}$ is the class of normalized analytic functions with $\mathcal{A}_1 = \mathcal{A}$. The purpose of the paper Alb Lupaş (2018b) is to derive certain subordination and superordination results involving the operator $IR_{\lambda,l}^{m,n}$ and we establish differential sandwich-type theorems.

If $f \in \mathcal{A}$ and $f(z) = z + \sum_{j=2}^{\infty} a_j z^j$, then

$$IR_{\lambda,l}^{m,n} f(z) = z + \sum_{j=2}^{\infty} \left(\frac{1 + \lambda(j-1) + l}{l+1} \right)^m \frac{(n+j-1)!}{n!\,(j-1)!} a_j^2 z^j, \quad z \in U.$$

Using simple computation, one obtains the next result.

Proposition 3.1 (Alb Lupaş, 2015i). *For $m, n \in \mathbb{N}$ and $\lambda, l \geq 0$, we have*

$$(n+1)\, IR_{\lambda,l}^{m,n+1} f(z) - n IR_{\lambda,l}^{m,n} f(z) = z \left(IR_{\lambda,l}^{m,n} f(z) \right)'. \tag{3.2.5.1}$$

We begin with the following

Theorem 3.112 (Alb Lupaş, 2018b). *Let $\frac{z\left(IR_{\lambda,l}^{m,n} f(z)\right)'}{IR_{\lambda,l}^{m,n} f(z)} \in \mathcal{H}(U)$ and let the function $q(z)$ be analytic and univalent in U such that $q(z) \neq 0$, for all $z \in U$. Suppose that $\frac{zq'(z)}{q(z)}$ is starlike univalent in U. Let*

$$\operatorname{Re}\left(\frac{\xi}{\beta} q(z) + \frac{2\mu}{\beta} q^2(z) + 1 + z\frac{q''(z)}{q(z)} - z\frac{q'(z)}{q(z)} \right) > 0, \tag{3.2.5.2}$$

for $\alpha, \xi, \beta, \mu \in \mathbb{C}$, $\beta \neq 0$, $z \in U$ and

$$\begin{aligned} \psi_{\lambda,l}^{m,n}(\alpha, \xi, \mu, \beta; z) &:= \left(\alpha - \xi n + \mu n^2\right) \\ &+ (n+1)(\xi - 2n\mu - \beta) \frac{IR_{\lambda,l}^{m,n+1} f(z)}{IR_{\lambda,l}^{m,n} f(z)} \\ &+ \mu (n+1)^2 \left(\frac{IR_{\lambda,l}^{m,n+1} f(z)}{IR_{\lambda,l}^{m,n} f(z)} \right)^2 \\ &+ \beta(n+1) \frac{(n+2)\, IR_{\lambda,l}^{m,n+2} f(z) - (n+1)\, IR_{\lambda,l}^{m,n+1} f(z)}{(n+1)\, IR_{\lambda,l}^{m,n+1} f(z) - n IR_{\lambda,l}^{m,n} f(z)}. \end{aligned} \tag{3.2.5.3}$$

If q satisfies the following subordination

$$\psi_{\lambda,l}^{m,n}(\alpha,\beta,\mu;z) \prec \alpha + \xi q(z) + \mu(q(z))^2 + \beta\frac{zq'(z)}{q(z)}, \tag{3.2.5.4}$$

for $\alpha,\xi,\beta,\mu\in\mathbb{C}$, $\beta\neq 0$, *then*

$$\frac{z\left(IR_{\lambda,l}^{m,n}f(z)\right)'}{IR_{\lambda,l}^{m,n}f(z)} \prec q(z), \tag{3.2.5.5}$$

and q is the best dominant.

Proof. Let the function p be defined by $p(z) := \frac{z\left(IR_{\lambda,l}^{m,n}f(z)\right)'}{IR_{\lambda,l}^{m,n}f(z)}$, $z\in U$, $z\neq 0$, $f\in\mathcal{A}$. We have $p'(z) = (n+1)\frac{\left(IR_{\lambda,l}^{m,n+1}f(z)\right)'}{IR_{\lambda,l}^{m,n}f(z)} - (n+1)\frac{IR_{\lambda,l}^{m,n+1}f(z)}{IR_{\lambda,l}^{m,n}f(z)}\cdot\frac{\left(IR_{\lambda,l}^{m,n}f(z)\right)'}{IR_{\lambda,l}^{m,n}f(z)}$.

By using the identity (3.2.5.1), we obtain

$$\begin{aligned}\frac{zp'(z)}{p(z)} &= (n+1)\frac{(n+2)IR_{\lambda,l}^{m,n+2}f(z)-(n+1)IR_{\lambda,l}^{m,n+1}f(z)}{(n+1)IR_{\lambda,l}^{m,n+1}f(z)-nIR_{\lambda,l}^{m,n}f(z)}\\ &\quad -(n+1)\frac{IR_{\lambda,l}^{m,n+1}f(z)}{IR_{\lambda,l}^{m,n}f(z)}.\end{aligned} \tag{3.2.5.6}$$

By setting $\theta(w) := \alpha+\xi w+\mu w^2$ and $\phi(w) := \frac{\beta}{w}$, it can be easily verified that θ is analytic in $\mathbb{C}$, ϕ is analytic in $\mathbb{C}\backslash\{0\}$ and that $\phi(w)\neq 0$, $w\in\mathbb{C}\backslash\{0\}$.

Also, by letting $Q(z) = zq'(z)\phi(q(z)) = \beta\frac{zq'(z)}{q(z)}$ and $h(z) = \theta(q(z)) + Q(z) = \alpha+\xi q(z)+\mu(q(z))^2+\beta\frac{zq'(z)}{q(z)}$, we find that $Q(z)$ is starlike univalent in U.

We have $h'(z) = \xi q'(z) + 2\mu q(z)q'(z) + \beta\frac{q'(z)}{q(z)} + \beta z\frac{q''(z)}{q(z)} - \beta z\left(\frac{q'(z)}{q(z)}\right)^2$ and $\frac{zh'(z)}{Q(z)} = \frac{\xi}{\beta}q(z) + \frac{2\mu}{\beta}q^2(z) + 1 + z\frac{q''(z)}{q(z)} - z\frac{q'(z)}{q(z)}$.

We deduce that

$$\operatorname{Re}\left(\frac{zh'(z)}{Q(z)}\right) = \operatorname{Re}\left(\frac{\xi}{\beta}q(z)+\frac{2\mu}{\beta}q^2(z)+1+z\frac{q''(z)}{q(z)}-z\frac{q'(z)}{q(z)}\right) > 0.$$

By using (3.2.5.6), we obtain

$$\alpha+\xi p(z)+\mu(p(z))^2+\beta\frac{zp'(z)}{p(z)}$$

$$=\left(\alpha-\xi n+\mu n^2\right)+(n+1)(\xi-2n\mu-\beta)\frac{IR_{\lambda,l}^{m,n+1}f(z)}{IR_{\lambda,l}^{m,n}f(z)}$$

$$+\mu(n+1)^2\left(\frac{IR_{\lambda,l}^{m,n+1}f(z)}{IR_{\lambda,l}^{m,n}f(z)}\right)^2$$

$$+\beta(n+1)\frac{(n+2)IR_{\lambda,l}^{m,n+2}f(z)-(n+1)IR_{\lambda,l}^{m,n+1}f(z)}{(n+1)IR_{\lambda,l}^{m,n+1}f(z)-nIR_{\lambda,l}^{m,n}f(z)}.$$

By using (3.2.5.4), we have $\alpha+\xi p(z)+\mu(p(z))^2+\beta\frac{zp'(z)}{p(z)}\prec\alpha+\xi q(z)+\mu(q(z))^2+\beta\frac{zq'(z)}{q(z)}$.

By an application of Lemma 1.3, we have $p(z)\prec q(z)$, $z\in U$, i.e. $\frac{z\left(IR_{\lambda,l}^{m,n}f(z)\right)'}{IR_{\lambda,l}^{m,n}f(z)}\prec q(z)$, $z\in U$ and q is the best dominant. □

Corollary 3.22 (Alb Lupaş, 2018b). *Let $m,n\in\mathbb{N}$, $\lambda,l\geq 0$. Assume that (3.2.5.2) holds. If $f\in\mathcal{A}$ and*

$$\psi_{\lambda,l}^{m,n}(\alpha,\beta,\mu;z)\prec\alpha+\xi\frac{1+Az}{1+Bz}+\mu\left(\frac{1+Az}{1+Bz}\right)^2+\frac{\beta(A-B)z}{(1+Az)(1+Bz)},$$

for $\alpha,\beta,\mu,\xi\in\mathbb{C}$, $\beta\neq 0$, $-1\leq B<A\leq 1$, where $\psi_{\lambda,l}^{m,n}$ is defined in (3.2.5.3), then

$$\frac{z\left(IR_{\lambda,l}^{m,n}f(z)\right)'}{IR_{\lambda,l}^{m,n}f(z)}\prec\frac{1+Az}{1+Bz},$$

and $\frac{1+Az}{1+Bz}$ is the best dominant.

Proof. For $q(z)=\frac{1+Az}{1+Bz}$, $-1\leq B<A\leq 1$ in Theorem 3.112, we get the corollary. □

Corollary 3.23 (Alb Lupaş, 2018b). *Let $m,n\in\mathbb{N}$, $\lambda,l\geq 0$. Assume that (3.2.5.2) holds. If $f\in\mathcal{A}$ and*

$$\psi_{\lambda,l}^{m,n}(\alpha,\beta,\mu;z)\prec\alpha+\xi\left(\frac{1+z}{1-z}\right)^\gamma+\mu\left(\frac{1+z}{1-z}\right)^{2\gamma}+\frac{2\beta\gamma z}{1-z^2},$$

for $\alpha, \beta, \mu, \xi \in \mathbb{C}$, $0 < \gamma \leq 1$, $\beta \neq 0$, *where* $\psi_{\lambda,l}^{m,n}$ *is defined in* (3.2.5.3), *then*

$$\frac{z\left(IR_{\lambda,l}^{m,n} f(z)\right)'}{IR_{\lambda,l}^{m,n} f(z)} \prec \left(\frac{1+z}{1-z}\right)^{\gamma},$$

and $\left(\frac{1+z}{1-z}\right)^{\gamma}$ *is the best dominant.*

Proof. Corollary follows by using Theorem 3.112 for $q(z) = \left(\frac{1+z}{1-z}\right)^{\gamma}$, $0 < \gamma \leq 1$. □

Theorem 3.113 (Alb Lupaş, 2018b). *Let* q *be analytic and univalent in* U *such that* $q(z) \neq 0$ *and* $\frac{zq'(z)}{q(z)}$ *be starlike univalent in* U. *Assume that*

$$\operatorname{Re}\left(\frac{\xi}{\beta} q(z) q'(z) + \frac{2\mu}{\beta} q^2(z) q'(z)\right) > 0, \quad \text{for } \xi, \beta, \mu \in \mathbb{C}, \quad \beta \neq 0. \tag{3.2.5.7}$$

If $f \in \mathcal{A}$, $\frac{z\left(IR_{\lambda,l}^{m,n} f(z)\right)'}{IR_{\lambda,l}^{m,n} f(z)} \in \mathcal{H}[q(0), 1] \cap Q$ *and* $\psi_{\lambda,l}^{m,n}(\alpha, \beta, \mu; z)$ *is univalent in* U, *where* $\psi_{\lambda,l}^{m,n}(\alpha, \beta, \mu; z)$ *is as defined in* (3.2.5.3), *then*

$$\alpha + \xi q(z) + \mu (q(z))^2 + \frac{\beta z q'(z)}{q(z)} \prec \psi_{\lambda,l}^{m,n}(\alpha, \beta, \mu; z) \tag{3.2.5.8}$$

implies

$$q(z) \prec \frac{z\left(IR_{\lambda,l}^{m,n} f(z)\right)'}{IR_{\lambda,l}^{m,n} f(z)}, \quad z \in U, \tag{3.2.5.9}$$

and q *is the best subordinant.*

Proof. Let the function p be defined by $p(z) := \frac{z\left(IR_{\lambda,l}^{m,n} f(z)\right)'}{IR_{\lambda,l}^{m,n} f(z)}$, $z \in U$, $z \neq 0$, $f \in \mathcal{A}$.

By setting $\nu(w) := \alpha + \xi w + \mu w^2$ and $\phi(w) := \frac{\beta}{w}$ it can be easily verified that ν is analytic in $\mathbb{C}$, ϕ is analytic in $\mathbb{C} \backslash \{0\}$ and that $\phi(w) \neq 0$, $w \in \mathbb{C} \backslash \{0\}$.

Since $\frac{\nu'(q(z))}{\phi(q(z))} = \frac{q'(z) q(z)[\xi + 2\mu q(z)]}{\beta}$, it follows that $\operatorname{Re}\left(\frac{\nu'(q(z))}{\phi(q(z))}\right) = \operatorname{Re}\left(\frac{\xi}{\beta} q(z) q'(z) + \frac{2\mu}{\beta} q^2(z) q'(z)\right) > 0$, for $\alpha, \beta, \mu \in \mathbb{C}$, $\mu \neq 0$.

By using (3.2.5.6) and (3.2.5.8), we obtain $\alpha + \xi q(z) + \mu (q(z))^2 + \frac{\beta z q'(z)}{q(z)} \prec \alpha + \xi p(z) + \mu (p(z))^2 + \frac{\beta z p'(z)}{p(z)}$.

Using Lemma 1.6, we have $q(z) \prec p(z) = \frac{z\left(IR_{\lambda,l}^{m,n} f(z)\right)'}{IR_{\lambda,l}^{m,n} f(z)}$, $z \in U$, and q is the best subordinant. □

Corollary 3.24 (Alb Lupaş, 2018b). *Let $m, n \in \mathbb{N}$, $\lambda, l \geq 0$. Assume that* (3.2.5.7) *holds. If $f \in \mathcal{A}$, $\frac{z\left(IR_{\lambda,l}^{m,n} f(z)\right)'}{IR_{\lambda,l}^{m,n} f(z)} \in \mathcal{H}[q(0), 1] \cap Q$ and*

$$\alpha + \xi \frac{1 + Az}{1 + Bz} + \mu \left(\frac{1 + Az}{1 + Bz}\right)^2 + \frac{\beta (A - B) z}{(1 + Az)(1 + Bz)} \prec \psi_{\lambda,l}^{m,n}(\alpha, \beta, \mu; z),$$

for $\alpha, \beta, \xi, \mu \in \mathbb{C}$, $\beta \neq 0$, $-1 \leq B < A \leq 1$, where $\psi_{\lambda,l}^{m,n}$ is defined in (3.2.5.3)*, then*

$$\frac{1 + Az}{1 + Bz} \prec \frac{z\left(IR_{\lambda,l}^{m,n} f(z)\right)'}{IR_{\lambda,l}^{m,n} f(z)},$$

and $\frac{1+Az}{1+Bz}$ is the best subordinant.

Proof. For $q(z) = \frac{1+Az}{1+Bz}$, $-1 \leq B < A \leq 1$ in Theorem 3.113, we get the corollary. □

Corollary 3.25 (Alb Lupaş, 2018b). *Let $m, n \in \mathbb{N}$, $\lambda, l \geq 0$. Assume that* (3.2.5.7) *holds. If $f \in \mathcal{A}$, $\frac{z\left(IR_{\lambda,l}^{m,n} f(z)\right)'}{IR_{\lambda,l}^{m,n} f(z)} \in \mathcal{H}[q(0), 1] \cap Q$ and*

$$\alpha + \xi \left(\frac{1 + z}{1 - z}\right)^\gamma + \mu \left(\frac{1 + z}{1 - z}\right)^{2\gamma} + \frac{2\beta\gamma z}{1 - z^2} \prec \psi_{\lambda,l}^{m,n}(\alpha, \beta, \mu; z),$$

for $\alpha, \beta, \mu, \xi \in \mathbb{C}$, $\beta \neq 0$, $0 < \gamma \leq 1$, where $\psi_{\lambda,l}^{m,n}$ is defined in (3.2.5.3)*, then*

$$\left(\frac{1 + z}{1 - z}\right)^\gamma \prec \frac{z\left(IR_{\lambda,l}^{m,n} f(z)\right)'}{IR_{\lambda,l}^{m,n} f(z)},$$

and $\left(\frac{1+z}{1-z}\right)^\gamma$ is the best subordinant.

Proof. For $q(z) = \left(\frac{1+z}{1-z}\right)^\gamma$, $0 < \gamma \leq 1$ in Theorem 3.113, we get the corollary. □

Combining Theorems 3.112 and 3.113, we state the following sandwich theorem.

Theorem 3.114 (Alb Lupaş, 2018b). *Let* q_1 *and* q_2 *be analytic and univalent in* U *such that* $q_1(z) \neq 0$ *and* $q_2(z) \neq 0$, *for all* $z \in U$, *with* $\frac{zq_1'(z)}{q_1(z)}$ *and* $\frac{zq_2'(z)}{q_2(z)}$ *being starlike univalent. Suppose that* q_1 *satisfies* (3.2.5.2) *and* q_2 *satisfies* (3.2.5.7). *If* $f \in \mathcal{A}$, $\frac{z\left(IR_{\lambda,l}^{m,n} f(z)\right)'}{IR_{\lambda,l}^{m,n} f(z)} \in \mathcal{H}[q(0),1] \cap Q$ *and* $\psi_{\lambda,l}^{m,n}(\alpha,\beta,\mu;z)$ *is as defined in* (3.2.5.3) *univalent in* U, *then*

$$\alpha + \xi q_1(z) + \mu (q_1(z))^2 + \frac{\beta z q_1'(z)}{q_1(z)} \prec \psi_{\lambda,l}^{m,n}(\alpha,\beta,\mu;z)$$

$$\prec \alpha + \xi q_2(z) + \mu (q_2(z))^2 + \frac{\beta z q_2'(z)}{q_2(z)},$$

for $\alpha,\beta,\mu,\xi \in \mathbb{C}$, $\beta \neq 0$, *implies*

$$q_1(z) \prec \frac{z\left(IR_{\lambda,l}^{m,n} f(z)\right)'}{IR_{\lambda,l}^{m,n} f(z)} \prec q_2(z),$$

and q_1 *and* q_2 *are respectively the best subordinant and the best dominant.*

For $q_1(z) = \frac{1+A_1 z}{1+B_1 z}$, $q_2(z) = \frac{1+A_2 z}{1+B_2 z}$, where $-1 \leq B_2 < B_1 < A_1 < A_2 \leq 1$, we have the following corollary.

Corollary 3.26 (Alb Lupaş, 2018b). *Let* $m,n \in \mathbb{N}$, $\lambda, l \geq 0$. *Assume that* (3.2.5.2) *and* (3.2.5.7) *hold. If* $f \in \mathcal{A}$, $\frac{z\left(IR_{\lambda,l}^{m,n} f(z)\right)'}{IR_{\lambda,l}^{m,n} f(z)} \in \mathcal{H}[q(0),1] \cap Q$ *and*

$$\alpha + \xi \frac{1+A_1 z}{1+B_1 z} + \mu \left(\frac{1+A_1 z}{1+B_1 z}\right)^2 + \frac{\beta (A_1 - B_1) z}{(1+A_1 z)(1+B_1 z)} \prec \psi_{\lambda,l}^{m,n}(\alpha,\beta,\mu;z)$$

$$\prec \alpha + \xi \frac{1+A_2 z}{1+B_2 z} + \mu \left(\frac{1+A_2 z}{1+B_2 z}\right)^2 + \frac{\beta (A_2 - B_2) z}{(1+A_2 z)(1+B_2 z)},$$

for $\alpha,\beta,\mu,\xi \in \mathbb{C}$, $\beta \neq 0$, $-1 \leq B_2 \leq B_1 < A_1 \leq A_2 \leq 1$, *where* $\psi_{\lambda,l}^{m,n}$ *is defined in* (3.2.5.3), *then*

$$\frac{1+A_1 z}{1+B_1 z} \prec \frac{z\left(IR_{\lambda,l}^{m,n} f(z)\right)'}{IR_{\lambda,l}^{m,n} f(z)} \prec \frac{1+A_2 z}{1+B_2 z},$$

hence $\frac{1+A_1 z}{1+B_1 z}$ *and* $\frac{1+A_2 z}{1+B_2 z}$ *are the best subordinant and the best dominant, respectively.*

For $q_1(z) = \left(\frac{1+z}{1-z}\right)^{\gamma_1}$, $q_2(z) = \left(\frac{1+z}{1-z}\right)^{\gamma_2}$, where $0 < \gamma_1 < \gamma_2 \leq 1$, we have the following corollary.

Corollary 3.27 (Alb Lupaş, 2018b). *Let $m, n \in \mathbb{N}$, $\lambda, l \geq 0$. Assume that (3.2.5.2) and (3.2.5.7) hold. If $f \in \mathcal{A}$, $\frac{z\left(IR_{\lambda,l}^{m,n} f(z)\right)'}{IR_{\lambda,l}^{m,n} f(z)} \in \mathcal{H}[q(0), 1] \cap Q$ and*

$$\alpha + \xi\left(\frac{1+z}{1-z}\right)^{\gamma_1} + \mu\left(\frac{1+z}{1-z}\right)^{2\gamma_1} + \frac{2\beta\gamma_1 z}{1-z^2} \prec \psi_{\lambda,l}^{m,n}(\alpha, \beta, \mu; z)$$
$$\prec \alpha + \xi\left(\frac{1+z}{1-z}\right)^{\gamma_2} + \mu\left(\frac{1+z}{1-z}\right)^{2\gamma_2} + \frac{2\beta\gamma_2 z}{1-z^2},$$

for $\alpha, \beta, \mu, \xi \in \mathbb{C}$, $\beta \neq 0$, $0 < \gamma_1 < \gamma_2 \leq 1$, where $\psi_{\lambda,l}^{m,n}$ is defined in (3.2.5.3), then

$$\left(\frac{1+z}{1-z}\right)^{\gamma_1} \prec \frac{z\left(IR_{\lambda,l}^{m,n} f(z)\right)'}{IR_{\lambda,l}^{m,n} f(z)} \prec \left(\frac{1+z}{1-z}\right)^{\gamma_2},$$

hence $\left(\frac{1+z}{1-z}\right)^{\gamma_1}$ and $\left(\frac{1+z}{1-z}\right)^{\gamma_2}$ are the best subordinant and the best dominant, respectively.

We have also

Theorem 3.115 (Alb Lupaş, 2018b). *Let $\frac{z\left(IR_{\lambda,l}^{m,n} f(z)\right)'}{IR_{\lambda,l}^{m,n} f(z)} \in \mathcal{H}(U), f \in \mathcal{A}$, $z \in U$, $m, n \in \mathbb{N}$, $\lambda, l \geq 0$ and let the function $q(z)$ be convex and univalent in U such that $q(0) = 1$, $z \in U$. Assume that*

$$\operatorname{Re}\left(\frac{\alpha+\beta}{\beta} + z\frac{q''(z)}{q'(z)}\right) > 0, \tag{3.2.5.10}$$

for $\alpha, \beta \in \mathbb{C}$, $\beta \neq 0$, $z \in U$, and

$$\psi_{\lambda,l}^{m,n}(\alpha, \beta; z) := \beta(n+1)(n+2)\frac{IR_{\lambda,l}^{m,n+2} f(z)}{IR_{\lambda,l}^{m,n} f(z)}$$
$$- \beta(n+1)^2\left(\frac{IR_{\lambda,l}^{m,n+1} f(z)}{IR_{\lambda,l}^{m,n} f(z)}\right)^2$$
$$+ (\alpha - \beta)(n+1)\frac{IR_{\lambda,l}^{m,n+1} f(z)}{IR_{\lambda,l}^{m,n} f(z)} - \alpha n. \tag{3.2.5.11}$$

If q satisfies the following subordination

$$\psi_{\lambda,l}^{m,n}(\alpha,\beta;z) \prec \alpha q(z) + \beta z q'(z), \tag{3.2.5.12}$$

for $\alpha,\beta \in \mathbb{C}$, $\beta \neq 0$, $z \in U$, then

$$\frac{z\left(IR_{\lambda,l}^{m,n} f(z)\right)'}{IR_{\lambda,l}^{m,n} f(z)} \prec q(z), \quad z \in U, \tag{3.2.5.13}$$

and q is the best dominant.

Proof. Let the function p be defined by $p(z) := \frac{z\left(IR_{\lambda,l}^{m,n} f(z)\right)'}{IR_{\lambda,l}^{m,n} f(z)}$, $z \in U$, $z \neq 0$, $f \in \mathcal{A}$. The function p is analytic in U and $p(0) = 1$. We have

$$p'(z) = (n+1)\frac{\left(IR_{\lambda,l}^{m,n+1} f(z)\right)'}{IR_{\lambda,l}^{m,n} f(z)} - (n+1)\frac{IR_{\lambda,l}^{m,n+1} f(z)}{IR_{\lambda,l}^{m,n} f(z)} \cdot \frac{\left(IR_{\lambda,l}^{m,n} f(z)\right)'}{IR_{\lambda,l}^{m,n} f(z)}.$$

By using the identity (3.2.5.1), we obtain

$$zp'(z) = (n+1)(n+2)\frac{IR_{\lambda,l}^{m,n+2} f(z)}{IR_{\lambda,l}^{m,n} f(z)} - (n+1)^2\left(\frac{IR_{\lambda,l}^{m,n+1} f(z)}{IR_{\lambda,l}^{m,n} f(z)}\right)^2 - (n+1)\frac{IR_{\lambda,l}^{m,n+1} f(z)}{IR_{\lambda,l}^{m,n} f(z)}. \tag{3.2.5.14}$$

By setting $\theta(w) := \alpha w$ and $\phi(w) := \beta$, it can be easily verified that θ is analytic in $\mathbb{C}$, ϕ is analytic in $\mathbb{C}\backslash\{0\}$ and that $\phi(w) \neq 0$, $w \in \mathbb{C}\backslash\{0\}$.

Also, by letting $Q(z) = zq'(z)\phi(q(z)) = \beta z q'(z)$, we find that $Q(z)$ is starlike univalent in U.

Let $h(z) = \theta(q(z)) + Q(z) = \alpha q(z) + \beta z q'(z)$.

We have $\operatorname{Re}\left(\frac{zh'(z)}{Q(z)}\right) = \operatorname{Re}\left(\frac{\alpha+\beta}{\beta} + z\frac{q''(z)}{q'(z)}\right) > 0$.

By using (3.2.5.14), we obtain

$$\alpha p(z) + \beta z p'(z) = \beta(n+1)(n+2)\frac{IR_{\lambda,l}^{m,n+2} f(z)}{IR_{\lambda,l}^{m,n} f(z)} - \beta(n+1)^2\left(\frac{IR_{\lambda,l}^{m,n+1} f(z)}{IR_{\lambda,l}^{m,n} f(z)}\right)^2 + (\alpha-\beta)(n+1)\frac{IR_{\lambda,l}^{m,n+1} f(z)}{IR_{\lambda,l}^{m,n} f(z)} - \alpha n.$$

By using (3.2.5.12), we have $\alpha p(z) + \beta z p'(z) \prec \alpha q(z) + \beta z q'(z)$.

From Lemma 1.3, we have $p(z) \prec q(z)$, $z \in U$, i.e. $\frac{z\left(IR_{\lambda,l}^{m,n} f(z)\right)'}{IR_{\lambda,l}^{m,n} f(z)} \prec q(z)$, $z \in U$, and q is the best dominant. $\square$

Corollary 3.28 (Alb Lupaş, 2018b). *Let* $q(z) = \frac{1+Az}{1+Bz}$, $z \in U$, $-1 \leq B < A \leq 1$, $m, n \in \mathbb{N}$, $\lambda, l \geq 0$. *Assume that* (3.2.5.10) *holds. If* $f \in \mathcal{A}$ *and*

$$\psi_{\lambda,l}^{m,n}(\alpha, \beta; z) \prec \alpha \frac{1+Az}{1+Bz} + \frac{\beta(A-B)z}{(1+Bz)^2},$$

for $\alpha, \beta \in \mathbb{C}$, $\beta \neq 0$, $-1 \leq B < A \leq 1$, *where* $\psi_{\lambda,l}^{m,n}$ *is defined in* (3.2.5.11), *then*

$$\frac{z\left(IR_{\lambda,l}^{m,n} f(z)\right)'}{IR_{\lambda,l}^{m,n} f(z)} \prec \frac{1+Az}{1+Bz},$$

and $\frac{1+Az}{1+Bz}$ *is the best dominant.*

Proof. For $q(z) = \frac{1+Az}{1+Bz}$, $-1 \leq B < A \leq 1$, in Theorem 3.115, we get the corollary. $\square$

Corollary 3.29 (Alb Lupaş, 2018b). *Let* $q(z) = \left(\frac{1+z}{1-z}\right)^{\gamma}$, $m, n \in \mathbb{N}$, $\lambda, l \geq 0$. *Assume that* (3.2.5.10) *holds. If* $f \in \mathcal{A}$ *and*

$$\psi_{\lambda,l}^{m,n}(\alpha, \beta; z) \prec \alpha \left(\frac{1+z}{1-z}\right)^{\gamma} + \frac{2\beta\gamma z}{1-z^2}\left(\frac{1+z}{1-z}\right)^{\gamma},$$

for $\alpha, \beta \in \mathbb{C}$, $0 < \gamma \leq 1$, $\beta \neq 0$, *where* $\psi_{\lambda,l}^{m,n}$ *is defined in* (3.2.5.11), *then*

$$\frac{z\left(IR_{\lambda,l}^{m,n} f(z)\right)'}{IR_{\lambda,l}^{m,n} f(z)} \prec \left(\frac{1+z}{1-z}\right)^{\gamma},$$

and $\left(\frac{1+z}{1-z}\right)^{\gamma}$ *is the best dominant.*

Proof. Corollary follows by using Theorem 3.115 for $q(z) = \left(\frac{1+z}{1-z}\right)^{\gamma}$, $0 < \gamma \leq 1$. $\square$

Theorem 3.116 (Alb Lupaş, 2018b). *Let* q *be convex and univalent in* U *such that* $q(0) = 1$. *Assume that*

$$\operatorname{Re}\left(\frac{\alpha}{\beta} q'(z)\right) > 0, \quad \text{for } \alpha, \beta \in \mathbb{C}, \quad \beta \neq 0. \tag{3.2.5.15}$$

If $f \in \mathcal{A}, \frac{z\left(IR_{\lambda,l}^{m,n} f(z)\right)'}{IR_{\lambda,l}^{m,n} f(z)} \in \mathcal{H}[q(0),1] \cap Q$ *and* $\psi_{\lambda,l}^{m,n}(\alpha,\beta;z)$ *is univalent in* U*, where* $\psi_{\lambda,l}^{m,n}(\alpha,\beta;z)$ *is as defined in* (3.2.5.11)*, then*

$$\alpha q(z) + \beta z q'(z) \prec \psi_{\lambda,l}^{m,n}(\alpha,\beta;z) \tag{3.2.5.16}$$

implies

$$q(z) \prec \frac{z\left(IR_{\lambda,l}^{m,n} f(z)\right)'}{IR_{\lambda,l}^{m,n} f(z)}, \quad \delta \in \mathbb{C}, \quad \delta \neq 0, \quad z \in U, \tag{3.2.5.17}$$

and q *is the best subordinant.*

Proof. Let the function p be defined by $p(z) := \frac{z\left(IR_{\lambda,l}^{m,n} f(z)\right)'}{IR_{\lambda,l}^{m,n} f(z)}$, $z \in U$, $z \neq 0$, $f \in \mathcal{A}$. The function p is analytic in U and $p(0) = 1$.

By setting $\nu(w) := \alpha w$ and $\phi(w) := \beta$ it can be easily verified that ν is analytic in $\mathbb{C}$, ϕ is analytic in $\mathbb{C}\backslash\{0\}$ and that $\phi(w) \neq 0$, $w \in \mathbb{C}\backslash\{0\}$.

Since $\frac{\nu'(q(z))}{\phi(q(z))} = \frac{\alpha}{\beta} q'(z)$, it follows that $\operatorname{Re}\left(\frac{\nu'(q(z))}{\phi(q(z))}\right) = \operatorname{Re}\left(\frac{\alpha}{\beta} q'(z)\right) > 0$, for $\alpha, \beta \in \mathbb{C}$, $\beta \neq 0$.

Now, by using (3.2.5.16), we obtain $\alpha q(z) + \beta z q'(z) \prec \alpha p(z) + \beta z p'(z)$, $z \in U$. From Lemma 1.6, we have $q(z) \prec p(z) = \frac{z\left(IR_{\lambda,l}^{m,n} f(z)\right)'}{IR_{\lambda,l}^{m,n} f(z)}$, $z \in U$, and q is the best subordinant. □

Corollary 3.30 (Alb Lupaş, 2018b). *Let* $q(z) = \frac{1+Az}{1+Bz}$, $-1 \leq B < A \leq 1$, $z \in U$, $m, n \in \mathbb{N}$, $\lambda, l \geq 0$. *Assume that* (3.2.5.15) *holds. If* $f \in \mathcal{A}$, $\frac{z\left(IR_{\lambda,l}^{m,n} f(z)\right)'}{IR_{\lambda,l}^{m,n} f(z)} \in \mathcal{H}[q(0),1] \cap Q$, *and*

$$\alpha \frac{1+Az}{1+Bz} + \frac{\beta(A-B)z}{(1+Bz)^2} \prec \psi_{\lambda,l}^{m,n}(\alpha,\beta;z),$$

for $\alpha, \beta \in \mathbb{C}$, $\beta \neq 0$, $-1 \leq B < A \leq 1$, *where* $\psi_{\lambda,l}^{m,n}$ *is defined in* (3.2.5.11)*, then*

$$\frac{1+Az}{1+Bz} \prec \frac{z\left(IR_{\lambda,l}^{m,n} f(z)\right)'}{IR_{\lambda,l}^{m,n} f(z)},$$

and $\frac{1+Az}{1+Bz}$ *is the best subordinant.*

Proof. For $q(z) = \frac{1+Az}{1+Bz}$, $-1 \leq B < A \leq 1$, in Theorem 3.116 we get the corollary. □

Corollary 3.31 (Alb Lupaş, 2018b). *Let* $q(z) = \left(\frac{1+z}{1-z}\right)^{\gamma}$, $m, n \in \mathbb{N}$, $\lambda, l \geq 0$. *Assume that* (3.2.5.15) *holds. If* $f \in \mathcal{A}$, $\frac{z(IR_{\lambda,l}^{m,n} f(z))'}{IR_{\lambda,l}^{m,n} f(z)} \in \mathcal{H}[q(0), 1] \cap Q$ *and*

$$\alpha\left(\frac{1+z}{1-z}\right)^{\gamma} + \frac{2\beta\gamma z}{1-z^2}\left(\frac{1+z}{1-z}\right)^{\gamma} \prec \psi_{\lambda,l}^{m,n}(\alpha, \beta; z),$$

for $\alpha, \beta \in \mathbb{C}$, $0 < \gamma \leq 1$, $\beta \neq 0$, *where* $\psi_{\lambda,l}^{m,n}$ *is defined in* (3.2.5.11), *then*

$$\left(\frac{1+z}{1-z}\right)^{\gamma} \prec \frac{z\left(IR_{\lambda,l}^{m,n} f(z)\right)'}{IR_{\lambda,l}^{m,n} f(z)},$$

and $\left(\frac{1+z}{1-z}\right)^{\gamma}$ *is the best subordinant.*

Proof. Corollary follows by using Theorem 3.116 for $q(z) = \left(\frac{1+z}{1-z}\right)^{\gamma}$, $0 < \gamma \leq 1$. □

Combining Theorems 3.115 and 3.116, we state the following sandwich theorem.

Theorem 3.117 (Alb Lupaş, 2018b). *Let* q_1 *and* q_2 *be convex and univalent in* U *such that* $q_1(z) \neq 0$ *and* $q_2(z) \neq 0$, *for all* $z \in U$. *Suppose that* q_1 *satisfies* (3.2.5.10) *and* q_2 *satisfies* (3.2.5.15). *If* $f \in \mathcal{A}$, $\frac{z\left(IR_{\lambda,l}^{m,n} f(z)\right)'}{IR_{\lambda,l}^{m,n} f(z)} \in \mathcal{H}[q(0), 1] \cap Q$, *and* $\psi_{\lambda,l}^{m,n}(\alpha, \beta; z)$ *is defined as in* (3.2.5.11) *univalent in* U, *then*

$$\alpha q_1(z) + \beta z q_1'(z) \prec \psi_{\lambda,l}^{m,n}(\alpha, \beta; z) \prec \alpha q_2(z) + \beta z q_2'(z),$$

for $\alpha, \beta \in \mathbb{C}$, $\beta \neq 0$, *implies*

$$q_1(z) \prec \frac{z\left(IR_{\lambda,l}^{m,n} f(z)\right)'}{IR_{\lambda,l}^{m,n} f(z)} \prec q_2(z), \quad z \in U,$$

and q_1 *and* q_2 *are respectively the best subordinant and the best dominant.*

For $q_1(z) = \frac{1+A_1 z}{1+B_1 z}$, $q_2(z) = \frac{1+A_2 z}{1+B_2 z}$, where $-1 \leq B_2 < B_1 < A_1 < A_2 \leq 1$, we have the following corollary.

Corollary 3.32 (Alb Lupaş, 2018b). *Let* $m, n \in \mathbb{N}$, $\lambda, l \geq 0$. *Assume that* (3.2.5.10) *and* (3.2.5.15) *hold for* $q_1(z) = \frac{1+A_1 z}{1+B_1 z}$ *and* $q_2(z) = \frac{1+A_2 z}{1+B_2 z}$, *respectively. If* $f \in \mathcal{A}$, $\frac{z\left(IR_{\lambda,l}^{m,n}f(z)\right)'}{IR_{\lambda,l}^{m,n}f(z)} \in \mathcal{H}\left[q(0),1\right] \cap Q$ *and*

$$\alpha \frac{1+A_1 z}{1+B_1 z} + \frac{\beta\left(A_1 - B_1\right) z}{\left(1+B_1 z\right)^2} \prec \psi_{\lambda,l}^{m,n}\left(\alpha,\beta;z\right) \prec \alpha \frac{1+A_2 z}{1+B_2 z} + \frac{\beta\left(A_2 - B_2\right) z}{\left(1+B_2 z\right)^2},$$

$z \in U$, *for* $\alpha, \beta \in \mathbb{C}$, $\beta \neq 0$, $-1 \leq B_2 \leq B_1 < A_1 \leq A_2 \leq 1$, *where* $\psi_{\lambda,l}^{m,n}$ *is defined in* (3.2.5.11), *then*

$$\frac{1+A_1 z}{1+B_1 z} \prec \frac{z\left(IR_{\lambda,l}^{m,n}f(z)\right)'}{IR_{\lambda,l}^{m,n}f(z)} \prec \frac{1+A_2 z}{1+B_2 z},$$

$z \in U$, *hence* $\frac{1+A_1 z}{1+B_1 z}$ *and* $\frac{1+A_2 z}{1+B_2 z}$ *are the best subordinant and the best dominant, respectively.*

For $q_1(z) = \left(\frac{1+z}{1-z}\right)^{\gamma_1}$, $q_2(z) = \left(\frac{1+z}{1-z}\right)^{\gamma_2}$, where $0 < \gamma_1 < \gamma_2 \leq 1$, we have the following corollary.

Corollary 3.33 (Alb Lupaş, 2018b). *Let* $m, n \in \mathbb{N}$, $\lambda, l \geq 0$. *Assume that* (3.2.5.10) *and* (3.2.5.15) *hold for* $q_1(z) = \left(\frac{1+z}{1-z}\right)^{\gamma_1}$ *and* $q_2(z) = \left(\frac{1+z}{1-z}\right)^{\gamma_2}$, *respectively. If* $f \in \mathcal{A}$, $\frac{z\left(IR_{\lambda,l}^{m,n}f(z)\right)'}{IR_{\lambda,l}^{m,n}f(z)} \in \mathcal{H}\left[q(0),1\right] \cap Q$ *and*

$$\alpha\left(\frac{1+z}{1-z}\right)^{\gamma_1} + \frac{2\beta\gamma_1 z}{1-z^2}\left(\frac{1+z}{1-z}\right)^{\gamma_1} \prec \psi_{\lambda,l}^{m,n}\left(\alpha,\beta;z\right) \prec \alpha\left(\frac{1+z}{1-z}\right)^{\gamma_2} + \frac{2\beta\gamma_2 z}{1-z^2}\left(\frac{1+z}{1-z}\right)^{\gamma_2},$$

$z \in U$, *for* $\alpha, \beta \in \mathbb{C}$, $\beta \neq 0$, $0 < \gamma_1 < \gamma_2 \leq 1$, *where* $\psi_{\lambda,l}^{m,n}$ *is defined in* (3.2.5.11), *then*

$$\left(\frac{1+z}{1-z}\right)^{\gamma_1} \prec \frac{z\left(IR_{\lambda,l}^{m,n}f(z)\right)'}{IR_{\lambda,l}^{m,n}f(z)} \prec \left(\frac{1+z}{1-z}\right)^{\gamma_2},$$

$z \in U$, *hence* $\left(\frac{1+z}{1-z}\right)^{\gamma_1}$ *and* $\left(\frac{1+z}{1-z}\right)^{\gamma_2}$ *are the best subordinant and the best dominant, respectively.*

In the paper Alb Lupaş (2017c), we obtain another subordination and superordination results for the operator $IR_{\lambda.l}^{m,n}$ and we establish differential sandwich-type theorems.

Using simple computation, we obtain the following relation.

Proposition 3.2 (Alb Lupaş, 2015h). *For $m, n \in \mathbb{N}$ and $\lambda \geq 0$, we have*

$$IR_{\lambda,l}^{m+1,n} f(z) = \frac{1+l-\lambda}{l+1} IR_{\lambda,l}^{m,n} f(z) + \frac{\lambda}{l+1} z \left(IR_{\lambda,l}^{m,n} f(z) \right)' \quad (3.2.5.18)$$

We intend to find sufficient conditions for certain normalized analytic functions f such that

$$q_1(z) \prec \frac{z^{\delta} IR_{\lambda,l}^{m+1,n} f(z)}{\left(IR_{\lambda,l}^{m,n} f(z) \right)^{1+\delta}} \prec q_2(z),$$

$z \in U$, $0 < \delta \leq 1$, where q_1 and q_2 are given univalent functions.

Theorem 3.118 (Alb Lupaş, 2017c). *Let* $\frac{z^{\delta} IR_{\lambda,l}^{m+1,n} f(z)}{\left(IR_{\lambda,l}^{m,n} f(z) \right)^{1+\delta}} \in \mathcal{H}(U)$ *and let the function $q(z)$ be analytic and univalent in U such that $q(z) \neq 0$, for all $z \in U$. Suppose that $\frac{zq'(z)}{q(z)}$ is starlike univalent in U. Let*

$$\operatorname{Re}\left(\frac{\xi}{\beta} q(z) + \frac{2\mu}{\beta} q^2(z) + 1 + z\frac{q''(z)}{q'(z)} - z\frac{q'(z)}{q(z)} \right) > 0, \quad (3.2.5.19)$$

for $\alpha, \xi, \beta, \mu \in \mathbb{C}$, $\beta \neq 0$, $z \in U$ and

$$\begin{aligned} \psi_{\lambda,l}^{m,n}(\alpha, \xi, \mu, \beta; z) := {} & \alpha + \beta\frac{(l+1)}{\lambda} + \beta\frac{(l+1)}{\lambda} \frac{IR_{\lambda,l}^{m+2,n} f(z)}{IR_{\lambda,l}^{m+1,n} f(z)} \\ & - \beta\frac{(l+1)(1+\delta)}{\lambda} \frac{IR_{\lambda,l}^{m+1,n} f(z)}{IR_{\lambda,l}^{m+1,n} f(z)} \\ & + \xi\frac{z^{\delta} IR_{\lambda,l}^{m+1,n} f(z)}{\left(IR_{\lambda,l}^{m,n} f(z) \right)^{1+\delta}} + \mu\frac{z^{2\delta} \left(IR_{\lambda,l}^{m+1,n} f(z) \right)^2}{\left(IR_{\lambda,l}^{m,n} f(z) \right)^{2+2\delta}}. \end{aligned} \quad (3.2.5.20)$$

If q satisfies the following subordination

$$\psi_{\lambda,l}^{m,n}(\alpha, \beta, \mu; z) \prec \alpha + \xi q(z) + \mu (q(z))^2 + \beta\frac{zq'(z)}{q(z)}, \quad (3.2.5.21)$$

for $\alpha, \xi, \beta, \mu \in \mathbb{C}$, $\beta \neq 0$, *then* $\frac{z^{\delta} IR_{\lambda,l}^{m+1,n} f(z)}{\left(IR_{\lambda,l}^{m,n} f(z)\right)^{1+\delta}} \prec q(z)$, *and* q *is the best dominant.*

Proof. Consider $p(z) := \frac{z^{\delta} IR_{\lambda,l}^{m+1,n} f(z)}{\left(IR_{\lambda,l}^{m,n} f(z)\right)^{1+\delta}}$, $z \in U$, $z \neq 0$, $f \in \mathcal{A}$. Differentiating, we obtain $p'(z) = \frac{\delta(1+l)}{\lambda} \frac{z^{\delta-1} IR_{\lambda,l}^{m+1,n} f(z)}{\left(IR_{\lambda,l}^{m,n} f(z)\right)^{1+\delta}} + \frac{l+1}{\lambda} \frac{z^{\delta-1} IR_{\lambda,l}^{m+2,n} f(z)}{\left(IR_{\lambda,l}^{m,n} f(z)\right)^{1+\delta}} - \frac{(l+1)(1+\delta)}{\lambda} \frac{z^{\delta-1} \left(IR_{\lambda,l}^{m+1,n} f(z)\right)^2}{\left(IR_{\lambda,l}^{m,n} f(z)\right)^{2+\delta}}$.

By using the identity (3.2.5.18), we obtain

$$\frac{zp'(z)}{p(z)} = \frac{\delta(l+1)}{\lambda} + \frac{l+1}{\lambda} \frac{IR_{\lambda,l}^{m+2,n} f(z)}{IR_{\lambda,l}^{m+1,n} f(z)} - \frac{(l+1)(1+\delta)}{\lambda} \frac{IR_{\lambda,l}^{m+1,n} f(z)}{IR_{\lambda,l}^{m+1,n} f(z)}. \tag{3.2.5.22}$$

By setting $\theta(w) := \alpha + \xi w + \mu w^2$ and $\phi(w) := \frac{\beta}{w}$, it can be easily verified that θ is analytic in $\mathbb{C}$, ϕ is analytic in $\mathbb{C}\backslash\{0\}$ and that $\phi(w) \neq 0$, $w \in \mathbb{C}\backslash\{0\}$.

Also, by letting $Q(z) = zq'(z)\phi(q(z)) = \beta\frac{zq'(z)}{q(z)}$ and $h(z) = \theta(q(z)) + Q(z) = \alpha + \xi q(z) + \mu(q(z))^2 + \beta\frac{zq'(z)}{q(z)}$, we find that $Q(z)$ is starlike univalent in U.

We get $h'(z) = \xi q'(z) + 2\mu q(z) q'(z) + \beta\frac{q'(z)}{q(z)} + \beta z\frac{q''(z)}{q(z)} - \beta z\left(\frac{q'(z)}{q(z)}\right)^2$ and $\frac{zh'(z)}{Q(z)} = \frac{\xi}{\beta} q(z) + \frac{2\mu}{\beta} q^2(z) + 1 + z\frac{q''(z)}{q(z)} - z\frac{q'(z)}{q(z)}$.

So we deduce that

$$\operatorname{Re}\left(\frac{zh'(z)}{Q(z)}\right) = \operatorname{Re}\left(\frac{\xi}{\beta} q(z) + \frac{2\mu}{\beta} q^2(z) + 1 + z\frac{q''(z)}{q(z)} - z\frac{q'(z)}{q(z)}\right) > 0.$$

By using (3.2.5.22), we obtain $\alpha + \xi p(z) + \mu(p(z))^2 + \beta\frac{zp'(z)}{p(z)} = \alpha + \beta\frac{(l+1)}{\lambda} + \beta\frac{(l+1)}{\lambda} \frac{IR_{\lambda,l}^{m+2,n} f(z)}{IR_{\lambda,l}^{m+1,n} f(z)} - \beta\frac{(l+1)(1+\delta)}{\lambda} \frac{IR_{\lambda,l}^{m+1,n} f(z)}{IR_{\lambda,l}^{m+1,n} f(z)} + \xi\frac{z^{\delta} IR_{\lambda,l}^{m+1,n} f(z)}{\left(IR_{\lambda,l}^{m,n} f(z)\right)^{1+\delta}} + \mu\frac{z^{2\delta}\left(IR_{\lambda,l}^{m+1,n} f(z)\right)^2}{\left(IR_{\lambda,l}^{m,n} f(z)\right)^{2+2\delta}}$.

By using (3.2.5.21), we have $\alpha + \xi p(z) + \mu(p(z))^2 + \beta\frac{zp'(z)}{p(z)} \prec \alpha + \xi q(z) + \mu(q(z))^2 + \beta\frac{zq'(z)}{q(z)}$.

Applying Lemma 1.3, we obtain $p(z) \prec q(z)$, $z \in U$, i.e. $\frac{z^{\delta} IR_{\lambda,l}^{m+1,n} f(z)}{\left(IR_{\lambda,l}^{m,n} f(z)\right)^{1+\delta}} \prec q(z)$, $z \in U$ and q is the best dominant. ☐

Corollary 3.34 (Alb Lupaş, 2017c). *Let $m, n \in \mathbb{N}$, $\lambda, l \geq 0$. Assume that* (3.2.5.19) *holds. If $f \in \mathcal{A}$ and*

$$\psi_{\lambda,l}^{m,n}(\alpha,\beta,\mu;z) \prec \alpha + \xi\frac{1+Az}{1+Bz} + \mu\left(\frac{1+Az}{1+Bz}\right)^2 + \frac{\beta(A-B)z}{(1+Az)(1+Bz)},$$

for $\alpha, \beta, \mu, \xi \in \mathbb{C}$, $\beta \neq 0$, $-1 \leq B < A \leq 1$, where $\psi_{\lambda,l}^{m,n}$ is defined in (3.2.5.20), *then*

$$\frac{z^{\delta} IR_{\lambda,l}^{m+1,n} f(z)}{\left(IR_{\lambda,l}^{m,n} f(z)\right)^{1+\delta}} \prec \frac{1+Az}{1+Bz},$$

and $\frac{1+Az}{1+Bz}$ is the best dominant.

Proof. For $q(z) = \frac{1+Az}{1+Bz}$, $-1 \leq B < A \leq 1$ in Theorem 3.118, we get the corollary. □

Corollary 3.35 (Alb Lupaş, 2017c). *Let $m, n \in \mathbb{N}$, $\lambda, l \geq 0$. Assume that* (3.2.5.19) *holds. If $f \in \mathcal{A}$ and*

$$\psi_{\lambda,l}^{m,n}(\alpha,\beta,\mu;z) \prec \alpha + \xi\left(\frac{1+z}{1-z}\right)^{\gamma} + \mu\left(\frac{1+z}{1-z}\right)^{2\gamma} + \frac{2\beta\gamma z}{1-z^2},$$

for $\alpha, \beta, \mu, \xi \in \mathbb{C}$, $0 < \gamma \leq 1$, $\beta \neq 0$, where $\psi_{\lambda,l}^{m,n}$ is defined in (3.2.5.20), *then*

$$\frac{z^{\delta} IR_{\lambda,l}^{m+1,n} f(z)}{\left(IR_{\lambda,l}^{m,n} f(z)\right)^{1+\delta}} \prec \left(\frac{1+z}{1-z}\right)^{\gamma},$$

and $\left(\frac{1+z}{1-z}\right)^{\gamma}$ is the best dominant.

Proof. Corollary follows by using Theorem 3.118 for $q(z) = \left(\frac{1+z}{1-z}\right)^{\gamma}$, $0 < \gamma \leq 1$. □

Theorem 3.119 (Alb Lupaş, 2017c). *Let q be analytic and univalent in U such that $q(z) \neq 0$ and $\frac{zq'(z)}{q(z)}$ be starlike univalent in U. Assume that*

$$\operatorname{Re}\left(\frac{\xi}{\beta}q(z)q'(z) + \frac{2\mu}{\beta}q^2(z)q'(z)\right) > 0, \quad \text{for } \xi, \beta, \mu \in \mathbb{C}, \quad \beta \neq 0. \tag{3.2.5.23}$$

If $f \in \mathcal{A}$, $\frac{z^{\delta} IR_{\lambda,l}^{m+1,n} f(z)}{\left(IR_{\lambda,l}^{m,n} f(z)\right)^{1+\delta}} \in \mathcal{H}[q(0),1] \cap Q$ *and* $\psi_{\lambda,l}^{m,n}(\alpha,\beta,\mu;z)$ *is univalent in* U, *where* $\psi_{\lambda,l}^{m,n}(\alpha,\beta,\mu;z)$ *is as defined in* (3.2.5.20), *then*

$$\alpha + \xi q(z) + \mu (q(z))^2 + \frac{\beta z q'(z)}{q(z)} \prec \psi_{\lambda,l}^{m,n}(\alpha,\beta,\mu;z) \qquad (3.2.5.24)$$

implies $q(z) \prec \frac{z^{\delta} IR_{\lambda,l}^{m+1,n} f(z)}{\left(IR_{\lambda,l}^{m,n} f(z)\right)^{1+\delta}}$, $z \in U$, *and* q *is the best subordinant.*

Proof. Consider $p(z) := \frac{z^{\delta} IR_{\lambda,l}^{m+1,n} f(z)}{\left(IR_{\lambda,l}^{m,n} f(z)\right)^{1+\delta}}$, $z \in U$, $z \neq 0$, $f \in \mathcal{A}$.

By setting $\theta(w) := \alpha + \xi w + \mu w^2$ and $\phi(w) := \frac{\beta}{w}$ it can be easily verified that θ is analytic in $\mathbb{C}$, ϕ is analytic in $\mathbb{C}\backslash\{0\}$ and that $\phi(w) \neq 0$, $w \in \mathbb{C}\backslash\{0\}$.

Since $\frac{\theta'(q(z))}{\phi(q(z))} = \frac{q'(z)q(z)[\xi+2\mu q(z)]}{\beta}$, it follows that $\operatorname{Re}\left(\frac{\theta'(q(z))}{\phi(q(z))}\right) = \operatorname{Re}\left(\frac{\xi}{\beta} q(z) q'(z) + \frac{2\mu}{\beta} q^2(z) q'(z)\right) > 0$, for $\alpha,\beta,\mu \in \mathbb{C}$, $\mu \neq 0$.

By using (3.2.5.22) and (3.2.5.24), we get $\alpha + \xi q(z) + \mu(q(z))^2 + \frac{\beta z q'(z)}{q(z)} \prec \alpha + \xi p(z) + \mu(p(z))^2 + \frac{\beta z p'(z)}{p(z)}$. Applying Lemma 1.6, we obtain $q(z) \prec p(z) = \frac{z^{\delta} IR_{\lambda,l}^{m+1,n} f(z)}{\left(IR_{\lambda,l}^{m,n} f(z)\right)^{1+\delta}}$, $z \in U$, and q is the best subordinant. □

Corollary 3.36 (Alb Lupaş, 2017c). *Let* $m,n \in \mathbb{N}$, $\lambda, l \geq 0$. *Assume that* (3.2.5.23) *holds. If* $f \in \mathcal{A}$, $\frac{z^{\delta} IR_{\lambda,l}^{m+1,n} f(z)}{\left(IR_{\lambda,l}^{m,n} f(z)\right)^{1+\delta}} \in \mathcal{H}[q(0),1] \cap Q$ *and*

$$\alpha + \xi \frac{1+Az}{1+Bz} + \mu \left(\frac{1+Az}{1+Bz}\right)^2 + \frac{\beta(A-B)z}{(1+Az)(1+Bz)} \prec \psi_{\lambda,l}^{m,n}(\alpha,\beta,\mu;z),$$

for $\alpha,\beta,\xi,\mu \in \mathbb{C}$, $\beta \neq 0$, $-1 \leq B < A \leq 1$, *where* $\psi_{\lambda,l}^{m,n}$ *is defined in* (3.2.5.20), *then*

$$\frac{1+Az}{1+Bz} \prec \frac{z^{\delta} IR_{\lambda,l}^{m+1,n} f(z)}{\left(IR_{\lambda,l}^{m,n} f(z)\right)^{1+\delta}},$$

and $\frac{1+Az}{1+Bz}$ *is the best subordinant.*

Proof. For $q(z) = \frac{1+Az}{1+Bz}$, $-1 \leq B < A \leq 1$ in Theorem 3.119, we get the corollary. □

Corollary 3.37 (Alb Lupaş, 2017c). *Let $m, n \in \mathbb{N}$, $\lambda, l \geq 0$. Assume that (3.2.5.23) holds. If $f \in \mathcal{A}$, $\frac{z^{\delta} IR_{\lambda,l}^{m+1,n} f(z)}{\left(IR_{\lambda,l}^{m,n} f(z)\right)^{1+\delta}} \in \mathcal{H}[q(0), 1] \cap Q$ and*

$$\alpha + \xi \left(\frac{1+z}{1-z}\right)^{\gamma} + \mu \left(\frac{1+z}{1-z}\right)^{2\gamma} + \frac{2\beta\gamma z}{1-z^2} \prec \psi_{\lambda,l}^{m,n}(\alpha, \beta, \mu; z),$$

for $\alpha, \beta, \mu, \xi \in \mathbb{C}$, $\beta \neq 0$, $0 < \gamma \leq 1$, where $\psi_{\lambda,l}^{m,n}$ is defined in (3.2.5.20), then

$$\left(\frac{1+z}{1-z}\right)^{\gamma} \prec \frac{z^{\delta} IR_{\lambda,l}^{m+1,n} f(z)}{\left(IR_{\lambda,l}^{m,n} f(z)\right)^{1+\delta}},$$

and $\left(\frac{1+z}{1-z}\right)^{\gamma}$ is the best subordinant.

Proof. For $q(z) = \left(\frac{1+z}{1-z}\right)^{\gamma}$, $0 < \gamma \leq 1$ in Theorem 3.119, we get the corollary. □

Combining Theorems 3.118 and 3.119, we state the following sandwich theorem.

Theorem 3.120 (Alb Lupaş, 2017c). *Let q_1 and q_2 be analytic and univalent in U such that $q_1(z) \neq 0$ and $q_2(z) \neq 0$, for all $z \in U$, with $\frac{zq_1'(z)}{q_1(z)}$ and $\frac{zq_2'(z)}{q_2(z)}$ being starlike univalent. Suppose that q_1 satisfies (3.2.5.19) and q_2 satisfies (3.2.5.23). If $f \in \mathcal{A}$, $\frac{z^{\delta} IR_{\lambda,l}^{m+1,n} f(z)}{\left(IR_{\lambda,l}^{m,n} f(z)\right)^{1+\delta}} \in \mathcal{H}[q(0), 1] \cap Q$ and $\psi_{\lambda,l}^{m,n}(\alpha, \beta, \mu; z)$ is as defined in (3.2.5.20) univalent in U, then*

$$\alpha + \xi q_1(z) + \mu (q_1(z))^2 + \frac{\beta z q_1'(z)}{q_1(z)} \prec \psi_{\lambda,l}^{m,n}(\alpha, \beta, \mu; z)$$
$$\prec \alpha + \xi q_2(z) + \mu (q_2(z))^2 + \frac{\beta z q_2'(z)}{q_2(z)},$$

for $\alpha, \beta, \mu, \xi \in \mathbb{C}$, $\beta \neq 0$, implies

$$q_1(z) \prec \frac{z^{\delta} IR_{\lambda,l}^{m+1,n} f(z)}{\left(IR_{\lambda,l}^{m,n} f(z)\right)^{1+\delta}} \prec q_2(z),$$

and q_1 and q_2 are respectively the best subordinant and the best dominant.

For $q_1(z) = \frac{1+A_1 z}{1+B_1 z}$, $q_2(z) = \frac{1+A_2 z}{1+B_2 z}$, where $-1 \le B_2 < B_1 < A_1 < A_2 \le 1$, we have the following corollary.

Corollary 3.38 (Alb Lupaş, 2017c). *Let* $m, n \in \mathbb{N}$, $\lambda, l \ge 0$. *Assume that* (3.2.5.19) *and* (3.2.5.23) *hold. If* $f \in \mathcal{A}$, $\frac{z^{\delta} IR_{\lambda,l}^{m+1,n} f(z)}{\left(IR_{\lambda,l}^{m,n} f(z)\right)^{1+\delta}} \in \mathcal{H}[q(0),1] \cap Q$ *and*

$$\alpha + \xi \frac{1+A_1 z}{1+B_1 z} + \mu \left(\frac{1+A_1 z}{1+B_1 z}\right)^2 + \frac{\beta (A_1 - B_1) z}{(1+A_1 z)(1+B_1 z)} \prec \psi_{\lambda,l}^{m,n}(\alpha, \beta, \mu; z)$$
$$\prec \alpha + \xi \frac{1+A_2 z}{1+B_2 z} + \mu \left(\frac{1+A_2 z}{1+B_2 z}\right)^2 + \frac{\beta (A_2 - B_2) z}{(1+A_2 z)(1+B_2 z)},$$

for $\alpha, \beta, \mu, \xi \in \mathbb{C}$, $\beta \ne 0$, $-1 \le B_2 \le B_1 < A_1 \le A_2 \le 1$, *where* $\psi_{\lambda,l}^{m,n}$ *is defined in* (3.2.5.20), *then*

$$\frac{1+A_1 z}{1+B_1 z} \prec \frac{z^{\delta} IR_{\lambda,l}^{m+1,n} f(z)}{\left(IR_{\lambda,l}^{m,n} f(z)\right)^{1+\delta}} \prec \frac{1+A_2 z}{1+B_2 z},$$

hence $\frac{1+A_1 z}{1+B_1 z}$ *and* $\frac{1+A_2 z}{1+B_2 z}$ *are the best subordinant and the best dominant, respectively.*

For $q_1(z) = \left(\frac{1+z}{1-z}\right)^{\gamma_1}$, $q_2(z) = \left(\frac{1+z}{1-z}\right)^{\gamma_2}$, where $0 < \gamma_1 < \gamma_2 \le 1$, we have the following corollary.

Corollary 3.39 (Alb Lupaş, 2017c). *Let* $m, n \in \mathbb{N}$, $\lambda, l \ge 0$. *Assume that* (3.2.5.19) *and* (3.2.5.23) *hold. If* $f \in \mathcal{A}$, $\frac{z^{\delta} IR_{\lambda,l}^{m+1,n} f(z)}{\left(IR_{\lambda,l}^{m,n} f(z)\right)^{1+\delta}} \in \mathcal{H}[q(0),1] \cap Q$ *and*

$$\alpha + \xi \left(\frac{1+z}{1-z}\right)^{\gamma_1} + \mu \left(\frac{1+z}{1-z}\right)^{2\gamma_1} + \frac{2\beta\gamma_1 z}{1-z^2} \prec \psi_{\lambda,l}^{m,n}(\alpha, \beta, \mu; z)$$
$$\prec \alpha + \xi \left(\frac{1+z}{1-z}\right)^{\gamma_2} + \mu \left(\frac{1+z}{1-z}\right)^{2\gamma_2} + \frac{2\beta\gamma_2 z}{1-z^2},$$

for $\alpha, \beta, \mu, \xi \in \mathbb{C}$, $\beta \ne 0$, $0 < \gamma_1 < \gamma_2 \le 1$, *where* $\psi_{\lambda,l}^{m,n}$ *is defined in* (3.2.5.20), *then*

$$\left(\frac{1+z}{1-z}\right)^{\gamma_1} \prec \frac{z^{\delta} IR_{\lambda,l}^{m+1,n} f(z)}{\left(IR_{\lambda,l}^{m,n} f(z)\right)^{1+\delta}} \prec \left(\frac{1+z}{1-z}\right)^{\gamma_2},$$

hence $\left(\frac{1+z}{1-z}\right)^{\gamma_1}$ *and* $\left(\frac{1+z}{1-z}\right)^{\gamma_2}$ *are the best subordinant and the best dominant, respectively.*

Changing the functions θ and ϕ, we obtain the following results.

Theorem 3.121 (Alb Lupaş, 2017c). *Let* $\frac{z^{\delta} IR_{\lambda,l}^{m+1,n} f(z)}{\left(IR_{\lambda,l}^{m,n} f(z)\right)^{1+\delta}} \in \mathcal{H}(U)$, $f \in \mathcal{A}$, $z \in U$, $m,n \in \mathbb{N}$, $\lambda, l \geq 0$ *and let the function* $q(z)$ *be convex and univalent in* U *such that* $q(0) = 1$, $z \in U$. *Assume that*

$$\operatorname{Re}\left(\frac{\alpha+\beta}{\beta} + z\frac{q''(z)}{q'(z)}\right) > 0, \tag{3.2.5.25}$$

for $\alpha, \beta \in \mathbb{C}$, $\beta \neq 0$, $z \in U$, *and*

$$\begin{aligned}\psi_{\lambda,l}^{m,n}(\alpha,\beta;z) :=\ & \frac{\beta(l+1)}{\lambda}\frac{z^{\delta} IR_{\lambda,l}^{m+2,n} f(z)}{\left(IR_{\lambda,l}^{m,n} f(z)\right)^{1+\delta}} \\ & + \left(\alpha + \frac{\beta\delta(l+1)}{\lambda}\right)\frac{z^{\delta} IR_{\lambda,l}^{m+1,n} f(z)}{\left(IR_{\lambda,l}^{m,n} f(z)\right)^{1+\delta}} \\ & - \frac{\beta(1+\delta)(l+1)}{\lambda}\frac{z^{\delta}\left(IR_{\lambda,l}^{m+1,n} f(z)\right)^{2}}{\left(IR_{\lambda,l}^{m,n} f(z)\right)^{2+\delta}}.\end{aligned} \tag{3.2.5.26}$$

If q *satisfies the following subordination*

$$\psi_{\lambda,l}^{m,n}(\alpha,\beta;z) \prec \alpha q(z) + \beta z q'(z), \tag{3.2.5.27}$$

for $\alpha, \beta \in \mathbb{C}$, $\beta \neq 0$, $z \in U$, *then*

$$\frac{z^{\delta} IR_{\lambda,l}^{m+1,n} f(z)}{\left(IR_{\lambda,l}^{m,n} f(z)\right)^{1+\delta}} \prec q(z), \quad z \in U,$$

and q *is the best dominant.*

Proof. Consider $p(z) := \frac{z^{\delta} IR_{\lambda,l}^{m+1,n} f(z)}{\left(IR_{\lambda,l}^{m,n} f(z)\right)^{1+\delta}}$, $z \in U$, $z \neq 0$, $f \in \mathcal{A}$. The function p is analytic in U and $p(0) = 1$.

Differentiating, we get

$$p'(z)=\frac{\delta(1+l)}{\lambda}\frac{z^{\delta-1}IR_{\lambda,l}^{m+1,n}f(z)}{\left(IR_{\lambda,l}^{m,n}f(z)\right)^{1+\delta}}+\frac{l+1}{\lambda}\frac{z^{\delta-1}IR_{\lambda,l}^{m+2,n}f(z)}{\left(IR_{\lambda,l}^{m,n}f(z)\right)^{1+\delta}}$$
$$-\frac{(l+1)(1+\delta)}{\lambda}\frac{z^{\delta-1}\left(IR_{\lambda,l}^{m+1,n}f(z)\right)^{2}}{\left(IR_{\lambda,l}^{m,n}f(z)\right)^{2+\delta}}.$$

By using the identity (3.2.5.18), we get

$$zp'(z)=\frac{l+1}{\lambda}\frac{z^{\delta}IR_{\lambda,l}^{m+2,n}f(z)}{\left(IR_{\lambda,l}^{m,n}f(z)\right)^{1+\delta}}+\frac{\delta(1+l)}{\lambda}\frac{z^{\delta}IR_{\lambda,l}^{m+1,n}f(z)}{\left(IR_{\lambda,l}^{m,n}f(z)\right)^{1+\delta}}$$
$$-\frac{(l+1)(1+\delta)}{\lambda}\frac{z^{\delta}\left(IR_{\lambda,l}^{m+1,n}f(z)\right)^{2}}{\left(IR_{\lambda,l}^{m,n}f(z)\right)^{2+\delta}}. \qquad (3.2.5.28)$$

By setting $\theta(w):=\alpha w$ and $\phi(w):=\beta$, it can be easily verified that θ is analytic in $\mathbb{C}$, ϕ is analytic in $\mathbb{C}\backslash\{0\}$ and that $\phi(w)\neq 0$, $w\in\mathbb{C}\backslash\{0\}$.

Also, by letting $Q(z)=zq'(z)\phi(q(z))=\beta zq'(z)$, we find that $Q(z)$ is starlike univalent in U.

Let $h(z)=\theta(q(z))+Q(z)=\alpha q(z)+\beta zq'(z)$. We have $\operatorname{Re}\left(\frac{zh'(z)}{Q(z)}\right)=\operatorname{Re}\left(\frac{\alpha+\beta}{\beta}+z\frac{q''(z)}{q'(z)}\right)>0$.

By using (3.2.5.28), we obtain

$$\alpha p(z)+\beta zp'(z)=\frac{\beta(l+1)}{\lambda}\frac{z^{\delta}IR_{\lambda,l}^{m+2,n}f(z)}{\left(IR_{\lambda,l}^{m,n}f(z)\right)^{1+\delta}}$$
$$+\left(\alpha+\frac{\beta\delta(l+1)}{\lambda}\right)\frac{z^{\delta}IR_{\lambda,l}^{m+1,n}f(z)}{\left(IR_{\lambda,l}^{m,n}f(z)\right)^{1+\delta}}$$
$$-\frac{\beta(1+\delta)(l+1)}{\lambda}\frac{z^{\delta}\left(IR_{\lambda,l}^{m+1,n}f(z)\right)^{2}}{\left(IR_{\lambda,l}^{m,n}f(z)\right)^{2+\delta}}.$$

By using (3.2.5.27), we have $\alpha p(z)+\beta zp'(z)\prec\alpha q(z)+\beta zq'(z)$. From Lemma 1.3, we have $p(z)\prec q(z)$, $z\in U$, i.e. $\frac{z^{\delta}IR_{\lambda,l}^{m+1,n}f(z)}{\left(IR_{\lambda,l}^{m,n}f(z)\right)^{1+\delta}}\prec q(z)$, $z\in U$, and q is the best dominant. $\square$

Corollary 3.40 (Alb Lupaş, 2017c). *Let* $q(z) = \frac{1+Az}{1+Bz}$, $z \in U$, $-1 \leq B < A \leq 1$, $m, n \in \mathbb{N}$, $\lambda, l \geq 0$. *Assume that* (3.2.5.25) *holds. If* $f \in \mathcal{A}$ *and*

$$\psi_{\lambda,l}^{m,n}(\alpha, \beta; z) \prec \alpha \frac{1+Az}{1+Bz} + \frac{\beta(A-B)z}{(1+Bz)^2},$$

for $\alpha, \beta \in \mathbb{C}$, $\beta \neq 0$, $-1 \leq B < A \leq 1$, *where* $\psi_{\lambda,l}^{m,n}$ *is defined in* (3.2.5.26), *then*

$$\frac{z^{\delta} IR_{\lambda,l}^{m+1,n} f(z)}{\left(IR_{\lambda,l}^{m,n} f(z)\right)^{1+\delta}} \prec \frac{1+Az}{1+Bz},$$

and $\frac{1+Az}{1+Bz}$ *is the best dominant.*

Proof. For $q(z) = \frac{1+Az}{1+Bz}$, $-1 \leq B < A \leq 1$, in Theorem 3.121, we get the corollary. □

Corollary 3.41 (Alb Lupaş, 2017c). *Let* $q(z) = \left(\frac{1+z}{1-z}\right)^{\gamma}$, $m, n \in \mathbb{N}$, $\lambda, l \geq 0$. *Assume that* (3.2.5.25) *holds. If* $f \in \mathcal{A}$ *and*

$$\psi_{\lambda,l}^{m,n}(\alpha, \beta; z) \prec \alpha \left(\frac{1+z}{1-z}\right)^{\gamma} + \frac{2\beta\gamma z}{1-z^2}\left(\frac{1+z}{1-z}\right)^{\gamma},$$

for $\alpha, \beta \in \mathbb{C}$, $0 < \gamma \leq 1$, $\beta \neq 0$, *where* $\psi_{\lambda,l}^{m,n}$ *is defined in* (3.2.5.26), *then*

$$\frac{z^{\delta} IR_{\lambda,l}^{m+1,n} f(z)}{\left(IR_{\lambda,l}^{m,n} f(z)\right)^{1+\delta}} \prec \left(\frac{1+z}{1-z}\right)^{\gamma},$$

and $\left(\frac{1+z}{1-z}\right)^{\gamma}$ *is the best dominant.*

Proof. Corollary follows by using Theorem 3.121 for $q(z) = \left(\frac{1+z}{1-z}\right)^{\gamma}$, $0 < \gamma \leq 1$. □

Theorem 3.122 (Alb Lupaş, 2017c). *Let* q *be convex and univalent in* U *such that* $q(0) = 1$. *Assume that*

$$\operatorname{Re}\left(\frac{\alpha}{\beta} q'(z)\right) > 0, \quad \text{for } \alpha, \beta \in \mathbb{C}, \quad \beta \neq 0. \tag{3.2.5.29}$$

If $f \in \mathcal{A}, \frac{z^{\delta} IR_{\lambda,l}^{m+1,n} f(z)}{\left(IR_{\lambda,l}^{m,n} f(z)\right)^{1+\delta}} \in \mathcal{H}\left[q(0), 1\right] \cap Q$ *and* $\psi_{\lambda,l}^{m,n}(\alpha, \beta; z)$ *is univalent in* U*, where* $\psi_{\lambda,l}^{m,n}(\alpha, \beta; z)$ *is as defined in* (3.2.5.16)*, then*

$$\alpha q(z) + \beta z q'(z) \prec \psi_{\lambda,l}^{m,n}(\alpha, \beta; z) \tag{3.2.5.30}$$

implies

$$q(z) \prec \frac{z^{\delta} IR_{\lambda,l}^{m+1,n} f(z)}{\left(IR_{\lambda,l}^{m,n} f(z)\right)^{1+\delta}},$$

$\delta \in \mathbb{C}$, $\delta \neq 0$, $z \in U$, *and* q *is the best subordinant.*

Proof. Consider $p(z) := \frac{z^{\delta} IR_{\lambda,l}^{m+1,n} f(z)}{\left(IR_{\lambda,l}^{m,n} f(z)\right)^{1+\delta}}$, $z \in U$, $z \neq 0$, $f \in \mathcal{A}$. The function p is analytic in U and $p(0) = 1$.

By setting $\theta(w) := \alpha w$ and $\phi(w) := \beta$ it can be easily verified that θ is analytic in $\mathbb{C}$, ϕ is analytic in $\mathbb{C} \backslash \{0\}$ and that $\phi(w) \neq 0$, $w \in \mathbb{C} \backslash \{0\}$.

Since $\frac{\theta'(q(z))}{\phi(q(z))} = \frac{\alpha}{\beta} q'(z)$, it follows that $\operatorname{Re}\left(\frac{\theta'(q(z))}{\phi(q(z))}\right) = \operatorname{Re}\left(\frac{\alpha}{\beta} q'(z)\right) > 0$, for $\alpha, \beta \in \mathbb{C}$, $\beta \neq 0$.

Now, by using (3.2.5.30), we obtain $\alpha q(z) + \beta z q'(z) \prec \alpha p(z) + \beta z p'(z)$, $z \in U$. From Lemma 1.6, we have $q(z) \prec p(z) = \frac{z^{\delta} IR_{\lambda,l}^{m+1,n} f(z)}{\left(IR_{\lambda,l}^{m,n} f(z)\right)^{1+\delta}}$, $z \in U$, and q is the best subordinant. □

Corollary 3.42 (Alb Lupaş, 2017c). *Let* $q(z) = \frac{1+Az}{1+Bz}$, $-1 \le B < A \le 1$, $z \in U$, $m, n \in \mathbb{N}$, $\lambda, l \ge 0$. *Assume that* (3.2.5.29) *holds. If* $f \in \mathcal{A}$, $\frac{z^{\delta} IR_{\lambda,l}^{m+1,n} f(z)}{\left(IR_{\lambda,l}^{m,n} f(z)\right)^{1+\delta}} \in \mathcal{H}\left[q(0), 1\right] \cap Q$*, and*

$$\alpha \frac{1 + Az}{1 + Bz} + \frac{\beta(A - B) z}{(1 + Bz)^2} \prec \psi_{\lambda,l}^{m,n}(\alpha, \beta; z),$$

for $\alpha, \beta \in \mathbb{C}$, $\beta \neq 0$, $-1 \le B < A \le 1$, *where* $\psi_{\lambda,l}^{m,n}$ *is defined in* (3.2.5.26)*, then*

$$\frac{1 + Az}{1 + Bz} \prec \frac{z^{\delta} IR_{\lambda,l}^{m+1,n} f(z)}{\left(IR_{\lambda,l}^{m,n} f(z)\right)^{1+\delta}},$$

and $\frac{1+Az}{1+Bz}$ *is the best subordinant.*

Proof. For $q(z) = \frac{1+Az}{1+Bz}$, $-1 \le B < A \le 1$, in Theorem 3.122, we get the corollary. □

Corollary 3.43 (Alb Lupaş, 2017c). *Let* $q(z) = \left(\frac{1+z}{1-z}\right)^{\gamma}$, $m, n \in \mathbb{N}$, $\lambda, l \geq 0$. *Assume that* (3.2.5.29) *holds. If* $f \in \mathcal{A}$, $\frac{z^{\delta} IR_{\lambda,l}^{m+1,n} f(z)}{\left(IR_{\lambda,l}^{m,n} f(z)\right)^{1+\delta}} \in \mathcal{H}[q(0), 1] \cap Q$ *and*

$$\alpha \left(\frac{1+z}{1-z}\right)^{\gamma} + \frac{2\beta\gamma z}{1-z^2} \left(\frac{1+z}{1-z}\right)^{\gamma} \prec \psi_{\lambda,l}^{m,n}(\alpha, \beta; z),$$

for $\alpha, \beta \in \mathbb{C}$, $0 < \gamma \leq 1$, $\beta \neq 0$, *where* $\psi_{\lambda,l}^{m,n}$ *is defined in* (3.2.5.26), *then*

$$\left(\frac{1+z}{1-z}\right)^{\gamma} \prec \frac{z^{\delta} IR_{\lambda,l}^{m+1,n} f(z)}{\left(IR_{\lambda,l}^{m,n} f(z)\right)^{1+\delta}},$$

and $\left(\frac{1+z}{1-z}\right)^{\gamma}$ *is the best subordinant.*

Proof. Corollary follows by using Theorem 3.122 for $q(z) = \left(\frac{1+z}{1-z}\right)^{\gamma}$, $0 < \gamma \leq 1$. □

Combining Theorems 3.121 and 3.122, we state the following sandwich theorem.

Theorem 3.123 (Alb Lupaş, 2017c). *Let* q_1 *and* q_2 *be convex and univalent in* U *such that* $q_1(z) \neq 0$ *and* $q_2(z) \neq 0$, *for all* $z \in U$. *Suppose that* q_1 *satisfies* (3.2.5.25) *and* q_2 *satisfies* (3.2.5.29). *If* $f \in \mathcal{A}$, $\frac{z^{\delta} IR_{\lambda,l}^{m+1,n} f(z)}{\left(IR_{\lambda,l}^{m,n} f(z)\right)^{1+\delta}} \in \mathcal{H}[q(0), 1] \cap Q$, *and* $\psi_{\lambda,l}^{m,n}(\alpha, \beta; z)$ *is as defined in* (3.2.5.26) *univalent in* U, *then*

$$\alpha q_1(z) + \beta z q_1'(z) \prec \psi_{\lambda,l}^{m,n}(\alpha, \beta; z) \prec \alpha q_2(z) + \beta z q_2'(z),$$

for $\alpha, \beta \in \mathbb{C}$, $\beta \neq 0$, *implies*

$$q_1(z) \prec \frac{z^{\delta} IR_{\lambda,l}^{m+1,n} f(z)}{\left(IR_{\lambda,l}^{m,n} f(z)\right)^{1+\delta}} \prec q_2(z),$$

$z \in U$, *and* q_1 *and* q_2 *are respectively the best subordinant and the best dominant.*

For $q_1(z)=\frac{1+A_1z}{1+B_1z}$, $q_2(z)=\frac{1+A_2z}{1+B_2z}$, where $-1\le B_2<B_1<A_1<A_2\le 1$, we have the following corollary.

Corollary 3.44 (Alb Lupaş, 2017c). *Let $m,n\in\mathbb{N}$, $\lambda,l\ge 0$. Assume that (3.2.5.25) and (3.2.5.29) hold for $q_1(z)=\frac{1+A_1z}{1+B_1z}$ and $q_2(z)=\frac{1+A_2z}{1+B_2z}$, respectively. If $f\in\mathcal{A}$, $\frac{z^\delta IR_{\lambda,l}^{m+1,n}f(z)}{\left(IR_{\lambda,l}^{m,n}f(z)\right)^{1+\delta}}\in\mathcal{H}[q(0),1]\cap Q$ and*

$$\alpha\frac{1+A_1z}{1+B_1z}+\frac{\beta(A_1-B_1)z}{(1+B_1z)^2}\prec\psi_{\lambda,l}^{m,n}(\alpha,\beta;z)\prec\alpha\frac{1+A_2z}{1+B_2z}+\frac{\beta(A_2-B_2)z}{(1+B_2z)^2},$$

$z\in U$, for $\alpha,\beta\in\mathbb{C}$, $\beta\ne 0$, $-1\le B_2\le B_1<A_1\le A_2\le 1$, where $\psi_{\lambda,l}^{m,n}$ is defined in (3.2.5.26), then

$$\frac{1+A_1z}{1+B_1z}\prec\frac{z^\delta IR_{\lambda,l}^{m+1,n}f(z)}{\left(IR_{\lambda,l}^{m,n}f(z)\right)^{1+\delta}}\prec\frac{1+A_2z}{1+B_2z},$$

$z\in U$, hence $\frac{1+A_1z}{1+B_1z}$ and $\frac{1+A_2z}{1+B_2z}$ are the best subordinant and the best dominant, respectively.

For $q_1(z)=\left(\frac{1+z}{1-z}\right)^{\gamma_1}$, $q_2(z)=\left(\frac{1+z}{1-z}\right)^{\gamma_2}$, where $0<\gamma_1<\gamma_2\le 1$, we have the following corollary.

Corollary 3.45 (Alb Lupaş, 2017c). *Let $m,n\in\mathbb{N}$, $\lambda,l\ge 0$. Assume that (3.2.5.25) and (3.2.5.29) hold for $q_1(z)=\left(\frac{1+z}{1-z}\right)^{\gamma_1}$ and $q_2(z)=\left(\frac{1+z}{1-z}\right)^{\gamma_2}$, respectively. If $f\in\mathcal{A}$, $\frac{z^\delta IR_{\lambda,l}^{m+1,n}f(z)}{\left(IR_{\lambda,l}^{m,n}f(z)\right)^{1+\delta}}\in\mathcal{H}[q(0),1]\cap Q$ and*

$$\alpha\left(\frac{1+z}{1-z}\right)^{\gamma_1}+\frac{2\beta\gamma_1z}{1-z^2}\left(\frac{1+z}{1-z}\right)^{\gamma_1}\prec\psi_{\lambda,l}^{m,n}(\alpha,\beta;z)$$
$$\prec\alpha\left(\frac{1+z}{1-z}\right)^{\gamma_2}+\frac{2\beta\gamma_2z}{1-z^2}\left(\frac{1+z}{1-z}\right)^{\gamma_2},$$

$z\in U$, for $\alpha,\beta\in\mathbb{C}$, $\beta\ne 0$, $0<\gamma_1<\gamma_2\le 1$, where $\psi_{\lambda,l}^{m,n}$ is defined in (3.2.5.26), then

$$\left(\frac{1+z}{1-z}\right)^{\gamma_1}\prec\frac{z^\delta IR_{\lambda,l}^{m+1,n}f(z)}{\left(IR_{\lambda,l}^{m,n}f(z)\right)^{1+\delta}}\prec\left(\frac{1+z}{1-z}\right)^{\gamma_2},$$

$z\in U$, hence $\left(\frac{1+z}{1-z}\right)^{\gamma_1}$ and $\left(\frac{1+z}{1-z}\right)^{\gamma_2}$ are the best subordinant and the best dominant, respectively.

3.2.6 *Differential Subordination and Superordination Results Using Fractional Integral of Confluent Hypergeometric Function*

Both the theory of differential subordination and its dual, the theory of differential superordination, introduced by Professors Miller and Mocanu are based on reinterpreting certain inequalities for real-valued functions for the case of complex-valued functions. Studying subordination and superordination properties using different types of operators is a technique that is still widely used, some studies resulting in sandwich-type theorems as it is the case of the present study. The fractional integral of confluent hypergeometric function is introduced in the paper Alb Lupaş and Oros (2021b) and certain subordination and superordination results are stated in theorems and corollaries, the study being completed by the statement of a sandwich-type theorem connecting the results obtained by using the two theories in Alb Lupaş and Oros (2021b) and Alb Lupaş (2021b).

Let a and c be complex numbers with $c \neq 0, -1, -2, \ldots$ and $\lambda > 0$. The fractional integral of confluent hypergeometric function defined in Definition 2.20 is

$$\begin{aligned} D_z^{-\lambda}\phi(a,c;z) &= \frac{1}{\Gamma(\lambda)}\int_0^z \frac{\phi(a,c;t)}{(z-t)^{1-\lambda}}dt \\ &= \frac{1}{\Gamma(\lambda)}\frac{\Gamma(c)}{\Gamma(a)}\sum_{k=0}^{\infty}\frac{\Gamma(a+k)}{\Gamma(c+k)\Gamma(k+1)}\int_0^z \frac{t^k}{(z-t)^{1-\lambda}}dt. \end{aligned} \tag{3.2.6.1}$$

After a simple calculation, the fractional integral of confluent hypergeometric function has the following form

$$D_z^{-\lambda}\phi(a,c;z) = \frac{\Gamma(c)}{\Gamma(a)}\sum_{k=0}^{\infty}\frac{\Gamma(a+k)}{\Gamma(c+k)\Gamma(\lambda+k+1)}z^{k+\lambda}. \tag{3.2.6.2}$$

We note that $D_z^{-\lambda}\phi(a,c;z) \in \mathcal{H}[0,\lambda]$.

The first subordination result obtained using the operator given by (3.2.6.1) is the following theorem.

Theorem 3.124 (Alb Lupaş and Oros, 2021b). *Let* $\left(\frac{D_z^{-\lambda}\phi(a,c;z)}{z}\right)^{\delta} \in \mathcal{H}(U)$ *and consider a function* q *analytic and univalent in* U *with* $q(z) \neq 0$, *when* $z \in U$, $a, c \in \mathbb{C}$, $c \neq 0, -1, -2, \ldots$ *and* $\lambda, \delta > 0$. *Suppose that* $\frac{zq'(z)}{q(z)}$ *is* *starlike and univalent function in* U*. Let*

$$\operatorname{Re}\left(1+\frac{\xi}{\beta}q(z)+\frac{2\mu}{\beta}(q(z))^2-\frac{zq'(z)}{q(z)}+\frac{zq''(z)}{q'(z)}\right)>0, \tag{3.2.6.3}$$

for $\alpha,\xi,\mu,\beta\in\mathbb{C}$, $\beta\neq 0$, $z\in U$ *and*

$$\begin{aligned}\psi_\lambda^{a,c}(\delta,\alpha,\xi,\mu,\beta;z) &:= \alpha+\xi\left[\frac{D_z^{-\lambda}\phi(a,c;z)}{z}\right]^\delta \\ &\quad +\mu\left[\frac{D_z^{-\lambda}\phi(a,c;z)}{z}\right]^{2\delta}+\beta\delta\left[\frac{z\left(D_z^{-\lambda}\phi(a,c;z)\right)'}{D_z^{-\lambda}\phi(a,c;z)}-1\right].\end{aligned} \tag{3.2.6.4}$$

If function q *satisfies*

$$\psi_\lambda^{a,c}(\delta,\alpha,\xi,\mu,\beta;z)\prec\alpha+\xi q(z)+\mu(q(z))^2+\beta\frac{zq'(z)}{q(z)}, \tag{3.2.6.5}$$

considering $\alpha,\xi,\mu,\beta\in\mathbb{C}$, $\beta\neq 0$, *then*

$$\left(\frac{D_z^{-\lambda}\phi(a,c;z)}{z}\right)^\delta\prec q(z),\quad z\in U, \tag{3.2.6.6}$$

and q *is the best dominant.*

Proof. Take function p of the form

$$p(z):=\left(\frac{D_z^{-\lambda}\phi(a,c;z)}{z}\right)^\delta,\quad z\in U,\ z\neq 0.$$

We compute

$$\begin{aligned}p'(z)&=\delta\left(\frac{D_z^{-\lambda}\phi(a,c;z)}{z}\right)^{\delta-1}\left[\frac{\left(D_z^{-\lambda}\phi(a,c;z)\right)'}{z}-\frac{D_z^{-\lambda}\phi(a,c;z)}{z^2}\right]\\ &=\delta\left(\frac{D_z^{-\lambda}\phi(a,c;z)}{z}\right)^{\delta-1}\frac{\left(D_z^{-\lambda}\phi(a,c;z)\right)'}{z}-\frac{\delta}{z}p(z).\end{aligned}$$

Then

$$\frac{zp'(z)}{p(z)}=\delta\left[\frac{z\left(D_z^{-\lambda}\phi(a,c;z)\right)'}{D_z^{-\lambda}\phi(a,c;z)}-1\right].$$

By setting

$$\theta(w):=\alpha+\xi w+\mu w^2$$

and

$$\phi(w) := \frac{\beta}{w}.$$

it is evidently that θ and ϕ are analytic in $\mathbb{C}$, respectively $\mathbb{C}\backslash\{0\}$ and $\phi(w) \neq 0$, $w \in \mathbb{C}\backslash\{0\}$.

Also, considering

$$Q(z) = zq'(z)\phi(q(z)) = \beta\frac{zq'(z)}{q(z)}$$

and

$$h(z) = \theta(q(z)) + Q(z) = \alpha + \xi q(z) + \mu(q(z))^2 + \beta\frac{zq'(z)}{q(z)},$$

$Q(z)$ is starlike univalent in U.

We have

$$h'(z) = \xi + q'(z) + 2\mu q(z)q'(z) + \beta\frac{(q'(z) + zq''(z))q(z) - z(q'(z))^2}{(q(z))^2}$$

and

$$\frac{zh'(z)}{Q(z)} = \frac{zh'(z)}{\beta\frac{zq'(z)}{q(z)}} = 1 + \frac{\xi}{\beta}q(z) + \frac{2\mu}{\beta}(q(z))^2 - \frac{zq'(z)}{q(z)} + \frac{zq''(z)}{q'(z)}.$$

We obtain

$$\operatorname{Re}\left(\frac{zh'(z)}{Q(z)}\right) = \operatorname{Re}\left(1 + \frac{\xi}{\beta}q(z) + \frac{2\mu}{\beta}(q(z))^2 - \frac{zq'(z)}{q(z)} + \frac{zq''(z)}{q'(z)}\right) > 0.$$

We get

$$\begin{aligned}
&\alpha + \xi p(z) + \mu(p(z))^2 + \beta\frac{zp'(z)}{p(z)} \\
&\quad = \alpha + \xi\left[\frac{D_z^{-\lambda}\phi(a,c;z)}{z}\right]^{\delta} + \mu\left[\frac{D_z^{-\lambda}\phi(a,c;z)}{z}\right]^{2\delta} \\
&\qquad + \beta\delta\left[\frac{z\left(D_z^{-\lambda}\phi(a,c;z)\right)'}{D_z^{-\lambda}\phi(a,c;z)} - 1\right].
\end{aligned}$$

By using (3.2.6.5), we get

$$\alpha + \xi p(z) + \mu(p(z))^2 + \beta\frac{zp'(z)}{p(z)} \prec \alpha + \beta q(z) + \mu(q(z))^2 + \beta\frac{zq'(z)}{q(z)}.$$

Applying Lemma 1.3, we obtain $p(z) \prec q(z)$, $z \in U$, i.e.

$$\left(\frac{D_z^{-\lambda}\phi(a,c;z)}{z}\right)^{\delta} \prec q(z), \quad z \in U$$

and q is the best dominant. □

Corollary 3.46 (Alb Lupaş and Oros, 2021b). *Let* $a, c \in \mathbb{C}$, $c \neq 0, -1, -2, \ldots$ *and* $\lambda, \delta > 0$. *Assume that* (3.2.6.3) *holds. If*

$$\psi_{\lambda}^{a,c}(\delta, \alpha, \xi, \mu, \beta; z) \prec \alpha + \xi\frac{1+Az}{1+Bz} + \mu\left(\frac{1+Az}{1+Bz}\right)^2 + \beta\frac{(A-B)z}{(1+Az)(1+Bz)},$$

for $\alpha, \xi, \mu, \beta \in \mathbb{C}$, $\beta \neq 0$, $-1 \leq B < A \leq 1$, *where* $\psi_{\lambda}^{a,c}$ *is defined in* (3.2.6.4), *then*

$$\left(\frac{D_z^{-\lambda}\phi(a,c;z)}{z}\right)^{\delta} \prec \frac{1+Az}{1+Bz}, \quad z \in U,$$

and $\frac{1+Az}{1+Bz}$ *is the best dominant.*

Proof. The results stated in this corollary are obtained using $q(z) = \frac{1+Az}{1+Bz}$, $-1 \leq B < A \leq 1$ in Theorem 3.124. □

Corollary 3.47 (Alb Lupaş and Oros, 2021b). *Let* $a, c \in \mathbb{C}$, $c \neq 0, -1, -2, \ldots$ *and* $\lambda, \delta > 0$. *Assume that* (3.2.6.3) *holds. If*

$$\psi_{\lambda}^{a,c}(\delta, \alpha, \xi, \mu, \beta; z) \prec \alpha + \xi\left(\frac{1+z}{1-z}\right)^{\gamma} + \mu\left(\frac{1+z}{1-z}\right)^{2\gamma} + \beta\frac{2\gamma z}{1-z^2},$$

for $\alpha, \xi, \mu, \beta \in \mathbb{C}$, $0 < \gamma \leq 1$, $\beta \neq 0$, *where* $\psi_{\lambda}^{a,c}$ *is defined in* (3.2.6.4), *then*

$$\left(\frac{D_z^{-\lambda}\phi(a,c;z)}{z}\right)^{\delta} \prec \left(\frac{1+z}{1-z}\right)^{\gamma}, \quad z \in U,$$

and $\left(\frac{1+z}{1-z}\right)^{\gamma}$ *is the best dominant.*

Proof. The conclusion of the corollary follows from Theorem 3.124 by taking $q(z) = \left(\frac{1+z}{1-z}\right)^{\gamma}$, $0 < \gamma \leq 1$. □

Theorem 3.125 (Alb Lupaş and Oros, 2021b). *Consider q an analytic function in U with $q(z) \neq 0$ and let $\frac{zq'(z)}{q(z)}$ be starlike and univalent in U. Assume that*

$$\operatorname{Re}\left(\frac{2\mu}{\beta}(q(z))^2 + \frac{\xi}{\beta}q(z)\right) > 0, \quad \text{for } \xi, \mu, \beta \in \mathbb{C}, \quad \beta \neq 0. \tag{3.2.6.7}$$

Let $a, c \in \mathbb{C}$, $c \neq 0, -1, -2, \ldots$ and $\lambda, \delta > 0$. If $\left(\frac{D_z^{-\lambda}\phi(a,c;z)}{z}\right)^{\delta} \in \mathcal{H}[0, (\lambda-1)\delta] \cap Q$ and $\psi_{\lambda}^{a,c}(\delta, \alpha, \xi, \mu, \beta; z)$ is univalent in U, where $\psi_{\lambda}^{a,c}(\delta, \alpha, \xi, \mu, \beta; z)$ is as defined in (3.2.6.4), then

$$\alpha + \xi q(z) + \mu (q(z))^2 + \beta \frac{zq'(z)}{q(z)} \prec \psi_{\lambda}^{a,c}(\delta, \alpha, \xi, \mu, \beta; z) \tag{3.2.6.8}$$

implies

$$q(z) \prec \left(\frac{D_z^{-\lambda}\phi(a,c;z)}{z}\right)^{\delta}, \quad z \in U, \tag{3.2.6.9}$$

and q is the best subordinant.

Proof. Define function p as

$$p(z) := \left(\frac{D_z^{-\lambda}\phi(a,c;z)}{z}\right)^{\delta}, \quad z \in U, z \neq 0.$$

By setting

$$\theta(w) := \alpha + \xi w + \mu w^2$$

and

$$\phi(w) := \frac{\beta}{w},$$

it is evidently that θ and ϕ are analytic in $\mathbb{C}$, respectively $\mathbb{C}\backslash\{0\}$ and $\phi(w) \neq 0$, $w \in \mathbb{C}\backslash\{0\}$.

Since

$$\frac{\theta'(q(z))}{\phi(q(z))} = \frac{q'(z)[\xi + 2\mu q(z)]q(z)}{\beta},$$

it follows that

$$\operatorname{Re}\left(\frac{\theta'(q(z))}{\phi(q(z))}\right)=\operatorname{Re}\left(\frac{2\mu}{\beta}(q(z))^2+\frac{\xi}{\beta}q(z)\right)>0,$$

for $\mu,\xi,\beta\in\mathbb{C}$, $\beta\neq 0$.

We get

$$\alpha+\xi q(z)+\mu(q(z))^2+\beta\frac{zq'(z)}{q(z)}\prec\alpha+\xi p(z)+\mu(p(z))^2+\beta\frac{zp'(z)}{p(z)}.$$

Using Lemma 1.6, we obtain

$$q(z)\prec p(z)=\left(\frac{D_z^{-\lambda}\phi(a,c;z)}{z}\right)^{\delta},\quad z\in U,$$

and q is the best subordinant. □

Corollary 3.48 (Alb Lupaş and Oros, 2021b). *Let* $a,c\in\mathbb{C}$, $c\neq 0,-1,-2,\ldots$ *and* $\lambda,\delta>0$. *Assume that* (3.2.6.7) *holds. If* $\left(\frac{D_z^{-\lambda}\phi(a,c;z)}{z}\right)^{\delta}\in\mathcal{H}[0,(\lambda-1)\delta]\cap Q$ *and*

$$\alpha+\xi\frac{1+Az}{1+Bz}+\mu\left(\frac{1+Az}{1+Bz}\right)^2+\beta\frac{(A-B)z}{(1+Az)(1+Bz)}\prec\psi_{\lambda}^{a,c}(\delta,\alpha,\xi,\mu,\beta;z),$$

for $\alpha,\xi,\mu,\beta\in\mathbb{C}$, $\beta\neq 0$, $-1\leq B<A\leq 1$, *where* $\psi_{\lambda}^{a,c}$ *is defined in* (3.2.6.4), *then*

$$\frac{1+Az}{1+Bz}\prec\left(\frac{D_z^{-\lambda}\phi(a,c;z)}{z}\right)^{\delta},\quad z\in U,$$

and $\frac{1+Az}{1+Bz}$ *is the best subordinant.*

Proof. Using Theorem 3.125, the conclusion of the corollary derives from setting $q(z)=\frac{1+Az}{1+Bz}$, $-1\leq B<A\leq 1$. □

Corollary 3.49 (Alb Lupaş and Oros, 2021b). *Let* $a,c\in\mathbb{C}$, $c\neq 0,-1,-2,\ldots$ *and* $\lambda,\delta>0$. *Assume that* (3.2.6.7) *holds. If* $\left(\frac{D_z^{-\lambda}\phi(a,c;z)}{z}\right)^{\delta}\in\mathcal{H}[0,(\lambda-1)\delta]\cap Q$ *and*

$$\alpha+\xi\left(\frac{1+z}{1-z}\right)^{\gamma}+\mu\left(\frac{1+z}{1-z}\right)^{2\gamma}+\beta\frac{2\gamma z}{1-z^2}\prec\psi_{\lambda}^{a,c}(\delta,\alpha,\xi,\mu,\beta;z),$$

for $\alpha, \xi, \mu, \beta \in \mathbb{C}$, $0 < \gamma \le 1$, $\beta \neq 0$, *where* $\psi_\lambda^{a,c}$ *is defined in* (3.2.6.4), *then*

$$\left(\frac{1+z}{1-z}\right)^\gamma \prec \left(\frac{D_z^{-\lambda}\phi(a,c;z)}{z}\right)^\delta, \quad z \in U,$$

and $\left(\frac{1+z}{1-z}\right)^\gamma$ *is the best subordinant.*

Proof. By using Theorem 3.125 for $q(z) = \left(\frac{1+z}{1-z}\right)^\gamma$, $0 < \gamma \le 1$ we obtain the result. □

Using the conclusions of Theorems 3.124 and 3.125 combined, a sandwich-type theorem can be stated as it follows:

Theorem 3.126 (Alb Lupaş and Oros, 2021b). *Consider two analytic and univalent functions* q_1 *and* q_2 *satisfying* $q_1(z) \neq 0$ *and* $q_2(z) \neq 0$, *when* $z \in U$, *such that* $\frac{zq_1'(z)}{q_1(z)}$ *and* $\frac{zq_2'(z)}{q_2(z)}$ *be starlike and univalent functions. Suppose that* q_1 *satisfies* (3.2.6.3) *and* q_2 *satisfies* (3.2.6.7). *Let* a, c *be complex numbers with* $c \neq 0, -1, -2, \ldots$ *and* $\lambda, \delta > 0$. *If* $\left(\frac{D_z^{-\lambda}\phi(a,c;z)}{z}\right)^\delta \in \mathcal{H}[0,(\lambda-1)\delta] \cap Q$ *and* $\psi_\lambda^{a,c}(\delta, \alpha, \xi, \mu, \beta; z)$ *is as defined in* (3.2.6.4) *univalent in* U, *then*

$$\alpha + \xi q_1(z) + \mu (q_1(z))^2 + \beta \frac{zq_1'(z)}{q_1(z)} \prec \psi_\lambda^{a,c}(\delta, \alpha, \xi, \mu, \beta; z)$$

$$\prec \alpha + \xi q_2(z) + \mu (q_2(z))^2 + \beta \frac{zq_2'(z)}{q_2(z)},$$

for $\alpha, \xi, \mu, \beta \in \mathbb{C}$, $\beta \neq 0$, *implies*

$$q_1(z) \prec \left(\frac{D_z^{-\lambda}\phi(a,c;z)}{z}\right)^\delta \prec q_2(z), \quad z \in U,$$

and q_1 *and* q_2 *are the best subordinant and the best dominant, respectively.*

For $q_1(z) = \frac{1+A_1 z}{1+B_1 z}$, $q_2(z) = \frac{1+A_2 z}{1+B_2 z}$, where $-1 \le B_2 < B_1 < A_1 < A_2 \le 1$, we have the following corollary.

Corollary 3.50 (Alb Lupaş and Oros, 2021b). *Let* $a, c \in \mathbb{C}$, $c \neq 0, -1, -2, \ldots$ *and* $\lambda, \delta > 0$. *Assume that* (3.2.6.3) *and* (3.2.6.7) *hold.*

If $\left(\frac{D_z^{-\lambda}\phi(a,c;z)}{z}\right)^{\delta} \in \mathcal{H}\left[0,(\lambda-1)\delta\right] \cap Q$ *and*

$$\begin{aligned}\alpha+\xi\frac{1+A_1z}{1+B_1z}+\mu\left(\frac{1+A_1z}{1+B_1z}\right)^2+\beta\frac{(A_1-B_1)z}{(1+A_1z)(1+B_1z)}\\ \prec \psi_\lambda^{a,c}(\delta,\alpha,\xi,\mu,\beta;z)\\ \prec \alpha+\xi\frac{1+A_2z}{1+B_2z}+\mu\left(\frac{1+A_2z}{1+B_2z}\right)^2+\frac{(A_2-B_2)z}{(1+A_2z)(1+B_2z)},\end{aligned}$$

for $\alpha,\xi,\mu,\beta\in\mathbb{C}$, $\beta\neq 0$, $-1\le B_2\le B_1<A_1\le A_2\le 1$, *where* $\psi_\lambda^{a,c}$ *is defined in* (3.2.6.4), *then*

$$\frac{1+A_1z}{1+B_1z}\prec\left(\frac{D_z^{-\lambda}\phi(a,c;z)}{z}\right)^{\delta}\prec\frac{1+A_2z}{1+B_2z},$$

hence $\frac{1+A_1z}{1+B_1z}$ *and* $\frac{1+A_2z}{1+B_2z}$ *are the best subordinant and the best dominant, respectively.*

Corollary 3.51 (Alb Lupaş and Oros, 2021b). *Let* $a,c\in\mathbb{C}$, $c\neq 0,-1,-2,\ldots$ *and* $\lambda,\delta>0$. *Assume that* (3.2.6.3) *and* (3.2.6.7) *hold. If* $\left(\frac{D_z^{-\lambda}\phi(a,c;z)}{z}\right)^{\delta}\in\mathcal{H}\left[0,(\lambda-1)\delta\right]\cap Q$ *and*

$$\begin{aligned}\alpha+\xi\left(\frac{1+z}{1-z}\right)^{\gamma_1}+\mu\left(\frac{1+z}{1-z}\right)^{2\gamma_1}+\beta\frac{2\gamma_1z}{1-z^2}\prec\psi_\lambda^{a,c}(\delta,\alpha,\xi,\mu,\beta;z)\\ \prec\alpha+\xi\left(\frac{1+z}{1-z}\right)^{\gamma_2}+\mu\left(\frac{1+z}{1-z}\right)^{2\gamma_2}+\beta\frac{2\gamma_2z}{1-z^2},\end{aligned}$$

for $\alpha,\xi,\mu,\beta\in\mathbb{C}$, $0<\gamma_1,\gamma_2\le 1$, $\beta\neq 0$,, *where* $\psi_\lambda^{a,c}$ *is defined in* (3.2.6.4), *then*

$$\left(\frac{1+z}{1-z}\right)^{\gamma_1}\prec\left(\frac{D_z^{-\lambda}\phi(a,c;z)}{z}\right)^{\delta}\prec\left(\frac{1+z}{1-z}\right)^{\gamma_2},$$

hence $\left(\frac{1+z}{1-z}\right)^{\gamma_1}$ *and* $\left(\frac{1+z}{1-z}\right)^{\gamma_2}$ *are the best subordinant and the best dominant, respectively.*

Theorem 3.127 (Alb Lupaş, 2021b). *Let* q *be an analytic and univalent function in* U *with* $q(z)\neq 0$, $\forall\ z\in U$, *and* $\frac{z\left(D_z^{-\lambda}\phi(a,c;z)\right)'}{D_z^{-\lambda}\phi(a,c;z)}\in\mathcal{H}(U)$, *where*

$\lambda > 0$ and a, $c \in \mathbb{C}$, $c \neq 0, -1, -2, \ldots$. Suppose that $\frac{zq'(z)}{q(z)}$ is starlike univalent in U. Consider

$$\operatorname{Re}\left(\frac{2\mu}{\beta}q^2(z) + \frac{\xi}{\beta}q(z) + 1 - z\frac{q'(z)}{q(z)} + z\frac{q''(z)}{q'(z)}\right) > 0, \quad z \in U, \quad (3.2.6.10)$$

with $\mu, \beta, \xi, \alpha \in \mathbb{C}$, $\beta \neq 0$, and

$$\psi_\lambda^{a,b}(\alpha, \beta, \xi, \mu; z) := \alpha + \beta + (\xi - \beta)\frac{z\left(D_z^{-\lambda}\phi(a, c; z)\right)'}{D_z^{-\lambda}\phi(a, c; z)} + \mu\left(\frac{z\left(D_z^{-\lambda}\phi(a, c; z)\right)'}{D_z^{-\lambda}\phi(a, c; z)}\right)^2 + \beta\frac{z\left(D_z^{-\lambda}\phi(a, c; z)\right)''}{\left(D_z^{-\lambda}\phi(a, c; z)\right)'}. \quad (3.2.6.11)$$

If the following subordination is satisfied by q,

$$\psi_\lambda^{a,b}(\alpha, \beta, \xi, \mu; z) \prec \alpha + \xi q(z) + \mu(q(z))^2 + \beta\frac{zq'(z)}{q(z)}, \quad (3.2.6.12)$$

then

$$\frac{z\left(D_z^{-\lambda}\phi(a, c; z)\right)'}{D_z^{-\lambda}\phi(a, c; z)} \prec q(z), \quad (3.2.6.13)$$

and q is the best dominant.

Proof. Consider

$$p(z) := \frac{z\left(D_z^{-\lambda}\phi(a, c; z)\right)'}{D_z^{-\lambda}\phi(a, c; z)}, \quad z \in U, \ z \neq 0.$$

Differentiating with respect to z, we get $p'(z) = \frac{\left(D_z^{-\lambda}\phi(a,c;z)\right)'}{D_z^{-\lambda}\phi(a,c;z)} + z\frac{\left(D_z^{-\lambda}\phi(a,c;z)\right)''}{D_z^{-\lambda}\phi(a,c;z)} - z\left(\frac{\left(D_z^{-\lambda}\phi(a,c;z)\right)'}{D_z^{-\lambda}\phi(a,c;z)}\right)^2$ and

$$\frac{zp'(z)}{p(z)} = 1 - z\frac{\left(D_z^{-\lambda}\phi(a, c; z)\right)'}{D_z^{-\lambda}\phi(a, c; z)} + z\frac{\left(D_z^{-\lambda}\phi(a, c; z)\right)''}{\left(D_z^{-\lambda}\phi(a, c; z)\right)'}. \quad (3.2.6.14)$$

By setting

$$\phi(w) := \frac{\beta}{w}$$

and

$$\theta(w) := \alpha + \xi w + \mu w^2,$$

evidently ϕ is analytic in $\mathbb{C}\backslash\{0\}$, $\phi(w) \neq 0$, $\forall\, w \in \mathbb{C}\backslash\{0\}$ and θ is analytic in $\mathbb{C}$.

Considering

$$Q(z) = zq'(z)\phi(q(z)) = \beta\frac{zq'(z)}{q(z)}$$

and

$$h(z) = \theta(q(z)) + Q(z) = \alpha + \xi q(z) + \mu(q(z))^2 + \beta\frac{zq'(z)}{q(z)},$$

it yields that Q is a starlike univalent function in U.

Differentiating, we obtain $h'(z) = \xi q'(z) + 2\mu q(z)q'(z) + \beta\frac{q'(z)}{q(z)} + \beta z\frac{q''(z)}{q(z)} - \beta z\left(\frac{q'(z)}{q(z)}\right)^2$ and $\frac{zh'(z)}{Q(z)} = \frac{2\mu}{\beta}q^2(z) + \frac{\xi}{\beta}q(z) + 1 - z\frac{q'(z)}{q(z)} + z\frac{q''(z)}{q(z)}$.

We deduce that

$$\operatorname{Re}\left(\frac{zh'(z)}{Q(z)}\right) = \operatorname{Re}\left(\frac{2\mu}{\beta}q^2(z) + \frac{\xi}{\beta}q(z) + 1 - z\frac{q'(z)}{q(z)} + z\frac{q''(z)}{q(z)}\right) > 0.$$

By using (3.2.6.14), we obtain $\alpha + \xi p(z) + \mu(p(z))^2 + \beta\frac{zp'(z)}{p(z)} = \alpha + \beta + (\xi - \beta)\frac{z\left(D_z^{-\lambda}\phi(a,c;z)\right)'}{D_z^{-\lambda}\phi(a,c;z)} + \mu\left(\frac{z\left(D_z^{-\lambda}\phi(a,c;z)\right)'}{D_z^{-\lambda}\phi(a,c;z)}\right)^2 + \beta\frac{z\left(D_z^{-\lambda}\phi(a,c;z)\right)''}{\left(D_z^{-\lambda}\phi(a,c;z)\right)'}$.

By using (3.2.6.12), we have $\alpha + \xi p(z) + \mu(p(z))^2 + \beta\frac{zp'(z)}{p(z)} \prec \alpha + \xi q(z) + \mu(q(z))^2 + \beta\frac{zq'(z)}{q(z)}$.

Applying Lemma 1.3, we get $p(z) \prec q(z)$, $\forall\, z \in U$, which means $\frac{z\left(D_z^{-\lambda}\phi(a,c;z)\right)'}{D_z^{-\lambda}\phi(a,c;z)} \prec q(z)$, $\forall\, z \in U$ and the function q is the best dominant. □

Corollary 3.52 (Alb Lupaş, 2021b). *Let* $\lambda > 0$, a, c, α, β, ξ, $\mu \in \mathbb{C}$, $c \neq 0, -1, -2, \ldots$, $\beta \neq 0$, *and relation* (3.2.6.10) *holds for* $q(z) = \frac{1+Az}{1+Bz}$, *with* $-1 \leq B < A \leq 1$. *If*

$$\psi_\lambda^{a,b}(\alpha, \beta, \xi, \mu; z) \prec \alpha + \xi\frac{1+Az}{1+Bz} + \mu\left(\frac{1+Az}{1+Bz}\right)^2 + \frac{\beta(A-B)z}{(1+Az)(1+Bz)},$$

with $\psi_\lambda^{a,b}$ *is defined in* (3.2.6.11), *then*

$$\frac{z\left(D_z^{-\lambda}\phi(a,c;z)\right)'}{D_z^{-\lambda}\phi(a,c;z)} \prec \frac{1+Az}{1+Bz},$$

and $\frac{1+Az}{1+Bz}$ *is the best dominant.*

Proof. We get the corollary considering $q(z) = \frac{1+Az}{1+Bz}$, $-1 \le B < A \le 1$ in Theorem 3.127. □

Corollary 3.53 (Alb Lupaş, 2021b). *Let* $\lambda > 0$, a, c, α, β, ξ, $\mu \in \mathbb{C}$, $c \neq 0, -1, -2, \ldots$, $\beta \neq 0$. *Assume that* (3.2.6.10) *holds for* $q(z) = \left(\frac{1+z}{1-z}\right)^{\gamma}$, $0 < \gamma \le 1$. *If*

$$\psi_{\lambda}^{a,b}(\alpha, \beta, \xi, \mu; z) \prec \alpha + \xi \left(\frac{1+z}{1-z}\right)^{\gamma} + \mu \left(\frac{1+z}{1-z}\right)^{2\gamma} + \frac{2\beta\gamma z}{1-z^2},$$

with $\psi_{\lambda}^{a,b}$ *is defined in* (3.2.6.11), *then*

$$\frac{z\left(D_z^{-\lambda}\phi(a, c; z)\right)'}{D_z^{-\lambda}\phi(a, c; z)} \prec \left(\frac{1+z}{1-z}\right)^{\gamma},$$

and $\left(\frac{1+z}{1-z}\right)^{\gamma}$ *is the best dominant.*

Proof. Put $q(z) = \left(\frac{1+z}{1-z}\right)^{\gamma}$, $0 < \gamma \le 1$ in Theorem 3.127 to obtain the corollary. □

Theorem 3.128 (Alb Lupaş, 2021b). *Let q be an analytic and univalent function in U with $q(z) \neq 0$, $\forall\, z \in U$, such that $\frac{zq'(z)}{q(z)}$ is a starlike univalent function in U and*

$$\operatorname{Re}\left(\frac{2\mu}{\beta} q^2(z)\, q'(z) + \frac{\xi}{\beta} q(z)\, q'(z)\right) > 0, \quad \text{for } \mu, \beta, \xi \in \mathbb{C}, \quad \beta \neq 0. \tag{3.2.6.15}$$

If $\frac{z\left(D_z^{-\lambda}\phi(a,c;z)\right)'}{D_z^{-\lambda}\phi(a,c;z)} \in Q \cap \mathcal{H}[q(0), 1]$ *and* $\psi_{\lambda}^{a,b}(\alpha, \beta, \xi, \mu; z)$ *is an univalent function in* U, *where* $\psi_{\lambda}^{a,b}(\alpha, \beta, \xi, \mu; z)$ *is defined by* (3.2.6.11) *and* $\lambda > 0$, a, $c \in \mathbb{C}$, $c \neq 0, -1, -2, \ldots$, *then*

$$\alpha + \xi q(z) + \mu (q(z))^2 + \frac{\beta z q'(z)}{q(z)} \prec \psi_{\lambda}^{a,b}(\alpha, \beta, \xi, \mu; z) \tag{3.2.6.16}$$

implies

$$q(z) \prec \frac{z\left(D_z^{-\lambda}\phi(a, c; z)\right)'}{D_z^{-\lambda}\phi(a, c; z)}, \quad z \in U, \tag{3.2.6.17}$$

and function q is the best subordinant.

Proof. Define the function p by

$$p(z) := \frac{z\left(D_z^{-\lambda}\phi(a,c;z)\right)'}{D_z^{-\lambda}\phi(a,c;z)}, \quad z \in U, \ z \neq 0.$$

Considering

$$\phi(w) := \frac{\beta}{w}$$

and

$$\theta(w) := \alpha + \xi w + \mu w^2,$$

it is easy to show that ϕ is analytic in $\mathbb{C}\backslash\{0\}$, $\phi(w) \neq 0$, $w \in \mathbb{C}\backslash\{0\}$ and θ is also analytic in $\mathbb{C}$.

Since $\frac{\theta'(q(z))}{\phi(q(z))} = \frac{q(z)[\xi+2\mu q(z)]q'(z)}{\beta}$, it yields $\operatorname{Re}\left(\frac{\theta'(q(z))}{\phi(q(z))}\right) = \operatorname{Re}\left(\frac{2\mu}{\beta}q^2(z)q'(z) + \frac{\xi}{\beta}q(z)q'(z)\right) > 0$, where $\mu, \beta, \alpha \in \mathbb{C}$, $\mu \neq 0$.

From (3.2.6.14) and (3.2.6.16), we get

$$\alpha + \xi q(z) + \mu(q(z))^2 + \frac{\beta z q'(z)}{q(z)} \prec \alpha + \xi p(z) + \mu(p(z))^2 + \frac{\beta z p'(z)}{p(z)}.$$

Applying Lemma 1.6, we obtain $q(z) \prec p(z)$, therefore

$$q(z) \prec \frac{z\left(D_z^{-\lambda}\phi(a,c;z)\right)'}{D_z^{-\lambda}\phi(a,c;z)}, \quad z \in U,$$

and the best subordinant is function q. □

Corollary 3.54 (Alb Lupaş, 2021b). *Consider $\lambda > 0$ and a, c, α, ξ, μ, $\beta \in \mathbb{C}$, $c \neq 0, -1, -2, \ldots$, $\beta \neq 0$. Assume that (3.2.6.15) holds for $q(z) = \frac{1+Az}{1+Bz}$, $-1 \leq B < A \leq 1$. If $\frac{z\left(D_z^{-\lambda}\phi(a,c;z)\right)'}{D_z^{-\lambda}\phi(a,c;z)} \in Q \cap \mathcal{H}[q(0),1]$ and*

$$\alpha + \xi\frac{1+Az}{1+Bz} + \mu\left(\frac{1+Az}{1+Bz}\right)^2 + \frac{\beta(A-B)z}{(1+Az)(1+Bz)} \prec \psi_\lambda^{a,b}(\alpha,\beta,\xi,\mu;z),$$

where $\psi_\lambda^{a,b}$ is defined in (3.2.6.11), then

$$\frac{1+Az}{1+Bz} \prec \frac{z\left(D_z^{-\lambda}\phi(a,c;z)\right)'}{D_z^{-\lambda}\phi(a,c;z)},$$

and the best subordinant is $\frac{1+Az}{1+Bz}$.

Proof. When $-1 \leq B < A \leq 1$, consider $q(z) = \frac{1+Az}{1+Bz}$ in Theorem 3.128 and obtain the corollary. □

Corollary 3.55 (Alb Lupaş, 2021b). *Let* $\lambda > 0$, a, c, α, ξ, μ, $\beta \in \mathbb{C}$, $c \neq 0, -1, -2, \ldots$, $\beta \neq 0$. *Assume that* (3.2.6.15) *holds for* $q(z) = \left(\frac{1+z}{1-z}\right)^{\gamma}$, $0 < \gamma \leq 1$. *If* $\frac{z\left(D_z^{-\lambda}\phi(a,c;z)\right)'}{D_z^{-\lambda}\phi(a,c;z)} \in Q \cap \mathcal{H}[q(0), 1]$ *and*

$$\alpha + \xi\left(\frac{1+z}{1-z}\right)^{\gamma} + \mu\left(\frac{1+z}{1-z}\right)^{2\gamma} + \frac{2\beta\gamma z}{1-z^2} \prec \psi_{\lambda}^{a,b}(\alpha, \beta, \xi, \mu; z),$$

where $\psi_{\lambda}^{a,b}$ *is defined in* (3.2.6.11), *then*

$$\left(\frac{1+z}{1-z}\right)^{\gamma} \prec \frac{z\left(D_z^{-\lambda}\phi(a,c;z)\right)'}{D_z^{-\lambda}\phi(a,c;z)},$$

and the best subordinant is $\left(\frac{1+z}{1-z}\right)^{\gamma}$.

Proof. Put in Theorem 3.128 $q(z) = \left(\frac{1+z}{1-z}\right)^{\gamma}$, when $0 < \gamma \leq 1$. □

The sandwich theorem is obtained combining Theorems 3.127 and 3.128.

Theorem 3.129 (Alb Lupaş, 2021b). *Consider* q_1 *and* q_2 *analytic and univalent functions in* U, *with* $q_1(z) \neq 0$ *and* $q_2(z) \neq 0$, $\forall\ z \in U$, *such that* $\frac{zq_1'(z)}{q_1(z)}$ *and* $\frac{zq_2'(z)}{q_2(z)}$ *are starlike univalent. Assume that* q_1 *satisfies relation* (3.2.6.10) *and* q_2 *satisfies relation* (3.2.6.15). *If* $\frac{z\left(D_z^{-\lambda}\phi(a,c;z)\right)'}{D_z^{-\lambda}\phi(a,c;z)} \in Q \cap \mathcal{H}[q(0), 1]$ *and* $\psi_{\lambda}^{a,b}(\alpha, \beta, \xi, \mu; z)$ *is defined by* (3.2.6.11) *and is univalent in* U, $\lambda > 0$ *and* a, $c \in \mathbb{C}$, $c \neq 0, -1, -2, \ldots$, *then*

$$\alpha + \xi q_1(z) + \mu(q_1(z))^2 + \frac{\beta z q_1'(z)}{q_1(z)} \prec \psi_{\lambda}^{a,b}(\alpha, \beta, \xi, \mu; z)$$

$$\prec \alpha + \xi q_2(z) + \mu(q_2(z))^2 + \frac{\beta z q_2'(z)}{q_2(z)},$$

for $\alpha, \beta, \mu, \xi \in \mathbb{C}$, $\beta \neq 0$, *implies*

$$q_1(z) \prec \frac{z\left(D_z^{-\lambda}\phi(a,c;z)\right)'}{D_z^{-\lambda}\phi(a,c;z)} \prec q_2(z),$$

and q_1 *and* q_2 *are respectively the best subordinant and the best dominant.*

For $q_1(z) = \frac{1+A_1z}{1+B_1z}$, $q_2(z) = \frac{1+A_2z}{1+B_2z}$, where $-1 \le B_2 < B_1 < A_1 < A_2 \le 1$, we have the following corollary.

Corollary 3.56 (Alb Lupaş, 2021b). *Let* $a, c, \alpha, \xi, \mu, \beta \in \mathbb{C}$, $c \neq 0, -1, -2, \ldots$, $\beta \neq 0$ *and* $\lambda > 0$. *Assume that* (3.2.6.10) *and* (3.2.6.15) *hold for* $q_1(z) = \frac{1+A_1z}{1+B_1z}$, *respectively* $q_2(z) = \frac{1+A_2z}{1+B_2z}$, $-1 \le B_2 \le B_1 < A_1 \le A_2 \le 1$. *If* $\frac{z\left(D_z^{-\lambda}\phi(a,c;z)\right)'}{D_z^{-\lambda}\phi(a,c;z)} \in \mathcal{H}[q(0),1] \cap Q$ *and*

$$\alpha + \xi\frac{1+A_1z}{1+B_1z} + \mu\left(\frac{1+A_1z}{1+B_1z}\right)^2 + \frac{\beta(A_1-B_1)z}{(1+A_1z)(1+B_1z)} \prec \psi_\lambda^{a,b}(\alpha,\beta,\xi,\mu;z)$$

$$\prec \alpha + \xi\frac{1+A_2z}{1+B_2z} + \mu\left(\frac{1+A_2z}{1+B_2z}\right)^2 + \frac{\beta(A_2-B_2)z}{(1+A_2z)(1+B_2z)},$$

with $\psi_\lambda^{a,b}$ *defined by* (3.2.6.11), *then*

$$\frac{1+A_1z}{1+B_1z} \prec \frac{z\left(D_z^{-\lambda}\phi(a,c;z)\right)'}{D_z^{-\lambda}\phi(a,c;z)} \prec \frac{1+A_2z}{1+B_2z},$$

hence $\frac{1+A_1z}{1+B_1z}$ *and* $\frac{1+A_2z}{1+B_2z}$ *are the best subordinant, respectively the best dominant.*

For $q_1(z) = \left(\frac{1+z}{1-z}\right)^{\gamma_1}$, $q_2(z) = \left(\frac{1+z}{1-z}\right)^{\gamma_2}$, where $0 < \gamma_1 < \gamma_2 \le 1$, we have the following corollary.

Corollary 3.57 (Alb Lupaş, 2021b). *Let* $\lambda > 0$, $a, c, \alpha, \beta, \mu, \xi \in \mathbb{C}$, $c \neq 0, -1, -2, \ldots$, $\beta \neq 0$. *Assume that* (3.2.6.10) *and* (3.2.6.15) *hold for* $q_1(z) = \left(\frac{1+z}{1-z}\right)^{\gamma_1}$, *respectively* $q_2(z) = \left(\frac{1+z}{1-z}\right)^{\gamma_2}$, $0 < \gamma_1 < \gamma_2 \le 1$. *If* $\frac{z\left(D_z^{-\lambda}\phi(a,c;z)\right)'}{D_z^{-\lambda}\phi(a,c;z)} \in Q \cap \mathcal{H}[q(0),1]$ *and*

$$\alpha + \xi\left(\frac{1+z}{1-z}\right)^{\gamma_1} + \mu\left(\frac{1+z}{1-z}\right)^{2\gamma_1} + \frac{2\beta\gamma_1 z}{1-z^2} \prec \psi_\lambda^{a,b}(\alpha,\beta,\xi,\mu;z)$$

$$\prec \alpha + \xi\left(\frac{1+z}{1-z}\right)^{\gamma_2} + \mu\left(\frac{1+z}{1-z}\right)^{2\gamma_2} + \frac{2\beta\gamma_2 z}{1-z^2},$$

where $\psi_\lambda^{a,b}$ *is defined in* (3.2.6.11), *then*

$$\left(\frac{1+z}{1-z}\right)^{\gamma_1} \prec \frac{z\left(D_z^{-\lambda}\phi(a,c;z)\right)'}{D_z^{-\lambda}\phi(a,c;z)} \prec \left(\frac{1+z}{1-z}\right)^{\gamma_2},$$

hence $\left(\frac{1+z}{1-z}\right)^{\gamma_1}$ *and* $\left(\frac{1+z}{1-z}\right)^{\gamma_2}$ *are the best subordinant and the best dominant, respectively.*

Changing the functions ϕ and θ in Theorem 3.127, we get

Theorem 3.130 (Alb Lupaş, 2021b). *Consider* $\lambda > 0$, $a, c \in \mathbb{C}$, $c \neq 0, -1, -2, \ldots$, *the convex and univalent function* q *in* U *with* $q(0) = \lambda$ *and* $\frac{z\left(D_z^{-\lambda}\phi(a,c;z)\right)'}{D_z^{-\lambda}\phi(a,c;z)} \in \mathcal{H}(U)$, $\forall\ z \in U$. *Suppose that*

$$\operatorname{Re}\left(z\frac{q''(z)}{q'(z)} + \frac{\alpha}{\beta} + 1\right) > 0, \quad z \in U, \tag{3.2.6.18}$$

where $\alpha, \beta \in \mathbb{C}$, $\beta \neq 0$, *and*

$$\psi_\lambda^{a,b}(\alpha, \beta; z) := (\alpha + \beta)\frac{z\left(D_z^{-\lambda}\phi(a,c;z)\right)'}{D_z^{-\lambda}\phi(a,c;z)} - \beta\left(\frac{z\left(D_z^{-\lambda}\phi(a,c;z)\right)'}{D_z^{-\lambda}\phi(a,c;z)}\right)^2 + \beta\frac{z^2\left(D_z^{-\lambda}\phi(a,c;z)\right)''}{D_z^{-\lambda}\phi(a,c;z)}. \tag{3.2.6.19}$$

If the following subordination is satisfied by q,

$$\psi_\lambda^{a,b}(\alpha, \beta; z) \prec \alpha q(z) + \beta z q'(z), \quad z \in U, \tag{3.2.6.20}$$

then

$$\frac{z\left(D_z^{-\lambda}\phi(a,c;z)\right)'}{D_z^{-\lambda}\phi(a,c;z)} \prec q(z), \quad z \in U, \tag{3.2.6.21}$$

and the best dominant is the function q.

Proof. Define

$$p(z) := \frac{z\left(D_z^{-\lambda}\phi(a,c;z)\right)'}{D_z^{-\lambda}\phi(a,c;z)}, \quad z \in U,\ z \neq 0$$

with $p(0) = \lambda$, an analytic function in U. Differentiating with respect to z, we get $p'(z) = \frac{\left(D_z^{-\lambda}\phi(a,c;z)\right)'}{D_z^{-\lambda}\phi(a,c;z)} - z\left(\frac{\left(D_z^{-\lambda}\phi(a,c;z)\right)'}{D_z^{-\lambda}\phi(a,c;z)}\right)^2 + z\frac{\left(D_z^{-\lambda}\phi(a,c;z)\right)''}{D_z^{-\lambda}\phi(a,c;z)}$ and

$$zp'(z) = \frac{z\left(D_z^{-\lambda}\phi(a,c;z)\right)'}{D_z^{-\lambda}\phi(a,c;z)} - \left(\frac{z\left(D_z^{-\lambda}\phi(a,c;z)\right)'}{D_z^{-\lambda}\phi(a,c;z)}\right)^2 + \frac{z^2\left(D_z^{-\lambda}\phi(a,c;z)\right)''}{D_z^{-\lambda}\phi(a,c;z)}. \tag{3.2.6.22}$$

Let

$$\phi(w) := \beta$$

analytic in $\mathbb{C}\backslash\{0\}$ with $\phi(w) \neq 0$, $w \in \mathbb{C}\backslash\{0\}$ and

$$\theta(w) := \alpha w$$

analytic in $\mathbb{C}$.

Consider

$$Q(z) = zq'(z)\phi(q(z)) = \beta z q'(z)$$

starlike univalent in U and

$$h(z) = \theta(q(z)) + Q(z) = \alpha q(z) + \beta z q'(z).$$

We obtain $\operatorname{Re}\left(\frac{zh'(z)}{Q(z)}\right) = \operatorname{Re}\left(z\frac{q''(z)}{q'(z)} + \frac{\alpha}{\beta} + 1\right) > 0$.

From (3.2.6.22), it yields $\alpha p(z) + \beta z p'(z) = (\alpha+\beta)\frac{z\left(D_z^{-\lambda}\phi(a,c;z)\right)'}{D_z^{-\lambda}\phi(a,c;z)} - \beta\left(\frac{z\left(D_z^{-\lambda}\phi(a,c;z)\right)'}{D_z^{-\lambda}\phi(a,c;z)}\right)^2 + \beta\frac{z^2\left(D_z^{-\lambda}\phi(a,c;z)\right)''}{D_z^{-\lambda}\phi(a,c;z)}$.

By using (3.2.6.20), we get $\alpha p(z) + \beta z p'(z) \prec \alpha q(z) + \beta z q'(z)$.

Lemma 1.3 gives $p(z) \prec q(z)$, $\forall\ z \in U$, so we obtain $\frac{z\left(D_z^{-\lambda}\phi(a,c;z)\right)'}{D_z^{-\lambda}\phi(a,c;z)} \prec q(z)$, $\forall\ z \in U$, and the best dominant is function q. □

Corollary 3.58 (Alb Lupaş, 2021b). *Let $-1 \leq B < A \leq 1$, $\lambda > 0$, a, c, α, $\beta \in \mathbb{C}$, $c \neq 0, -1, -2, \ldots$, $\beta \neq 0$, and $q(z) = \frac{1+Az}{1+Bz}$, $z \in U$, fulfilling the relation* (3.2.6.18). *If*

$$\psi_\lambda^{a,b}(\alpha,\beta;z) \prec \frac{\beta(A-B)z}{(1+Bz)^2} + \alpha\frac{1+Az}{1+Bz},$$

with $\psi_\lambda^{a,b}$ defined by (3.2.6.19), *then*

$$\frac{z\left(D_z^{-\lambda}\phi(a,c;z)\right)'}{D_z^{-\lambda}\phi(a,c;z)} \prec \frac{1+Az}{1+Bz},$$

and the best dominant is $\frac{1+Az}{1+Bz}$.

Proof. In Theorem 3.130 consider $q(z) = \frac{1+Az}{1+Bz}$, with $-1 \leq B < A \leq 1$. □

Corollary 3.59 (Alb Lupaş, 2021b). *Let* $\lambda > 0$, a, c, α, $\beta \in \mathbb{C}$, $c \neq 0, -1, -2, \ldots, \beta \neq 0$, *and* $q(z) = \left(\frac{1+z}{1-z}\right)^{\gamma}$, $0 < \gamma \leq 1$, *fulfilling the relation* (3.2.6.18). *If*

$$\psi_{\lambda}^{a,b}(\alpha, \beta; z) \prec \frac{2\beta\gamma z}{1-z^2}\left(\frac{1+z}{1-z}\right)^{\gamma} + \alpha\left(\frac{1+z}{1-z}\right)^{\gamma},$$

with $\psi_{\lambda}^{a,b}$ *defined by* (3.2.6.19), *then*

$$\frac{z\left(D_z^{-\lambda}\phi(a, c; z)\right)'}{D_z^{-\lambda}\phi(a, c; z)} \prec \left(\frac{1+z}{1-z}\right)^{\gamma},$$

and $\left(\frac{1+z}{1-z}\right)^{\gamma}$ *is the best dominant.*

Proof. In Theorem 3.130, put $q(z) = \left(\frac{1+z}{1-z}\right)^{\gamma}$, with $0 < \gamma \leq 1$. □

Changing the functions ϕ and θ in Theorem 3.128 to be the same as in Theorem 3.130, we get

Theorem 3.131 (Alb Lupaş, 2021b). *Let* $\lambda > 0$, a, $c \in \mathbb{C}$, $c \neq 0, -1, -2, \ldots$ *and* q *be a convex and univalent function in* U *with* $q(0) = \lambda$. *Suppose that*

$$\operatorname{Re}\left(\frac{\alpha}{\beta}q'(z)\right) > 0, \quad \alpha, \beta \in \mathbb{C}, \quad \beta \neq 0. \tag{3.2.6.23}$$

If $\frac{z\left(D_z^{-\lambda}\phi(a,c;z)\right)'}{D_z^{-\lambda}\phi(a,c;z)} \in Q \cap \mathcal{H}[q(0), 1]$ *and* $\psi_{\lambda}^{a,b}(\alpha, \beta; z)$ *defined in* (3.2.6.19) *is univalent in* U, *then*

$$\alpha q(z) + \beta z q'(z) \prec \psi_{\lambda}^{a,b}(\alpha, \beta; z) \tag{3.2.6.24}$$

implies

$$q(z) \prec \frac{z\left(D_z^{-\lambda}\phi(a, c; z)\right)'}{D_z^{-\lambda}\phi(a, c; z)}, \quad \forall\, z \in U, \tag{3.2.6.25}$$

and the best subordinant is the function q.

Proof. Define the analytic function

$$p(z) := \frac{z\left(D_z^{-\lambda}\phi(a,c;z)\right)'}{D_z^{-\lambda}\phi(a,c;z)}, \quad z\in U, \ z\neq 0$$

and $p(0)=\lambda$.

Set

$$\phi(w) := \beta$$

analytic in $\mathbb{C}\backslash\{0\}$ with $\phi(w)\neq 0$, $\forall\ w\in\mathbb{C}\backslash\{0\}$ and

$$\theta(w) := \alpha w$$

analytic in $\mathbb{C}$.

Since $\frac{\theta'(q(z))}{\phi(q(z))} = \frac{\alpha}{\beta}q'(z)$, from (3.2.6.23), we get $\operatorname{Re}\left(\frac{\theta'(q(z))}{\phi(q(z))}\right) = \operatorname{Re}\left(\frac{\alpha}{\beta}q'(z)\right) > 0$, with $\alpha,\beta\in\mathbb{C}$, $\beta\neq 0$.

Relation (3.2.6.24) gives the differential superordination

$$\alpha q(z) + \beta z q'(z) \prec \alpha p(z) + \beta z p'(z), \quad z\in U,$$

and applying Lemma 1.6, we obtain $q(z)\prec p(z)$, which means

$$q(z) \prec \frac{z\left(D_z^{-\lambda}\phi(a,c;z)\right)'}{D_z^{-\lambda}\phi(a,c;z)}, \quad z\in U,$$

and the best subordinant is the function q. □

Corollary 3.60 (Alb Lupaş, 2021b). *Assume that* (3.2.6.23) *holds for* $q(z) = \frac{1+Az}{1+Bz}$, $z\in U$, *with* $-1\leq B < A\leq 1$, $\lambda>0$, a, $c\in\mathbb{C}$, $c\neq 0,-1,-2,\ldots$. *If* $\frac{z\left(D_z^{-\lambda}\phi(a,c;z)\right)'}{D_z^{-\lambda}\phi(a,c;z)} \in Q\cap\mathcal{H}[q(0),1]$, *and*

$$\alpha\frac{1+Az}{1+Bz} + \frac{\beta(A-B)z}{(1+Bz)^2} \prec \psi_\lambda^{a,b}(\alpha,\beta;z),$$

where $\psi_\lambda^{a,b}$ *is defined in* (3.2.6.19), $\alpha,\beta\in\mathbb{C}$, $\beta\neq 0$, *then*

$$\frac{1+Az}{1+Bz} \prec \frac{z\left(D_z^{-\lambda}\phi(a,c;z)\right)'}{D_z^{-\lambda}\phi(a,c;z)},$$

and the best subordinant is $\frac{1+Az}{1+Bz}$.

Proof. In Theorem 3.131, consider $q(z) = \frac{1+Az}{1+Bz}$, with $-1 \leq B < A \leq 1$. $\square$

Corollary 3.61 (Alb Lupaş, 2021b). *Assume that* (3.2.6.23) *holds for* $q(z) = \left(\frac{1+z}{1-z}\right)^{\gamma}$, $0 < \gamma \leq 1$, $\lambda > 0$, a, $c \in \mathbb{C}$, $c \neq 0, -1, -2, \ldots$. *If* $\frac{z\left(D_z^{-\lambda}\phi(a,c;z)\right)'}{D_z^{-\lambda}\phi(a,c;z)} \in Q \cap \mathcal{H}[q(0),1]$ *and*

$$\alpha\left(\frac{1+z}{1-z}\right)^{\gamma} + \frac{2\beta\gamma z}{1-z^2}\left(\frac{1+z}{1-z}\right)^{\gamma} \prec \psi_{\lambda}^{a,b}(\alpha,\beta;z),$$

where $\psi_{\lambda}^{a,b}$ *is defined in* (3.2.6.19), $\alpha, \beta \in \mathbb{C}$, $\beta \neq 0$, *then*

$$\left(\frac{1+z}{1-z}\right)^{\gamma} \prec \frac{z\left(D_z^{-\lambda}\phi(a,c;z)\right)'}{D_z^{-\lambda}\phi(a,c;z)},$$

and the best subordinant is $\left(\frac{1+z}{1-z}\right)^{\gamma}$.

Proof. In Theorem 3.131, consider $q(z) = \left(\frac{1+z}{1-z}\right)^{\gamma}$, with $0 < \gamma \leq 1$. $\square$

The sandwich theorem is obtained combining Theorems 3.130 and 3.131.

Theorem 3.132 (Alb Lupaş, 2021b). *Consider* q_1 *and* q_2 *convex and univalent functions in* U *with* $q_1(z) \neq 0$ *and* $q_2(z) \neq 0$, $\forall\ z \in U$. *Assume that relation* (3.2.6.18) *is satisfied by* q_1 *and relation* (3.2.6.23) *is satisfied by* q_2. *If* $\frac{z\left(D_z^{-\lambda}\phi(a,c;z)\right)'}{D_z^{-\lambda}\phi(a,c;z)} \in Q \cap \mathcal{H}[q(0),1]$ *and* $\psi_{\lambda}^{a,b}(\alpha,\beta;z)$ *is univalent in* U *defined by* (3.2.6.19), $\lambda > 0$, a, c, α, $\beta \in \mathbb{C}$, $c \neq 0, -1, -2, \ldots$, $\beta \neq 0$, *then*

$$\alpha q_1(z) + \beta z q_1'(z) \prec \psi_{\lambda}^{a,b}(\alpha,\beta;z) \prec \alpha q_2(z) + \beta z q_2'(z),$$

implies

$$q_1(z) \prec \frac{z\left(D_z^{-\lambda}\phi(a,c;z)\right)'}{D_z^{-\lambda}\phi(a,c;z)} \prec q_2(z), \quad \forall\ z \in U,$$

and the best subordinant is q_1 *and the best dominant is* q_2.

Letting $q_1(z) = \frac{1+A_1z}{1+B_1z}$, $q_2(z) = \frac{1+A_2z}{1+B_2z}$, where $-1 \leq B_2 < B_1 < A_1 < A_2 \leq 1$, in Theorem 3.132, we get

Corollary 3.62 (Alb Lupaş, 2021b). *Assume that* (3.2.6.18) *and* (3.2.6.23) *hold for* $q_1(z) = \frac{1+A_1z}{1+B_1z}$ *and* $q_2(z) = \frac{1+A_2z}{1+B_2z}$, *respectively,* $\lambda > 0$,

a, c, α, $\beta \in \mathbb{C}$, $c \neq 0, -1, -2, \ldots$, $\beta \neq 0$, $-1 \leq B_2 \leq B_1 < A_1 \leq A_2 \leq 1$. If $\frac{z\left(D_z^{-\lambda}\phi(a,c;z)\right)'}{D_z^{-\lambda}\phi(a,c;z)} \in Q \cap \mathcal{H}[q(0),1]$ and

$$\alpha \frac{1+A_1 z}{1+B_1 z} + \frac{\beta (A_1 - B_1) z}{(1+B_1 z)^2} \prec \psi_\lambda^{a,b}(\alpha, \beta; z)$$

$$\prec \alpha \frac{1+A_2 z}{1+B_2 z} + \frac{\beta (A_2 - B_2) z}{(1+B_2 z)^2}, \quad \forall\, z \in U,$$

where $\psi_\lambda^{a,b}$ is defined in (3.2.6.19), $\alpha, \beta \in \mathbb{C}$, $\beta \neq 0$, then

$$\frac{1+A_1 z}{1+B_1 z} \prec \frac{z\left(D_z^{-\lambda}\phi(a,c;z)\right)'}{D_z^{-\lambda}\phi(a,c;z)} \prec \frac{1+A_2 z}{1+B_2 z}, \quad z \in U,$$

hence the best subordinant is $\frac{1+A_1 z}{1+B_1 z}$ and the best dominant is $\frac{1+A_2 z}{1+B_2 z}$.

Setting $q_1(z) = \left(\frac{1+z}{1-z}\right)^{\gamma_1}$, $q_2(z) = \left(\frac{1+z}{1-z}\right)^{\gamma_2}$, where $0 < \gamma_1 < \gamma_2 \leq 1$, in Theorem 3.132, we obtain

Corollary 3.63 (Alb Lupaş, 2021b). *Assume that (3.2.6.18) and (3.2.6.23) hold for $q_1(z) = \left(\frac{1+z}{1-z}\right)^{\gamma_1}$ and $q_2(z) = \left(\frac{1+z}{1-z}\right)^{\gamma_2}$, respectively, $\lambda > 0$, a, c, α, $\beta \in \mathbb{C}$, $c \neq 0, -1, -2, \ldots$, $\beta \neq 0$, $0 < \gamma_1 < \gamma_2 \leq 1$. If $\frac{z\left(D_z^{-\lambda}\phi(a,c;z)\right)'}{D_z^{-\lambda}\phi(a,c;z)} \in Q \cap \mathcal{H}[q(0),1]$ and*

$$\alpha \left(\frac{1+z}{1-z}\right)^{\gamma_1} + \frac{2\beta\gamma_1 z}{1-z^2}\left(\frac{1+z}{1-z}\right)^{\gamma_1} \prec \psi_\lambda^{a,b}(\alpha, \beta; z)$$

$$\prec \alpha \left(\frac{1+z}{1-z}\right)^{\gamma_2} + \frac{2\beta\gamma_2 z}{1-z^2}\left(\frac{1+z}{1-z}\right)^{\gamma_2}, \quad z \in U,$$

where $\psi_\lambda^{a,b}$ is defined by (3.2.6.19), then

$$\left(\frac{1+z}{1-z}\right)^{\gamma_1} \prec \frac{z\left(D_z^{-\lambda}\phi(a,c;z)\right)'}{D_z^{-\lambda}\phi(a,c;z)} \prec \left(\frac{1+z}{1-z}\right)^{\gamma_2}, \quad \forall\, z \in U,$$

hence the best subordinant is $\left(\frac{1+z}{1-z}\right)^{\gamma_1}$ and the best dominant is $\left(\frac{1+z}{1-z}\right)^{\gamma_2}$.

Chapter 4

Recent Developments in the Strong Differential Subordinations and Superordinations Theory

In this chapter, we expose some of the results obtained by the author regarding few operators presented in the previous chapter using the strong differential subordinations and superordinations theory.

4.1 Applications of Differential Operators in the Theory of Strong Differential Subordinations

In this section, we present several strong differential subordination results involving few of the defined operators presented in Chapter 2.

4.1.1 *Certain Strong Differential Subordinations Regarding the Hadamard Product of Multiplier Transformation and Ruscheweyh Derivative*

In the paper Alb Lupaş (2011g), we established several strong differential subordinations regarding the new operator $IR_{\lambda,l}^{m}$ defined by the Hadamard product of the multiplier transformation $I(m,\lambda,l)$ and the Ruscheweyh operator R^m, considering in Definition 2.10 $n = m$, given by $IR_{\lambda,l}^{m}$: $\mathcal{A}_{\zeta}^{*} \to \mathcal{A}_{\zeta}^{*}$,

$$IR_{\lambda,l}^{m} f(z,\zeta) = (I(m,\lambda,l) * R^m) f(z,\zeta),$$

where $\mathcal{A}_{n\zeta}^{*} = \{f \in \mathcal{H}(U \times \overline{U}),\ f(z,\zeta) = z + a_{n+1}(\zeta) z^{n+1} + \cdots,\ z \in U,\ \zeta \in \overline{U}\}$ is the class of normalized analytic functions with $\mathcal{A}_{1\zeta}^{*} = \mathcal{A}_{\zeta}^{*}$.

If $f \in \mathcal{A}_{\zeta}^{*}$, $f(z,\zeta) = z + \sum_{j=2}^{\infty} a_j(\zeta) z^j$, then $IR_{\lambda,l}^{m} f(z,\zeta) = z + \sum_{j=2}^{\infty} \left(\frac{1+\lambda(j-1)+l}{l+1}\right)^m C_{m+j-1}^{m} a_j^2(\zeta) z^j$, $z \in U, \zeta \in \overline{U}$.

Theorem 4.1 (Alb Lupaş, 2011g). *Let $g(z,\zeta)$ be a convex function such that $g(0,\zeta)=1$ and let h be the function $h(z,\zeta)=g(z,\zeta)+zg'(z,\zeta)$, $z\in U$, $\zeta\in\overline{U}$. If $\lambda, l\geq 0$, $m\in\mathbb{N}$, $f\in\mathcal{A}_{\zeta}^{*}$ and the strong differential subordination*

$$\frac{1}{z}\left(\frac{m+1}{\lambda}IR_{\lambda,l}^{m+1}f(z,\zeta)-\frac{m-2}{\lambda}IR_{\lambda,l}^{m}f(z,\zeta)\right)$$
$$+\frac{\lambda(m-1)-(l+1)}{\lambda(l+1)}\left(IR_{\lambda,l}^{m}f(z,\zeta)\right)_{z}'+\left(1-\frac{m-1}{l+1}-\frac{2}{\lambda}\right)$$
$$-\frac{2(l+1)(m-1)-2\lambda m}{\lambda(l+1)}\int_{0}^{z}\frac{IR_{\lambda,l}^{m}f(t,\zeta)-t}{t^{2}}dt\prec\prec h(z,\zeta), \tag{4.1.1.1}$$

$z\in U$, $\zeta\in\overline{U}$, holds, then

$$\left(IR_{\lambda,l}^{m}f(z,\zeta)\right)_{z}'\prec\prec g(z,\zeta),\quad z\in U,\quad \zeta\in\overline{U}$$

and this result is sharp.

Proof. With notation $p(z,\zeta)=\left(IR_{\lambda,l}^{m}f(z,\zeta)\right)'=1+\sum_{j=2}^{\infty}\left(\frac{1+\lambda(j-1)+l}{l+1}\right)^{m}C_{m+j-1}^{m}ja_{j}^{2}(\zeta)z^{j-1}$ and $p(0,\zeta)=1$, we obtain for $f(z,\zeta)=z+\sum_{j=2}^{\infty}a_{j}(\zeta)z^{j}$,

$p(z,\zeta)+zp_{z}'(z,\zeta)$

$$=1+\sum_{j=2}^{\infty}\left(\frac{1+\lambda(j-1)+l}{l+1}\right)^{m}C_{m+j-1}^{m}ja_{j}^{2}(\zeta)z^{j-1}$$
$$+\sum_{j=2}^{\infty}\left(\frac{1+\lambda(j-1)+l}{l+1}\right)^{m}C_{m+j-1}^{m}j(j-1)a_{j}^{2}(\zeta)z^{j-1}$$
$$=1+\sum_{j=2}^{\infty}\left(\frac{1+\lambda(j-1)+l}{l+1}\right)^{m}C_{m+j-1}^{m}j^{2}a_{j}^{2}(\zeta)z^{j-1}$$
$$=\frac{1}{z}\left(z+\sum_{j=2}^{\infty}\left(\frac{1+\lambda(j-1)+l}{l+1}\right)^{m+1}C_{m+j}^{m+1}\frac{m+1}{\lambda}a_{j}^{2}(\zeta)z^{j}\right.$$
$$-\sum_{j=2}^{\infty}\left(\frac{1+\lambda(j-1)+l}{l+1}\right)^{m}C_{m+j-1}^{m}\frac{\lambda(m-1)-(l+1)}{\lambda(l+1)}ja_{j}^{2}(\zeta)z^{j}$$
$$-\sum_{j=2}^{\infty}\left(\frac{1+\lambda(j-1)+l}{l+1}\right)^{m}C_{m+j-1}^{m}\frac{m-2}{\lambda}a_{j}^{2}(\zeta)z^{j}$$

$$- \sum_{j=2}^{\infty} \left(\frac{1+\lambda(j-1)+l}{l+1} \right)^m C_{m+j-1}^m \frac{1}{j-1}$$

$$\times \frac{2(l+1)(m-1) - 2\lambda m}{\lambda(l+1)} a_j^2(\zeta) z^j \Bigg)$$

$$= \frac{1}{z} \left[\frac{m+1}{\lambda} \left(z + \sum_{j=2}^{\infty} \left(\frac{1+\lambda(j-1)+l}{l+1} \right)^{m+1} C_{m+j}^{m+1} a_j^2(\zeta) z^j \right) \right.$$

$$\left. - \frac{m-2}{\lambda} \left(z + \sum_{j=2}^{\infty} \left(\frac{1+\lambda(j-1)+l}{l+1} \right)^m C_{m+j-1}^m a_j^2(\zeta) z^j \right) \right]$$

$$+ \left(1 - \frac{m+1}{\lambda} - \frac{m-2}{\lambda} \right)$$

$$+ \left(1 + \sum_{j=2}^{\infty} \left(\frac{1+\lambda(j-1)+l}{l+1} \right)^m C_{m+j-1}^m a_j^2(\zeta) j z^{j-1} \right) \frac{\lambda(m-1)-(l+1)}{\lambda(l+1)}$$

$$- \frac{\lambda(m-1)-(l+1)}{\lambda(l+1)}$$

$$- \sum_{j=2}^{\infty} \left(\frac{1+\lambda(j-1)+l}{l+1} \right)^m C_{m+j-1}^m \frac{1}{j-1}$$

$$\times \frac{2(l+1)(m-1) - 2\lambda m}{\lambda(l+1)} a_j^2(\zeta) z^{j-1}$$

$$= \frac{1}{z} \left(\frac{m+1}{\lambda} IR_{\lambda,l}^{m+1} f(z,\zeta) - \frac{m-2}{\lambda} IR_{\lambda,l}^m f(z,\zeta) \right)$$

$$+ \frac{\lambda(m-1)-(l+1)}{\lambda(l+1)} \left(IR_{\lambda,l}^m f(z,\zeta) \right)'_z$$

$$+ \frac{\lambda l - \lambda m + 2\lambda - 2l - 2}{\lambda(l+1)}$$

$$- \frac{2(l+1)(m-1) - 2\lambda m}{\lambda(l+1)} \sum_{j=2}^{\infty} \left(\frac{1+\lambda(j-1)+l}{l+1} \right)^m$$

$$\times C_{m+j-1}^m \frac{1}{j-1} a_j^2(\zeta) z^{j-1}$$

$$= \frac{1}{z}\left(\frac{m+1}{\lambda} IR_{\lambda,l}^{m+1} f(z,\zeta) - \frac{m-2}{\lambda} IR_{\lambda,l}^{m} f(z,\zeta)\right)$$
$$+ \frac{\lambda(m-1)-(l+1)}{\lambda(l+1)} \left(IR_{\lambda,l}^{m} f(z,\zeta)\right)_z' + \left(1 - \frac{m-1}{l+1} - \frac{2}{\lambda}\right)$$
$$- \frac{2(l+1)(m-1)-2\lambda m}{\lambda(l+1)} \int_0^z \frac{IR_{\lambda,l}^{m} f(t,\zeta) - t}{t^2} dt.$$

We have $p(z,\zeta) + zp_z'(z,\zeta) \prec\prec h(z,\zeta) = g(z,\zeta) + zg_z'(z,\zeta)$, $z \in U$, $\zeta \in \overline{U}$. By using Lemma 1.8, we obtain $p(z,\zeta) \prec\prec g(z,\zeta)$, $z \in U$, $\zeta \in \overline{U}$, i.e. $\left(IR_{\lambda,l}^{m} f(z,\zeta)\right)_z' \prec\prec g(z,\zeta)$, $z \in U$, $\zeta \in \overline{U}$ and this result is sharp. □

Theorem 4.2 (Alb Lupaş, 2011g). *Let $h(z,\zeta)$ be a convex function such that $h(0,\zeta) = 1$. If $\lambda, l \geq 0$, $m \in \mathbb{N}$, $f \in \mathcal{A}_\zeta^*$ and the strong differential subordination*

$$\frac{1}{z}\left(\frac{m+1}{\lambda} IR_{\lambda,l}^{m+1} f(z,\zeta) - \frac{m-2}{\lambda} IR_{\lambda,l}^{m} f(z,\zeta)\right)$$
$$+ \frac{\lambda(m-1)-(l+1)}{\lambda(l+1)} \left(IR_{\lambda,l}^{m} f(z,\zeta)\right)_z' + \left(1 - \frac{m-1}{l+1} - \frac{2}{\lambda}\right)$$
$$- \frac{2(l+1)(m-1)-2\lambda m}{\lambda(l+1)} \int_0^z \frac{IR_{\lambda,l}^{m} f(t,\zeta) - t}{t^2} dt \prec\prec h(z,\zeta),$$

$z \in U$, $\zeta \in \overline{U}$, holds, then

$$\left(IR_{\lambda,l}^{m} f(z,\zeta)\right)_z' \prec\prec g(z,\zeta) \prec\prec h(z,\zeta), \quad z \in U, \quad \zeta \in \overline{U},$$

where $g(z,\zeta) = \frac{1}{z}\int_0^z h(t,\zeta)\,dt$ is convex and it is the best dominant.

Proof. With notation $p(z,\zeta) = \left(IR_{\lambda,l}^{m} f(z,\zeta)\right)_z' = 1 + \sum_{j=2}^{\infty} \left(\frac{1+\lambda(j-1)+l}{l+1}\right)^m C_{m+j-1}^m j a_j^2(\zeta) z^{j-1}$ and $p(0,\zeta) = 1$, we obtain for $f(z,\zeta) = z + \sum_{j=2}^{\infty} a_j(\zeta) z^j$, $p(z,\zeta) + zp_z'(z,\zeta) = \frac{1}{z}\left(\frac{m+1}{\lambda} IR_{\lambda,l}^{m+1} f(z,\zeta) - \frac{m-2}{\lambda} IR_{\lambda,l}^{m} f(z,\zeta)\right) + \frac{\lambda(m-1)-(l+1)}{\lambda(l+1)} \left(IR_{\lambda,l}^{m} f(z,\zeta)\right)_z' + \left(1 - \frac{m-1}{l+1} - \frac{2}{\lambda}\right) - \frac{2(l+1)(m-1)-2\lambda m}{\lambda(l+1)} \int_0^z \frac{IR_{\lambda,l}^{m} f(t,\zeta)-t}{t^2} dt.$

We have $p(z,\zeta) + zp_z'(z,\zeta) \prec\prec h(z,\zeta)$, $z \in U$, $\zeta \in \overline{U}$. Since $p(z,\zeta) \in \mathcal{H}^*[1,1,\zeta]$, using Lemma 1.7, for $n = 1$ and $\gamma = 1$, we obtain $p(z,\zeta) \prec\prec g(z,\zeta) \prec\prec h(z,\zeta)$, $z \in U$, $\zeta \in \overline{U}$, i.e. $\left(IR_{\lambda,l}^{m} f(z,\zeta)\right)_z' \prec\prec g(z,\zeta) = \frac{1}{z}\int_0^z h(t,\zeta)\,dt \prec\prec h(z,\zeta)$, $z \in U$, $\zeta \in \overline{U}$, and $g(z,\zeta)$ is convex and it is the best dominant. □

Theorem 4.3 (Alb Lupaş, 2011g). *Let $g(z,\zeta)$ be a convex function, $g(0,\zeta)=1$ and let h be the function $h(z,\zeta)=g(z,\zeta)+zg'_z(z,\zeta)$, $z\in U$, $\zeta\in\overline{U}$. If $\lambda,l\geq 0$, $m\in\mathbb{N}$, $f\in\mathcal{A}^*_\zeta$ and verifies the strong differential subordination*

$$\left(IR^m_{\lambda,l}f(z,\zeta)\right)'_z \prec\prec h(z,\zeta), \quad z\in U,\ \zeta\in\overline{U}, \tag{4.1.1.2}$$

then

$$\frac{IR^m_{\lambda,l}f(z,\zeta)}{z}\prec\prec g(z,\zeta),\ \ z\in U,\ \ \zeta\in\overline{U},$$

and this result is sharp.

Proof. For $f\in\mathcal{A}^*_\zeta$, $f(z,\zeta)=z+\sum_{j=2}^\infty a_j(\zeta)z^j$, we have

$$IR^m_{\lambda,l}f(z,\zeta) = z+\sum_{j=2}^{\infty}\left(\frac{1+\lambda(j-1)+l}{l+1}\right)^m C^m_{m+j-1}a_j^2(\zeta)z^j, \quad z\in U,\ \ \zeta\in\overline{U}.$$

Consider $p(z,\zeta)=\frac{IR^m_{\lambda,l}f(z,\zeta)}{z}=\frac{z+\sum_{j=2}^\infty\left(\frac{1+\lambda(j-1)+l}{l+1}\right)^m C^m_{m+j-1}a_j^2(\zeta)z^j}{z}=1+\sum_{j=2}^\infty\left(\frac{1+\lambda(j-1)+l}{l+1}\right)^m C^m_{m+j-1}a_j^2(\zeta)z^{j-1}$.

We have $p(z,\zeta)+zp'_z(z,\zeta)=\left(IR^m_{\lambda,l}f(z,\zeta)\right)'_z$, $z\in U$, $\zeta\in\overline{U}$.

Then $\left(IR^m_{\lambda,l}f(z,\zeta)\right)'_z\prec\prec h(z,\zeta)$, $z\in U$, $\zeta\in\overline{U}$ becomes $p(z,\zeta)+zp'_z(z,\zeta)\prec\prec h(z,\zeta)=g(z,\zeta)+zg'_z(z,\zeta)$, $z\in U$, $\zeta\in\overline{U}$. By using Lemma 1.8, we obtain $p(z,\zeta)\prec\prec g(z,\zeta)$, $z\in U$, $\zeta\in\overline{U}$, i.e. $\frac{IR^m_{\lambda,l}f(z,\zeta)}{z}\prec\prec g(z,\zeta)$, $z\in U$, $\zeta\in\overline{U}$. □

Theorem 4.4 (Alb Lupaş, 2011g). *Let $h(z,\zeta)$ be a convex function, $h(0,\zeta)=1$. If $\lambda,l\geq 0$, $m\in\mathbb{N}$, $f\in\mathcal{A}^*_\zeta$ and verifies the strong differential subordination*

$$\left(IR^m_{\lambda,l}f(z,\zeta)\right)'_z \prec\prec h(z,\zeta), \quad z\in U,\ \ \zeta\in\overline{U},$$

then

$$\frac{IR^m_{\lambda,l}f(z,\zeta)}{z}\prec\prec g(z,\zeta)\prec\prec h(z,\zeta), \quad z\in U,\ \ \zeta\in\overline{U},$$

where $g(z,\zeta)=\frac{1}{z}\int_0^z h(t,\zeta)\,dt$ is convex and it is the best dominant.

Proof. The proof is the same with the proof of Theorem 4.3, using Lemma 1.7, for $n = 1$ and $\gamma = 1$. $\square$

Theorem 4.5 (Alb Lupaş, 2011g). *Let $g(z,\zeta)$ be a convex function such that $g(0,\zeta) = 1$ and let h be the function $h(z,\zeta) = g(z,\zeta) + zg'_z(z,\zeta)$, $z \in U$, $\zeta \in \overline{U}$. If $\lambda, l \geq 0$, $m \in \mathbb{N}$, $f \in \mathcal{A}^*_\zeta$ and verifies the strong differential subordination*

$$\left(\frac{zIR^{m+1}_{\lambda,l} f(z,\zeta)}{IR^m_{\lambda,l} f(z,\zeta)}\right)'_z \prec\prec h(z,\zeta), \quad z \in U, \ \zeta \in \overline{U}, \tag{4.1.1.3}$$

then

$$\frac{IR^{m+1}_{\lambda,l} f(z,\zeta)}{IR^m_{\lambda,l} f(z,\zeta)} \prec\prec g(z,\zeta), \quad z \in U, \ \zeta \in \overline{U}$$

and this result is sharp.

Proof. For $f \in \mathcal{A}^*_\zeta$, $f(z,\zeta) = z + \sum_{j=2}^{\infty} a_j(\zeta) z^j$, we have

$$IR^m_{\lambda,l} f(z,\zeta) = z + \sum_{j=2}^{\infty} \left(\frac{1+\lambda(j-1)+l}{l+1}\right)^m C^m_{m+j-1} a_j^2(\zeta) z^j, \quad z \in U, \ \zeta \in \overline{U}.$$

Consider $p(z,\zeta) = \frac{IR^{m+1}_{\lambda,l} f(z,\zeta)}{IR^m_{\lambda,l} f(z,\zeta)} = \frac{z+\sum_{j=2}^{\infty}\left(\frac{1+\lambda(j-1)+l}{l+1}\right)^{m+1} C^{m+1}_{m+j} a_j^2(\zeta) z^j}{z+\sum_{j=2}^{\infty}\left(\frac{1+\lambda(j-1)+l}{l+1}\right)^{m} C^{m}_{m+j-1} a_j^2(\zeta) z^j} = \frac{1+\sum_{j=2}^{\infty}\left(\frac{1+\lambda(j-1)+l}{l+1}\right)^{m+1} C^{m+1}_{m+j} a_j^2(\zeta) z^{j-1}}{1+\sum_{j=2}^{\infty}\left(\frac{1+\lambda(j-1)+l}{l+1}\right)^{m} C^{m}_{m+j-1} a_j^2(\zeta) z^{j-1}}$.

We have $p'_z(z,\zeta) = \frac{\left(IR^{m+1}_{\lambda,l} f(z,\zeta)\right)'}{IR^m_{\lambda,l} f(z,\zeta)} - p(z,\zeta) \cdot \frac{\left(IR^m_{\lambda,l} f(z,\zeta)\right)'}{IR^m_{\lambda,l} f(z,\zeta)}$.

Then, $p(z,\zeta) + zp'_z(z,\zeta) = \left(\frac{zIR^{m+1}_{\lambda,l} f(z,\zeta)}{IR^m_{\lambda,l} f(z,\zeta)}\right)'$.

Relation (4.1.1.3) becomes $p(z,\zeta) + zp'_z(z,\zeta) \prec\prec h(z,\zeta) = g(z,\zeta) + zg'_z(z,\zeta)$, $z \in U$, $\zeta \in \overline{U}$ and by using Lemma 1.8, we obtain $p(z,\zeta) \prec\prec g(z,\zeta)$, $z \in U$, $\zeta \in \overline{U}$, i.e. $\frac{IR^{m+1}_{\lambda,l} f(z,\zeta)}{IR^m_{\lambda,l} f(z,\zeta)} \prec\prec g(z,\zeta)$, $z \in U$, $\zeta \in \overline{U}$. $\square$

Theorem 4.6 (Alb Lupaş, 2011g). *Let $h(z,\zeta)$ be a convex function, $h(0,\zeta) = 1$. If $\lambda, l \geq 0$, $m \in \mathbb{N}$, $f \in \mathcal{A}^*_\zeta$ and verifies the strong differential subordination*

$$\left(\frac{zIR^{m+1}_{\lambda,l} f(z,\zeta)}{IR^m_{\lambda,l} f(z,\zeta)}\right)'_z \prec\prec h(z,\zeta), \quad z \in U, \ \zeta \in \overline{U},$$

then,

$$\frac{IR^{m+1}_{\lambda,l} f(z,\zeta)}{IR^{m}_{\lambda,l} f(z,\zeta)} \prec\prec g(z,\zeta) \prec\prec h(z,\zeta), \quad z \in U, \ \zeta \in \overline{U},$$

where $g(z,\zeta) = \frac{1}{z} \int_0^z h(t,\zeta)\, dt$ *is convex and it is the best dominant.*

Proof. The proof is the same with the proof of Theorem 4.4, using Lemma 1.7, for $n = 1$ and $\gamma = 1$. □

4.1.2 *On Special Strong Differential Subordinations Regarding an Operator Defined by Multiplier Transformation and Ruscheweyh Derivative*

In the paper Alb Lupaş (2012d), we establish several strong differential subordinations regarding the extended new operator $RI^{\alpha}_{m,\lambda,l}$ defined in Definition 2.7, given by $RI^{\alpha}_{m,\lambda,l} : \mathcal{A}^*_{\zeta} \to \mathcal{A}^*_{\zeta}$,

$$RI^{\alpha}_{m,\lambda,l} f(z,\zeta) = (1-\alpha) R^m f(z,\zeta) + \alpha I(m,\lambda,l) f(z,\zeta),$$

where $R^m f(z,\zeta)$ denote the extended Ruscheweyh derivative, $I(m,\lambda,l)$ is the extended multiplier transformation and $\mathcal{A}^*_{n\zeta} = \{f \in \mathcal{H}(U \times \overline{U}),\ f(z,\zeta) = z + a_{n+1}(\zeta) z^{n+1} + \cdots,\ z \in U,\ \zeta \in \overline{U}\}$, with $\mathcal{A}^*_{1\zeta} = \mathcal{A}^*_{\zeta}$ is the class of normalized analytic functions. A certain subclass, denoted by $\mathcal{RI}_m(\delta,\lambda,l,\alpha,\zeta)$, of analytic functions is introduced by means of the new operator.

If $f \in \mathcal{A}^*_{\zeta}$, $f(z,\zeta) = z + \sum_{j=2}^{\infty} a_j(\zeta) z^j$, then $RI^{\alpha}_{m,\lambda,l} f(z,\zeta) = z + \sum_{j=2}^{\infty} \left\{\alpha \left(\frac{1+\lambda(j-1)+l}{l+1}\right)^m + (1-\alpha) C^m_{m+j-1}\right\} a_j(\zeta) z^j$, $z \in U$, $\zeta \in \overline{U}$.

Definition 4.1 (Alb Lupaş, 2012d). Let $\delta \in [0,1)$, $\alpha, \lambda, l \geq 0$ and $m \in \mathbb{N}$. A function $f(z,\zeta) \in \mathcal{A}^*_{\zeta}$ is said to be in the subclass $\mathcal{RI}_m(\delta,\lambda,l,\alpha,\zeta)$ if it satisfies the inequality

$$\operatorname{Re}\left(RI^{\alpha}_{m,\lambda,l} f(z,\zeta)\right)'_z > \delta, \quad z \in U, \ \zeta \in \overline{U}.$$

Theorem 4.7 (Alb Lupaş, 2012d). *The set* $\mathcal{RI}_m(\delta,\lambda,l,\alpha,\zeta)$ *is convex.*

Proof. Let the functions

$$f_j(z,\zeta) = z + \sum_{j=2}^{\infty} a_{jk}(\zeta) z^j, \quad k = 1,2, \quad z \in U, \ \zeta \in \overline{U},$$

be in the subclass $\mathcal{RI}_m(\delta,\lambda,l,\alpha,\zeta)$. It is sufficient to show that the function

$$h(z,\zeta)=\eta_1 f_1(z,\zeta)+\eta_2 f_2(z,\zeta)$$

is in the subclass $\mathcal{RI}_m(\delta,\lambda,l,\alpha,\zeta)$, with η_1 and η_2 nonnegative such that $\eta_1+\eta_2=1$.

Since $h(z,\zeta)=z+\sum_{j=2}^{\infty}(\eta_1 a_{j1}(\zeta)+\eta_2 a_{j2}(\zeta))z^j$, $z\in U$, $\zeta\in\overline{U}$, then

$$RI_{m,\lambda,l}^{\alpha}h(z,\zeta)=z+ \sum_{j=2}^{\infty}\left\{\alpha\left(\frac{1+\lambda(j-1)+l}{l+1}\right)^m+(1-\alpha)C_{m+j-1}^m\right\}(\eta_1 a_{j1}(\zeta)+\eta_2 a_{j2}(\zeta))z^j, \tag{4.1.2.1}$$

$z\in U$, $\zeta\in\overline{U}$.

Differentiating (4.1.2.1) with respect to z, we obtain $\left(RI_{m,\lambda,l}^{\alpha}h(z,\zeta)\right)_z'=1+\sum_{j=2}^{\infty}\left\{\alpha\left(\frac{1+\lambda(j-1)+l}{l+1}\right)^m+(1-\alpha)C_{m+j-1}^m\right\}(\eta_1 a_{j1}(\zeta)+\eta_2 a_{j2}(\zeta))$ jz^{j-1}, $z\in U$, $\zeta\in\overline{U}$.

Hence,

$$\begin{aligned}\operatorname{Re}\left(RI_{m,\lambda,l}^{\alpha}h(z,\zeta)\right)_z'&=1\\&+\operatorname{Re}\left(\eta_1\sum_{j=2}^{\infty}j\left\{\alpha\left(\frac{1+\lambda(j-1)+l}{l+1}\right)^m+(1-\alpha)C_{m+j-1}^m\right\}a_{j1}(\zeta)z^{j-1}\right)\\&+\operatorname{Re}\left(\eta_2\sum_{j=2}^{\infty}j\left\{\alpha\left(\frac{1+\lambda(j-1)+l}{l+1}\right)^m+(1-\alpha)C_{m+j-1}^m\right\}a_{j2}(\zeta)z^{j-1}\right).\end{aligned} \tag{4.1.2.2}$$

Taking into account that $f_1,f_2\in\mathcal{RI}_m(\delta,\lambda,l,\alpha,\zeta)$, we deduce

$$\operatorname{Re}\left(\eta_k\sum_{j=2}^{\infty}j\left\{\alpha\left(\frac{1+\lambda(j-1)+l}{l+1}\right)^m+(1-\alpha)C_{m+j-1}^m\right\}a_{jk}(\zeta)z^{j-1}\right) \tag{4.1.2.3}$$

$$>\eta_k(\delta-1),\qquad k=1,2.$$

Using (4.1.2.3), we get from (4.1.2.2)

$$\operatorname{Re}\left(RI_{m,\lambda,l}^{\alpha}h(z,\zeta)\right)_z'>1+\eta_1(\delta-1)+\eta_2(\delta-1)=\delta,\quad z\in U,\ \zeta\in\overline{U},$$

which is equivalent that $\mathcal{RI}_m(\delta,\lambda,l,\alpha,\zeta)$ is convex. □

Theorem 4.8 (Alb Lupaş, 2012d). *Let $g(z,\zeta)$ be a convex function such that $g(0,\zeta)=1$ and let $h(z,\zeta)=g(z,\zeta)+\frac{1}{c+2}zg'_z(z,\zeta)$, where $z\in U$, $\zeta\in\overline{U}$, $c>0$.*

If $\alpha,\lambda,l\geq 0$, $m\in\mathbb{N}$, $f\in\mathcal{RI}_m(\delta,\lambda,l,\alpha,\zeta)$ and $F(z,\zeta)=I_c(f)(z,\zeta)=\frac{c+2}{z^{c+1}}\int_0^z t^c f(t,\zeta)\,dt$, $z\in U$, $\zeta\in\overline{U}$, then

$$\left(RI^{\alpha}_{m,\lambda,l}f(z,\zeta)\right)'_z \prec\prec h(z,\zeta),\quad z\in U,\ \zeta\in\overline{U}, \tag{4.1.2.4}$$

implies

$$\left(RI^{\alpha}_{m,\lambda,l}F(z,\zeta)\right)'_z \prec\prec g(z,\zeta),\quad z\in U,\ \zeta\in\overline{U},$$

and this result is sharp.

Proof. We obtain that

$$z^{c+1}F(z,\zeta)=(c+2)\int_0^z t^c f(t,\zeta)\,dt. \tag{4.1.2.5}$$

Differentiating (4.1.2.5), with respect to z, we have $(c+1)F(z,\zeta)+zF'_z(z,\zeta)=(c+2)f(z,\zeta)$ and

$$(c+1)\,RI^{\alpha}_{m,\lambda,l}F(z,\zeta)+z\left(RI^{\alpha}_{m,\lambda,l}F(z,\zeta)\right)'_z=(c+2)\,RI^{\alpha}_{m,\lambda,l}f(z,\zeta),\quad z\in U,\ \zeta\in\overline{U}. \tag{4.1.2.6}$$

Differentiating (4.1.2.6) with respect to z, we have

$$\left(RI^{\alpha}_{m,\lambda,l}F(z,\zeta)\right)'_z+\frac{1}{c+2}z\left(RI^{\alpha}_{m,\lambda,l}F(z,\zeta)\right)''_{z^2}=\left(RI^{\alpha}_{m,\lambda,l}f(z,\zeta)\right)'_z,\quad z\in U,\ \zeta\in\overline{U}. \tag{4.1.2.7}$$

Using (4.1.2.7), the strong differential subordination (4.1.2.4) becomes

$$\left(RI^{\alpha}_{m,\lambda,l}F(z,\zeta)\right)'_z+\frac{1}{c+2}z\left(RI^{\alpha}_{m,\lambda,l}F(z,\zeta)\right)''_{z^2}\prec g(z)+\frac{1}{c+2}zg'_z(z). \tag{4.1.2.8}$$

Denote

$$p(z,\zeta)=\left(RI^{\alpha}_{m,\lambda,l}F(z,\zeta)\right)'_z,\quad z\in U,\ \zeta\in\overline{U}. \tag{4.1.2.9}$$

Replacing (4.1.2.9) in (4.1.2.8), we obtain

$$p(z,\zeta)+\frac{1}{c+2}zp'_z(z,\zeta)\prec\prec g(z,\zeta)+\frac{1}{c+2}zg'_z(z,\zeta),\quad z\in U,\ \zeta\in\overline{U}.$$

Using Lemma 1.8, we have

$$p(z,\zeta) \prec\prec g(z,\zeta), \quad z \in U, \ \zeta \in \overline{U}, \quad \text{i.e.} \ \left(RI^{\alpha}_{m,\lambda,l}F(z,\zeta)\right)'_z \prec\prec g(z,\zeta),$$

$z \in U$, $\zeta \in \overline{U}$, and this result is sharp. □

Theorem 4.9 (Alb Lupaş, 2012d). *Let* $h(z,\zeta) = \frac{\zeta+(2\delta-\zeta)z}{1+z}$, $z \in U$, $\zeta \in \overline{U}$, $\delta \in [0,1)$ *and* $c > 0$. *If* $\alpha, \lambda, l \geq 0$, $m \in \mathbb{N}$ *and* I_c *is given by Theorem* 4.8, *then*

$$I_c\left[\mathcal{RI}_m(\delta,\lambda,l,\alpha,\zeta)\right] \subset \mathcal{RI}_m(\delta^*,\lambda,l,\alpha,\zeta), \tag{4.1.2.10}$$

where $\delta^* = 2\delta - \zeta + \frac{2(c+2)(\zeta-\delta)}{n}\beta(c+1)$ *and* $\beta(x) = \int_0^1 \frac{t^x}{t+1}dt$.

Proof. The function h is convex and using the same steps as in the proof of Theorem 4.8, we get from the hypothesis of Theorem 4.9 that

$$p(z,\zeta) + \frac{1}{c+2}zp'_z(z,\zeta) \prec\prec h(z,\zeta),$$

where $p(z,\zeta)$ is defined in (4.1.2.9).

Using Lemma 1.7, we deduce that

$$p(z,\zeta) \prec\prec g(z,\zeta) \prec\prec h(z,\zeta),$$

that is

$$\left(RI^{\alpha}_{m,\lambda,l}F(z,\zeta)\right)'_z \prec\prec g(z,\zeta) \prec\prec h(z,\zeta),$$

where

$$\begin{aligned} g(z,\zeta) &= \frac{c+2}{z^{c+2}}\int_0^z t^{c+1}\frac{\zeta+(2\delta-\zeta)t}{1+t}dt \\ &= (2\delta-\zeta) + \frac{2(c+2)(\zeta-\delta)}{z^{c+2}}\int_0^z \frac{t^{c+1}}{1+t}dt. \end{aligned}$$

Since g is convex and $g\left(U \times \overline{U}\right)$ is symmetric with respect to the real axis, we deduce

$$\begin{aligned} \operatorname{Re}\left(RI^{\alpha}_{m,\lambda,l}F(z,\zeta)\right)'_z &\geq \min_{|z|=1}\operatorname{Re} g(z,\zeta) = \operatorname{Re} g(1,\zeta) = \delta^* \\ &= 2\delta - \zeta + 2(c+2)(\zeta-\delta)\beta(c). \end{aligned} \tag{4.1.2.11}$$

From (4.1.2.11), we deduce inclusion (4.1.2.10). □

Theorem 4.10 (Alb Lupaş, 2012d). *Let $g(z,\zeta)$ be a convex function such that $g(0,\zeta)=1$ and let h be the function $h(z,\zeta)=g(z,\zeta)+zg'_z(z,\zeta)$, $z\in U$, $\zeta\in\overline{U}$.*

If $\alpha,\lambda,l\geq 0$, $m\in\mathbb{N}$, $f\in\mathcal{A}^*_\zeta$ and satisfies the strong differential subordination

$$\left(RI^{\alpha}_{m,\lambda,l}f(z,\zeta)\right)'_z \prec\prec h(z,\zeta),\quad z\in U,\ \zeta\in\overline{U}, \tag{4.1.2.12}$$

then

$$\frac{RI^{\alpha}_{m,\lambda,l}f(z,\zeta)}{z}\prec\prec g(z,\zeta),\quad z\in U,\ \zeta\in\overline{U},$$

and this result is sharp.

Proof. By using the properties of operator $RI^{\alpha}_{m,\lambda,l}$, we have

$$\begin{aligned}&RI^{\alpha}_{m,\lambda,l}f(z,\zeta)\\&\quad=z+\sum_{j=2}^{\infty}\left\{\alpha\left(\frac{1+\lambda(j-1)+l}{l+1}\right)^m+(1-\alpha)\,C^m_{m+j-1}\right\}a_j(\zeta)\,z^j,\end{aligned}$$

$z\in U$, $\zeta\in\overline{U}$.

Consider

$$\begin{aligned}p(z,\zeta)&=\frac{RI^{\alpha}_{m,\lambda,l}f(z,\zeta)}{z}\\&=\frac{z+\sum_{j=2}^{\infty}\left\{\alpha\left(\frac{1+\lambda(j-1)+l}{l+1}\right)^m+(1-\alpha)\,C^m_{m+j-1}\right\}a_j(\zeta)\,z^j}{z}\\&=1+p_1(\zeta)\,z+p_2(\zeta)\,z^2+\cdots,\quad z\in U,\ \zeta\in\overline{U}.\end{aligned}$$

Let $RI^{\alpha}_{m,\lambda,l}f(z,\zeta)=zp(z,\zeta)$, $z\in U$, $\zeta\in\overline{U}$. Differentiating with respect to z, we obtain $\left(RI^{\alpha}_{m,\lambda,l}f(z,\zeta)\right)'_z=p(z,\zeta)+zp'_z(z,\zeta)$, $\quad z\in U,\ \zeta\in\overline{U}$.

Then (4.1.2.12) becomes

$$p(z,\zeta)+zp'_z(z,\zeta)\prec\prec h(z,\zeta)=g(z,\zeta)+zg'_z(z,\zeta),\quad z\in U,\ \zeta\in\overline{U}.$$

By using Lemma 1.8, we have

$$p(z,\zeta)\prec\prec g(z,\zeta),\quad \text{i.e.}\ \frac{RI^{\alpha}_{m,\lambda,l}f(z,\zeta)}{z}\prec\prec g(z,\zeta),\quad z\in U,\ \zeta\in\overline{U}.$$

□

Theorem 4.11 (Alb Lupaş, 2012d). *Let $h(z,\zeta)$ be a convex function such that $h(0,\zeta)=1$.*

If $\alpha,\lambda,l\geq 0$, $m\in\mathbb{N}$, $f\in\mathcal{A}^*_\zeta$ and satisfies the strong differential subordination

$$\left(RI^\alpha_{m,\lambda,l}f(z,\zeta)\right)'_z \prec\prec h(z,\zeta),\quad z\in U,\ \zeta\in\overline{U}, \tag{4.1.2.13}$$

then

$$\frac{RI^\alpha_{m,\lambda,l}f(z,\zeta)}{z}\prec\prec g(z,\zeta)\prec\prec h(z,\zeta),\ \ z\in U,\ \ \zeta\in\overline{U},$$

where $g(z,\zeta)=\frac{1}{z}\int_0^z h(t,\zeta)\,dt$ is convex and it is the best dominant.

Proof. Let

$$\begin{aligned}p(z,\zeta)&=\frac{RI^\alpha_{m,\lambda,l}f(z,\zeta)}{z}\\&=\frac{z+\sum_{j=2}^\infty\left\{\alpha\left(\frac{1+\lambda(j-1)+l}{l+1}\right)^m+(1-\alpha)C^m_{m+j-1}\right\}a_j(\zeta)z^j}{z}\\&=1+\sum_{j=2}^\infty\left\{\alpha\left(\frac{1+\lambda(j-1)+l}{l+1}\right)^m+(1-\alpha)C^m_{m+j-1}\right\}a_j(\zeta)z^{j-1}\\&=1+\sum_{j=2}^\infty p_j(\zeta)z^{j-1},\quad z\in U,\ \zeta\in\overline{U}.\end{aligned}$$

Differentiating with respect to z, we obtain $\left(RI^\alpha_{m,\lambda,l}f(z,\zeta)\right)'_z = p(z,\zeta)+zp'_z(z,\zeta)$, $z\in U$, $\zeta\in\overline{U}$, and (4.1.2.13) becomes

$$p(z,\zeta)+zp'_z(z,\zeta)\prec\prec h(z,\zeta),\quad z\in U,\ \zeta\in\overline{U}.$$

Since $p(z,\zeta)\in\mathcal{H}^*[1,1,\zeta]$, using Lemma 1.7 for $n=1$ and $\gamma=1$, we have

$$p(z,\zeta)\prec\prec g(z,\zeta)\prec\prec h(z,\zeta),\quad z\in U,\ \zeta\in\overline{U},\quad \text{i.e.}$$

$$\frac{RI^\alpha_{m,\lambda,l}f(z,\zeta)}{z}\prec\prec g(z,\zeta)=\frac{1}{z}\int_0^z h(t,\zeta)\,dt\prec\prec h(z,\zeta),\quad z\in U,\ \zeta\in\overline{U},$$

and $g(z,\zeta)$ is convex and it is the best dominant. □

Corollary 4.1 (Alb Lupaş, 2012d). *Let* $h(z,\zeta) = \frac{\zeta+(2\beta-\zeta)z}{1+z}$ *a convex function in* $U \times \overline{U}$, $0 \le \beta < 1$. *If* $\alpha \ge 0$, $m \in \mathbb{N}$, $f \in \mathcal{A}^*_\zeta$ *and verifies the strong differential subordination*

$$\left(RI^\alpha_{m,\lambda,l} f(z,\zeta)\right)'_z \prec\prec h(z,\zeta), \quad z \in U,\ \zeta \in \overline{U}, \tag{4.1.2.14}$$

then

$$\frac{RI^\alpha_{m,\lambda,l} f(z,\zeta)}{z} \prec\prec g(z,\zeta) \prec\prec h(z,\zeta), \quad z \in U,\ \zeta \in \overline{U},$$

where g *is given by* $g(z,\zeta) = 2\beta - \zeta + \frac{2(\zeta-\beta)}{z}\ln(1+z)$, $z \in U$, $\zeta \in \overline{U}$. *The function* g *is convex and it is the best dominant.*

Proof. Following the same steps as in the proof of Theorem 4.11 and considering $p(z,\zeta) = \frac{RI^\alpha_{m,\lambda,l} f(z,\zeta)}{z}$, the strong differential subordination (4.1.2.14) becomes

$$p(z,\zeta) + zp'_z(z,\zeta) \prec\prec h(z,\zeta) = \frac{\zeta + (2\beta-\zeta)z}{1+z}, \quad z \in U,\ \zeta \in \overline{U}.$$

By using Lemma 1.7 for $n = 1$ and $\gamma = 1$, we have $p(z,\zeta) \prec\prec g(z,\zeta) \prec\prec h(z,\zeta)$, $z \in U$, $\zeta \in \overline{U}$, i.e.

$$\frac{RI^\alpha_{m,\lambda,l} f(z,\zeta)}{z} \prec\prec g(z,\zeta) = \frac{1}{z}\int_0^z h(t,\zeta)\,dt = \frac{1}{z}\int_0^z \frac{\zeta + (2\beta-\zeta)t}{1+t}dt$$
$$= 2\beta - \zeta + \frac{2(\zeta-\beta)}{z}\ln(1+z), \quad z \in U,\ \zeta \in \overline{U}.$$

□

Theorem 4.12 (Alb Lupaş, 2012d). *Let* $g(z,\zeta)$ *be a convex function such that* $g(0,\zeta) = 1$ *and let* h *be the function* $h(z,\zeta) = g(z,\zeta) + zg'_z(z,\zeta)$, $z \in U$, $\zeta \in \overline{U}$.

If $\alpha, \lambda, l \ge 0$, $m \in \mathbb{N}$, $f \in \mathcal{A}^*_\zeta$ *and the strong differential subordination*

$$\left(\frac{zRI^\alpha_{m+1,\lambda,l} f(z,\zeta)}{RI^\alpha_{m,\lambda,l} f(z,\zeta)}\right)'_z \prec\prec h(z,\zeta), \quad z \in U,\ \zeta \in \overline{U}, \tag{4.1.2.15}$$

holds, then

$$\frac{RI^\alpha_{m+1,\lambda,l} f(z,\zeta)}{RI^\alpha_{m,\lambda,l} f(z,\zeta)} \prec\prec g(z,\zeta), \quad z \in U,\ \zeta \in \overline{U},$$

and this result is sharp.

Proof. For $f \in \mathcal{A}_{\zeta}^{*}$, $f(z,\zeta) = z + \sum_{j=2}^{\infty} a_j(\zeta) z^j$, we have $RI_{m,\lambda,l}^{\alpha} f(z,\zeta) = z + \sum_{j=2}^{\infty} \left\{ \alpha \left(\frac{1+\lambda(j-1)+l}{l+1} \right)^m + (1-\alpha) C_{m+j-1}^{m} \right\} a_j(\zeta) z^j$, $z \in U$, $\zeta \in \overline{U}$.

Consider

$$p(z,\zeta) = \frac{RI_{m+1,\lambda,l}^{\alpha} f(z,\zeta)}{RI_{m,\lambda,l}^{\alpha} f(z,)}$$

$$= \frac{z + \sum_{j=2}^{\infty} \left\{ \alpha \left(\frac{1+\lambda(j-1)+l}{l+1} \right)^{m+1} + (1-\alpha) C_{m+j}^{m+1} \right\} a_j(\zeta) z^j}{z + \sum_{j=2}^{\infty} \left\{ \alpha \left(\frac{1+\lambda(j-1)+l}{l+1} \right)^{m} + (1-\alpha) C_{m+j-1}^{m} \right\} a_j(\zeta) z^j}.$$

We have $p_z'(z,\zeta) = \frac{\left(RI_{m+1,\lambda,l}^{\alpha} f(z,\zeta)\right)_z'}{RI_{m,\lambda,l}^{\alpha} f(z,\zeta)} - p(z,\zeta) \cdot \frac{\left(RI_{m,\lambda,l}^{\alpha} f(z,\zeta)\right)_z'}{RI_{m,\lambda,l}^{\alpha} f(z,\zeta)}$ and we obtain $p(z,\zeta) + z \cdot p_z'(z,\zeta) = \left(\frac{z RI_{m+1,\lambda,l}^{\alpha} f(z,\zeta)}{RI_{m,\lambda,l}^{\alpha} f(z,\zeta)} \right)_z'$.

Relation (4.1.2.15) becomes

$$p(z,\zeta) + zp_z'(z,\zeta) \prec\prec h(z,\zeta) = g(z,\zeta) + zg_z'(z,\zeta), \quad z \in U, \ \zeta \in \overline{U}.$$

By using Lemma 1.8, we have

$$p(z,\zeta) \prec\prec g(z,\zeta), \text{ i.e. } \frac{RI_{m+1,\lambda,l}^{\alpha} f(z,\zeta)}{RI_{m,\lambda,l}^{\alpha} f(z,\zeta)} \prec\prec g(z,\zeta), \quad z \in U, \ \zeta \in \overline{U}.$$

□

Theorem 4.13 (Alb Lupaş, 2012d). *Let $g(z,\zeta)$ be a convex function such that $g(0,\zeta) = 1$ and let h be the function $h(z,\zeta) = g(z,\zeta) + zg_z'(z,\zeta)$, $z \in U$, $\zeta \in \overline{U}$.*

If $\alpha, \lambda, l \geq 0$, $m \in \mathbb{N}$, $f \in \mathcal{A}_{\zeta}^{*}$ and the strong differential subordination

$$\begin{aligned} &\frac{(m+1)(m+2)}{z} RI_{m+2,\lambda,l}^{\alpha} f(z,\zeta) \\ &- \frac{(m+1)(2m+1)}{z} RI_{m+1,\lambda,l}^{\alpha} f(z,\zeta) + \frac{m^2}{z} RI_{m,\lambda,l}^{\alpha} f(z,\zeta) \\ &+ \frac{\alpha}{z} \left[\frac{(l+1)^2}{\lambda^2} - (m+1)(m+2) \right] I(m+2,\lambda,l) f(z,\zeta) \\ &- \frac{\alpha}{z} \left[\frac{2(l+1-\lambda)(l+1)}{\lambda^2} - (m+1)(2m+1) \right] I(m+1,\lambda,l) f(z,\zeta) \\ &+ \frac{\alpha}{z} \left[\frac{(l+1-\lambda)^2}{\lambda^2} - m^2 \right] I(m,\lambda,l) f(z,\zeta) \prec\prec h(z,\zeta), \quad z \in U, \ \zeta \in \overline{U}, \end{aligned} \tag{4.1.2.16}$$

holds, then

$$[RI^{\alpha}_{m,\lambda,l}f(z,\zeta)]'_z \prec\prec g(z,\zeta), \quad z \in U, \ \zeta \in \overline{U}.$$

This result is sharp.

Proof. Let

$$\begin{aligned}
p(z,\zeta) &= \left(RI^{\alpha}_{m,\lambda,l}f(z,\zeta)\right)'_z = (1-\alpha)\left(R^m f(z,\zeta)\right)'_z + \alpha\left(I(m,\lambda,l)f(z,\zeta)\right)'_z \\
&= 1 + \sum_{j=2}^{\infty}\left\{\alpha\left(\frac{1+\lambda(j-1)+l}{l+1}\right)^m + (1-\alpha)C^m_{m+j-1}\right\} ja_j(\zeta)z^{j-1} \\
&= 1 + p_1(\zeta)z + p_2(\zeta)z^2 + \cdots. \qquad (4.1.2.17)
\end{aligned}$$

By using the properties of operators $RI^{\alpha}_{m,\lambda,l}$, R^m and $I(m,\lambda,l)$, after a short calculation, we obtain

$$\begin{aligned}
&p(z,\zeta) + zp'_z(z,\zeta) \\
&= \frac{(m+1)(m+2)}{z}RI^{\alpha}_{m+2,\lambda,l}f(z,\zeta) - \frac{(m+1)(2m+1)}{z}RI^{\alpha}_{m+1,\lambda,l}f(z,\zeta) \\
&\quad + \frac{m^2}{z}RI^{\alpha}_{m,\lambda,l}f(z,\zeta) + \frac{\alpha}{z}\left[\frac{(l+1)^2}{\lambda^2} - (m+1)(m+2)\right] \\
&\quad \times I(m+2,\lambda,l)f(z,\zeta) - \frac{\alpha}{z}\left[\frac{2(l+1-\lambda)(l+1)}{\lambda^2} - (m+1)(2m+1)\right] \\
&\quad \times I(m+1,\lambda,l)f(z,\zeta) + \frac{\alpha}{z}\left[\frac{(l+1-\lambda)^2}{\lambda^2} - m^2\right] I(m,\lambda,l)f(z,\zeta).
\end{aligned}$$

Using the notation in (4.1.2.17), the differential subordination becomes

$$p(z,\zeta) + zp'_z(z,\zeta) \prec\prec h(z,\zeta) = g(z,\zeta) + zg'_z(z,\zeta).$$

By using Lemma 1.8, we have

$$p(z,\zeta) \prec\prec g(z,\zeta), \text{ i.e. } \left(RI^{\alpha}_{m,\lambda,l}f(z,\zeta)\right)'_z \prec\prec g(z,\zeta), \ z \in U, \ \zeta \in \overline{U},$$

and this result is sharp. □

Theorem 4.14 (Alb Lupaş, 2012d). *Let $h(z,\zeta)$ be a convex function such that $h(0,\zeta) = 1$.*

If $\alpha, \lambda, l \geq 0$, $m \in \mathbb{N}$, $f \in \mathcal{A}_\zeta^*$ and satisfies the strong differential subordination

$$\begin{aligned}
&\frac{(m+1)(m+2)}{z} RI^{\alpha}_{m+2,\lambda,l} f(z,\zeta) \\
&\quad - \frac{(m+1)(2m+1)}{z} RI^{\alpha}_{m+1,\lambda,l} f(z,\zeta) + \frac{m^2}{z} RI^{\alpha}_{m,\lambda,l} f(z,\zeta) \\
&\quad + \frac{\alpha}{z}\left[\frac{(l+1)^2}{\lambda^2} - (m+1)(m+2)\right] I(m+2,\lambda,l) f(z,\zeta) \\
&\quad - \frac{\alpha}{z}\left[\frac{2(l+1-\lambda)(l+1)}{\lambda^2} - (m+1)(2m+1)\right] I(m+1,\lambda,l) f(z,\zeta) \\
&\quad + \frac{\alpha}{z}\left[\frac{(l+1-\lambda)^2}{\lambda^2} - m^2\right] I(m,\lambda,l) f(z,\zeta) \prec\prec h(z,\zeta), \quad z \in U, \ \zeta \in \overline{U},
\end{aligned} \tag{4.1.2.18}$$

then

$$\left(RI^{\alpha}_{m,\lambda,l} f(z,\zeta)\right)'_z \prec\prec g(z,\zeta) \prec\prec h(z,\zeta), \quad z \in U, \ \zeta \in \overline{U},$$

where $g(z,\zeta) = \frac{1}{z}\int_0^z h(t,\zeta)\, dt$ is convex and it is the best dominant.

Proof. Using the properties of operator $RI^{\alpha}_{m,\lambda,l}$ and considering $p(z,\zeta) = \left(RI^{\alpha}_{m,\lambda,l} f(z,\zeta)\right)'_z$, we obtain

$$\begin{aligned}
&p(z,\zeta) + z p'_z(z,\zeta) \\
&\quad = \frac{(m+1)(m+2)}{z} RI^{\alpha}_{m+2,\lambda,l} f(z,\zeta) - \frac{(m+1)(2m+1)}{z} RI^{\alpha}_{m+1,\lambda,l} f(z,\zeta) \\
&\qquad + \frac{m^2}{z} RI^{\alpha}_{m,\lambda,l} f(z,\zeta) + \frac{\alpha}{z}\left[\frac{(l+1)^2}{\lambda^2} - (m+1)(m+2)\right] \\
&\qquad \times I(m+2,\lambda,l) f(z,\zeta) - \frac{\alpha}{z}\left[\frac{2(l+1-\lambda)(l+1)}{\lambda^2} - (m+1)(2m+1)\right] \\
&\qquad \times I(m+1,\lambda,l) f(z,\zeta) + \frac{\alpha}{z}\left[\frac{(l+1-\lambda)^2}{\lambda^2} - m^2\right] \\
&\qquad \times I(m,\lambda,l) f(z,\zeta), \quad z \in U, \ \zeta \in \overline{U}.
\end{aligned}$$

Then (4.1.2.18) becomes

$$p(z,\zeta) + zp'_z(z,\zeta) \prec\prec h(z,\zeta), \quad z \in U, \ \zeta \in \overline{U}.$$

By using Lemma 1.7, for $n = 1$ and $\gamma = 1$, we obtain $p(z,\zeta) \prec\prec g(z,\zeta) \prec\prec h(z,\zeta)$, $z \in U$, $\zeta \in \overline{U}$, i.e. $\left(RI^{\alpha}_{m,\lambda,l} f(z,\zeta)\right)'_z \prec\prec g(z,\zeta) = \frac{1}{z}\int_0^z h(t,\zeta)\,dt \prec\prec h(z,\zeta)$, $z \in U$, $\zeta \in \overline{U}$, and $g(z,\zeta)$ is convex and it is the best dominant. □

Corollary 4.2 (Alb Lupaş, 2012d). *Let* $h(z,\zeta) = \frac{\zeta+(2\beta-\zeta)z}{1+z}$ *a convex function in* $U \times \overline{U}$, $0 \le \beta < 1$. *If* $\alpha \ge 0$, $m \in \mathbb{N}$, $f \in \mathcal{A}^*_\zeta$ *and verifies the strong differential subordination*

$$\begin{aligned}
&\frac{(m+1)(m+2)}{z} RI^{\alpha}_{m+2,\lambda,l} f(z,\zeta) \\
&\quad - \frac{(m+1)(2m+1)}{z} RI^{\alpha}_{m+1,\lambda,l} f(z,\zeta) + \frac{m^2}{z} RI^{\alpha}_{m,\lambda,l} f(z,\zeta) \\
&\quad + \frac{\alpha}{z}\left[\frac{(l+1)^2}{\lambda^2} - (m+1)(m+2)\right] I(m+2,\lambda,l) f(z,\zeta) \\
&\quad - \frac{\alpha}{z}\left[\frac{2(l+1-\lambda)(l+1)}{\lambda^2} - (m+1)(2m+1)\right] I(m+1,\lambda,l) f(z,\zeta) \\
&\quad + \frac{\alpha}{z}\left[\frac{(l+1-\lambda)^2}{\lambda^2} - m^2\right] I(m,\lambda,l) f(z,\zeta) \prec\prec h(z,\zeta), \\
&\quad z \in U, \ \zeta \in \overline{U},
\end{aligned} \tag{4.1.2.19}$$

then

$$\left(RI^{\alpha}_{m,\lambda,l} f(z,\zeta)\right)'_z \prec\prec g(z,\zeta) \prec\prec h(z,\zeta), \quad z \in U, \ \zeta \in \overline{U},$$

where g *is given by* $g(z,\zeta) = 2\beta - \zeta + \frac{2(\zeta-\beta)}{z}\ln(1+z)$, $z \in U$, $\zeta \in \overline{U}$. *The function* g *is convex and it is the best dominant.*

Proof. Following the same steps as in the proof of Theorem 4.14 and considering $p(z,\zeta) = \left(RI^{\alpha}_{m,\lambda,l} f(z,\zeta)\right)'_z$, the strong differential subordination (4.1.2.19) becomes

$$p(z,\zeta) + zp'_z(z,\zeta) \prec\prec h(z,\zeta) = \frac{\zeta + (2\beta-\zeta)z}{1+z}, \quad z \in U, \ \zeta \in \overline{U}.$$

By using Lemma 1.7 for $n = 1$ and $\gamma = 1$, we have $p(z,\zeta) \prec\prec g(z,\zeta) \prec\prec h(z,\zeta)$, $z \in U$, $\zeta \in \overline{U}$, i.e.

$$\left(RI^{\alpha}_{m,\lambda,l}f(z,\zeta)\right)'_z \prec\prec g(z,\zeta) = \frac{1}{z}\int_0^z h(t,\zeta)\,dt$$

$$= \frac{1}{z}\int_0^z \frac{\zeta + (2\beta-\zeta)t}{1+t}dt = 2\beta - \zeta + \frac{2(\zeta-\beta)}{z}\ln(1+z), \quad z \in U, \ \zeta \in \overline{U}.$$

□

4.2 Applications of Differential Operators in the Theory of Strong Differential Superordinations

In this section, we present several strong differential superordination results involving few of the defined operators presented in Chapter 2.

4.2.1 *A Note on Special Strong Differential Superordinations Regarding an Operator Defined by Multiplier Transformation and Ruscheweyh Derivative*

In the paper Alb Lupaş (2012f), we established several strong differential superordinations regarding the extended new operator $RI^{\alpha}_{m,\lambda,l}$ defined in Alb Lupaş (2012d) and exposed in Definition 2.7 by using the extended Ruscheweyh derivative and the extended multiplier transformation, $RI^{\alpha}_{m,\lambda,l} : \mathcal{A}^*_{n\zeta} \to \mathcal{A}^*_{n\zeta}$,

$$RI^{\alpha}_{m,\lambda,l}f(z,\zeta) = (1-\alpha)R^m f(z,\zeta) + \alpha I(m,\lambda,l)\,f(z,\zeta), \quad z \in U, \ \zeta \in \overline{U},$$

where $R^m f(z,\zeta)$ denotes the extended Ruscheweyh derivative, $I(m,\lambda,l)$ is the extended multiplier transformation and $\mathcal{A}^*_{n\zeta} = \{f \in \mathcal{H}(U \times \overline{U}),\ f(z,\zeta) = z + a_{n+1}(\zeta)\,z^{n+1} + \cdots,\ z \in U,\ \zeta \in \overline{U}\}$ is the class of normalized analytic functions.

If $f \in \mathcal{A}^*_{n\zeta}$, $f(z,\zeta) = z + \sum_{j=n+1}^{\infty} a_j(\zeta)\,z^j$, then

$$RI^{\alpha}_{m,\lambda,l}f(z,\zeta)$$

$$= z + \sum_{j=n+1}^{\infty}\left\{\alpha\left(\frac{1+\lambda(j-1)+l}{l+1}\right)^m + (1-\alpha)\,C^m_{m+j-1}\right\}$$

$$\times\, a_j(\zeta)\,z^j, \quad z \in U, \ \zeta \in \overline{U}.$$

Theorem 4.15 (Alb Lupaş, 2012f). *Let $h(z,\zeta)$ be a convex function in $U \times \overline{U}$ with $h(0,\zeta) = 1$. Let $m \in \mathbb{N}$, $\lambda, \alpha, l \geq 0$, $f(z,\zeta) \in \mathcal{A}^*_{n\zeta}$, $F(z,\zeta) = I_c(f)(z,\zeta) = \frac{c+2}{z^{c+1}} \int_0^z t^c f(t,\zeta)\,dt$, $z \in U$, $\zeta \in \overline{U}$,* Re $c > -2$, *and suppose that $\left(RI^{\alpha}_{m,\lambda,l} f(z,\zeta)\right)'_z$ is univalent in $U \times \overline{U}$, $\left(RI^{\alpha}_{m,\lambda,l} F(z,\zeta)\right)'_z \in \mathcal{H}^*[1,n,\zeta] \cap Q^*$ and*

$$h(z,\zeta) \prec\prec \left(RI^{\alpha}_{m,\lambda,l} f(z,\zeta)\right)'_z, \quad z \in U, \ \zeta \in \overline{U}. \tag{4.2.1.1}$$

Then

$$q(z,\zeta) \prec\prec \left(RI^{\alpha}_{m,\lambda,l} F(z,\zeta)\right)'_z, \quad z \in U, \ \zeta \in \overline{U},$$

where $q(z,\zeta) = \frac{c+2}{nz^{\frac{c+2}{n}}} \int_0^z h(t,\zeta) t^{\frac{c+2}{n}-1} dt$. The function q is convex and it is the best subordinant.

Proof. We have

$$z^{c+1} F(z,\zeta) = (c+2) \int_0^z t^c f(t,\zeta)\,dt$$

and differentiating it, with respect to z, we obtain $(c+1) F(z,\zeta) + zF'_z(z,\zeta) = (c+2) f(z,\zeta)$ and

$$(c+1) RI^{\alpha}_{m,\lambda,l} F(z,\zeta) + z \left(RI^{\alpha}_{m,\lambda,l} F(z,\zeta)\right)'_z = (c+2) RI^{\alpha}_{m,\lambda,l} f(z,\zeta),$$
$$z \in U, \ \zeta \in \overline{U}.$$

Differentiating the last relation with respect to z, we have

$$\left(RI^{\alpha}_{m,\lambda,l} F(z,\zeta)\right)'_z + \frac{1}{c+2} z \left(RI^{\alpha}_{m,\lambda,l} F(z,\zeta)\right)''_{z^2} = \left(RI^{\alpha}_{m,\lambda,l} f(z,\zeta)\right)'_z,$$
$$z \in U, \ \zeta \in \overline{U}. \tag{4.2.1.2}$$

Using (4.2.1.2), the strong differential superordination (4.2.1.1) becomes

$$h(z,\zeta) \prec\prec \left(RI^{\alpha}_{m,\lambda,l} F(z,\zeta)\right)'_z + \frac{1}{c+2} z \left(RI^{\alpha}_{m,\lambda,l} F(z,\zeta)\right)''_{z^2}. \tag{4.2.1.3}$$

Denote

$$p(z,\zeta) = \left(RI^{\alpha}_{m,\lambda,l} F(z,\zeta)\right)'_z, \quad z \in U, \ \zeta \in \overline{U}. \tag{4.2.1.4}$$

Replacing (4.2.1.4) in (4.2.1.3), we obtain

$$h(z,\zeta) \prec\prec p(z,\zeta) + \frac{1}{c+2} z p'_z(z,\zeta), \quad z \in U, \ \zeta \in \overline{U}.$$

Using Lemma 1.10 for $\gamma = c+2$, we have

$$q(z,\zeta) \prec\prec p(z,\zeta), \text{ i.e. } q(z,\zeta) \prec\prec \left(RI^{\alpha}_{m,\lambda,l}F(z,\zeta)\right)'_z, \quad z \in U, \ \zeta \in \overline{U},$$

where $q(z,\zeta) = \frac{c+2}{nz^{\frac{c+2}{n}}}\int_0^z h(t,\zeta)t^{\frac{c+2}{n}-1}dt$. The function q is convex and it is the best subordinant. □

Corollary 4.3 (Alb Lupaş, 2012f). *Let* $h(z,\zeta) = \frac{1+(2\beta-\zeta)z}{1+z}$, *where* $\beta \in [0,1)$. *Let* $m \in \mathbb{N}$, $\lambda, \alpha, l \geq 0$, $f(z,\zeta) \in \mathcal{A}^*_{n\zeta}$, $F(z,\zeta) = I_c(f)(z,\zeta) = \frac{c+2}{z^{c+1}}\int_0^z t^c f(t,\zeta)\,dt$, $z \in U$, $\zeta \in \overline{U}$, Re $c > -2$, *and suppose that* $\left(RI^{\alpha}_{m,\lambda,l}f(z,\zeta)\right)'_z$ *is univalent in* $U \times \overline{U}$, $\left(RI^{\alpha}_{m,\lambda,l}F(z,\zeta)\right)'_z \in \mathcal{H}^*[1,n,\zeta] \cap Q^*$ *and*

$$h(z,\zeta) \prec\prec \left(RI^{\alpha}_{m,\lambda,l}f(z,\zeta)\right)'_z, \quad z \in U, \ \zeta \in \overline{U}, \tag{4.2.1.5}$$

then

$$q(z,\zeta) \prec\prec \left(RI^{\alpha}_{m,\lambda,l}F(z,\zeta)\right)'_z, \quad z \in U, \ \zeta \in \overline{U},$$

where q *is given by* $q(z,\zeta) = 2\beta - \zeta + \frac{(c+2)(1+\zeta-2\beta)}{nz^{\frac{c+2}{n}}}\int_0^z \frac{t^{\frac{c+2}{n}-1}}{t+1}dt$, $z \in U$, $\zeta \in \overline{U}$. *The function* q *is convex and it is the best subordinant.*

Proof. Following the same steps as in the proof of Theorem 4.15 and considering $p(z,\zeta) = \left(RI^{\alpha}_{m,\lambda,l}F(z,\zeta)\right)'_z$, the strong differential superordination (4.2.1.5) becomes

$$h(z,\zeta) = \frac{1+(2\beta-\zeta)z}{1+z} \prec\prec p(z,\zeta) + \frac{1}{c+2}zp'_z(z,\zeta), \quad z \in U, \ \zeta \in \overline{U}.$$

By using Lemma 1.10 for $\gamma = c+2$, we have $q(z,\zeta) \prec\prec p(z,\zeta)$, i.e.

$$\begin{aligned} q(z,\zeta) &= \frac{c+2}{nz^{\frac{c+2}{n}}}\int_0^z h(t,\zeta)t^{\frac{c+2}{n}-1}dt = \frac{c+2}{nz^{\frac{c+2}{n}}}\int_0^z \frac{1+(2\beta-\zeta)t}{1+t}t^{\frac{c+2}{n}-1}dt \\ &= 2\beta-\zeta+\frac{(c+2)(1+\zeta-2\beta)}{nz^{\frac{c+2}{n}}}\int_0^z \frac{t^{\frac{c+2}{n}-1}}{t+1}dt \prec\prec \left(RI^{\alpha}_{m,\lambda,l}F(z,\zeta)\right)'_z, \\ & z \in U, \ \zeta \in \overline{U}. \end{aligned}$$

The function q is convex and it is the best subordinant. □

Theorem 4.16 (Alb Lupaş, 2012f). *Let $q(z,\zeta)$ be a convex function in $U \times \overline{U}$ and let $h(z,\zeta) = q(z,\zeta) + \frac{1}{c+2} z q_z'(z,\zeta)$, where $z \in U$, $\zeta \in \overline{U}$,* $\operatorname{Re} c > -2$.

Let $m \in \mathbb{N}$, $\lambda, \alpha, l \geq 0$, $f(z,\zeta) \in \mathcal{A}_{n\zeta}^*$, $F(z,\zeta) = I_c(f)(z,\zeta) = \frac{c+2}{z^{c+1}} \int_0^z t^c f(t,\zeta)\, dt$, $z \in U$, $\zeta \in \overline{U}$, and suppose that $\left(RI_{m,\lambda,l}^{\alpha} f(z,\zeta)\right)_z'$ is univalent in $U \times \overline{U}$, $\left(RI_{m,\lambda,l}^{\alpha} F(z,\zeta)\right)_z' \in \mathcal{H}^*[1,n,\zeta] \cap Q^*$ and

$$h(z,\zeta) \prec\prec \left(RI_{m,\lambda,l}^{\alpha} f(z,\zeta)\right)_z', \quad z \in U, \ \zeta \in \overline{U}, \tag{4.2.1.6}$$

then

$$q(z,\zeta) \prec\prec \left(RI_{m,\lambda,l}^{\alpha} F(z,\zeta)\right)_z', \quad z \in U, \ \zeta \in \overline{U},$$

where $q(z,\zeta) = \frac{c+2}{nz^{\frac{c+2}{n}}} \int_0^z h(t,\zeta) t^{\frac{c+2}{n}-1} dt$. The function q is the best subordinant.

Proof. Following the same steps as in the proof of Theorem 4.15 and considering $p(z,\zeta) = \left(RI_{m,\lambda,l}^{\alpha} F(z,\zeta)\right)_z'$, $z \in U$, $\zeta \in \overline{U}$, the strong differential superordination (4.2.1.6) becomes

$$h(z,\zeta) = q(z,\zeta) + \frac{1}{c+2} z q_z'(z,\zeta) \prec\prec p(z,\zeta) + \frac{1}{c+2} z p_z'(z,\zeta),$$
$$z \in U, \ \zeta \in \overline{U}.$$

Using Lemma 1.11 for $\gamma = c+2$, we have

$$q(z,\zeta) \prec\prec p(z,\zeta), \text{ i.e. } q(z,\zeta) \prec\prec \left(RI_{m,\lambda,l}^{\alpha} F(z,\zeta)\right)_z', \quad z \in U, \ \zeta \in \overline{U},$$

where $q(z,\zeta) = \frac{c+2}{nz^{\frac{c+2}{n}}} \int_0^z h(t,\zeta) t^{\frac{c+2}{n}-1} dt$. The function q is the best subordinant. □

Theorem 4.17 (Alb Lupaş, 2012f). *Let $h(z,\zeta)$ be a convex function, $h(0,\zeta) = 1$. Let $m \in \mathbb{N}$, $\lambda, \alpha, l \geq 0$, $f(z,\zeta) \in \mathcal{A}_{n\zeta}^*$ and suppose that $\left(RI_{m,\lambda,l}^{\alpha} f(z,\zeta)\right)_z'$ is univalent and $\frac{RI_{m,\lambda,l}^{\alpha} f(z,\zeta)}{z} \in \mathcal{H}^*[1,n,\zeta] \cap Q^*$. If*

$$h(z,\zeta) \prec\prec \left(RI_{m,\lambda,l}^{\alpha} f(z,\zeta)\right)_z', \quad z \in U, \ \zeta \in \overline{U}, \tag{4.2.1.7}$$

then

$$q(z,\zeta) \prec\prec \frac{RI_{m,\lambda,l}^{\alpha} f(z,\zeta)}{z}, \quad z \in U, \ \zeta \in \overline{U},$$

where $q(z,\zeta) = \frac{1}{nz^{\frac{1}{n}}} \int_0^z h(t,\zeta) t^{\frac{1}{n}-1} dt$. The function q is convex and it is the best subordinant.

Proof. By using the properties of operator $RI^{\alpha}_{m,\lambda,l}$, we have

$$RI^{\alpha}_{m,\lambda,l}f(z,\zeta) = z + \sum_{j=n+1}^{\infty}\left\{\alpha\left(\frac{1+\lambda(j-1)+l}{l+1}\right)^{m} + (1-\alpha)C^{m}_{m+j-1}\right\} \times a_j(\zeta)z^j, \quad z\in U,\ \zeta\in\overline{U}.$$

Consider

$$p(z,\zeta) = \frac{RI^{\alpha}_{m,\lambda,l}f(z,\zeta)}{z} = \frac{z+\sum_{j=n+1}^{\infty}\left\{\alpha\left(\frac{1+\lambda(j-1)+l}{l+1}\right)^{m}+(1-\alpha)C^{m}_{m+j-1}\right\}a_j(\zeta)z^j}{z} = 1+p_n(\zeta)z^n+p_{n+1}(\zeta)z^{n+1}+\cdots,\quad z\in U,\ \zeta\in\overline{U}.$$

We deduce that $p\in\mathcal{H}^*[1,n,\zeta]$.

Let $RI^{\alpha}_{m,\lambda,l}f(z,\zeta)=zp(z,\zeta)$, $z\in U$, $\zeta\in\overline{U}$. Differentiating with respect to z, we obtain $\left(RI^{\alpha}_{m,\lambda,l}f(z,\zeta)\right)'_z = p(z,\zeta)+zp'_z(z,\zeta)$, $z\in U$, $\zeta\in\overline{U}$.

Then (4.2.1.7) becomes

$$h(z,\zeta)\prec\prec p(z,\zeta)+zp'_z(z,\zeta),\quad z\in U,\ \zeta\in\overline{U}.$$

By using Lemma 1.10 for $\gamma=1$, we have

$$q(z,\zeta)\prec\prec p(z,\zeta),\quad \text{i.e.}\quad q(z,\zeta)\prec\prec\frac{RI^{\alpha}_{m,\lambda,l}f(z,\zeta)}{z},\quad z\in U,\ \zeta\in\overline{U},$$

where $q(z,\zeta)=\frac{1}{nz^{\frac{1}{n}}}\int_0^z h(t,\zeta)t^{\frac{1}{n}-1}dt$. The function q is convex and it is the best subordinant. □

Corollary 4.4 (Alb Lupaş, 2012f). *Let $h(z,\zeta)=\frac{1+(2\beta-\zeta)z}{1+z}$ be a convex function in $U\times\overline{U}$, where $0\le\beta<1$. Let $m\in\mathbb{N}$, $\lambda,\alpha,l\ge 0$, $f(z,\zeta)\in\mathcal{A}^*_{n\zeta}$ and suppose that $\left(RI^{\alpha}_{m,\lambda,l}f(z,\zeta)\right)'_z$ is univalent and $\frac{RI^{\alpha}_{m,\lambda,l}f(z,\zeta)}{z}\in\mathcal{H}^*[1,n,\zeta]\cap Q^*$. If*

$$h(z,\zeta)\prec\prec\left(RI^{\alpha}_{m,\lambda,l}f(z,\zeta)\right)'_z,\quad z\in U,\ \zeta\in\overline{U},\tag{4.2.1.8}$$

then

$$q(z,\zeta)\prec\prec\frac{RI^{\alpha}_{m,\lambda,l}f(z,\zeta)}{z},\quad z\in U,\ \zeta\in\overline{U},$$

where q is given by $q(z,\zeta)=2\beta-\zeta+\frac{1+\zeta-2\beta}{nz^{\frac{1}{n}}}\int_0^z\frac{t^{\frac{1}{n}-1}}{t+1}dt$, $z\in U$, $\zeta\in\overline{U}$. The function q is convex and it is the best subordinant.

Proof. Following the same steps as in the proof of Theorem 4.17 and considering $p(z,\zeta) = \frac{RI^{\alpha}_{m,\lambda,l}f(z,\zeta)}{z}$, the strong differential superordination (4.2.1.8) becomes

$$h(z,\zeta) = \frac{1+(2\beta-\zeta)z}{1+z} \prec\prec p(z,\zeta) + zp'_z(z,\zeta), \quad z \in U, \ \zeta \in \overline{U}.$$

By using Lemma 1.10 for $\gamma = 1$, we have $q(z,\zeta) \prec\prec p(z,\zeta)$, i.e.

$$\begin{aligned} q(z,\zeta) &= \frac{1}{nz^{\frac{1}{n}}} \int_0^z h(t,\zeta) t^{\frac{1}{n}-1} dt = \frac{1}{nz^{\frac{1}{n}}} \int_0^z \frac{1+(2\beta-\zeta)t}{1+t} t^{\frac{1}{n}-1} dt \\ &= 2\beta - \zeta + \frac{1+\zeta-2\beta}{nz^{\frac{1}{n}}} \int_0^z \frac{t^{\frac{1}{n}-1}}{t+1} dt \prec\prec \frac{RI^{\alpha}_{m,\lambda,l}f(z,\zeta)}{z}, \\ & \quad z \in U, \ \zeta \in \overline{U}. \end{aligned}$$

The function q is convex and it is the best subordinant. □

Theorem 4.18 (Alb Lupaş, 2012f). *Let $q(z,\zeta)$ be convex in $U \times \overline{U}$ and let h be defined by $h(z,\zeta) = q(z,\zeta) + zq'_z(z,\zeta)$. If $m \in \mathbb{N}$, $\lambda, \alpha, l \geq 0$, $f(z,\zeta) \in \mathcal{A}^*_{n\zeta}$, suppose that $\left(RI^{\alpha}_{m,\lambda,l}f(z)\right)'_z$ is univalent and $\frac{RI^{\alpha}_{m,\lambda,l}f(z,\zeta)}{z} \in \mathcal{H}^*[1,n,\zeta] \cap Q^*$ and satisfies the strong differential superordination*

$$h(z,\zeta) = q(z,\zeta) + zq'_z(z,\zeta) \prec\prec \left(RI^{\alpha}_{m,\lambda,l}f(z,\zeta)\right)'_z, \quad z \in U, \ \zeta \in \overline{U}, \tag{4.2.1.9}$$

then

$$q(z,\zeta) \prec\prec \frac{RI^{\alpha}_{m,\lambda,l}f(z,\zeta)}{z}, \quad z \in U, \ \zeta \in \overline{U},$$

where $q(z,\zeta) = \frac{1}{nz^{\frac{1}{n}}} \int_0^z h(t,\zeta) t^{\frac{1}{n}-1} dt$. The function q is the best subordinant.

Proof. Following the same steps as in the proof of Theorem 4.17 and considering $p(z,\zeta) = \frac{RI^{\alpha}_{m,\lambda,l}f(z,\zeta)}{z}$, the strong differential superordination (4.2.1.9) becomes

$$q(z,\zeta) + zq'_z(z,\zeta) \prec\prec p(z,\zeta) + zp'_z(z,\zeta), \quad z \in U, \ \zeta \in \overline{U}.$$

Using Lemma 1.11 for $\gamma = 1$, we have

$$q(z,\zeta) \prec\prec p(z,\zeta), \quad z \in U, \ \zeta \in \overline{U}, \quad \text{i.e.}$$

$$q(z,\zeta) = \frac{1}{nz^{\frac{1}{n}}} \int_0^z h(t,\zeta) t^{\frac{1}{n}-1} dt \prec\prec \frac{RI^{\alpha}_{m,\lambda,l}f(z,\zeta)}{z}, \quad z \in U, \ \zeta \in \overline{U},$$

and q is the best subordinant. □

Theorem 4.19 (Alb Lupaş, 2012f). *Let $h(z,\zeta)$ be a convex function, $h(0,\zeta)=1$. Let $m\in\mathbb{N}$, $\lambda,\alpha,l\geq 0$, $f(z,\zeta)\in\mathcal{A}^*_{n\zeta}$ and suppose that $\left(\frac{zRI^{\alpha}_{m+1,\lambda,l}f(z,\zeta)}{RI^{\alpha}_{m,\lambda,l}f(z,\zeta)}\right)'_z$ is univalent and $\frac{RI^{\alpha}_{m+1,\lambda,l}f(z,\zeta)}{RI^{\alpha}_{m,\lambda,l}f(z,\zeta)}\in\mathcal{H}^*[1,n,\zeta]\cap Q^*$. If*

$$h(z,\zeta)\prec\prec\left(\frac{zRI^{\alpha}_{m+1,\lambda,l}f(z,\zeta)}{RI^{\alpha}_{m,\lambda,l}f(z,\zeta)}\right)'_z,\quad z\in U,\ \zeta\in\overline{U},\tag{4.2.1.10}$$

then

$$q(z,\zeta)\prec\prec\frac{RI^{\alpha}_{m+1,\lambda,l}f(z,\zeta)}{RI^{\alpha}_{m,\lambda,l}f(z,\zeta)},\quad z\in U,\ \zeta\in\overline{U},$$

where $q(z,\zeta)=\frac{1}{nz^{\frac{1}{n}}}\int_0^z h(t,\zeta)t^{\frac{1}{n}-1}dt$. The function q is convex and it is the best subordinant.

Proof. For $f\in\mathcal{A}^*_{n\zeta}$, $f(z,\zeta)=z+\sum_{j=n+1}^{\infty}a_j(\zeta)z^j$, we have $RI^{\alpha}_{m,\lambda,l}f(z,\zeta)=z+\sum_{j=n+1}^{\infty}\left\{\alpha\left(\frac{1+\lambda(j-1)+l}{l+1}\right)^m+(1-\alpha)C^m_{m+j-1}\right\}a_j(\zeta)z^j$, $z\in U$, $\zeta\in\overline{U}$.

Consider

$$p(z,\zeta)=\frac{RI^{\alpha}_{m+1,\lambda,l}f(z,\zeta)}{RI^{\alpha}_{m,\lambda,l}f(z,\zeta)}$$
$$=\frac{z+\sum_{j=n+1}^{\infty}\left\{\alpha\left(\frac{1+\lambda(j-1)+l}{l+1}\right)^{m+1}+(1-\alpha)C^{m+1}_{m+j}\right\}a_j(\zeta)z^j}{z+\sum_{j=n+1}^{\infty}\left\{\alpha\left(\frac{1+\lambda(j-1)+l}{l+1}\right)^{m}+(1-\alpha)C^{m}_{m+j-1}\right\}a_j(\zeta)z^j}.$$

We have $p'_z(z,\zeta)=\frac{\left(RI^{\alpha}_{m+1,\lambda,l}f(z,\zeta)\right)'_z}{RI^{\alpha}_{m,\lambda,l}f(z,\zeta)}-p(z,\zeta)\cdot\frac{\left(RI^{\alpha}_{m,\lambda,l}f(z,\zeta)\right)'_z}{RI^{\alpha}_{m,\lambda,l}f(z,\zeta)}$ and we obtain $p(z,\zeta)+z\cdot p'_z(z,\zeta)=\left(\frac{zRI^{\alpha}_{m+1,\lambda,l}f(z,\zeta)}{RI^{\alpha}_{m,\lambda,l}f(z,\zeta)}\right)'_z$.

Relation (4.2.1.10) becomes

$$h(z,\zeta)\prec\prec p(z,\zeta)+zp'_z(z,\zeta),\quad z\in U,\ \zeta\in\overline{U}.$$

By using Lemma 1.10 for $\gamma=1$, we have

$$q(z,\zeta)\prec\prec p(z,\zeta),\ \text{i.e.}\ q(z,\zeta)\prec\prec\frac{RI^{\alpha}_{m+1,\lambda,l}f(z,\zeta)}{RI^{\alpha}_{m,\lambda,l}f(z,\zeta)},\quad z\in U,\ \zeta\in\overline{U},$$

where $q(z,\zeta)=\frac{1}{nz^{\frac{1}{n}}}\int_0^z h(t,\zeta)t^{\frac{1}{n}-1}dt$. The function q is convex and it is the best subordinant. □

Corollary 4.5 (Alb Lupaş, 2012f). *Let* $h(z,\zeta) = \frac{1+(2\beta-\zeta)z}{1+z}$ *be a convex function in* $U \times \overline{U}$, *where* $0 \leq \beta < 1$. *Let* $\lambda, \alpha, l \geq 0$, $m \in \mathbb{N}$, $f(z,\zeta) \in \mathcal{A}^*_{n\zeta}$ *and suppose that* $\left(\frac{zRI^{\alpha}_{m+1,\lambda,l}f(z,)}{RI^{\alpha}_{m,\lambda,l}f(z,\zeta)}\right)'_z$ *is univalent and* $\frac{RI^{\alpha}_{m+1,\lambda,l}f(z,\zeta)}{RI^{\alpha}_{m,\lambda,l}f(z,\zeta)} \in \mathcal{H}^*[1,n,\zeta] \cap Q^*$. *If*

$$h(z,\zeta) \prec\prec \left(\frac{zRI^{\alpha}_{m+1,\lambda,l}f(z,\zeta)}{RI^{\alpha}_{m,\lambda,l}f(z,\zeta)}\right)'_z, \quad z \in U,\ \zeta \in \overline{U}, \tag{4.2.1.11}$$

then

$$q(z,\zeta) \prec\prec \frac{RI^{\alpha}_{m+1,\lambda,l}f(z,\zeta)}{RI^{\alpha}_{m,\lambda,l}f(z,\zeta)}, \quad z \in U,\ \zeta \in \overline{U},$$

where q *is given by* $q(z,\zeta) = 2\beta - \zeta + \frac{1+\zeta-2\beta}{nz^{\frac{1}{n}}}\int_0^z \frac{t^{\frac{1}{n}-1}}{t+1}dt$, $z \in U$, $\zeta \in \overline{U}$. *The function* q *is convex and it is the best subordinant.*

Proof. Following the same steps as in the proof of Theorem 4.19 and considering $p(z,\zeta) = \frac{RI^{\alpha}_{m+1,\lambda,l}f(z,\zeta)}{RI^{\alpha}_{m,\lambda,l}f(z,\zeta)}$, the strong differential superordination (4.2.1.11) becomes

$$h(z,\zeta) = \frac{1+(2\beta-\zeta)z}{1+z} \prec\prec p(z,\zeta) + zp'_z(z,\zeta), \quad z \in U,\ \zeta \in \overline{U}.$$

By using Lemma 1.10 for $\gamma = 1$, we have $q(z,\zeta) \prec\prec p(z,\zeta)$, i.e.

$$\begin{aligned} q(z,\zeta) &= \frac{1}{nz^{\frac{1}{n}}}\int_0^z h(t,\zeta)t^{\frac{1}{n}-1}dt = \frac{1}{nz^{\frac{1}{n}}}\int_0^z \frac{1+(2\beta-\zeta)t}{1+t}t^{\frac{1}{n}-1}dt \\ &= 2\beta - \zeta + \frac{1+\zeta-2\beta}{nz^{\frac{1}{n}}}\int_0^z \frac{t^{\frac{1}{n}-1}}{t+1}dt \prec\prec \frac{RI^{\alpha}_{m+1,\lambda,l}f(z,\zeta)}{RI^{\alpha}_{m,\lambda,l}f(z,\zeta)}, \\ & z \in U,\ \zeta \in \overline{U}. \end{aligned}$$

The function q is convex and it is the best subordinant. □

Theorem 4.20 (Alb Lupaş, 2012f). *Let* $q(z,\zeta)$ *be a convex function and* h *be defined by* $h(z,\zeta) = q(z,\zeta) + zq'_z(z,\zeta)$. *Let* $\lambda, \alpha, l \geq 0$, $m \in \mathbb{N}$, $f(z,\zeta) \in \mathcal{A}^*_{n\zeta}$ *and suppose that* $\left(\frac{zRI^{\alpha}_{m+1,\lambda,l}f(z,\zeta)}{RI^{\alpha}_{m,\lambda,l}f(z,\zeta)}\right)'_z$ *is univalent and* $\frac{RI^{\alpha}_{m+1,\lambda,l}f(z,\zeta)}{RI^{\alpha}_{m,\lambda,l}f(z,\zeta)} \in \mathcal{H}^*[1,n,\zeta] \cap Q^*$. *If*

$$h(z,\zeta) = q(z,\zeta) + zq'_z(z,\zeta) \prec\prec \left(\frac{zRI^{\alpha}_{m+1,\lambda,l}f(z,\zeta)}{RI^{\alpha}_{m,\lambda,l}f(z,\zeta)}\right)'_z, \quad z \in U,\ \zeta \in \overline{U}, \tag{4.2.1.12}$$

then

$$q(z,\zeta) \prec\prec \frac{RI^{\alpha}_{m+1,\lambda,l}f(z,\zeta)}{RI^{\alpha}_{m,\lambda,l}f(z,\zeta)}, \quad z \in U,\ \zeta \in \overline{U},$$

where $q(z,\zeta) = \frac{1}{nz^{\frac{1}{n}}}\int_0^z h(t,\zeta)t^{\frac{1}{n}-1}dt$. *The function* q *is the best subordinant.*

Proof. Following the same steps as in the proof of Theorem 4.19 and considering $p(z,\zeta) = \frac{RI^{\alpha}_{m+1,\lambda,l}f(z,\zeta)}{RI^{\alpha}_{m,\lambda,l}f(z,\zeta)}$, the strong differential superordination (4.2.1.12) becomes

$$h(z,\zeta) = q(z,\zeta) + zq'_z(z,\zeta) \prec\prec p(z,\zeta) + zp'_z(z,\zeta), \quad z \in U,\ \zeta \in \overline{U}.$$

By using Lemma 1.11 for $\gamma = 1$, we have

$$q(z,\zeta) \prec\prec p(z,\zeta), \quad z \in U,\ \zeta \in \overline{U}, \quad \text{i.e.}$$

$$q(z,\zeta) = \frac{1}{nz^{\frac{1}{n}}}\int_0^z h(t,\zeta)t^{\frac{1}{n}-1}dt \prec\prec \frac{RI^{\alpha}_{m+1,\lambda,l}f(z,\zeta)}{RI^{\alpha}_{m,\lambda,l}f(z,\zeta)}, \quad z \in U,\ \zeta \in \overline{U},$$

and q is the best subordinant. □

Theorem 4.21 (Alb Lupaş, 2012f). *Let* $h(z,\zeta)$ *be a convex function in* $U \times \overline{U}$ *with* $h(0,\zeta) = 1$ *and let* $\lambda,\alpha,l \geq 0$, $m \in \mathbb{N}$, $f(z,\zeta) \in \mathcal{A}^*_{n\zeta}$, $\frac{(m+1)(m+2)}{z}RI^{\alpha}_{m+2,\lambda,l}f(z,\zeta) - \frac{(m+1)(2m+1)}{z}RI^{\alpha}_{m+1,\lambda,l}f(z,\zeta) + \frac{m^2}{z}RI^{\alpha}_{m,\lambda,l}f(z,\zeta) + \frac{\alpha}{z}\left[\frac{(l+1)^2}{\lambda^2} - (m+1)(m+2)\right]I(m+2,\lambda,l)f(z,\zeta) - \frac{\alpha}{z}\left[\frac{2(l+1-\lambda)(l+1)}{\lambda^2} - (m+1)(2m+1)\right]I(m+1,\lambda,l)f(z,\zeta) + \frac{\alpha}{z}\left[\frac{(l+1-\lambda)^2}{\lambda^2} - m^2\right]I(m,\lambda,l)f(z,\zeta)$ *is univalent and* $[RD^m_{\lambda,\alpha}f(z,\zeta)]'_z \in \mathcal{H}^*[1,n,\zeta] \cap Q^*$. *If*

$$\begin{aligned} h(z,\zeta) \prec\prec &\frac{(m+1)(m+2)}{z}RI^{\alpha}_{m+2,\lambda,l}f(z,\zeta) \\ &- \frac{(m+1)(2m+1)}{z}RI^{\alpha}_{m+1,\lambda,l}f(z,\zeta) + \frac{m^2}{z}RI^{\alpha}_{m,\lambda,l}f(z,\zeta) \\ &+ \frac{\alpha}{z}\left[\frac{(l+1)^2}{\lambda^2} - (m+1)(m+2)\right]I(m+2,\lambda,l)f(z,\zeta) \\ &- \frac{\alpha}{z}\left[\frac{2(l+1-\lambda)(l+1)}{\lambda^2} - (m+1)(2m+1)\right]I(m+1,\lambda,l)f(z,\zeta) \\ &+ \frac{\alpha}{z}\left[\frac{(l+1-\lambda)^2}{\lambda^2} - m^2\right]I(m,\lambda,l)f(z,\zeta), \quad z \in U,\ \zeta \in \overline{U}, \end{aligned} \tag{4.2.1.13}$$

holds, then

$$q(z,\zeta) \prec\prec [RI^{\alpha}_{m,\lambda,l}f(z,\zeta)]'_z, \quad z \in U, \ \zeta \in \overline{U},$$

where $q(z,\zeta) = \frac{1}{nz^{\frac{1}{n}}}\int_0^z h(t,\zeta)t^{\frac{1}{n}-1}dt$. *The function q is convex and it is the best subordinant.*

Proof. For $f \in \mathcal{A}^*_{n\zeta}$, $f(z,\zeta) = z + \sum_{j=n+1}^{\infty} a_j(\zeta) z^j$, we have

$$\begin{aligned} &RI^{\alpha}_{m,\lambda,l}f(z,\zeta) \\ &\quad = z + \sum_{j=n+1}^{\infty}\left\{\alpha\left(\frac{1+\lambda(j-1)+l}{l+1}\right)^m + (1-\alpha)C^m_{m+j-1}\right\}a_j(\zeta)z^j, \\ &\qquad z \in U, \ \zeta \in \overline{U}. \end{aligned}$$

Let

$$\begin{aligned} p(z,\zeta) &= \left(RI^{\alpha}_{m,\lambda,l}f(z,\zeta)\right)'_z \\ &= 1 + \sum_{j=n+1}^{\infty}\left\{\alpha\left(\frac{1+\lambda(j-1)+l}{l+1}\right)^m + (1-\alpha)C^m_{m+j-1}\right\}ja_j(\zeta)z^{j-1} \\ &= 1 + p_n(\zeta)z^n + p_{n+1}(\zeta)z^{n+1} + \cdots. \qquad (4.2.1.14) \end{aligned}$$

By using the properties of operators $RI^{\alpha}_{m,\lambda,l}$, R^m and $I(m,\lambda,l)$, after a short calculation, we obtain

$$\begin{aligned} &p(z,\zeta) + zp'_z(z,\zeta) \\ &\quad = \frac{(m+1)(m+2)}{z}RI^{\alpha}_{m+2,\lambda,l}f(z,\zeta) - \frac{(m+1)(2m+1)}{z}RI^{\alpha}_{m+1,\lambda,l}f(z,\zeta) \\ &\qquad + \frac{m^2}{z}RI^{\alpha}_{m,\lambda,l}f(z,\zeta) + \frac{\alpha}{z}\left[\frac{(l+1)^2}{\lambda^2} - (m+1)(m+2)\right] \\ &\qquad \times I(m+2,\lambda,l)f(z,\zeta) - \frac{\alpha}{z}\left[\frac{2(l+1-\lambda)(l+1)}{\lambda^2} - (m+1)(2m+1)\right] \\ &\qquad \times I(m+1,\lambda,l)f(z,\zeta) + \frac{\alpha}{z}\left[\frac{(l+1-\lambda)^2}{\lambda^2} - m^2\right]I(m,\lambda,l)f(z,\zeta). \end{aligned}$$

Using the notation in (4.2.1.14), the strong differential superordination becomes

$$h(z,\zeta) \prec\prec p(z,\zeta) + zp'_z(z,\zeta).$$

By using Lemma 1.10 for $\gamma = 1$, we have

$$q(z,\zeta) \prec\prec p(z,\zeta), \quad z \in U, \ \zeta \in \overline{U}, \text{ i.e. } q(z,\zeta) \prec\prec \left(RI^{\alpha}_{m,\lambda,l}f(z,\zeta)\right)'_z, \quad z \in U, \ \zeta \in \overline{U},$$

where $q(z,\zeta) = \frac{1}{nz^{\frac{1}{n}}}\int_0^z h(t,\zeta)t^{\frac{1}{n}-1}dt$. The function q is convex and it is the best subordinant. □

Corollary 4.6 (Alb Lupaş, 2012f). *Let* $h(z,\zeta) = \frac{1+(2\beta-\zeta)z}{1+z}$ *be a convex function in* $U \times \overline{U}$, *where* $0 \le \beta < 1$. *Let* $\lambda, \alpha, l \ge 0$, $m \in \mathbb{N}$, $f(z,\zeta) \in \mathcal{A}^*_{n\zeta}$, *suppose that* $\frac{(m+1)(m+2)}{z}RI^{\alpha}_{m+2,\lambda,l}f(z,\zeta) - \frac{(m+1)(2m+1)}{z}RI^{\alpha}_{m+1,\lambda,l}f(z,\zeta) + \frac{m^2}{z}RI^{\alpha}_{m,\lambda,l}f(z,\zeta) + \frac{\alpha}{z}\left[\frac{(l+1)^2}{\lambda^2} - (m+1)(m+2)\right]I(m+2,\lambda,l)f(z,\zeta) - \frac{\alpha}{z}\left[\frac{2(l+1-\lambda)(l+1)}{\lambda^2} - (m+1)(2m+1)\right]I(m+1,\lambda,l)f(z,\zeta) + \frac{\alpha}{z}\left[\frac{(l+1-\lambda)^2}{\lambda^2} - m^2\right]I(m,\lambda,l)f(z,\zeta)$ *is univalent in* $U \times \overline{U}$ *and* $[RI^{\alpha}_{m,\lambda,l}f(z,\zeta)]'_z \in \mathcal{H}^*[1,n,\zeta] \cap Q^*$. *If*

$$\begin{aligned} h(z,\zeta) \prec\prec\ & \frac{(m+1)(m+2)}{z}RI^{\alpha}_{m+2,\lambda,l}f(z,\zeta) \\ & - \frac{(m+1)(2m+1)}{z}RI^{\alpha}_{m+1,\lambda,l}f(z,\zeta) + \frac{m^2}{z}RI^{\alpha}_{m,\lambda,l}f(z,\zeta) \\ & + \frac{\alpha}{z}\left[\frac{(l+1)^2}{\lambda^2} - (m+1)(m+2)\right]I(m+2,\lambda,l)f(z,\zeta) \\ & - \frac{\alpha}{z}\left[\frac{2(l+1-\lambda)(l+1)}{\lambda^2} - (m+1)(2m+1)\right] \\ & \times I(m+1,\lambda,l)f(z,\zeta) \\ & + \frac{\alpha}{z}\left[\frac{(l+1-\lambda)^2}{\lambda^2} - m^2\right]I(m,\lambda,l)f(z,\zeta), \quad z \in U, \ \zeta \in \overline{U}, \end{aligned} \tag{4.2.1.15}$$

then

$$q(z,\zeta) \prec\prec \left(RI^{\alpha}_{m,\lambda,l}f(z,\zeta)\right)'_z, \quad z \in U, \ \zeta \in \overline{U},$$

where q *is given by* $q(z,\zeta) = 2\beta - \zeta + \frac{1+\zeta-\beta}{nz^{\frac{1}{n}}}\int_0^z \frac{t^{\frac{1}{n}-1}}{t+1}dt$, $z \in U$, $\zeta \in \overline{U}$. *The function* q *is convex and it is the best subordinant.*

Proof. Following the same steps as in the proof of Theorem 4.21 and considering $p(z,\zeta) = \left(RI^{\alpha}_{m,\lambda,l}f(z,\zeta)\right)'_z$, the strong differential superordination

(4.2.1.15) becomes

$$h(z,\zeta)=\frac{1+(2\beta-\zeta)z}{1+z}\prec\prec p(z,\zeta)+zp'_z(z,\zeta),\quad z\in U,\ \zeta\in\overline{U}.$$

By using Lemma 1.10 for $\gamma=1$, we have $q(z,\zeta)\prec\prec p(z,\zeta)$, i.e.

$$\begin{aligned}q(z,\zeta)&=\frac{1}{nz^{\frac{1}{n}}}\int_0^z h(t,\zeta)t^{\frac{1}{n}-1}dt=\frac{1}{nz^{\frac{1}{n}}}\int_0^z\frac{1+(2\beta-\zeta)t}{1+t}t^{\frac{1}{n}-1}dt\\&=2\beta-\zeta+\frac{1+\zeta-2\beta}{nz^{\frac{1}{n}}}\int_0^z\frac{t^{\frac{1}{n}-1}}{t+1}dt\prec\prec\left(RI^{\alpha}_{m,\lambda,l}f(z,\zeta)\right)'_z,\\&\quad z\in U,\ \zeta\in\overline{U}.\end{aligned}$$

The function q is convex and it is the best subordinant. □

Theorem 4.22 (Alb Lupaş, 2012f). *Let $q(z,\zeta)$ be a convex function in $U\times\overline{U}$ and $h(z,\zeta)=q(z,\zeta)+zq'_z(z,\zeta)$.*

Let $\lambda,\alpha,l\geq 0$, $m\in\mathbb{N}$, $f(z,\zeta)\in\mathcal{A}^*_{n\zeta}$, suppose that $\frac{(m+1)(m+2)}{z}RI^{\alpha}_{m+2,\lambda,l}f(z,\zeta)-\frac{(m+1)(2m+1)}{z}RI^{\alpha}_{m+1,\lambda,l}f(z,\zeta)+\frac{m^2}{z}RI^{\alpha}_{m,\lambda,l}f(z,\zeta)+\frac{\alpha}{z}\left[\frac{(l+1)^2}{\lambda^2}-(m+1)(m+2)\right]I(m+2,\lambda,l)f(z,\zeta)-\frac{\alpha}{z}\left[\frac{2(l+1-\lambda)(l+1)}{\lambda^2}-(m+1)(2m+1)\right]I(m+1,\lambda,l)f(z,\zeta)+\frac{\alpha}{z}\left[\frac{(l+1-\lambda)^2}{\lambda^2}-m^2\right]I(m,\lambda,l)f(z,\zeta)$ is univalent in $U\times\overline{U}$ and $[RI^{\alpha}_{m,\lambda,l}f(z,\zeta)]'_z\in\mathcal{H}^*[1,n,\zeta]\cap Q^*$. If

$$\begin{aligned}h(z,\zeta)&=q(z,\zeta)+zq'_z(z,\zeta)\\&\prec\prec\frac{(m+1)(m+2)}{z}RI^{\alpha}_{m+2,\lambda,l}f(z,\zeta)\\&\quad-\frac{(m+1)(2m+1)}{z}RI^{\alpha}_{m+1,\lambda,l}f(z,\zeta)+\frac{m^2}{z}RI^{\alpha}_{m,\lambda,l}f(z,\zeta)\\&\quad+\frac{\alpha}{z}\left[\frac{(l+1)^2}{\lambda^2}-(m+1)(m+2)\right]I(m+2,\lambda,l)f(z,\zeta)\\&\quad-\frac{\alpha}{z}\left[\frac{2(l+1-\lambda)(l+1)}{\lambda^2}-(m+1)(2m+1)\right]\\&\quad\times I(m+1,\lambda,l)f(z,\zeta)\\&\quad+\frac{\alpha}{z}\left[\frac{(l+1-\lambda)^2}{\lambda^2}-m^2\right]I(m,\lambda,l)f(z,\zeta),\quad z\in U,\ \zeta\in\overline{U},\end{aligned}\tag{4.2.1.16}$$

then

$$q(z,\zeta) \prec\prec \left(RI^{\alpha}_{m,\lambda,l}f(z,\zeta)\right)'_z, \quad z\in U,\ \zeta\in\overline{U},$$

where $q(z,\zeta) = \frac{1}{nz^{\frac{1}{n}}}\int_0^z h(t,\zeta)t^{\frac{1}{n}-1}dt$. *The function* q *is the best subordinant.*

Proof. Following the same steps as in the proof of Theorem 4.21 and considering $p(z,\zeta) = \left(RI^{\alpha}_{m,\lambda,l}f\,(z,\zeta)\right)'_z$, the strong differential superordination (4.2.1.16) becomes

$$h(z,\zeta) = q\,(z,\zeta) + zq'_z\,(z,\zeta) \prec\prec p(z,\zeta) + zp'_z(z,\zeta), \quad z\in U,\ \zeta\in\overline{U}.$$

By using Lemma 1.11 for $\gamma = 1$, we have $q(z,\zeta) \prec\prec p(z,\zeta)$, i.e.

$$q(z,\zeta) = \frac{1}{nz^{\frac{1}{n}}}\int_0^z h(t,\zeta)t^{\frac{1}{n}-1}dt \prec\prec \left(RI^{\alpha}_{m,\lambda,l}f(z,\zeta)\right)'_z, \quad z\in U,\ \zeta\in\overline{U}.$$

The function q is the best subordinant. □

4.2.2 *Certain Strong Differential Superordinations for Convolution Product of Multiplier Transformation and Ruscheweyh Operator*

In the paper Alb Lupaş (2012i), we established several strong differential superordinations regarding the new operator $IR^m_{\lambda,l}$ defined in Alb Lupaş (2011g) and exposed in Definition 2.10 for $n = m$, by convolution product of the extended multiplier transformation and the extended Ruscheweyh derivative, $IR^m_{\lambda,l} : \mathcal{A}^*_\zeta \to \mathcal{A}^*_\zeta$,

$$IR^m_{\lambda,l}f\,(z,\zeta) = \left(I\,(m,\lambda,l) * R^m\right)f\,(z,\zeta), \quad z\in U, \quad \zeta\in\overline{U},$$

where $R^m f(z,\zeta)$ denote the extended Ruscheweyh derivative, $I\,(m,\lambda,l)$ $f\,(z,\zeta)$ is the extended multiplier transformation and $\mathcal{A}^*_{n\zeta} = \{f \in \mathcal{H}$ $(U\times\overline{U}),\ f(z,\zeta) = z + a_{n+1}\,(\zeta)\,z^{n+1} + \cdots,\ z\in U, \zeta\in\overline{U}\}$, with $\mathcal{A}^*_{1\zeta} = \mathcal{A}^*_\zeta$, is the class of normalized analytic functions.

If $f\in\mathcal{A}^*_\zeta$, $f(z,\zeta) = z + \sum_{j=2}^{\infty} a_j\,(\zeta)\,z^j$, then

$$IR^m_{\lambda,l}f\,(z,\zeta) = z + \sum_{j=2}^{\infty}\left(\frac{1+\lambda\,(j-1)+l}{l+1}\right)^m C^m_{m+j-1}a^2_j\,(\zeta)\,z^j, \quad z\in U,\ \zeta\in\overline{U}.$$

Theorem 4.23 (Alb Lupaş, 2012i). *Let* $h(z,\zeta)$ *be a convex function in* $U\times\overline{U}$ *with* $h(0,\zeta)=1$. *Let* $m\in\mathbb{N}$, $\lambda,l\geq 0$, $f(z,\zeta)\in\mathcal{A}^{*}_{\zeta}$, $F(z,\zeta)=I_c(f)(z,\zeta)=\frac{c+2}{z^{c+1}}\int_0^z t^c f(t,\zeta)\,dt$, $z\in U$, $\zeta\in\overline{U}$, Re $c>-2$, *and suppose that* $\left(IR^m_{\lambda,l}f(z,\zeta)\right)'_z$ *is univalent in* $U\times\overline{U}$, $\left(IR^m_{\lambda,l}F(z,\zeta)\right)'_z\in\mathcal{H}^*[1,1,\zeta]\cap Q^*$ *and*

$$h(z,\zeta)\prec\prec\left(IR^m_{\lambda,l}f(z,\zeta)\right)'_z,\quad z\in U,\ \zeta\in\overline{U},\tag{4.2.2.1}$$

then

$$q(z,\zeta)\prec\prec\left(IR^m_{\lambda,l}F(z,\zeta)\right)'_z,\quad z\in U,\ \zeta\in\overline{U},$$

where $q(z,\zeta)=\frac{c+2}{z^{c+2}}\int_0^z h(t,\zeta)t^{c+1}dt$. *The function* q *is convex and it is the best subordinant.*

Proof. We have

$$z^{c+1}F(z,\zeta)=(c+2)\int_0^z t^c f(t,\zeta)\,dt$$

and differentiating it, with respect to z, we obtain $(c+1)F(z,\zeta)+zF'_z(z,\zeta)=(c+2)f(z,\zeta)$ and

$$\begin{aligned}&(c+1)IR^m_{\lambda,l}F(z,\zeta)+z\left(IR^m_{\lambda,l}F(z,\zeta)\right)'_z\\&=(c+2)IR^m_{\lambda,l}f(z,\zeta),\quad z\in U,\ \zeta\in\overline{U}.\end{aligned}$$

Differentiating the last relation with respect to z, we have

$$\begin{aligned}&\left(IR^m_{\lambda,l}F(z,\zeta)\right)'_z+\frac{1}{c+2}z\left(IR^m_{\lambda,l}F(z,\zeta)\right)''_{z^2}\\&=\left(IR^m_{\lambda,l}f(z,\zeta)\right)'_z,\quad z\in U,\ \zeta\in\overline{U}.\end{aligned}\tag{4.2.2.2}$$

Using (4.2.2.2), the strong differential superordination (4.2.2.1) becomes

$$h(z,\zeta)\prec\prec\left(IR^m_{\lambda,l}F(z,\zeta)\right)'_z+\frac{1}{c+2}z\left(IR^m_{\lambda,l}F(z,\zeta)\right)''_{z^2}.\tag{4.2.2.3}$$

Denote

$$p(z,\zeta)=\left(IR^m_{\lambda,l}F(z,\zeta)\right)'_z,\quad z\in U,\ \zeta\in\overline{U}.\tag{4.2.2.4}$$

Replacing (4.2.2.4) in (4.2.2.3), we obtain

$$h(z,\zeta)\prec\prec p(z,\zeta)+\frac{1}{c+2}zp'_z(z,\zeta),\quad z\in U,\ \zeta\in\overline{U}.$$

Using Lemma 1.10 for $n = 1$ and $\gamma = c + 2$, we have

$$q(z,\zeta) \prec\prec p(z,\zeta), \text{ i.e. } q(z,\zeta) \prec\prec \left(IR^m_{\lambda,l}F(z,\zeta)\right)'_z, \quad z \in U, \ \zeta \in \overline{U},$$

where $q(z,\zeta) = \frac{c+2}{z^{c+2}} \int_0^z h(t,\zeta)t^{c+1}dt$. The function q is convex and it is the best subordinant. □

Corollary 4.7 (Alb Lupaş, 2012i). *Let* $h(z,\zeta) = \frac{\zeta+(2\beta-\zeta)z}{1+z}$, *where* $\beta \in [0,1)$. *Let* $m \in \mathbb{N}$, $\lambda, l \geq 0$, $f(z,\zeta) \in \mathcal{A}^*_\zeta$, $F(z,\zeta) = I_c(f)(z,\zeta) = \frac{c+2}{z^{c+1}} \int_0^z t^c f(t,\zeta)\,dt$, $z \in U$, $\zeta \in \overline{U}$, $\operatorname{Re} c > -2$, *and suppose that* $\left(IR^m_{\lambda,l}f(z,\zeta)\right)'_z$ *is univalent in* $U \times \overline{U}$, $\left(IR^m_{\lambda,l}F(z,\zeta)\right)'_z \in \mathcal{H}^*[1,1,\zeta] \cap Q^*$ *and*

$$h(z,\zeta) \prec\prec \left(IR^m_{\lambda,l}f(z,\zeta)\right)'_z, \quad z \in U, \ \zeta \in \overline{U}, \tag{4.2.2.5}$$

then

$$q(z,\zeta) \prec\prec \left(IR^m_{\lambda,l}F(z,\zeta)\right)'_z, \quad z \in U, \ \zeta \in \overline{U},$$

where q *is given by* $q(z,\zeta) = 2\beta - \zeta + \frac{2(c+2)(\zeta-\beta)}{z^{c+2}} \int_0^z \frac{t^{c+1}}{t+1}dt$, $z \in U$, $\zeta \in \overline{U}$. *The function* q *is convex and it is the best subordinant.*

Proof. Following the same steps as in the proof of Theorem 4.23 and considering $p(z,\zeta) = \left(IR^m_{\lambda,l}F(z,\zeta)\right)'_z$, the strong differential superordination (4.2.2.5) becomes

$$h(z,\zeta) = \frac{\zeta + (2\beta-\zeta)z}{1+z} \prec\prec p(z,\zeta) + \frac{1}{c+2} z p'_z(z,\zeta), \quad z \in U, \ \zeta \in \overline{U}.$$

By using Lemma 1.10 for $n = 1$ and $\gamma = c + 2$, we have $q(z,\zeta) \prec\prec p(z,\zeta)$, i.e.

$$\begin{aligned} q(z,\zeta) &= \frac{c+2}{z^{c+2}} \int_0^z h(t,\zeta)t^{c+1}dt = \frac{c+2}{z^{c+2}} \int_0^z \frac{\zeta+(2\beta-\zeta)t}{1+t} t^{c+1}dt \\ &= 2\beta - \zeta + \frac{2(c+2)(\zeta-\beta)}{z^{c+2}} \int_0^z \frac{t^{c+1}}{t+1}dt \prec\prec \left(IR^m_{\lambda,l}F(z,\zeta)\right)'_z, \end{aligned}$$

$z \in U$, $\zeta \in \overline{U}$.

The function q is convex and it is the best subordinant. □

Theorem 4.24 (Alb Lupaş, 2012i). *Let* $q(z,\zeta)$ *be a convex function in* $U \times \overline{U}$ *and let* $h(z,\zeta) = q(z,\zeta) + \frac{1}{c+2} z q'_z(z,\zeta)$, *where* $z \in U$, $\zeta \in \overline{U}$, $\operatorname{Re} c > -2$.

Let $m \in \mathbb{N}$, $\lambda, l \geq 0$, $f(z,\zeta) \in \mathcal{A}_\zeta^*$, $F(z,\zeta) = I_c(f)(z,\zeta) = \frac{c+2}{z^{c+1}} \int_0^z t^c f(t,\zeta)\,dt$, $z \in U$, $\zeta \in \overline{U}$, *and suppose that* $\left(IR_{\lambda,l}^m f(z,\zeta)\right)_z'$ *is univalent in* $U \times \overline{U}$, $\left(IR_{\lambda,l}^m F(z,\zeta)\right)_z' \in \mathcal{H}^*[1,1,\zeta] \cap Q^*$ *and*

$$h(z,\zeta) \prec\prec \left(IR_{\lambda,l}^m f(z,\zeta)\right)_z', \quad z \in U, \ \zeta \in \overline{U}, \tag{4.2.2.6}$$

then

$$q(z,\zeta) \prec\prec \left(IR_{\lambda,l}^m F(z,\zeta)\right)_z', \quad z \in U, \ \zeta \in \overline{U},$$

where $q(z,\zeta) = \frac{c+2}{z^{c+2}} \int_0^z h(t,\zeta)t^{c+1}dt$. *The function* q *is the best subordinant.*

Proof. Following the same steps as in the proof of Theorem 4.23 and considering $p(z,\zeta) = \left(IR_{\lambda,l}^m F(z,\zeta)\right)'$, $z \in U$, $\zeta \in \overline{U}$, the strong differential superordination (4.2.2.6) becomes

$$h(z,\zeta) = q(z,\zeta) + \frac{1}{c+2} z q_z'(z,\zeta) \prec\prec p(z,\zeta) + \frac{1}{c+2} z p_z'(z,\zeta),$$

$z \in U$, $\zeta \in \overline{U}$.

Using Lemma 1.11 for $n = 1$ and $\gamma = c+2$, we have

$$q(z,\zeta) \prec\prec p(z,\zeta), \quad \text{i.e. } q(z,\zeta) \prec\prec \left(IR_{\lambda,l}^m F(z,\zeta)\right)_z', \quad z \in U, \ \zeta \in \overline{U},$$

where $q(z,\zeta) = \frac{c+2}{z^{c+2}} \int_0^z h(t,\zeta)t^{c+1}dt$. The function q is the best subordinant. □

Theorem 4.25 (Alb Lupaş, 2012i). *Let* $h(z,\zeta)$ *be a convex function,* $h(0,\zeta) = 1$. *Let* $\lambda, l \geq 0$, $m \in \mathbb{N}$, $f(z,\zeta) \in \mathcal{A}_\zeta^*$ *and suppose that* $\left(IR_{\lambda,l}^m f(z,\zeta)\right)_z'$ *is univalent and* $\frac{IR_{\lambda,l}^m f(z,\zeta)}{z} \in \mathcal{H}^*[1,1,\zeta] \cap Q^*$. *If*

$$h(z,\zeta) \prec\prec \left(IR_{\lambda,l}^m f(z,\zeta)\right)_z', \quad z \in U, \ \zeta \in \overline{U}, \tag{4.2.2.7}$$

then

$$q(z,\zeta) \prec\prec \frac{IR_{\lambda,l}^m f(z,\zeta)}{z}, \quad z \in U, \ \zeta \in \overline{U},$$

where $q(z,\zeta) = \frac{1}{z} \int_0^z h(t,\zeta)dt$. *The function* q *is convex and it is the best subordinant.*

Proof. Consider

$$p(z,\zeta)=\frac{IR_{\lambda,l}^{m}f(z,\zeta)}{z}=\frac{z+\sum_{j=2}^{\infty}\left(\frac{1+\lambda(j-1)+l}{l+1}\right)^{m}C_{m+j-1}^{m}a_j^2(\zeta)z^j}{z}$$
$$=1+\sum_{j=2}^{\infty}\left(\frac{1+\lambda(j-1)+l}{l+1}\right)^{m}C_{m+j-1}^{m}a_j^2(\zeta)z^{j-1}.$$

Evidently, $p\in\mathcal{H}^*[1,1,\zeta]$.

We have $p(z,\zeta)+zp_z'(z,\zeta)=\left(IR_{\lambda,l}^{m}f(z,\zeta)\right)_z'$, $z\in U$, $\zeta\in\overline{U}$.

Then (4.2.2.7) becomes

$$h(z,\zeta)\prec\prec p(z,\zeta)+zp_z'(z,\zeta),\quad z\in U,\ \zeta\in\overline{U}.$$

By using Lemma 1.10 for $n=1$ and $\gamma=1$, we have

$$q(z,\zeta)\prec\prec p(z,\zeta),\quad \text{i.e.}\quad q(z,\zeta)\prec\prec\frac{IR_{\lambda,l}^{m}f(z,\zeta)}{z},\quad z\in U,\ \zeta\in\overline{U},$$

where $q(z,\zeta)=\frac{1}{z}\int_0^z h(t,\zeta)dt$. The function q is convex and it is the best subordinant. □

Corollary 4.8 (Alb Lupaş, 2012i). *Let $h(z,\zeta)=\frac{\zeta+(2\beta-\zeta)z}{1+z}$ be a convex function in $U\times\overline{U}$, where $0\le\beta<1$. Let $\lambda,l\ge0$, $m\in\mathbb{N}$, $f(z,\zeta)\in\mathcal{A}_\zeta^*$ and suppose that $\left(IR_{\lambda,l}^{m}f(z,\zeta)\right)_z'$ is univalent and $\frac{IR_{\lambda,l}^{m}f(z,\zeta)}{z}\in\mathcal{H}^*[1,1,\zeta]\cap Q^*$. If*

$$h(z,\zeta)\prec\prec\left(IR_{\lambda,l}^{m}f(z,\zeta)\right)_z',\quad z\in U,\ \zeta\in\overline{U},\tag{4.2.2.8}$$

then

$$q(z,\zeta)\prec\prec\frac{IR_{\lambda,l}^{m}f(z,\zeta)}{z},\quad z\in U,\ \zeta\in\overline{U},$$

where q is given by $q(z,\zeta)=2\beta-\zeta+\frac{2(\zeta-\beta)}{z}\ln(1+z)$, $z\in U$, $\zeta\in\overline{U}$. The function q is convex and it is the best subordinant.

Proof. Following the same steps as in the proof of Theorem 4.25 and considering $p(z,\zeta)=\frac{IR_{\lambda,l}^{m}f(z,\zeta)}{z}$, the strong differential superordination (4.2.2.8) becomes

$$h(z,\zeta)=\frac{\zeta+(2\beta-\zeta)z}{1+z}\prec\prec p(z,\zeta)+zp_z'(z,\zeta),\quad z\in U,\ \zeta\in\overline{U}.$$

By using Lemma 1.10 for $n = 1$ and $\gamma = 1$, we have $q(z,\zeta) \prec\prec p(z,\zeta)$, i.e.

$$q(z,\zeta) = \frac{1}{z}\int_0^z h(t,\zeta)\,dt = \frac{1}{z}\int_0^z \frac{\zeta + (2\beta - \zeta)\,t}{1+t}dt$$
$$= 2\beta - \zeta + \frac{2(\zeta - \beta)}{z}\ln(1+z) \prec\prec \frac{IR^m_{\lambda,l}f(z,\zeta)}{z}, \quad z \in U,\ \zeta \in \overline{U}.$$

The function q is convex and it is the best subordinant. □

Theorem 4.26 (Alb Lupaş, 2012i). *Let $q(z,\zeta)$ be convex in $U \times \overline{U}$ and let h be defined by $h(z,\zeta) = q(z,\zeta) + zq'_z(z,\zeta)$. If $\lambda, l \geq 0$, $m \in \mathbb{N}$, $f(z,\zeta) \in \mathcal{A}^*_\zeta$, suppose that $\left(IR^m_{\lambda,l}f(z,\zeta)\right)'_z$ is univalent, $\frac{IR^m_{\lambda,l}f(z,\zeta)}{z} \in \mathcal{H}^*[1,1,\zeta] \cap Q^*$ and satisfies the strong differential superordination*

$$h(z,\zeta) = q(z,\zeta) + zq'_z(z,\zeta) \prec\prec \left(IR^m_{\lambda,l}f(z,\zeta)\right)'_z, \quad z \in U,\ \zeta \in \overline{U}, \tag{4.2.2.9}$$

then

$$q(z,\zeta) \prec\prec \frac{IR^m_{\lambda,l}f(z,\zeta)}{z}, \quad z \in U,\ \zeta \in \overline{U},$$

where $q(z,\zeta) = \frac{1}{z}\int_0^z h(t,\zeta)dt$. The function q is the best subordinant.

Proof. Let $p(z,\zeta) = \frac{IR^m_{\lambda,l}f(z,\zeta)}{z} = \frac{z + \sum_{j=2}^\infty \left(\frac{1+\lambda(j-1)+l}{l+1}\right)^m C^m_{m+j-1} a_j^2(\zeta) z^j}{z} = 1 + \sum_{j=2}^\infty \left(\frac{1+\lambda(j-1)+l}{l+1}\right)^m C^m_{m+j-1} a_j^2(\zeta) z^{j-1}$. Evidently, $p \in \mathcal{H}^*[1,1,\zeta]$.

Differentiating with respect to z, we obtain $p(z,\zeta) + zp'_z(z,\zeta) = \left(IR^m_{\lambda,l}f(z,\zeta)\right)'_z$, $z \in U$, $\zeta \in \overline{U}$, and (4.2.2.9) becomes

$$q(z,\zeta) + zq'_z(z,\zeta) \prec\prec p(z,\zeta) + zp'_z(z,\zeta), \quad z \in U,\ \zeta \in \overline{U}.$$

Using Lemma 1.11 for $n = 1$ and $\gamma = 1$, we have

$$q(z,\zeta) \prec\prec p(z,\zeta), \ \ z \in U,\ \ \zeta \in \overline{U}, \text{ i.e.}$$
$$q(z,\zeta) = \frac{1}{z}\int_0^z h(t,\zeta)dt \prec\prec \frac{IR^m_{\lambda,l}f(z,\zeta)}{z}, \ \ z \in U,\ \ \zeta \in \overline{U},$$

and q is the best subordinant. □

Theorem 4.27 (Alb Lupaş, 2012i). *Let $h(z,\zeta)$ be a convex function, $h(0,\zeta)=1$. Let $\lambda, l \geq 0$, $m \in \mathbb{N}$, $f(z,\zeta) \in \mathcal{A}_\zeta^*$ and suppose that $\left(\frac{zIR_{\lambda,l}^{m+1}f(z,\zeta)}{IR_{\lambda,l}^{m}f(z,\zeta)}\right)_z'$ is univalent and $\frac{IR_{\lambda,l}^{m+1}f(z,\zeta)}{IR_{\lambda,l}^{m}f(z,\zeta)} \in \mathcal{H}^*[1,1,\zeta] \cap Q^*$. If*

$$h(z,\zeta) \prec\prec \left(\frac{zIR_{\lambda,l}^{m+1}f(z,\zeta)}{IR_{\lambda,l}^{m}f(z,\zeta)}\right)_z', \quad z \in U,\ \zeta \in \overline{U}, \tag{4.2.2.10}$$

then

$$q(z,\zeta) \prec\prec \frac{IR_{\lambda,l}^{m+1}f(z,\zeta)}{IR_{\lambda,l}^{m}f(z,\zeta)}, \quad z \in U,\ \zeta \in \overline{U},$$

where $q(z,\zeta) = \frac{1}{z}\int_0^z h(t,\zeta)\,dt$. The function q is convex and it is the best subordinant.

Proof. Consider

$$\begin{aligned} p(z,\zeta) &= \frac{IR_{\lambda,l}^{m+1}f(z,\zeta)}{IR_{\lambda,l}^{m}f(z,\zeta)} \\ &= \frac{z+\sum_{j=2}^{\infty}\left(\frac{1+\lambda(j-1)+l}{l+1}\right)^{m+1} C_{m+j}^{m+1} a_j^2(\zeta)\, z^j}{z+\sum_{j=2}^{\infty}\left(\frac{1+\lambda(j-1)+l}{l+1}\right)^{m} C_{m+j-1}^{m} a_j^2(\zeta)\, z^j} \\ &= \frac{1+\sum_{j=2}^{\infty}\left(\frac{1+\lambda(j-1)+l}{l+1}\right)^{m+1} C_{m+j}^{m+1} a_j^2(\zeta)\, z^{j-1}}{1+\sum_{j=2}^{\infty}\left(\frac{1+\lambda(j-1)+l}{l+1}\right)^{m} C_{m+j-1}^{m} a_j^2(\zeta)\, z^{j-1}}. \end{aligned}$$

Evidently, $p \in \mathcal{H}^*[1,1,\zeta]$.
We have $p_z'(z,\zeta) = \frac{\left(IR_{\lambda,l}^{m+1}f(z,\zeta)\right)_z'}{IR_{\lambda,l}^{m}f(z,\zeta)} - p(z,\zeta)\cdot\frac{\left(IR_{\lambda,l}^{m}f(z,\zeta)\right)_z'}{IR_{\lambda,l}^{m}f(z,\zeta)}$. Then $p(z,\zeta) + zp_z'(z,\zeta) = \left(\frac{zIR_{\lambda,l}^{m+1}f(z,\zeta)}{IR_{\lambda,l}^{m}f(z,\zeta)}\right)_z'$.

Then (4.2.2.10) becomes

$$h(z,\zeta) \prec\prec p(z,\zeta) + zp_z'(z,\zeta), \quad z \in U,\ \zeta \in \overline{U}.$$

By using Lemma 1.10 for $n=1$ and $\gamma=1$, we have

$$q(z,\zeta) \prec\prec p(z,\zeta), \quad \text{i.e.} \quad q(z,\zeta) \prec\prec \frac{IR_{\lambda,l}^{m+1}f(z,\zeta)}{IR_{\lambda,l}^{m}f(z,\zeta)}, \quad z \in U,\ \zeta \in \overline{U},$$

where $q(z,\zeta) = \frac{1}{z}\int_0^z h(t,\zeta)\,dt$. The function q is convex and it is the best subordinant. □

Corollary 4.9 (Alb Lupaş, 2012i). *Let $h(z,\zeta) = \frac{\zeta+(2\beta-\zeta)z}{1+z}$ be a convex function in $U \times \overline{U}$, where $0 \leq \beta < 1$. Let $\lambda, l \geq 0$, $m \in \mathbb{N}$, $f(z,\zeta) \in \mathcal{A}_\zeta^*$ and suppose that $\left(\frac{zIR_{\lambda,l}^{m+1} f(z,\zeta)}{IR_{\lambda,l}^m f(z,\zeta)}\right)_z'$ is univalent, $\frac{IR_{\lambda,l}^{m+1} f(z,\zeta)}{IR_{\lambda,l}^m f(z,\zeta)} \in \mathcal{H}^*[1,1,\zeta] \cap Q^*$. If*

$$h(z,\zeta) \prec\prec \left(\frac{zIR_{\lambda,l}^{m+1} f(z,\zeta)}{IR_{\lambda,l}^m f(z,\zeta)}\right)_z', \quad z \in U, \ \zeta \in \overline{U}, \tag{4.2.2.11}$$

then

$$q(z,\zeta) \prec\prec \frac{IR_{\lambda,l}^{m+1} f(z,\zeta)}{IR_{\lambda,l}^m f(z,\zeta)}, \quad z \in U, \ \zeta \in \overline{U},$$

where q is given by $q(z,\zeta) = 2\beta - \zeta + \frac{2(\zeta-\beta)}{z} \ln(1+z)$, $z \in U$, $\zeta \in \overline{U}$. The function q is convex and it is the best subordinant.

Proof. Following the same steps as in the proof of Theorem 4.27 and considering $p(z,\zeta) = \frac{IR_{\lambda,l}^{m+1} f(z,\zeta)}{IR_{\lambda,l}^m f(z,\zeta)}$, the strong differential superordination (4.2.2.11) becomes

$$h(z,\zeta) = \frac{\zeta + (2\beta-\zeta)z}{1+z} \prec\prec p(z,\zeta) + zp_z'(z,\zeta), \quad z \in U, \ \zeta \in \overline{U}.$$

By using Lemma 1.10 for $n = 1$ and $\gamma = 1$, we have $q(z,\zeta) \prec\prec p(z,\zeta)$, i.e.

$$q(z,\zeta) = \frac{1}{z}\int_0^z h(t,\zeta)\,dt = \frac{1}{z}\int_0^z \frac{\zeta + (2\beta-\zeta)t}{1+t}dt$$

$$= 2\beta - \zeta + \frac{2(\zeta-\beta)}{z}\ln(1+z) \prec\prec \frac{IR_{\lambda,l}^{m+1} f(z,\zeta)}{IR_{\lambda,l}^m f(z,\zeta)}, \quad z \in U, \ \zeta \in \overline{U}.$$

The function q is convex and it is the best subordinant. □

Theorem 4.28 (Alb Lupaş, 2012i). *Let $q(z,\zeta)$ be convex in $U \times \overline{U}$ and let h be defined by $h(z,\zeta) = q(z,\zeta) + zq_z'(z,\zeta)$. If $\lambda, l \geq 0$, $m \in \mathbb{N}$, $f(z,\zeta) \in \mathcal{A}_\zeta$, suppose that $\left(\frac{zIR_{\lambda,l}^{m+1} f(z,\zeta)}{IR_{\lambda,l}^m f(z,\zeta)}\right)_z'$ is univalent, $\frac{IR_{\lambda,l}^{m+1} f(z,\zeta)}{IR_{\lambda,l}^m f(z,\zeta)} \in \mathcal{H}^*[1,1,\zeta] \cap Q^*$ and satisfies the strong differential superordination*

$$h(z,\zeta) = q(z,\zeta) + zq_z'(z,\zeta) \prec\prec \left(\frac{zIR_{\lambda,l}^{m+1} f(z,\zeta)}{IR_{\lambda,l}^m f(z,\zeta)}\right)_z', \quad z \in U, \ \zeta \in \overline{U}, \tag{4.2.2.12}$$

then

$$q(z,\zeta) \prec\prec \frac{IR_{\lambda,l}^{m+1} f(z,\zeta)}{IR_{\lambda,l}^{m} f(z,\zeta)}, \quad z \in U, \ \zeta \in \overline{U},$$

where $q(z,\zeta) = \frac{1}{z}\int_0^z h(t,\zeta)\, dt$. *The function* q *is the best subordinant.*

Proof. Let

$$p(z,\zeta) = \frac{IR_{\lambda,l}^{m+1} f(z,\zeta)}{IR_{\lambda,l}^{m} f(z,\zeta)}$$

$$= \frac{z + \sum_{j=2}^{\infty} \left(\frac{1+\lambda(j-1)+l}{l+1}\right)^{m+1} C_{m+j}^{m+1} a_j^2(\zeta)\, z^j}{z + \sum_{j=2}^{\infty} \left(\frac{1+\lambda(j-1)+l}{l+1}\right)^{m} C_{m+j-1}^{m} a_j^2(\zeta)\, z^j}$$

$$= \frac{1 + \sum_{j=2}^{\infty} \left(\frac{1+\lambda(j-1)+l}{l+1}\right)^{m+1} C_{m+j}^{m+1} a_j^2(\zeta)\, z^{j-1}}{1 + \sum_{j=2}^{\infty} \left(\frac{1+\lambda(j-1)+l}{l+1}\right)^{m} C_{m+j-1}^{m} a_j^2(\zeta)\, z^{j-1}}.$$

Evidently, $p \in \mathcal{H}^*[1,1,\zeta]$.

Differentiating with respect to z, we obtain $p(z,\zeta) + zp_z'(z,\zeta) = \left(\frac{zIR_{\lambda,l}^{m+1} f(z,\zeta)}{IR_{\lambda,l}^{m} f(z,\zeta)}\right)_z'$, $z \in U$, $\zeta \in \overline{U}$, and (4.2.2.12) becomes

$$q(z,\zeta) + zq_z'(z,\zeta) \prec\prec p(z,\zeta) + zp_z'(z,\zeta), \quad z \in U, \ \zeta \in \overline{U}.$$

Using Lemma 1.11 for $n = 1$ and $\gamma = 1$, we have

$$q(z,\zeta) \prec\prec p(z,\zeta), \quad z \in U, \ \zeta \in \overline{U}, \text{ i.e.}$$

$$q(z,\zeta) = \frac{1}{z}\int_0^z h(t,\zeta)\, dt \prec\prec \frac{IR_{\lambda,l}^{m+1} f(z,\zeta)}{IR_{\lambda,l}^{m} f(z,\zeta)}, \ z \in U, \ \zeta \in \overline{U},$$

and q is the best subordinant. □

Theorem 4.29 (Alb Lupaş, 2012i). *Let* $h(z,\zeta)$ *be a convex function,* $h(0,\zeta) = 1$. *Let* $\lambda, l \geq 0$, $m \in \mathbb{N}$, $f(z,\zeta) \in \mathcal{A}_\zeta^*$ *and suppose that* $\frac{l+1}{[\lambda(l-m+2)-(l+1)]z} \cdot \left[(m+1)\, IR_{\lambda,l}^{m+1} f(z,\zeta) - (m-2)\, IR_{\lambda,l}^{m} f(z,\zeta)\right] + \left(1 - \frac{l+1}{\lambda(l-m+2)-(l+1)}\right) - \frac{2(l+1)(m-1)-2\lambda m}{\lambda(l-m+2)-(l+1)} \int_0^z \frac{IR_{\lambda,l}^{m} f(t,\zeta)-t}{t^2} dt$ *is univalent and*

$\left(IR_{\lambda,l}^{m}f(z,\zeta)\right)_z' \in \mathcal{H}^*[1,1,\zeta] \cap Q^*$. *If*

$$
\begin{aligned}
&h(z,\zeta)\\
&\quad \prec\prec \frac{l+1}{[\lambda(l-m+2)-(l+1)]\,z}\\
&\quad\quad \times \left[(m+1)\,IR_{\lambda,l}^{m+1}f(z,\zeta)-(m-2)\,IR_{\lambda,l}^{m}f(z,\zeta)\right]\\
&\quad\quad + \left(1-\frac{l+1}{\lambda(l-m+2)-(l+1)}\right)\\
&\quad\quad - \frac{2(l+1)(m-1)-2\lambda m}{\lambda(l-m+2)-(l+1)}\int_0^z \frac{IR_{\lambda,l}^{m}f(t,\zeta)-t}{t^2}\,dt, \quad z\in U,\ \zeta\in\overline{U},
\end{aligned}
\tag{4.2.2.13}
$$

then

$$q(z,\zeta) \prec\prec \left(IR_{\lambda,l}^{m}f(z,\zeta)\right)_z', \quad z\in U,\ \zeta\in\overline{U},$$

where $q(z,\zeta) = \frac{\lambda(l-m+2)-(l+1)}{\lambda(l+1)z^{\frac{\lambda(l-m+2)-(l+1)}{\lambda(l+1)}}}\int_0^z h(t,\zeta)\,t^{\frac{\lambda(1-m)-(l+1)}{\lambda(l+1)}}dt$. *The function q is convex and it is the best subordinant.*

Proof. With notation

$$
\begin{aligned}
p(z,\zeta) &= \left(IR_{\lambda,l}^{m}f(z,\zeta)\right)_z'\\
&= 1+\sum_{j=2}^{\infty}\left(\frac{1+\lambda(j-1)+l}{l+1}\right)^m C_{m+j-1}^{m}\, j a_j^2(\zeta)\, z^{j-1}
\end{aligned}
$$

and $p(0,\zeta)=1$, we obtain for $f(z,\zeta) = z+\sum_{j=2}^{\infty} a_j(\zeta)\,z^j$,

$$
\begin{aligned}
&p(z,\zeta)+zp_z'(z,\zeta)\\
&\quad = 1+\sum_{j=2}^{\infty}\left(\frac{1+\lambda(j-1)+l}{l+1}\right)^m C_{m+j-1}^{m}\, j a_j^2(\zeta)\, z^{j-1}\\
&\quad\quad + \sum_{j=2}^{\infty}\left(\frac{1+\lambda(j-1)+l}{l+1}\right)^m C_{m+j-1}^{m}\, j(j-1)\, a_j^2(\zeta)\, z^{j-1}\\
&\quad = \frac{1}{z}\left(\frac{m+1}{\lambda} IR_{\lambda,l}^{m+1}f(z,\zeta) - \frac{m-2}{\lambda} IR_{\lambda,l}^{m}f(z,\zeta)\right)
\end{aligned}
$$

$$+\frac{\lambda(m-1)-(l+1)}{\lambda(l+1)}\left(IR_{\lambda,l}^{m}f(z,\zeta)\right)_{z}'+\left(1-\frac{m-1}{l+1}-\frac{2}{\lambda}\right)$$
$$-\frac{2(l+1)(m-1)-2\lambda m}{\lambda(l+1)}\int_{0}^{z}\frac{IR_{\lambda,l}^{m}f(t,\zeta)-t}{t^{2}}dt.$$

Therefore,

$$p(z,\zeta)+\frac{\lambda(l+1)}{\lambda(l-m+2)-(l+1)}zp_{z}'(z,\zeta)$$
$$=\frac{l+1}{[\lambda(l-m+2)-(l+1)]z}\Big[(m+1)IR_{\lambda,l}^{m+1}f(z,\zeta)$$
$$-(m-2)IR_{\lambda,l}^{m}f(z,\zeta)\Big]+\left(1-\frac{l+1}{\lambda(l-m+2)-(l+1)}\right)$$
$$-\frac{2(l+1)(m-1)-2\lambda m}{\lambda(l-m+2)-(l+1)}\int_{0}^{z}\frac{IR_{\lambda,l}^{m}f(t,\zeta)-t}{t^{2}}dt.$$

Then (4.2.2.13) becomes

$$h(z,\zeta)\prec\prec p(z,\zeta)+\frac{\lambda(l+1)}{\lambda(l-m+2)-(l+1)}zp_{z}'(z,\zeta),\quad z\in U,\ \zeta\in\overline{U}.$$

By using Lemma 1.10 for $n=1$ and $\gamma=1-\frac{m-1}{l+1}-\frac{1}{\lambda}$, we have

$$q(z,\zeta)\prec\prec p(z,\zeta),\ \ z\in U,\ \ \zeta\in\overline{U},\quad \text{i.e. } q(z,\zeta)\prec\prec\left(IR_{\lambda,l}^{m}f(z,\zeta)\right)_{z}',$$
$$z\in U,\ \ \zeta\in\overline{U},$$

where $q(z,\zeta)=\frac{\lambda(l-m+2)-(l+1)}{\lambda(l+1)z^{\frac{\lambda(l-m+2)-(l+1)}{\lambda(l+1)}}}\int_0^z h(t,\zeta)\,t^{\frac{\lambda(1-m)-(l+1)}{\lambda(l+1)}}dt$. The function q is convex and it is the best subordinant. □

Corollary 4.10 (Alb Lupaş, 2012i). *Let* $h(z,\zeta)=\frac{\zeta+(2\beta-\zeta)z}{1+z}$ *be a convex function in* $U\times\overline{U}$*, where* $0\le\beta<1$*. Let* $\lambda,l\ge 0$*,* $m\in\mathbb{N}$*,* $f(z,\zeta)\in\mathcal{A}_{\zeta}^{*}$ *and suppose that* $\frac{l+1}{[\lambda(l-m+2)-(l+1)]z}\cdot\Big[(m+1)IR_{\lambda,l}^{m+1}f(z,\zeta)-(m-2)IR_{\lambda,l}^{m}f(z,\zeta)\Big]+\left(1-\frac{l+1}{\lambda(l-m+2)-(l+1)}\right)-\frac{2(l+1)(m-1)-2\lambda m}{\lambda(l-m+2)-(l+1)}\int_0^z\frac{IR_{\lambda,l}^{m}f(t,\zeta)-t}{t^2}dt$ *is univalent,* $\left(IR_{\lambda,l}^{m}f(z,\zeta)\right)_{z}'\in\mathcal{H}^{*}[1,1,\zeta]\cap Q^{*}$*. If*

$$h(z,\zeta)$$
$$\prec\prec\frac{l+1}{[\lambda(l-m+2)-(l+1)]z}\Big[(m+1)IR_{\lambda,l}^{m+1}f(z,\zeta)$$

$$- (m-2)\, IR^m_{\lambda,l} f(z,\zeta)\Big] + \left(1 - \frac{l+1}{\lambda(l-m+2)-(l+1)}\right)$$

$$- \frac{2(l+1)(m-1)-2\lambda m}{\lambda(l-m+2)-(l+1)} \int_0^z \frac{IR^m_{\lambda,l} f(t,\zeta) - t}{t^2} dt, \quad z \in U, \ \zeta \in \overline{U}, \tag{4.2.2.14}$$

then

$$q(z,\zeta) \prec\prec \left(IR^m_{\lambda,l} f(z,\zeta)\right)'_z, \quad z \in U, \ \zeta \in \overline{U},$$

where q is given by $q(z,\zeta) = 2\beta - \zeta + 2(\zeta-\beta) \frac{\lambda(l-m+2)-(l+1)}{\lambda(l+1) z^{\frac{\lambda(l-m+2)-(l+1)}{\lambda(l+1)}}} \int_0^z \frac{t^{\frac{\lambda(1-m)-(l+1)}{\lambda(l+1)}}}{1+t} dt$, $z \in U$, $\zeta \in \overline{U}$. *The function q is convex and it is the best subordinant.*

Proof. Following the same steps as in the proof of Theorem 4.29 and considering $p(z,\zeta) = \left(IR^m_{\lambda,l} f(z,\zeta)\right)'_z$, the strong differential superordination (4.2.2.14) becomes

$$h(z,\zeta) = \frac{\zeta + (2\beta-\zeta)z}{1+z}$$

$$\prec\prec p(z,\zeta) + \frac{\lambda(l+1)}{\lambda(l-m+2)-(l+1)} z p'_z(z,\zeta), \quad z \in U, \ \zeta \in \overline{U}.$$

By using Lemma 1.10 for $n = 1$ and $\gamma = \frac{\lambda(l-m+2)-(l+1)}{\lambda(l+1)}$, we have $q(z,\zeta) \prec\prec p(z,\zeta)$, i.e.

$$q(z,\zeta) = \frac{\lambda(l-m+2)-(l+1)}{\lambda(l+1) z^{\frac{\lambda(l-m+2)-(l+1)}{\lambda(l+1)}}} \int_0^z h(t,\zeta)\, t^{\frac{\lambda(1-m)-(l+1)}{\lambda(l+1)}} dt$$

$$= \frac{\lambda(l-m+2)-(l+1)}{\lambda(l+1) z^{\frac{\lambda(l-m+2)-(l+1)}{\lambda(l+1)}}} \int_0^z t^{\frac{\lambda(1-m)-(l+1)}{\lambda(l+1)}} \frac{\zeta + (2\beta-\zeta)t}{1+t} dt$$

$$= 2\beta - \zeta + 2(\zeta-\beta) \frac{\lambda(l-m+2)-(l+1)}{\lambda(l+1) z^{\frac{\lambda(l-m+2)-(l+1)}{\lambda(l+1)}}} \int_0^z \frac{t^{\frac{\lambda(1-m)-(l+1)}{\lambda(l+1)}}}{1+t} dt$$

$$\prec\prec \left(IR^m_{\lambda,l} f(z,\zeta)\right)'_z, \quad z \in U, \zeta \in \overline{U}.$$

The function q is convex and it is the best subordinant. □

Theorem 4.30 (Alb Lupaş, 2012i). *Let $q(z,\zeta)$ be convex in $U \times \overline{U}$ and let h be defined by $h(z,\zeta) = q(z,\zeta) + \frac{\lambda(l+1)}{\lambda(l-m+2)-(l+1)} z q_z'(z,\zeta)$, $\lambda, l \geq 0$, $m \in \mathbb{N}$. If $f(z,\zeta) \in \mathcal{A}_\zeta^*$, suppose that $\frac{l+1}{[\lambda(l-m+2)-(l+1)]z}\Big[(m+1) IR_{\lambda,l}^{m+1} f(z,\zeta) - (m-2) IR_{\lambda,l}^m f(z,\zeta)\Big] + \left(1 - \frac{l+1}{\lambda(l-m+2)-(l+1)}\right) - \frac{2(l+1)(m-1)-2\lambda m}{\lambda(l-m+2)-(l+1)} \int_0^z \frac{IR_{\lambda,l}^m f(t,\zeta)-t}{t^2} dt$ is univalent, $\left(IR_{\lambda,l}^m f(z,\zeta)\right)_z' \in \mathcal{H}^*[1,1,\zeta] \cap Q^*$ and satisfies the strong differential superordination*

$$\begin{aligned} h(z,\zeta) &= q(z,\zeta) + \frac{\lambda(l+1)}{\lambda(l-m+2)-(l+1)} z q_z'(z,\zeta) \\ &\prec\prec \frac{l+1}{[\lambda(l-m+2)-(l+1)]z}\Big[(m+1) IR_{\lambda,l}^{m+1} f(z,\zeta) \\ &\quad - (m-2) IR_{\lambda,l}^m f(z,\zeta)\Big] + \left(1 - \frac{l+1}{\lambda(l-m+2)-(l+1)}\right) \\ &\quad - \frac{2(l+1)(m-1)-2\lambda m}{\lambda(l-m+2)-(l+1)} \int_0^z \frac{IR_{\lambda,l}^m f(t,\zeta)-t}{t^2} dt, \quad z \in U,\ \zeta \in \overline{U}, \end{aligned} \tag{4.2.2.15}$$

then

$$q(z,\zeta) \prec\prec \left(IR_{\lambda,l}^m f(z,\zeta)\right)_z', \quad z \in U,\ \zeta \in \overline{U},$$

where $q(z,\zeta) = \frac{\lambda(l-m+2)-(l+1)}{\lambda(l+1) z^{\frac{\lambda(l-m+2)-(l+1)}{\lambda(l+1)}}} \int_0^z h(t,\zeta) t^{\frac{\lambda(1-m)-(l+1)}{\lambda(l+1)}} dt$. The function q is the best subordinant.

Proof. Let

$$\begin{aligned} p(z,\zeta) &= \left(IR_{\lambda,l}^m f(z,\zeta)\right)_z' \\ &= 1 + \sum_{j=2}^{\infty} \left(\frac{1+\lambda(j-1)+l}{l+1}\right)^m C_{m+j-1}^m j a_j^2(\zeta) z^{j-1}. \end{aligned}$$

Differentiating with respect to z, we obtain

$$\begin{aligned} &p(z,\zeta) + z p_z'(z,\zeta) \\ &= \frac{1}{z}\left(\frac{m+1}{\lambda} IR_{\lambda,l}^{m+1} f(z,\zeta) - \frac{m-2}{\lambda} IR_{\lambda,l}^m f(z,\zeta)\right) + \frac{\lambda(m-1)-(l+1)}{\lambda(l+1)} \\ &\quad \times \left(IR_{\lambda,l}^m f(z,\zeta)\right)_z' + \left(1 - \frac{m-1}{l+1} - \frac{2}{\lambda}\right) - \frac{2(l+1)(m-1)-2\lambda m}{\lambda(l+1)} \\ &\quad \times \int_0^z \frac{IR_{\lambda,l}^m f(t,\zeta)-t}{t^2} dt \end{aligned}$$

and

$$p(z,\zeta)+\frac{\lambda(l+1)}{\lambda(l-m+2)-(l+1)}zp'_z(z,\zeta)$$
$$=\frac{l+1}{[\lambda(l-m+2)-(l+1)]z}\Big[(m+1)IR^{m+1}_{\lambda,l}f(z,\zeta)$$
$$-(m-2)IR^m_{\lambda,l}f(z,\zeta)\Big]+\left(1-\frac{l+1}{\lambda(l-m+2)-(l+1)}\right)$$
$$-\frac{2(l+1)(m-1)-2\lambda m}{\lambda(l-m+2)-(l+1)}\int_0^z\frac{IR^m_{\lambda,l}f(t,\zeta)-t}{t^2}dt,\quad z\in U,\ \zeta\in\overline{U},$$

and (4.2.2.15) becomes

$$q(z,\zeta)+\frac{\lambda(l+1)}{\lambda(l-m+2)-(l+1)}zq'_z(z,\zeta)$$
$$\prec\prec p(z,\zeta)+\frac{\lambda(l+1)}{\lambda(l-m+2)-(l+1)}zp'_z(z,\zeta),\quad z\in U,\ \zeta\in\overline{U}.$$

Using Lemma 1.11 for $n=1$ and $\gamma=1-\frac{m-1}{l+1}-\frac{1}{\lambda}$, we have $q(z,\zeta)\prec\prec p(z,\zeta)$, $z\in U$, $\zeta\in\overline{U}$, i.e.

$$q(z,\zeta)=\frac{\lambda(l-m+2)-(l+1)}{\lambda(l+1)z^{\frac{\lambda(l-m+2)-(l+1)}{\lambda(l+1)}}}\int_0^z h(t,\zeta)t^{\frac{\lambda(1-m)-(l+1)}{\lambda(l+1)n}}dt$$
$$\prec\prec\left(IR^m_{\lambda,l}f(z,\zeta)\right)'_z,\quad z\in U,\ \zeta\in\overline{U},$$

and q is the best subordinant. □

4.3 Applications of Differential Operators in the Theory of Strong Differential Subordinations and Superordinations

In this section, we present several sandwich-type results involving few of the defined operators presented in Chapter 2.

4.3.1 *Strong Differential Subordinations and Superordinations for Riemann–Liouville Fractional Integral of Extended q-Hypergeometric Function*

The notions of strong differential subordination and its dual, strong differential superordination, have been introduced as extensions of the classical differential subordination and superordination concepts, respectively. The dual theories have developed beautifully, important results being obtained involving different types of operators and certain hypergeometric functions. In the paper Alb Lupaş and Oros (2023a), quantum calculus and fractional calculus aspects are added to the study. The well-known q-hypergeometric function is given an extended form to fit the study of classes of functions specific to strong differential subordination and superordination theories. Riemann–Liouville fractional integral of extended q-hypergeometric function is defined here and it is involved in the investigation of strong differential subordinations and superordinations. The best dominants and the best subordinants are provided in the theorems proved for the strong differential subordinations and superordinations, respectively. For particular functions considered due to their remarkable geometric properties as best dominant or best subordinant, interesting corollaries are stated. The study is concluded by connecting the results obtained using the dual theories through sandwich-type theorems and corollaries.

The Riemann–Liouville fractional integral of the extended q-confluent hypergeometric function introduced in Definition 2.21 is

$$
\begin{aligned}
D_z^{-\alpha}\phi\left(m\left(\zeta\right),n\left(\zeta\right);q,z,\zeta\right) &= \frac{1}{\Gamma\left(\alpha\right)}\int_0^z \frac{\phi\left(m\left(\zeta\right),n\left(\zeta\right);q,t,\zeta\right)}{\left(z-t\right)^{1-\alpha}}dt \\
&= \frac{1}{\Gamma\left(\alpha\right)}\sum_{j=0}^{\infty}\frac{\left(m\left(\zeta\right),q\right)_j}{\left(q,q\right)_j\left(n\left(\zeta\right),q\right)_j}\int_0^z \frac{t^j}{\left(z-t\right)^{1-\alpha}}dt,
\end{aligned}
\tag{4.3.1.1}
$$

where the q-hypergeometric function $\phi\left(m\left(\zeta\right),n\left(\zeta\right);q,z,\zeta\right)$ is defined by

$$
\phi\left(m\left(\zeta\right),n\left(\zeta\right);q,z,\zeta\right) = \sum_{j=0}^{\infty}\frac{\left(m\left(\zeta\right),q\right)_j}{\left(q,q\right)_j\left(n\left(\zeta\right),q\right)_j}z^j,
$$

with

$$
\left(m\left(\zeta\right),q\right)_j = \begin{cases} 1, \; j=0, \\ \left(1-m\left(\zeta\right)\right)\left(1-m\left(\zeta\right)q\right)\left(1-m\left(\zeta\right)q^2\right)\ldots\left(1-m\left(\zeta\right)q^{j-1}\right), \\ \quad j\in\mathbb{N}, \end{cases}
$$

$$
\left(d\right)_k = \frac{\Gamma\left(d+k\right)}{\Gamma\left(d\right)} = d\left(d+1\right)\left(d+2\right)\ldots\left(d+k-1\right) \quad \text{and} \quad \left(d\right)_0 = 1,
$$

and $m\left(\zeta\right),n\left(\zeta\right)$ are holomorphic functions depending on the parameter $\zeta\in\overline{U}$, $\alpha>0$, $0<q<1$.

After some calculations, it can be written using the following form

$$D_z^{-\alpha}\phi\left(m\left(\zeta\right),n\left(\zeta\right);q,z,\zeta\right)=\sum_{j=0}^{\infty}\frac{\left(m\left(\zeta\right),q\right)_j}{\left(q,q\right)_j\left(n\left(\zeta\right),q\right)_j\left(j+1\right)_\alpha}z^{\alpha+j}, \tag{4.3.1.2}$$

and $D_z^{-\alpha}\phi\left(m\left(\zeta\right),n\left(\zeta\right);q,z,\zeta\right)\in\mathcal{H}\left[0,\alpha,\zeta\right]$.

The first new result obtained concerns the study of a strong differential subordination obtained by using the Riemann–Liouville fractional integral of the extended q-hypergeometric function for which the best dominant is provided.

Theorem 4.31 (Alb Lupaş and Oros, 2023a). *Let $g\left(z,\zeta\right)$ be a univalent function in $U\times\overline{U}$ such that $g\left(z,\zeta\right)\neq 0$, for all $z\in U\backslash\{0\}$, $\zeta\in\overline{U}$ and $\frac{z\left(D_z^{-\alpha}\phi(m(\zeta),n(\zeta);q,z,\zeta)\right)'}{D_z^{-\alpha}\phi(m(\zeta),n(\zeta);q,z,\zeta)}\in\mathcal{H}\left(U\times\overline{U}\right)$, where $\alpha>0$, $0<q<1$. Assume that $\frac{zg_z'(z,\zeta)}{g(z,\zeta)}$ is starlike univalent function in $U\times\overline{U}$ and*

$$\mathrm{Re}\left(\frac{\beta}{\psi}g\left(z,\zeta\right)+\frac{2\delta}{\psi}g^2\left(z,\zeta\right)+1-z\frac{g_z'\left(z,\zeta\right)}{g\left(z,\zeta\right)}+z\frac{g_{z^2}''\left(z,\zeta\right)}{g_z'\left(z,\zeta\right)}\right)>0, \tag{4.3.1.3}$$

for $\varepsilon,\beta,\psi,\delta\in\mathbb{C}$, $\psi\neq 0$, $z\in U\backslash\{0\}$, $\zeta\in\overline{U}$ and

$$\begin{aligned}\Psi_\alpha^q\left(\varepsilon,\beta,\delta,\psi;z,\zeta\right):=&\ \varepsilon+\psi+\left(\beta-\psi\right)\frac{z\left(D_z^{-\alpha}\phi\left(m\left(\zeta\right),n\left(\zeta\right);q,z,\zeta\right)\right)_z'}{D_z^{-\alpha}\phi\left(m\left(\zeta\right),n\left(\zeta\right);q,z,\zeta\right)}\\&+\delta\left(\frac{z\left(D_z^{-\alpha}\phi\left(m\left(\zeta\right),n\left(\zeta\right);q,z,\zeta\right)\right)_z'}{D_z^{-\alpha}\phi\left(m\left(\zeta\right),n\left(\zeta\right);q,z,\zeta\right)}\right)^2\\&+\delta\frac{z\left(D_z^{-\alpha}\phi\left(m\left(\zeta\right),n\left(\zeta\right);q,z,\zeta\right)\right)_{z^2}''}{\left(D_z^{-\alpha}\phi\left(m\left(\zeta\right),n\left(\zeta\right);q,z,\zeta\right)\right)_z'}.\end{aligned} \tag{4.3.1.4}$$

When g verifies the strong differential subordination

$$\Psi_\alpha^q\left(\varepsilon,\beta,\delta,\psi;z,\zeta\right)\prec\prec\varepsilon+\beta g\left(z,\zeta\right)+\delta\left(g\left(z,\zeta\right)\right)^2+\psi\frac{zg_z'\left(z,\zeta\right)}{g\left(z,\zeta\right)}, \tag{4.3.1.5}$$

for $\varepsilon,\beta,\psi,\delta\in\mathbb{C}$, $\psi\neq 0$, then

$$\frac{z\left(D_z^{-\alpha}\phi\left(m\left(\zeta\right),n\left(\zeta\right);q,z,\zeta\right)\right)_z'}{D_z^{-\alpha}\phi\left(m\left(\zeta\right),n\left(\zeta\right);q,z,\zeta\right)}\prec\prec g\left(z,\zeta\right),\quad\left(z,\zeta\right)\in U\times\overline{U}, \tag{4.3.1.6}$$

and the best dominant is the function g.

Proof. Define $p(z,\zeta) := \frac{z\left(D_z^{-\alpha}\phi(m(\zeta),n(\zeta);q,z,\zeta)\right)'_z}{D_z^{-\alpha}\phi(m(\zeta),n(\zeta);q,z,\zeta)}$, $(z,\zeta) \in (U\backslash\{0\}) \times \overline{U}$, and differentiating it with respect to z, we get

$$p'_z(z,\zeta) = \frac{\left(D_z^{-\alpha}\phi(m(\zeta),n(\zeta);q,z,\zeta)\right)'_z}{D_z^{-\alpha}\phi(m(\zeta),n(\zeta);q,z,\zeta)} - z\left(\frac{\left(D_z^{-\alpha}\phi(m(\zeta),n(\zeta);q,z,\zeta)\right)'_z}{D_z^{-\alpha}\phi(m(\zeta),n(\zeta);q,z,\zeta)}\right)^2 + z\frac{\left(D_z^{-\alpha}\phi(m(\zeta),n(\zeta);q,z,\zeta)\right)''_{z^2}}{D_z^{-\alpha}\phi(m(\zeta),n(\zeta);q,z,\zeta)}$$

and

$$\frac{zp'_z(z,\zeta)}{p(z,\zeta)} = 1 - z\frac{\left(D_z^{-\alpha}\phi(m(\zeta),n(\zeta);q,z,\zeta)\right)'_z}{D_z^{-\alpha}\phi(m(\zeta),n(\zeta);q,z,\zeta)} + z\frac{\left(D_z^{-\alpha}\phi(m(\zeta),n(\zeta);q,z,\zeta)\right)''_{z^2}}{\left(D_z^{-\alpha}\phi(m(\zeta),n(\zeta);q,z,\zeta)\right)'_z}. \tag{4.3.1.7}$$

Let $\theta(u) = \delta u^2 + \beta u + \varepsilon$ analytic in $\mathbb{C}$, and $\eta(u) = \frac{\psi}{u}$, analytic in $\mathbb{C}\backslash\{0\}$ with $\eta(u) \neq 0$, $u \in \mathbb{C}\backslash\{0\}$, we consider the starlike univalent functions $G(z,\zeta) = z\eta(g(z,\zeta))g'_z(z,\zeta) = \psi\frac{zg'_z(z,\zeta)}{g(z,\zeta)}$ and $h(z,\zeta) = G(z,\zeta) + \theta(g(z,\zeta)) = \varepsilon + \beta g(z,\zeta) + \delta(g(z,\zeta))^2 + \psi\frac{zg'_z(z,\zeta)}{g(z,\zeta)}$.

Differentiating it with respect to z, we get $h'_z(z,\zeta) = \beta g'_z(z,\zeta) + 2\delta g(z,\zeta)g'_z(z,\zeta) + \psi\frac{g'_z(z,\zeta)}{g(z,\zeta)} - \psi z\left(\frac{g'_z(z,\zeta)}{g(z,\zeta)}\right)^2 + \psi z\frac{g''_{z^2}(z,\zeta)}{g(z,\zeta)}$ and $\frac{zh'_z(z,\zeta)}{G(z,\zeta)} = \frac{\beta}{\psi}g(z,\zeta) + \frac{2\delta}{\psi}g^2(z,\zeta) + 1 - z\frac{g'_z(z,\zeta)}{g(z,\zeta)} + z\frac{g''_{z^2}(z,\zeta)}{g(z,\zeta)}$, therefore we have

$$\operatorname{Re}\left(\frac{zh'_z(z,\zeta)}{G(z,\zeta)}\right) = \operatorname{Re}\left(\frac{\beta}{\psi}g(z,\zeta) + \frac{2\delta}{\psi}g^2(z,\zeta) + 1 - z\frac{g'_z(z,\zeta)}{g(z,\zeta)} + z\frac{g''_{z^2}(z,\zeta)}{g(z,\zeta)}\right) > 0$$

by relation (4.3.1.3).

Using relation (4.3.1.7), we can write

$$\varepsilon + \beta p(z,\zeta) + \delta(p(z,\zeta))^2 + \psi\frac{zp'_z(z,\zeta)}{p(z,\zeta)} = \varepsilon + \psi + (\beta - \psi)\frac{z\left(D_z^{-\alpha}\phi(m(\zeta),n(\zeta);q,z,\zeta)\right)'_z}{D_z^{-\alpha}\phi(m(\zeta),n(\zeta);q,z,\zeta)}$$

$$+\delta\left(\frac{z\left(D_z^{-\alpha}\phi\left(m\left(\zeta\right),n\left(\zeta\right);q,z,\zeta\right)\right)_z'}{D_z^{-\alpha}\phi\left(m\left(\zeta\right),n\left(\zeta\right);q,z,\zeta\right)}\right)^2$$

$$+\delta\frac{z\left(D_z^{-\alpha}\phi\left(m\left(\zeta\right),n\left(\zeta\right);q,z,\zeta\right)\right)_{z^2}''}{\left(D_z^{-\alpha}\phi\left(m\left(\zeta\right),n\left(\zeta\right);q,z,\zeta\right)\right)_z'}.$$

Taking account the strong differential subordination (4.3.1.5), we get $\varepsilon+\beta p\left(z,\zeta\right)+\delta\left(p\left(z,\zeta\right)\right)^2+\psi\frac{zp_z'(z,\zeta)}{p(z,\zeta)} \prec\prec \varepsilon+\beta g\left(z,\zeta\right)+\delta\left(g\left(z,\zeta\right)\right)^2+\psi\frac{zg_z'(z,\zeta)}{g(z,\zeta)}$ and applying Lemma 1.9, we obtain $p\left(z,\zeta\right) \prec\prec g\left(z,\zeta\right)$, $(z,\zeta) \in U \times \overline{U}$, equivalently with $\frac{z\left(D_z^{-\alpha}\phi(m(\zeta),n(\zeta);q,z,\zeta)\right)_z'}{D_z^{-\alpha}\phi(m(\zeta),n(\zeta);q,z,\zeta)} \prec\prec g\left(z,\zeta\right)$ with g the best dominant. □

Corollary 4.11 (Alb Lupaş and Oros, 2023a). *Assume that relation* (4.3.1.3) *is true. If*

$$\Psi_\alpha^q\left(\varepsilon,\beta,\delta,\psi;z,\zeta\right) \prec\prec \varepsilon+\beta\frac{Mz+\zeta}{Nz+\zeta}+\delta\left(\frac{Mz+\zeta}{Nz+\zeta}\right)^2+\frac{\psi\left(M-N\right)\zeta z}{\left(Mz+\zeta\right)\left(Nz+\zeta\right)},$$

where $\alpha>0$, $0<q<1$, $\varepsilon,\beta,\psi,\delta\in\mathbb{C}$, $\psi\neq 0$, $-1\leq N<M\leq 1$ *and* Ψ_α^q *is defined by relation* (4.3.1.4), *then*

$$\frac{z\left(D_z^{-\alpha}\phi\left(m\left(\zeta\right),n\left(\zeta\right);q,z,\zeta\right)\right)_z'}{D_z^{-\alpha}\phi\left(m\left(\zeta\right),n\left(\zeta\right);q,z,\zeta\right)} \prec\prec \frac{Mz+\zeta}{Nz+\zeta},\quad (z,\zeta)\in U\times\overline{U},$$

with the best dominant $\frac{Mz+\zeta}{Nz+\zeta}$.

Corollary 4.12 (Alb Lupaş and Oros, 2023a). *Assume the relation* (4.3.1.3) *is true. If*

$$\Psi_\alpha^q\left(\varepsilon,\beta,\delta,\psi;z,\zeta\right) \prec\prec \varepsilon+\beta\left(\frac{z+\zeta}{\zeta-z}\right)^k+\delta\left(\frac{z+\zeta}{\zeta-z}\right)^{2k}+\frac{2k\psi\zeta z}{\zeta^2-z^2},$$

where $\alpha>0$, $0<q<1$, $\varepsilon,\beta,\psi,\delta\in\mathbb{C}$, $\psi\neq 0$, $0<k\leq 1$ *and* Ψ_α^q *is defined by relation* (4.3.1.4), *then*

$$\frac{z\left(D_z^{-\alpha}\phi\left(m\left(\zeta\right),n\left(\zeta\right);q,z,\zeta\right)\right)_z'}{D_z^{-\alpha}\phi\left(m\left(\zeta\right),n\left(\zeta\right);q,z,\zeta\right)} \prec\prec \left(\frac{z+\zeta}{\zeta-z}\right)^k,\quad (z,\zeta)\in U\times\overline{U},$$

with the best dominant $\left(\frac{z+\zeta}{\zeta-z}\right)^k$.

In the following theorem, the best subordinant of a strong differential superordination investigated in correlation to the Riemann–Liouville fractional integral of the extended q-hypergeometric function is obtained.

Theorem 4.32 (Alb Lupaş and Oros, 2023a). *Let g be an analytic and univalent function in $U\times\overline{U}$ with the properties $g(z,\zeta)\neq 0$ and $\frac{zg'_z(z,\zeta)}{g(z,\zeta)}$ starlike univalent. Assume that*

$$\operatorname{Re}\left(\frac{\beta}{\psi}g(z,\zeta)\,g'_z(z,\zeta)+\frac{2\delta}{\psi}g^2(z,\zeta)\,g'_z(z,\zeta)\right)>0,\ \text{for } \beta,\psi,\delta\in\mathbb{C},\ \ \psi\neq 0. \tag{4.3.1.8}$$

When $\frac{z\left(D_z^{-\alpha}\phi(m(\zeta),n(\zeta);q,z,\zeta)\right)'_z}{D_z^{-\alpha}\phi(m(\zeta),n(\zeta);q,z,\zeta)}\in\mathcal{H}\left[g(0,\zeta),1,\zeta\right]\cap Q^*$ and $\Psi^q_\alpha(\varepsilon,\beta,\delta,\psi;z,\zeta)$ defined by relation (4.3.1.4) is univalent in $U\times\overline{U}$, then

$$\varepsilon+\beta g(z,\zeta)+\delta\left(g(z,\zeta)\right)^2+\frac{\psi zg'_z(z,\zeta)}{g(z,\zeta)}\prec\prec\Psi^q_\alpha(\varepsilon,\beta,\delta,\psi;z,\zeta) \tag{4.3.1.9}$$

implies

$$g(z,\zeta)\prec\prec\frac{z\left(D_z^{-\alpha}\phi(m(\zeta),n(\zeta);q,z,\zeta)\right)'_z}{D_z^{-\alpha}\phi(m(\zeta),n(\zeta);q,z,\zeta)},\quad (z,\zeta)\in U\times\overline{U}, \tag{4.3.1.10}$$

and g is the best subordinant.

Proof. Define $p(z,\zeta):=\frac{z\left(D_z^{-\alpha}\phi(m(\zeta),n(\zeta);q,z,\zeta)\right)'_z}{D_z^{-\alpha}\phi(m(\zeta),n(\zeta);q,z,\zeta)}$, $(z,\zeta)\in(U\backslash\{0\})\times\overline{U}$ and consider the analytic functions $\theta(u)=\delta u^2+\beta u+\varepsilon$ in $\mathbb{C}$, and $\eta(u)=\frac{\psi}{u}$, respectively in $\mathbb{C}\backslash\{0\}$ with $\eta(u)\neq 0$, $u\in\mathbb{C}\backslash\{0\}$.

Differentiating it with respect to z, we can write

$$\frac{\theta'_z(g(z,\zeta))}{\eta(g(z,\zeta))}=\frac{\left[\beta+2\delta g(z,\zeta)\right]g(z,\zeta)\,g'_z(z,\zeta)}{\psi},$$

and

$$\operatorname{Re}\left(\frac{\theta'_z(g(z,\zeta))}{\eta(g(z,\zeta))}\right)=\operatorname{Re}\left(\frac{\beta}{\psi}g(z,\zeta)\,g'_z(z,\zeta)+\frac{2\delta}{\psi}g^2(z,\zeta)\,g'_z(z,\zeta)\right)>0,$$

for $\beta,\psi,\delta\in\mathbb{C}$, $\psi\neq 0$, taking account of relation (4.3.1.8).

Strong differential superordination (4.3.1.9) can be written using relation (4.3.1.7) as follows:

$$\begin{aligned}&\varepsilon+\beta g(z,\zeta)+\delta\left(g(z,\zeta)\right)^2+\frac{\psi zg'_z(z,\zeta)}{g(z,\zeta)}\\&\quad\prec\prec\varepsilon+\beta p(z,\zeta)+\delta\left(p(z,\zeta)\right)^2+\frac{\psi zp'_z(z,\zeta)}{p(z,\zeta)},\end{aligned}$$

and applying Lemma 1.12, we obtain

$$g(z,\zeta) \prec\prec p(z,\zeta) = \frac{z\left(D_z^{-\alpha}\phi\left(m\left(\zeta\right),n\left(\zeta\right);q,z,\zeta\right)\right)_z'}{D_z^{-\alpha}\phi\left(m\left(\zeta\right),n\left(\zeta\right);q,z,\zeta\right)}, \quad (z,\zeta) \in U \times \overline{U},$$

and g is the best subordinant. □

Corollary 4.13 (Alb Lupaş and Oros, 2023a). *Assume that relation* (4.3.1.8) *is true. If* $\frac{z\left(D_z^{-\alpha}\phi(m(\zeta),n(\zeta);q,z,\zeta)\right)_z'}{D_z^{-\alpha}\phi(m(\zeta),n(\zeta);q,z,\zeta)} \in \mathcal{H}\left[g\left(0,\zeta\right),1,\zeta\right] \cap Q^*$ *and*

$$\varepsilon+\beta\frac{Mz+\zeta}{Nz+\zeta}+\delta\left(\frac{Mz+\zeta}{Nz+\zeta}\right)^2+\frac{\psi\left(M-N\right)\zeta z}{\left(Mz+\zeta\right)\left(Nz+\zeta\right)} \prec\prec \Psi_\alpha^q\left(\varepsilon,\beta,\delta,\psi;z,\zeta\right),$$

where $\alpha > 0$, $0 < q < 1$, $\varepsilon,\beta,\psi,\delta \in \mathbb{C}$, $\psi \neq 0$, $-1 \leq N < M \leq 1$ *and* Ψ_α^q *defined by relation* (4.3.1.4), *then*

$$\frac{Mz+\zeta}{Nz+\zeta} \prec\prec \frac{z\left(D_z^{-\alpha}\phi\left(m\left(\zeta\right),n\left(\zeta\right);q,z,\zeta\right)\right)_z'}{D_z^{-\alpha}\phi\left(m\left(\zeta\right),n\left(\zeta\right);q,z,\zeta\right)}, \quad (z,\zeta) \in U \times \overline{U},$$

with the best subordinant $\frac{Mz+\zeta}{Nz+\zeta}$.

Corollary 4.14 (Alb Lupaş and Oros, 2023a). *Assume the relation* (4.3.1.8) *is true. If* $\frac{z\left(D_z^{-\alpha}\phi(m(\zeta),n(\zeta);q,z,\zeta)\right)_z'}{D_z^{-\alpha}\phi(m(\zeta),n(\zeta);q,z,\zeta)} \in \mathcal{H}\left[g\left(0,\zeta\right),1,\zeta\right] \cap Q^*$ *and*

$$\varepsilon+\beta\left(\frac{z+\zeta}{\zeta-z}\right)^k+\delta\left(\frac{z+\zeta}{\zeta-z}\right)^{2k}+\frac{2k\psi\zeta z}{\zeta^2-z^2} \prec\prec \Psi_\alpha^q\left(\varepsilon,\beta,\delta,\psi;z,\zeta\right),$$

where $\alpha > 0$, $0 < q < 1$, $\varepsilon,\beta,\psi,\delta \in \mathbb{C}$, $\psi \neq 0$, $0 < k \leq 1$ *and* Ψ_α^q *defined by relation* (4.3.1.4), *then*

$$\left(\frac{z+\zeta}{\zeta-z}\right)^k \prec\prec \frac{z\left(D_z^{-\alpha}\phi\left(m\left(\zeta\right),n\left(\zeta\right);q,z,\zeta\right)\right)_z'}{D_z^{-\alpha}\phi\left(m\left(\zeta\right),n\left(\zeta\right);q,z,\zeta\right)}, \quad (z,\zeta) \in U \times \overline{U},$$

with the best subordinant $\left(\frac{z+\zeta}{\zeta-z}\right)^k$.

Now, as an application of the results obtained so far, a sandwich-type theorem connects the dual results presented in Theorems 4.31 and 4.32. The corresponding corollaries follow naturally.

Theorem 4.33 (Alb Lupaş and Oros, 2023a). *Let* g_1, g_2 *analytic and univalent functions in* $U \times \overline{U}$ *with the properties* $g_1\left(z,\zeta\right) \neq 0$, $g_2\left(z,\zeta\right) \neq 0$, *for all* $(z,\zeta) \in U \times \overline{U}$, *and* $\frac{z(g_1)_z'(z,\zeta)}{g_1(z,\zeta)}$, $\frac{z(g_2)_z'(z,\zeta)}{g_2(z,\zeta)}$ *are starlike univalent. Assume that* g_1 *verifies* (4.3.1.3) *and* g_2 *verifies* (4.3.1.8).

If $\frac{z\left(D_z^{-\alpha}\phi(m(\zeta),n(\zeta);q,z,\zeta)\right)_z'}{D_z^{-\alpha}\phi(m(\zeta),n(\zeta);q,z,\zeta)} \in \mathcal{H}\left[g\left(0,\zeta\right),1,\zeta\right] \cap Q^*$ *and* $\Psi_{\alpha}^{q}\left(\varepsilon,\beta,\delta,\psi;z,\zeta\right)$ *defined by relation* (4.3.1.4) *is univalent in* $U \times \overline{U}$, $\alpha > 0$, $0 < q < 1$, $\varepsilon,\beta,\psi,\delta \in \mathbb{C}$, $\psi \neq 0$, *then*

$$\varepsilon + \beta g_1\left(z,\zeta\right) + \delta\left(g_1\left(z,\zeta\right)\right)^2 + \frac{\psi z\left(g_1\right)_z'\left(z,\zeta\right)}{g_1\left(z,\zeta\right)} \prec\prec \Psi_{\alpha}^{q}\left(\varepsilon,\beta,\delta,\psi;z,\zeta\right)$$

$$\prec\prec \varepsilon + \beta g_2\left(z,\zeta\right) + \delta\left(g_2\left(z,\zeta\right)\right)^2 + \frac{\psi z\left(g_2\right)_z'\left(z,\zeta\right)}{g_2\left(z,\zeta\right)},$$

implies

$$g_1\left(z,\zeta\right) \prec\prec \frac{z\left(D_z^{-\alpha}\phi\left(m\left(\zeta\right),n\left(\zeta\right);q,z,\zeta\right)\right)_z'}{D_z^{-\alpha}\phi\left(m\left(\zeta\right),n\left(\zeta\right);q,z,\zeta\right)} \prec\prec g_2\left(z,\zeta\right), \quad (z,\zeta) \in U \times \overline{U},$$

and g_1 *and* g_2 *are respectively the best subordinant and the best dominant.*

Corollary 4.15 (Alb Lupaş and Oros, 2023a). *Assume that relations* (4.3.1.3) *and* (4.3.1.8) *are true.*
If $\frac{z\left(D_z^{-\alpha}\phi(m(\zeta),n(\zeta);q,z,\zeta)\right)_z'}{D_z^{-\alpha}\phi(m(\zeta),n(\zeta);q,z,\zeta)} \in \mathcal{H}\left[g\left(0,\zeta\right),1,\zeta\right] \cap Q^*$ *and*

$$\varepsilon + \beta\frac{M_1 z + \zeta}{N_1 z + \zeta} + \delta\left(\frac{M_1 z + \zeta}{N_1 z + \zeta}\right)^2 + \frac{\psi\left(M_1 - N_1\right)\zeta z}{\left(M_1 z + \zeta\right)\left(N_1 z + \zeta\right)}$$

$$\prec\prec \Psi_{\alpha}^{q}\left(\varepsilon,\beta,\delta,\psi;z,\zeta\right) \prec\prec \varepsilon + \beta\frac{M_2 z + \zeta}{N_2 z + \zeta} + \delta\left(\frac{M_2 z + \zeta}{N_2 z + \zeta}\right)^2$$

$$+ \frac{\psi\left(M_2 - N_2\right)\zeta z}{\left(M_2 z + \zeta\right)\left(N_2 z + \zeta\right)},$$

where $\alpha > 0$, $0 < q < 1$, $\varepsilon,\beta,\psi,\delta \in \mathbb{C}$, $\psi \neq 0$, $-1 \leq N_2 < N_1 < M_1 < M_2 \leq 1$, *and* Ψ_{α}^{q} *defined by relation* (4.3.1.4), *then*

$$\frac{M_1 z + \zeta}{N_1 z + \zeta} \prec\prec \frac{z\left(D_z^{-\alpha}\phi\left(m\left(\zeta\right),n\left(\zeta\right);q,z,\zeta\right)\right)_z'}{D_z^{-\alpha}\phi\left(m\left(\zeta\right),n\left(\zeta\right);q,z,\zeta\right)} \prec\prec \frac{M_2 z + \zeta}{N_2 z + \zeta}, \quad (z,\zeta) \in U\times\overline{U},$$

with the best dominant $\frac{M_2 z+\zeta}{N_2 z+\zeta}$ *and the best subordinant* $\frac{M_1 z+\zeta}{N_1 z+\zeta}$.

Corollary 4.16 (Alb Lupaş and Oros, 2023a). *Assume that relations* (4.3.1.3) *and* (4.3.1.8) *are true.*
If $\frac{z\left(D_z^{-\alpha}\phi(m(\zeta),n(\zeta);q,z,\zeta)\right)_z'}{D_z^{-\alpha}\phi(m(\zeta),n(\zeta);q,z,\zeta)} \in \mathcal{H}\left[g\left(0,\zeta\right),1,\zeta\right] \cap Q^*$ *and*

$$\varepsilon + \beta\left(\frac{z+\zeta}{\zeta - z}\right)^{k_1} + \delta\left(\frac{z+\zeta}{\zeta - z}\right)^{2k_1} + \frac{2k_1\psi\zeta z}{\zeta^2 - z^2} \prec\prec \Psi_{\alpha}^{q}\left(\varepsilon,\beta,\delta,\psi;z,\zeta\right)$$

$$\prec\prec \varepsilon + \beta\left(\frac{z+\zeta}{\zeta - z}\right)^{k_2} + \delta\left(\frac{z+\zeta}{\zeta - z}\right)^{2k_2} + \frac{2k_2\psi\zeta z}{\zeta^2 - z^2},$$

where $\alpha > 0$, $0 < q < 1$, $\varepsilon, \beta, \psi, \delta \in \mathbb{C}$, $\psi \neq 0$, $0 < k_1 < k_2 \leq 1$, *and* Ψ_α^q *defined by relation* (4.3.1.4), *then*

$$\left(\frac{z+\zeta}{\zeta - z}\right)^{k_1} \prec\prec \frac{z\left(D_z^{-\alpha}\phi\left(m\left(\zeta\right), n\left(\zeta\right); q, z, \zeta\right)\right)_z'}{D_z^{-\alpha}\phi\left(m\left(\zeta\right), n\left(\zeta\right); q, z, \zeta\right)} \prec\prec \left(\frac{z+\zeta}{\zeta - z}\right)^{k_2},$$

$(z, \zeta) \in U \times \overline{U}$, *with the best dominant* $\left(\frac{z+\zeta}{\zeta-z}\right)^{k_2}$ *and the best subordinant* $\left(\frac{z+\zeta}{\zeta-z}\right)^{k_1}$.

Considering the functions $\theta(u) = \varepsilon u$ and $\eta(u) = \psi$, $u \in U$, we obtain other strong subordination and superordination theorems and corollaries.

Theorem 4.34 (Alb Lupaş and Oros, 2023a). *Consider* g *a convex and univalent function in* $U \times \overline{U}$ *with* $g(0, \zeta) = \alpha$, $\zeta \in \overline{U}$ *and* $\frac{z\left(D_z^{-\alpha}\phi(m(\zeta), n(\zeta); q, z, \zeta)\right)_z'}{D_z^{-\alpha}\phi(m(\zeta), n(\zeta); q, z, \zeta)} \in \mathcal{H}\left(U \times \overline{U}\right)$, *where* $\alpha > 0$, $0 < q < 1$. *Assume that*

$$\operatorname{Re}\left(\frac{\varepsilon + \psi}{\psi} + z\frac{g_{z^2}''(z, \zeta)}{g_z'(z, \zeta)}\right) > 0, \tag{4.3.1.11}$$

for $\varepsilon, \psi \in \mathbb{C}$, $\psi \neq 0$ *and*

$$\begin{aligned}\Psi_\alpha^q(\varepsilon, \psi; z, \zeta) := (\varepsilon + \psi) &\frac{z\left(D_z^{-\alpha}\phi\left(m\left(\zeta\right), n\left(\zeta\right); q, z, \zeta\right)\right)_z'}{D_z^{-\alpha}\phi\left(m\left(\zeta\right), n\left(\zeta\right); q, z, \zeta\right)} \\ &- \psi\left(\frac{z\left(D_z^{-\alpha}\phi\left(m\left(\zeta\right), n\left(\zeta\right); q, z, \zeta\right)\right)_z'}{D_z^{-\alpha}\phi\left(m\left(\zeta\right), n\left(\zeta\right); q, z, \zeta\right)}\right)^2 \\ &+ \psi\frac{z^2\left(D_z^{-\alpha}\phi\left(m\left(\zeta\right), n\left(\zeta\right); q, z, \zeta\right)\right)_{z^2}''}{D_z^{-\alpha}\phi\left(m\left(\zeta\right), n\left(\zeta\right); q, z, \zeta\right)}.\end{aligned} \tag{4.3.1.12}$$

When g *verifies the strong differential subordination*

$$\Psi_\alpha^q(\varepsilon, \psi; z, \zeta) \prec\prec \varepsilon g(z, \zeta) + \psi z g_z'(z, \zeta), \tag{4.3.1.13}$$

then

$$\frac{z\left(D_z^{-\alpha}\phi\left(m\left(\zeta\right), n\left(\zeta\right); q, z, \zeta\right)\right)_z'}{D_z^{-\alpha}\phi\left(m\left(\zeta\right), n\left(\zeta\right); q, z, \zeta\right)} \prec\prec g(z, \zeta), \quad (z, \zeta) \in U \times \overline{U}, \tag{4.3.1.14}$$

and g *is the best dominant.*

Proof. Define $p(z,\zeta) := \frac{z\left(D_z^{-\alpha}\phi(m(\zeta),n(\zeta);q,z,\zeta)\right)'_z}{D_z^{-\alpha}\phi(m(\zeta),n(\zeta);q,z,\zeta)}$, with $p(0,\zeta)=\alpha$. Differentiating it with respect to z, we get

$$\begin{aligned} p'_z(z,\zeta) &= \frac{\left(D_z^{-\alpha}\phi(m(\zeta),n(\zeta);q,z,\zeta)\right)'_z}{D_z^{-\alpha}\phi(m(\zeta),n(\zeta);q,z,\zeta)} \\ &\quad - z\left(\frac{\left(D_z^{-\alpha}\phi(m(\zeta),n(\zeta);q,z,\zeta)\right)'_z}{D_z^{-\alpha}\phi(m(\zeta),n(\zeta);q,z,\zeta)}\right)^2 \\ &\quad + z\frac{\left(D_z^{-\alpha}\phi(m(\zeta),n(\zeta);q,z,\zeta)\right)''_{z^2}}{D_z^{-\alpha}\phi(m(\zeta),n(\zeta);q,z,\zeta)}. \end{aligned} \tag{4.3.1.15}$$

Considering the analytic functions $\theta(u) = \varepsilon u$ in $\mathbb{C}$ and $\eta(u) = \psi \neq 0$ in $\mathbb{C}\backslash\{0\}$, we define the starlike univalent function $G(z,\zeta) = z\eta(g(z,\zeta))g'_z(z,\zeta) = \psi z g'_z(z,\zeta)$ in $U \times \overline{U}$ and $h(z,\zeta) = G(z,\zeta) + \theta(g(z,\zeta)) = \varepsilon g(z,\zeta) + \psi z g'_z(z,\zeta)$. Relation (4.3.1.11) can be written $\operatorname{Re}\left(\frac{zh'_z(z,\zeta)}{G(z,\zeta)}\right) = \operatorname{Re}\left(\frac{\varepsilon+\psi}{\psi} + z\frac{g''_{z^2}(z,\zeta)}{g'_z(z,\zeta)}\right) > 0$ and using relation (4.3.1.15), we obtain

$$\begin{aligned} \varepsilon p(z,\zeta) + \psi z p'_z(z,\zeta) &= (\varepsilon+\psi)\frac{z\left(D_z^{-\alpha}\phi(m(\zeta),n(\zeta);q,z,\zeta)\right)'_z}{D_z^{-\alpha}\phi(m(\zeta),n(\zeta);q,z,\zeta)} \\ &\quad - \psi\left(\frac{z\left(D_z^{-\alpha}\phi(m(\zeta),n(\zeta);q,z,\zeta)\right)'_z}{D_z^{-\alpha}\phi(m(\zeta),n(\zeta);q,z,\zeta)}\right)^2 \\ &\quad + \psi\frac{z^2\left(D_z^{-\alpha}\phi(m(\zeta),n(\zeta);q,z,\zeta)\right)''_{z^2}}{D_z^{-\alpha}\phi(m(\zeta),n(\zeta);q,z,\zeta)}. \end{aligned}$$

The strong differential subordination (4.3.1.13) can be written $\varepsilon p(z,\zeta) + \psi z p'_z(z,\zeta) \prec\prec \varepsilon g(z,\zeta) + \psi z g'_z(z,\zeta)$ and applying Lemma 1.9, we get $p(z,\zeta) \prec\prec g(z,\zeta)$, equivalently with $\frac{z\left(D_z^{-\alpha}\phi(m(\zeta),n(\zeta);q,z,\zeta)\right)'_z}{D_z^{-\alpha}\phi(m(\zeta),n(\zeta);q,z,\zeta)} \prec\prec g(z,\zeta)$, $(z,\zeta) \in U \times \overline{U}$, and g is the best dominant. □

Corollary 4.17 (Alb Lupaş and Oros, 2023a). *Assume that relation (4.3.1.11) is true. If*

$$\Psi_\alpha^q(\varepsilon,\psi;z,\zeta) \prec\prec \varepsilon\frac{Mz+\zeta}{Nz+\zeta} + \frac{\psi(M-N)\zeta z}{(Nz+\zeta)^2},$$

where $\alpha > 0$, $0 < q < 1$, $\varepsilon, \psi \in \mathbb{C}$, $\psi \neq 0$, $-1 \leq N < M \leq 1$ *and* Ψ_{α}^{q} *defined by relation* (4.3.1.12), *then*

$$\frac{z\left(D_z^{-\alpha}\phi\left(m\left(\zeta\right), n\left(\zeta\right); q, z, \zeta\right)\right)_z'}{D_z^{-\alpha}\phi\left(m\left(\zeta\right), n\left(\zeta\right); q, z, \zeta\right)} \prec\prec \frac{Mz+\zeta}{Nz+\zeta}, \quad (z, \zeta) \in U \times \overline{U},$$

with the best dominant $\frac{Mz+\zeta}{Nz+\zeta}$.

Corollary 4.18 (Alb Lupaş and Oros, 2023a). *Assume that relation* (4.3.1.11) *is true. If*

$$\Psi_{\alpha}^{q}\left(\varepsilon, \psi; z, \zeta\right) \prec\prec \varepsilon\left(\frac{z+\zeta}{\zeta-z}\right)^{k} + \frac{2k\psi\zeta z}{\zeta^2 - z^2}\left(\frac{z+\zeta}{\zeta-z}\right)^{k},$$

where $\alpha > 0$, $0 < q < 1$, $\varepsilon, \psi \in \mathbb{C}$, $\psi \neq 0$, $0 < k \leq 1$ *and* Ψ_{α}^{q} *defined by relation* (4.3.1.12), *then*

$$\frac{z\left(D_z^{-\alpha}\phi\left(m\left(\zeta\right), n\left(\zeta\right); q, z, \zeta\right)\right)_z'}{D_z^{-\alpha}\phi\left(m\left(\zeta\right), n\left(\zeta\right); q, z, \zeta\right)} \prec\prec \left(\frac{z+\zeta}{\zeta-z}\right)^{k}, \quad (z, \zeta) \in U \times \overline{U},$$

with the best dominant $\left(\frac{z+\zeta}{\zeta-z}\right)^{k}$.

Theorem 4.35 (Alb Lupaş and Oros, 2023a). *Let* g *a convex and univalent function in* $U \times \overline{U}$ *with* $g\left(0, \zeta\right) = \alpha$, $\zeta \in \overline{U}$, *and* $\alpha > 0$, $0 < q < 1$. *Assume that*

$$\operatorname{Re}\left(\frac{\varepsilon}{\psi} g_z'\left(z, \zeta\right)\right) > 0, \text{ for } \varepsilon, \psi \in \mathbb{C}, \psi \neq 0. \tag{4.3.1.16}$$

When $\frac{z\left(D_z^{-\alpha}\phi(m(\zeta), n(\zeta); q, z, \zeta)\right)_z'}{D_z^{-\alpha}\phi(m(\zeta), n(\zeta); q, z, \zeta)} \in \mathcal{H}\left[g\left(0, \zeta\right), 1, \zeta\right] \cap Q^*$ *and* $\Psi_{\alpha}^{q}\left(\varepsilon, \psi; z, \zeta\right)$ *defined by relation* (4.3.1.12) *is univalent in* $U \times \overline{U}$, *then*

$$\varepsilon g\left(z, \zeta\right) + \psi z g_z'\left(z, \zeta\right) \prec\prec \Psi_{\alpha}^{q}\left(\varepsilon, \psi; z, \zeta\right) \tag{4.3.1.17}$$

implies

$$g\left(z, \zeta\right) \prec\prec \frac{z\left(D_z^{-\alpha}\phi\left(m\left(\zeta\right), n\left(\zeta\right); q, z, \zeta\right)\right)_z'}{D_z^{-\alpha}\phi\left(m\left(\zeta\right), n\left(\zeta\right); q, z, \zeta\right)}, \quad (z, \zeta) \in U \times \overline{U}, \tag{4.3.1.18}$$

and g *is the best subordinant.*

Proof. Define $p(z,\zeta) = \frac{z\left(D_z^{-\alpha}\phi(m(\zeta),n(\zeta);q,z,\zeta)\right)'_z}{D_z^{-\alpha}\phi(m(\zeta),n(\zeta);q,z,\zeta)}$, $(z,\zeta) \in U \times \overline{U}$, with $p(0,\zeta) = \alpha$, $\zeta \in \overline{U}$ and consider the analytic functions $\theta(u) = \varepsilon u$ in $\mathbb{C}$ and $\eta(u) = \psi \neq 0$ in $\mathbb{C}\backslash\{0\}$.

Differentiating it with respect to z, we obtain $\frac{\theta'_z(g(z,\zeta))}{\eta(g(z,\zeta))} = \frac{\varepsilon}{\psi} g'_z(z,\zeta)$, and $\operatorname{Re}\left(\frac{\theta'_z(g(z,\zeta))}{\eta(g(z,\zeta))}\right) = \operatorname{Re}\left(\frac{\varepsilon}{\psi} g'_z(z,\zeta)\right) > 0$, for $\varepsilon, \psi \in \mathbb{C}, \psi \neq 0$, taking account of relation (4.3.1.16).

Applying Lemma 1.12 for the strong differential superordination (4.3.1.17) written in the following form

$$\varepsilon g(z,\zeta) + \psi z g'_z(z,\zeta) \prec\prec \varepsilon p(z,\zeta) + \psi z p'_z(z,\zeta),$$

we obtain

$$g(z,\zeta) \prec\prec p(z,\zeta) = \frac{z\left(D_z^{-\alpha}\phi(m(\zeta),n(\zeta);q,z,\zeta)\right)'_z}{D_z^{-\alpha}\phi(m(\zeta),n(\zeta);q,z,\zeta)}, \quad (z,\zeta) \in U \times \overline{U},$$

and g is the best subordinant. □

Corollary 4.19 (Alb Lupaş and Oros, 2023a). *Assume that relation (4.3.1.16) is true. If* $\frac{z\left(D_z^{-\alpha}\phi(m(\zeta),n(\zeta);q,z,\zeta)\right)'_z}{D_z^{-\alpha}\phi(m(\zeta),n(\zeta);q,z,\zeta)} \in \mathcal{H}[g(0,\zeta),1,\zeta] \cap Q^*$ *and*

$$\varepsilon\frac{Mz+\zeta}{Nz+\zeta} + \frac{\psi(M-N)\zeta z}{(Nz+\zeta)^2} \prec\prec \Psi_\alpha^q(\varepsilon,\psi;z,\zeta),$$

where $\alpha > 0$, $0 < q < 1$, $\varepsilon, \psi \in \mathbb{C}$, $\psi \neq 0$, $-1 \le N < M \le 1$ *and* Ψ_α^q *defined by relation* (4.3.1.12), *then*

$$\frac{Mz+\zeta}{Nz+\zeta} \prec\prec \frac{z\left(D_z^{-\alpha}\phi(m(\zeta),n(\zeta);q,z,\zeta)\right)'_z}{D_z^{-\alpha}\phi(m(\zeta),n(\zeta);q,z,\zeta)}, \quad (z,\zeta) \in U \times \overline{U},$$

with the best subordinant $\frac{Mz+\zeta}{Nz+\zeta}$.

Corollary 4.20 (Alb Lupaş and Oros, 2023a). *Assume that relation (4.3.1.16) is true. If* $\frac{z\left(D_z^{-\alpha}\phi(m(\zeta),n(\zeta);q,z,\zeta)\right)'_z}{D_z^{-\alpha}\phi(m(\zeta),n(\zeta);q,z,\zeta)} \in \mathcal{H}[g(0,\zeta),1,\zeta] \cap Q^*$ *and*

$$\varepsilon\left(\frac{z+\zeta}{\zeta-z}\right)^k + \frac{2k\psi\zeta z}{\zeta^2-z^2}\left(\frac{z+\zeta}{\zeta-z}\right)^k \prec\prec \Psi_\alpha^q(\varepsilon,\psi;z,\zeta),$$

where $\alpha > 0$, $0 < q < 1$, $\varepsilon, \psi \in \mathbb{C}$, $\psi \neq 0$, $0 < k \le 1$ *and* Ψ_α^q *defined by relation* (4.3.1.12), *then*

$$\left(\frac{z+\zeta}{\zeta-z}\right)^k \prec\prec \frac{z\left(D_z^{-\alpha}\phi(m(\zeta),n(\zeta);q,z,\zeta)\right)'_z}{D_z^{-\alpha}\phi(m(\zeta),n(\zeta);q,z,\zeta)}, \quad (z,\zeta) \in U \times \overline{U},$$

with the best subordinant $\left(\frac{z+\zeta}{\zeta-z}\right)^k$.

Together, Theorems 4.34 and 4.35 imply the following sandwich theorem.

Theorem 4.36 (Alb Lupaş and Oros, 2023a). *Let g_1, g_2 be convex and univalent functions in $U \times \overline{U}$ such that $g_1(z,\zeta) \neq 0$, $g_2(z,\zeta) \neq 0$, for all $(z,\zeta) \in U \times \overline{U}$, and $\frac{z(g_1)'_z(z,\zeta)}{g_1(z,\zeta)}$, $\frac{z(g_2)'_z(z,\zeta)}{g_2(z,\zeta)}$ are starlike univalent. Assuming that g_1 satisfies (4.3.1.11) and g_2 satisfies (4.3.1.16). If $\frac{z\left(D_z^{-\alpha}\phi(m(\zeta),n(\zeta);q,z,\zeta)\right)'_z}{D_z^{-\alpha}\phi(m(\zeta),n(\zeta);q,z,\zeta)} \in \mathcal{H}[g(0,\zeta),1,\zeta] \cap Q^*$, and $\Psi_\alpha^q(\varepsilon,\psi;z,\zeta)$ defined by relation (4.3.1.12) is univalent in $U \times \overline{U}$, $\alpha > 0$, $0 < q < 1$, then*

$$\varepsilon g_1(z,\zeta) + \psi z (g_1)'_z(z,\zeta) \prec\prec \Psi_\alpha^q(\varepsilon,\psi;z,\zeta) \prec\prec \varepsilon g_2(z,\zeta) + \psi z (g_2)'_z(z,\zeta),$$

for $\varepsilon, \psi \in \mathbb{C}$, $\psi \neq 0$, implies

$$g_1(z,\zeta) \prec\prec \frac{z\left(D_z^{-\alpha}\phi(m(\zeta),n(\zeta);q,z,\zeta)\right)'_z}{D_z^{-\alpha}\phi(m(\zeta),n(\zeta);q,z,\zeta)} \prec\prec g_2(z,\zeta), \quad (z,\zeta) \in U \times \overline{U},$$

and g_1 and g_2 are respectively the best subordinant and the best dominant.

Corollary 4.21 (Alb Lupaş and Oros, 2023a). *Assume that relations (4.3.1.11) and (4.3.1.16) are true.*
If $\frac{z\left(D_z^{-\alpha}\phi(m(\zeta),n(\zeta);q,z,\zeta)\right)'_z}{D_z^{-\alpha}\phi(m(\zeta),n(\zeta);q,z,\zeta)} \in \mathcal{H}[g(0,\zeta),1,\zeta] \cap Q^*$ and

$$\varepsilon\frac{M_1 z + \zeta}{N_1 z + \zeta} + \frac{\psi(M_1 - N_1)\zeta z}{(N_1 z + \zeta)^2} \prec\prec \Psi_\alpha^q(\varepsilon,\psi;z,\zeta)$$
$$\prec\prec \varepsilon\frac{M_2 z + \zeta}{N_2 z + \zeta} + \frac{\psi(M_2 - N_2)\zeta z}{(N_2 z + \zeta)^2},$$

where $\alpha > 0$, $0 < q < 1$, $\varepsilon, \psi \in \mathbb{C}$, $\psi \neq 0$, $-1 \leq N_2 < N_1 < M_1 < M_2 \leq 1$ and Ψ_α^q defined by relation (4.3.1.12), then

$$\frac{M_1 z + \zeta}{N_1 z + \zeta} \prec\prec \frac{z\left(D_z^{-\alpha}\phi(m(\zeta),n(\zeta);q,z,\zeta)\right)'_z}{D_z^{-\alpha}\phi(m(\zeta),n(\zeta);q,z,\zeta)} \prec\prec \frac{M_2 z + \zeta}{N_2 z + \zeta},$$

$(z,\zeta) \in U \times \overline{U}$, with the best dominant $\frac{M_2 z+\zeta}{N_2 z+\zeta}$ and the best subordinant $\frac{M_1 z+\zeta}{N_1 z+\zeta}$.

Corollary 4.22 (Alb Lupaş and Oros, 2023a). *Assume that relations (4.3.1.11) and (4.3.1.16) are true.*

If $\frac{z\left(D_z^{-\alpha}\phi(m(\zeta),n(\zeta);q,z,\zeta)\right)_z'}{D_z^{-\alpha}\phi(m(\zeta),n(\zeta);q,z,\zeta)} \in \mathcal{H}\left[g\left(0,\zeta\right),1,\zeta\right] \cap Q^*$ *and*

$$\begin{aligned}\varepsilon\left(\frac{z+\zeta}{\zeta-z}\right)^{k_1}+\frac{2k_1\psi\zeta z}{\zeta^2-z^2}\left(\frac{z+\zeta}{\zeta-z}\right)^{k_1} &\prec\prec \Psi_\alpha^q\left(\varepsilon,\psi;z,\zeta\right)\\ &\prec\prec \varepsilon\left(\frac{z+\zeta}{\zeta-z}\right)^{k_2}+\frac{2k_2\psi\zeta z}{\zeta^2-z^2}\left(\frac{z+\zeta}{\zeta-z}\right)^{k_2},\end{aligned}$$

where $\alpha > 0$, $0 < q < 1$, $\varepsilon, \psi \in \mathbb{C}$, $\psi \neq 0$, $0 < k_1 < k_2 \leq 1$ *and* Ψ_α^q *defined by relation* (4.3.1.12), *then*

$$\left(\frac{z+\zeta}{\zeta-z}\right)^{k_1} \prec\prec \frac{z\left(D_z^{-\alpha}\phi\left(m\left(\zeta\right),n\left(\zeta\right);q,z,\zeta\right)\right)_z'}{D_z^{-\alpha}\phi\left(m\left(\zeta\right),n\left(\zeta\right);q,z,\zeta\right)} \prec\prec \left(\frac{z+\zeta}{\zeta-z}\right)^{k_2},$$

$(z,\zeta) \in U \times \overline{U}$, *with the best dominant* $\left(\frac{z+\zeta}{\zeta-z}\right)^{k_2}$ *and the best subordinant* $\left(\frac{z+\zeta}{\zeta-z}\right)^{k_1}$.

4.3.2 *Strong Sandwich Results Involving Riemann–Liouville Fractional Integral of Extended q-Hypergeometric Function*

The classical theories of differential subordination and superordination have been extended to strong differential subordination and its dual, strong differential superordination. The two new theories have progressed well, revealing significant findings when various operators and specific hypergeometric functions have been included in the studies. The research revealed by the paper Alb Lupaş (2024) expands the topic of the investigation by incorporating aspects of fractional calculus and quantum calculus. An extended version of the well-known q-hypergeometric function is introduced to correspond to the study of functions belonging to classes that were previously described and that are particularly defined for strong differential subordination and superordination theories. The paper Alb Lupaş and Oros (2023a) defines the Riemann–Liouville fractional integral of the extended q-hypergeometric function, which is used to study strong differential subordinations and superordinations in Alb Lupaş (2024). The theorems established for the strong differential subordinations and superordinations, establish the best dominants and best subordinants, respectively. Interesting corollaries are provided for certain functions that are regarded as best dominant or best subordinant due to their particular geometric characteristics. Sandwich-type theorems and corollaries conclude the study, stated to connect the outcomes obtained by applying the dual theories.

The Riemann–Liouville fractional integral applied to the extended q-confluent hypergeometric function from Definition 2.21 is

$$\begin{aligned} D_t^{-\alpha}\phi\left(a\left(\tau\right),b\left(\tau\right);q,t,\tau\right) &= \frac{1}{\Gamma\left(\alpha\right)}\int_0^t \frac{\phi\left(a\left(\tau\right),b\left(\tau\right);q,y,\tau\right)}{\left(t-y\right)^{1-\alpha}}dy \\ &= \frac{1}{\Gamma\left(\alpha\right)}\sum_{j=0}^{\infty}\frac{\left(a\left(\tau\right),q\right)_j}{\left(q,q\right)_j\left(b\left(\tau\right),q\right)_j}\int_0^t \frac{y^j}{\left(t-y\right)^{1-\alpha}}dy, \end{aligned} \tag{4.3.2.1}$$

with $a\left(\tau\right),b\left(\tau\right)$ holomorphic functions depending on the parameter $\tau\in\overline{\Delta}$, $\alpha>0$, $0<q<1$.

After a simple calculation, it takes the following form

$$D_t^{-\alpha}\phi\left(a\left(\tau\right),b\left(\tau\right);q,t,\tau\right)=\sum_{j=0}^{\infty}\frac{\left(a\left(\tau\right),q\right)_j}{\left(q,q\right)_j\left(b\left(\tau\right),q\right)_j\left(j+1\right)_\alpha}t^{\alpha+j}, \tag{4.3.2.2}$$

and $D_t^{-\alpha}\phi\left(a\left(\tau\right),b\left(\tau\right);q,t,\tau\right)\in\mathcal{H}\left[0,\alpha,\tau\right]$.

Next assume that $0<q<1$, $\alpha>0$ and $a\left(\tau\right)$, $b\left(\tau\right)$ are holomorphic functions.

The strong differential subordination result obtained using the operator given by (4.3.2.2) is the next theorem:

Theorem 4.37 (Alb Lupaş, 2024). *Consider* $\left(\frac{D_t^{-\alpha}\phi(a(\tau),b(\tau);q,t,\tau)}{t}\right)^p\in\mathcal{H}\left(\Delta\times\overline{\Delta}\right)$ *and a univalent function* $w\left(t,\tau\right)$ *in* $U\times\overline{U}$ *with the property* $w\left(t,\tau\right)\neq 0$, $\forall\ t\in\Delta\backslash\{0\}$, $\tau\in\overline{\Delta}$. *Assuming that the function* $\frac{tw_t'(t,\tau)}{w(t,\tau)}$ *is starlike univalent in* $\Delta\times\overline{\Delta}$ *and*

$$\operatorname{Re}\left(1+\frac{n}{s}w\left(t,\tau\right)+\frac{2r}{s}\left(w\left(t,\tau\right)\right)^2-\frac{tw_t'\left(t,\tau\right)}{w\left(t,\tau\right)}+\frac{tw_{t^2}''\left(t,\tau\right)}{w_t'\left(t,\tau\right)}\right)>0, \tag{4.3.2.3}$$

for $m,n,r,s\in\mathbb{C}$, $s\neq 0$, $t\in\Delta\backslash\{0\}$, $\tau\in\overline{\Delta}$ *and*

$$\begin{aligned} &H_\alpha^q\left(p,m,n,r,s;t,\tau\right) \\ &\quad =: m+n\left[\frac{D_t^{-\alpha}\phi\left(a\left(\tau\right),b\left(\tau\right);q,t,\tau\right)}{z}\right]^p+r\left[\frac{D_t^{-\alpha}\phi\left(a\left(\tau\right),b\left(\tau\right);q,t,\tau\right)}{z}\right]^{2p} \\ &\qquad + sp\left[\frac{t\left(D_t^{-\alpha}\phi\left(a\left(\tau\right),b\left(\tau\right);q,t,\tau\right)\right)_t'}{D_t^{-\alpha}\phi\left(a\left(\tau\right),b\left(\tau\right);q,t,\tau\right)}-1\right]. \end{aligned} \tag{4.3.2.4}$$

If w is a solution of the strong subordination

$$H_{\alpha}^{q}(p,m,n,r,s;t,\tau) \prec\prec m+nw(t,\tau)+r(w(t,\tau))^{2}+s\frac{tw_{t}'(t,\tau)}{w(t,\tau)}, \tag{4.3.2.5}$$

then w is the best dominant of the strong subordination

$$\left(\frac{D_{t}^{-\alpha}\phi(a(\tau),b(\tau);q,t,\tau)}{t}\right)^{p} \prec\prec w(t,\tau), \quad (t,\tau)\in\Delta\times\overline{\Delta}. \tag{4.3.2.6}$$

Proof. Considering the function $u(t,\tau) = \left(\frac{D_{t}^{-\alpha}\phi(a(\tau),b(\tau);q,t,\tau)}{t}\right)^{p}$, $t \in \Delta\backslash\{0\}$, $\tau\in\overline{\Delta}$, differentiating it with respect to t, we get

$$u_{t}'(t,\tau)=p\left(\frac{D_{t}^{-\alpha}\phi(a(\tau),b(\tau);q,t,\tau)}{t}\right)^{p-1}\left[\frac{\left(D_{t}^{-\alpha}\phi(a(\tau),b(\tau);q,t,\tau)\right)_{t}'}{t}\right.$$
$$\left.-\frac{D_{t}^{-\alpha}\phi(a(\tau),b(\tau);q,t,\tau)}{t^{2}}\right]=p\left(\frac{D_{t}^{-\alpha}\phi(a(\tau),b(\tau);q,t,\tau)}{t}\right)^{p-1}$$
$$\times\frac{\left(D_{t}^{-\alpha}\phi(a(\tau),b(\tau);q,t,\tau)\right)_{t}'}{t}-\frac{p}{t}u(t,\tau)$$

and yields $\frac{tu_{t}'(t,\tau)}{u(t,\tau)}=p\left[\frac{t\left(D_{t}^{-\alpha}\phi(a(\tau),b(\tau);q,t,\tau)\right)'}{D_{t}^{-\alpha}\phi(a(\tau),b(\tau);q,t,\tau)}-1\right]$.

Define the analytic functions $f(z)=m+nz+rz^{2}$ and $g(z)=\frac{s}{z}$, with $g(z)\neq 0$, $z\in\mathbb{C}\backslash\{0\}$.

Define also the functions $F(t,\tau)=tw_{t}'(t,\tau)g(w(t,\tau))=s\frac{tw_{t}'(t,\tau)}{w(t,\tau)}$ and $G(t,\tau)=f(w(t,\tau))+F(t,\tau)=m+nw(t,\tau)+r(w(t,\tau))^{2}+s\frac{tw_{t}'(t,\tau)}{w(t,\tau)}$.

We will check the conditions from Lemma 1.9. It is evidently that $F(t,\tau)$ is starlike univalent.

Differentiating the function G with respect to t, we get $G_{t}'(t,\tau) = s+w_{t}'(t,\tau)+2rw(t,\tau)w_{t}'(t,\tau)+s\frac{\left(w_{t}'(t,\tau)+tw_{t^{2}}''(t,\tau)\right)w(t,\tau)-t\left(w_{t}'(t,\tau)\right)^{2}}{(w(t,\tau))^{2}}$ and $\frac{tG_{t}'(t,\tau)}{F(t,\tau)}=\frac{tG_{t}'(t,\tau)}{s\frac{tw_{t}'(t,\tau)}{w(t,\tau)}}=1+\frac{n}{s}w(t,\tau)+\frac{2r}{s}(w(t,\tau))^{2}-\frac{tw_{t}'(t,\tau)}{w(t,\tau)}+\frac{tw_{t^{2}}''(t,\tau)}{w_{t}'(t,\tau)}$.

The second condition

$$\operatorname{Re}\left(\frac{tG_{t}'(t,\tau)}{F(t,\tau)}\right)$$
$$=\operatorname{Re}\left(1+\frac{n}{s}w(t,\tau)+\frac{2r}{s}(w(t,\tau))^{2}-\frac{tw_{t}'(t,\tau)}{w(t,\tau)}+\frac{tw_{t^{2}}''(t,\tau)}{w_{t}'(t,\tau)}\right)>0$$

is true form the relation (4.3.2.3).

We get the function

$$\begin{aligned}
&m + nu(t,\tau) + r(u(t,\tau))^2 + s\frac{tu_t'(t,\tau)}{u(t,\tau)}\\
&\quad = m + n\left[\frac{D_t^{-\alpha}\phi(a(\tau),b(\tau);q,t,\tau)}{t}\right]^p + r\left[\frac{D_t^{-\alpha}\phi(a(\tau),b(\tau);q,t,\tau)}{t}\right]^{2p}\\
&\qquad + sp\left[\frac{t\left(D_t^{-\alpha}\phi(a(\tau),b(\tau);q,t,\tau)\right)_t'}{D_t^{-\alpha}\phi(a(\tau),b(\tau);q,t,\tau)} - 1\right].
\end{aligned}$$

Strong differential subordination (4.3.2.5) can be written in the following form $m+nu(t,\tau)+r(u(t,\tau))^2+s\frac{tu_t'(t,\tau)}{u(t,\tau)} \prec\prec m+nw(t,\tau)+r(w(t,\tau))^2+s\frac{tw_t'(t,\tau)}{w(t,\tau)}$.

The conditions from Lemma 1.9 being fulfilled, we obtain $u(t,\tau) \prec\prec w(t,\tau)$, $(t,\tau)\in\Delta\times\overline{\Delta}$, written as $\left(\frac{D_t^{-\alpha}\phi(a(\tau),b(\tau);q,t,\tau)}{t}\right)^p \prec\prec w(t,\tau)$ with w the best dominant. □

Corollary 4.23 (Alb Lupaş, 2024). *Assuming that relation* (4.3.2.3) *takes place, if the strong differential subordination*

$$\begin{aligned}
&H_\alpha^q(p,m,n,r,s;t,\tau)\\
&\quad \prec\prec m + n\frac{Mt+\tau}{Nt+\tau} + r\left(\frac{Mt+\tau}{Nt+\tau}\right)^2 + s\frac{(M-N)t\tau}{(Mt+\tau)(Nt+\tau)},
\end{aligned}$$

is fulfilled for $m,n,r,s\in\mathbb{C}$, $s\neq 0$, $-1\le N<M\le 1$, *and the function* $H_\alpha^q(p,m,n,r,s;t,\tau)$ *is defined by relation* (4.3.2.4), *then* $\frac{Mt+\tau}{Nt+\tau}$ *is the best dominant for the strong differential subordination*

$$\left(\frac{D_t^{-\alpha}\phi(a(\tau),b(\tau);q,t,\tau)}{t}\right)^p \prec\prec \frac{Mt+\tau}{Nt+\tau},\quad (t,\tau)\in\Delta\times\Delta.$$

Corollary 4.24 (Alb Lupaş, 2024). *Assuming that relation* (4.3.2.3) *takes place, if the strong differential subordination*

$$H_\alpha^q(p,m,n,r,s;t,\tau) \prec\prec m + n\left(\frac{\tau+t}{\tau-t}\right)^k + r\left(\frac{\tau+t}{\tau-t}\right)^{2k} + s\frac{2kt\tau}{\tau^2-t^2},$$

is fulfilled for $m,n,r,s\in\mathbb{C}$, $s\neq 0$, $0<k\le 1$, *and the function* $H_\alpha^q(p,m,n,r,s;t,\tau)$ *is defined by relation* (4.3.2.4), *then* $\left(\frac{\tau+t}{\tau-t}\right)^k$ *is the best*

dominant for the strong differential subordination

$$\left(\frac{D_t^{-\alpha}\phi\left(a\left(\tau\right),b\left(\tau\right);q,t,\tau\right)}{t}\right)^p \prec\prec \left(\frac{\tau+t}{\tau-t}\right)^k, \quad (t,\tau)\in\Delta\times\overline{\Delta}.$$

The strong differential superordination result obtained using the operator given by (4.3.2.2) is the next theorem:

Theorem 4.38 (Alb Lupaş, 2024). *Consider the analytic and univalent function w in $\Delta\times\overline{\Delta}$ with the properties $w(t,\tau)\neq 0$ and $\frac{tw_t'(t,\tau)}{w(t,\tau)}$ is starlike univalent. Suppose that*

$$\operatorname{Re}\left(\frac{2r}{s}\left(w\left(t,\tau\right)\right)^2+\frac{n}{s}w\left(t,\tau\right)\right)>0, \text{ for } n,r,s\in\mathbb{C}, \quad s\neq 0. \quad (4.3.2.7)$$

If $\left(\frac{D_t^{-\alpha}\phi(a(\tau),b(\tau);q,t,\tau)}{t}\right)^p \in \mathcal{H}\left[w\left(0,\tau\right),\left(\alpha-1\right)p,\tau\right]\cap Q^*$, *the function* $H_\alpha^q\left(p,m,n,r,s;t,\tau\right)$ *defined by the relation* (4.3.2.4) *is univalent in* $\Delta\times\overline{\Delta}$, *then the strong differential subordination*

$$m+nw\left(t,\tau\right)+r\left(w\left(t,\tau\right)\right)^2+s\frac{tw_t'\left(t,\tau\right)}{w\left(t,\tau\right)} \prec\prec H_\alpha^q\left(p,m,n,r,s;t,\tau\right) \quad (4.3.2.8)$$

is endowed for $m,n,r,s\in\mathbb{C}$, $s\neq 0$, then w is the best subordinant for the following strong differential superordination

$$w\left(t,\tau\right) \prec\prec \left(\frac{D_t^{-\alpha}\phi\left(a\left(\tau\right),b\left(\tau\right);q,t,\tau\right)}{t}\right)^p, \quad (t,\tau)\in\Delta\times\overline{\Delta}. \quad (4.3.2.9)$$

Proof. Considering again the function $u\left(t,\tau\right)=\left(\frac{D_t^{-\alpha}\phi(a(\tau),b(\tau);q,t,\tau)}{t}\right)^p$, $(t,\tau)\in(\Delta\backslash\{0\})\times\overline{\Delta}$, and the analytic functions $f\left(z\right)=m+nz+rz^2$ and $g\left(z\right)=\frac{s}{z}$, with $g\left(z\right)\neq 0$, $z\in\mathbb{C}\backslash\{0\}$, we verify the conditions from Lemma 1.12.

Taking account that $\frac{f_t'(w(t,\tau))}{g(w(t,\tau))}=\frac{w_t'(t,\tau)[n+2rw(t,\tau)]w(t,\tau)}{s}$, it follows that $\operatorname{Re}\left(\frac{f_t'(w(t,\tau))}{g(w(t,\tau))}\right)=\operatorname{Re}\left(\frac{2r}{s}\left(w\left(t,\tau\right)\right)^2+\frac{n}{s}w\left(t,\tau\right)\right)>0$, for $n,r,s\in\mathbb{C}$, $s\neq 0$, by relation (4.3.2.7).

Strong differential superordination (4.3.2.8) can be written as

$$m+nw\left(t,\tau\right)+r\left(w\left(t,\tau\right)\right)^2+s\frac{tw_t'\left(t,\tau\right)}{w\left(t,\tau\right)}$$

$$\prec\prec m+nu\left(t,\tau\right)+r\left(u\left(t,\tau\right)\right)^2+s\frac{tu_t'\left(t,\tau\right)}{u\left(t,\tau\right)}.$$

The conditions from Lemma 1.12 being fulfilled, we get

$$w(t,\tau) \prec\prec u(t,\tau) = \left(\frac{D_t^{-\alpha}\phi(a(\tau),b(\tau);q,t,\tau)}{t}\right)^p, \quad (t,\tau) \in \Delta \times \overline{\Delta},$$

and w is the best subordinant. □

Corollary 4.25 (Alb Lupaş, 2024). *Assuming that relation* (4.3.2.7) *takes place and* $\left(\frac{D_t^{-\alpha}\phi(a(\tau),b(\tau);q,t,\tau)}{t}\right)^p \in \mathcal{H}[w(0,\tau),(\alpha-1)p,\tau] \cap Q^*$, *if the strong differential superordination*

$$m + n\frac{Mt+\tau}{Nt+\tau} + r\left(\frac{Mt+\tau}{Nt+\tau}\right)^2 + s\frac{(M-N)t\tau}{(Mt+\tau)(Nt+\tau)}$$
$$\prec\prec H_\alpha^q(p,m,n,r,s;t,\tau),$$

is fulfilled for $m,n,r,s \in \mathbb{C}$, $s \neq 0$, $-1 \leq N < M \leq 1$, *and the function* $H_\alpha^q(p,m,n,r,s;t,\tau)$ *is defined by relation* (4.3.2.4), *then* $\frac{Mt+\tau}{Nt+\tau}$ *is the best subordinant for the strong differential superordination*

$$\frac{Mt+\tau}{Nt+\tau} \prec\prec \left(\frac{D_t^{-\alpha}\phi(a(\tau),b(\tau);q,t,\tau)}{t}\right)^p, \quad (t,\tau) \in \Delta \times \Delta.$$

Corollary 4.26 (Alb Lupaş, 2024). *Assuming that relation* (4.3.2.7) *takes place and* $\left(\frac{D_t^{-\alpha}\phi(a(\tau),b(\tau);q,t,\tau)}{t}\right)^p \in \mathcal{H}[w(0,\tau),(\alpha-1)p,\tau] \cap Q^*$, *if the strong differential subordination*

$$m + n\left(\frac{\tau+t}{\tau-t}\right)^k + r\left(\frac{\tau+t}{\tau-t}\right)^{2k} + s\frac{2kt\tau}{\tau^2-t^2} \prec\prec H_\alpha^q(p,m,n,r,s;t,\tau),$$

is fulfilled for $m,n,r,s \in \mathbb{C}$, $s \neq 0$, $0 < k \leq 1$, *and the function* $H_\alpha^q(p,m,n,r,s;t,\tau)$ *is defined by relation* (4.3.2.4), *then* $\left(\frac{\tau+t}{\tau-t}\right)^k$ *is the best subordinant for the strong differential superordination*

$$\left(\frac{\tau+t}{\tau-t}\right)^k \prec\prec \left(\frac{D_t^{-\alpha}\phi(a(\tau),b(\tau);q,t,\tau)}{t}\right)^p, \quad (t,\tau) \in \Delta \times \overline{\Delta}.$$

Looking at Theorems 4.37 and 4.38 together, they generate a sandwich-type result.

Theorem 4.39 (Alb Lupaş, 2024). *Consider the analytic and univalent functions w_1 and w_2 in $\Delta \times \overline{\Delta}$ with the properties $w_1(t,\tau) \neq 0$, $w_2(t,\tau\zeta) \neq 0$, $\forall\,(t,\tau) \in \Delta\times\overline{\Delta}$. Assuming the functions $\frac{t(w_1)'_t(t,\tau)}{w_1(t,\tau)}$, $\frac{t(w_2)'_t(t,\tau)}{w_2(t,\tau)}$ are starlike univalent in $\Delta \times \overline{\Delta}$ and w_1 satisfies relation (4.3.2.3) and w_2 satisfies relation (4.3.2.7) if $\left(\frac{D_t^{-\alpha}\phi(a(\tau),b(\tau);q,t,\tau)}{t}\right)^p \in \mathcal{H}[w(0,\tau),(\alpha-1)p,\tau] \cap Q^*$, the function $H_\alpha^q(p,m,n,r,s;t,\tau)$ defined in (4.3.2.4) is univalent in $\Delta \times \overline{\Delta}$ and the sandwich-type result*

$$m + nw_1(t,\tau) + r(w_1(t,\tau))^2 + s\frac{t(w_1)'_t(t,\tau)}{w_1(t,\tau)} \prec\prec H_\alpha^q(p,m,n,r,s;t,\tau)$$
$$\prec\prec m + nw_2(t,\tau) + r(w_2(t,\tau))^2 + s\frac{t(w_2)'_t(t,\tau)}{w_2(t,\tau)},$$

is endowed for $m,n,r,s \in \mathbb{C}$, $s \neq 0$, then w_1 and w_2 are respectively the best subordinant and the best dominant for the following sandwich-type result

$$w_1(t,\tau) \prec\prec \left(\frac{D_t^{-\alpha}\phi(a(\tau),b(\tau);q,t,\tau)}{t}\right)^p \prec\prec w_2(t,\tau), \quad (t,\tau) \in \Delta \times \overline{\Delta}.$$

Corollary 4.27 (Alb Lupaş, 2024). *Assuming that relations (4.3.2.3) and (4.3.2.7) take place and $\left(\frac{D_t^{-\alpha}\phi(a(\tau),b(\tau);q,t,\tau)}{t}\right)^p \in \mathcal{H}[w(0,\tau),(\alpha-1)p,\tau] \cap Q^*$, if the sandwich-type result*

$$m + n\frac{M_1t+\tau}{N_1t+\tau} + r\left(\frac{M_1t+\tau}{N_1t+\tau}\right)^2 + s\frac{(M_1-N_1)t\tau}{(M_1t+\tau)(N_1t+\tau)}$$
$$\prec\prec H_\alpha^q(p,m,n,r,s;t,\tau) \prec\prec m + n\frac{M_2t+\tau}{N_2t+\tau} + r\left(\frac{M_2t+\tau}{N_2t+\tau}\right)^2$$
$$+ s\frac{(M_2-N_2)t\tau}{(M_2t+\tau)(N_2t+\tau)},$$

is fulfilled for $m,n,r,s \in \mathbb{C}$, $s \neq 0$, $-1 \leq N_2 < N_1 < M_1 < M_2 \leq 1$, and the function $H_\alpha^q(p,m,n,r,s;t,\tau)$ is defined by relation (4.3.2.4), then $\frac{M_1t+\tau}{N_1t+\tau}$ and $\frac{M_2t+\tau}{N_2t+\tau}$ are respectively the best subordinant and the best dominant for the following sandwich-type result

$$\frac{M_1t+\tau}{N_1t+\tau} \prec\prec \left(\frac{D_t^{-\alpha}\phi(a(\tau),b(\tau);q,t,\tau)}{t}\right)^p \prec\prec \frac{M_2t+\tau}{N_2t+\tau}.$$

Corollary 4.28 (Alb Lupaş, 2024). *Assuming that relations* (4.3.2.3) *and* (4.3.2.7) *take place and* $\left(\frac{D_t^{-\alpha}\phi(a(\tau),b(\tau);q,t,\tau)}{t}\right)^p \in \mathcal{H}\left[w\left(0,\tau\right),(\alpha-1)\right.$ $\left.p,\tau\right] \cap Q^*$, *if the sandwich-type result*

$$m+n\left(\frac{\tau+t}{\tau-t}\right)^{k_1}+r\left(\frac{\tau+t}{\tau-t}\right)^{2k_1}+s\frac{2k_1t\tau}{\tau^2-t^2} \prec\prec H_\alpha^q\left(p,m,n,r,s;t,\tau\right)$$

$$\prec\prec m+n\left(\frac{\tau+t}{\tau-t}\right)^{k_2}+r\left(\frac{\tau+t}{\tau-t}\right)^{2k_2}+s\frac{2k_2t\tau}{\tau^2-t^2},$$

is fulfilled for $m,n,r,s\in\mathbb{C}$, $s\neq 0$, $-1\leq N_2<N_1<M_1<M_2\leq 1$, *and the function* $H_\alpha^q\left(p,m,n,r,s;t,\tau\right)$ *is defined by relation* (4.3.2.4), *then* $\left(\frac{\tau+t}{\tau-t}\right)^{k_1}$ *and* $\left(\frac{\tau+t}{\tau-t}\right)^{k_2}$ *are respectively the best subordinant and the best dominant for the following sandwich-type result*

$$\left(\frac{\tau+t}{\tau-t}\right)^{k_1} \prec\prec \left(\frac{D_t^{-\alpha}\phi\left(a\left(\tau\right),b\left(\tau\right);q,t,\tau\right)}{t}\right)^p \prec\prec \left(\frac{\tau+t}{\tau-t}\right)^{k_2}.$$

4.3.3 *Strong Differential Subordination and Superordination Results for Extended q-Analogue of Multiplier Transformation*

The results obtained in the paper Alb Lupaş and Ghanim (2023) refer to quantum calculus applications regarding the theories of strong differential subordination and superordination involving the extended q-analogue of the multiplier transformation. Interesting strong differential subordination and superordination results are derived.

The extended q-analogue of multiplier transformation to the new class of analytic functions $\mathcal{A}_\zeta^*$ defined in Definition 2.13 has the following form

$$\mathcal{I}_q^{m,l}f\left(z,\zeta\right)=z+\sum_{k=2}^{\infty}\left(\frac{\left[k+l\right]_q}{\left[1+l\right]_q}\right)^m a_k\left(\zeta\right)z^k, \quad z\in U,\ \zeta\in\overline{U},$$

with $l>-1$, $q\in(0,1)$, m a real number and $f(z,\zeta)=z+\sum_{k=2}^{\infty}a_k\left(\zeta\right)z^k\in\mathcal{A}_\zeta^*$.

Using the properties of q-calculus, we obtain

$$z\mathcal{D}_q\left(\mathcal{I}_q^{m,l}f\left(z,\zeta\right)\right)=\left(1+\frac{\left[l\right]_q}{q^l}\right)\mathcal{I}_q^{m+1,l}f\left(z,\zeta\right)-\frac{\left[l\right]_q}{q^l}\mathcal{I}_q^{m,l}f\left(z,\zeta\right).$$

We define a new subclass of normalized analytic functions using the extended q-analogue of the multiplier transformation $\mathcal{I}_q^{m,l}$ defined below.

Definition 4.2 (Alb Lupaş and Ghanim, 2023). Let $\alpha \in [0,1)$. The subclass $\mathcal{S}_{m,l,\zeta}^{q}(\alpha)$ consists the functions $f \in \mathcal{A}_{\zeta}^{*}$ with property

$$\operatorname{Re}\left(\mathcal{I}_q^{m,l} f(z,\zeta)\right)_z' > \alpha, \quad z \in U, \ \zeta \in \overline{U}.$$

The first result concerning the subclass $\mathcal{S}_{m,l,\zeta}^{q}(\alpha)$ establishes its convexity.

Theorem 4.40 (Alb Lupaş and Ghanim, 2023). *$\mathcal{S}_{m,l,\zeta}^{q}(\alpha)$ is a convex set.*

Proof. Consider the functions

$$f_j(z,\zeta) = z + \sum_{k=2}^{\infty} a_{jk}(\zeta) z^k, \quad z \in U, \quad \zeta \in \overline{U}, \ j = 1,2,$$

belonging to the subclass $\mathcal{S}_{m,l,\zeta}^{q}(\alpha)$. It is enough to prove that the function

$$f(z,\zeta) = \lambda_1 f_1(z,\zeta) + \lambda_2 f_2(z,\zeta)$$

belongs to the subclass $\mathcal{S}_{m,l,\zeta}^{q}(\alpha)$, with λ_1 and λ_2 positive real numbers such that $\lambda_1 + \lambda_2 = 1$.

The function f has the following form

$$f(z,\zeta) = z + \sum_{k=2}^{\infty} \left(\lambda_1 a_{1k}(\zeta) + \lambda_2 a_{2k}(\zeta)\right) z^k, \quad z \in U, \ \zeta \in \overline{U},$$

and

$$\mathcal{I}_q^{m,l} f(z,\zeta) = z + \sum_{k=2}^{\infty} \left(\frac{[k+l]_q}{[1+l]_q}\right)^m (\lambda_1 a_{1k}(\zeta) + \lambda_2 a_{2k}(\zeta)) z^k, \quad z \in U, \ \zeta \in \overline{U}. \tag{4.3.3.1}$$

Differentiating relation (4.3.3.1) with respect to z, we obtain

$$\left(\mathcal{I}_q^{m,l} f(z,\zeta)\right)_z' = 1 + \sum_{k=2}^{\infty} \left(\frac{[k+l]_q}{[1+l]_q}\right)^m \left(\lambda_1 a_{1k}(\zeta) + \lambda_2 a_{2k}(\zeta)\right) k z^{k-1},$$

$z \in U$, $\zeta \in \overline{U}$.

Hence,

$$\operatorname{Re}\left(\mathcal{I}_q^{m,l} f(z,\zeta)\right)_z' = 1 + \operatorname{Re}\left(\lambda_1 \sum_{k=2}^{\infty} k \left(\frac{[k+l]_q}{[1+l]_q}\right)^m a_{1k}(\zeta) z^{k-1}\right) + \operatorname{Re}\left(\lambda_2 \sum_{k=2}^{\infty} k \left(\frac{[k+l]_q}{[1+l]_q}\right)^m a_{2k}(\zeta) z^{k-1}\right). \tag{4.3.3.2}$$

Taking account that $f_1, f_2 \in \mathcal{S}^q_{m,l,\zeta}(\alpha)$ we can write

$$\operatorname{Re}\left(\lambda_j \sum_{k=2}^{\infty} k \left(\frac{[k+l]_q}{[1+l]_q}\right)^m a_{jk}(\zeta) z^{k-1}\right) > \lambda_k (\alpha - 1), \quad j = 1, 2. \tag{4.3.3.3}$$

Using relation (4.3.3.3), we get from (4.3.3.2)

$$\operatorname{Re}\left(\mathcal{I}_q^{m,l} f(z,\zeta)\right)'_z > 1 + \lambda_1 (\alpha - 1) + \lambda_2 (\alpha - 1) = \alpha, \quad z \in U,$$

which proved that the set $\mathcal{S}^q_{m,l,\zeta}(\alpha)$ is convex. □

We next investigate a series of strong differential subordinations involving the convex functions of the subclass $\mathcal{S}^q_{m,l,\zeta}(\alpha)$ and the extended q-analogue of the multiplier transformation $\mathcal{I}_q^{m,l}$.

Theorem 4.41 (Alb Lupaş and Ghanim, 2023). *Considering a convex function $g(z,\zeta)$, we define the function*

$$h(z,\zeta) = \frac{z g'_z(z,\zeta)}{a+2} + g(z,\zeta), \quad a > 0, \ z \in U, \ \zeta \in \overline{U}. \tag{4.3.3.4}$$

For $f \in \mathcal{S}^q_{m,l,\zeta}(\alpha)$, consider

$$F(z,\zeta) = \frac{a+2}{z^{a+1}} \int_0^z t^a f(t,\zeta)\, dt, \quad z \in U, \ \zeta \in \overline{U}. \tag{4.3.3.5}$$

Then the strong differential subordination

$$\left(\mathcal{I}_q^{m,l} f(z,\zeta)\right)'_z \prec\prec h(z,\zeta), \quad z \in U, \ \zeta \in \overline{U}, \tag{4.3.3.6}$$

implies the sharp strong differential subordination

$$\left(\mathcal{I}_q^{m,l} F(z,\zeta)\right)'_z \prec\prec g(z,\zeta), \quad z \in U, \ \zeta \in \overline{U}.$$

Proof. Relation (4.3.3.5) can be written as follows

$$z^{a+1} F(z,\zeta) = (a+2) \int_0^z t^a f(t,\zeta)\, dt, \tag{4.3.3.7}$$

and differentiating it with respect to z, we have

$$z F'_z(z,\zeta) + (a+1) F(z,\zeta) = (a+2) f(z,\zeta) \tag{4.3.3.8}$$

and

$$z\left(\mathcal{I}_q^{m,l}F(z,\zeta)\right)_z' + (a+1)\,\mathcal{I}_q^{m,l}F(z,\zeta)$$
$$= (a+2)\,\mathcal{I}_q^{m,l}f(z,\zeta), \quad z\in U, \quad \zeta\in\overline{U}. \tag{4.3.3.9}$$

Differentiating the last relation with respect to z, we get

$$\frac{z\left(\mathcal{I}_q^{m,l}F(z,\zeta)\right)_{z^2}''}{a+2} + \left(\mathcal{I}_q^{m,l}F(z,\zeta)\right)_z' = \left(\mathcal{I}_q^{m,l}f(z,\zeta)\right)_z', \quad z\in U,\ \zeta\in\overline{U}, \tag{4.3.3.10}$$

and the strong differential subordination (4.3.3.6) can be written in the form

$$\frac{z\left(\mathcal{I}_q^{m,l}F(z,\zeta)\right)_{z^2}''}{a+2} + \left(\mathcal{I}_q^{m,l}F(z,\zeta)\right)_z' \prec\prec \frac{zg_z'(z,\zeta)}{a+2} + g(z,\zeta). \tag{4.3.3.11}$$

Denoting

$$p(z,\zeta) = \left(\mathcal{I}_q^{m,l}F(z,\zeta)\right)_z', \quad z\in U,\ \zeta\in\overline{U}, \tag{4.3.3.12}$$

strong differential subordination (4.3.3.11) has the following form

$$\frac{zp_z'(z,\zeta)}{a+2} + p(z,\zeta) \prec\prec \frac{zg_z'(z,\zeta)}{a+2} + g(z,\zeta), \quad z\in U,\ \zeta\in\overline{U}.$$

Applying Lemma 1.8, we get

$$p(z,\zeta) \prec\prec g(z,\zeta),$$

that means

$$\left(\mathcal{I}_q^{m,l}F(z,\zeta)\right)_z' \prec\prec g(z,\zeta), \quad z\in U,\ \zeta\in\overline{U},$$

and this result is sharp. □

Theorem 4.42 (Alb Lupaş and Ghanim, 2023). *Let* $h(z,\zeta) = \frac{\zeta+(2\delta-\zeta)z}{1+z}$, $z\in U$, $\zeta\in\overline{U}$. *If we denote by*

$$I_a(f)(z,\zeta) = \frac{a+2}{z^{a+1}}\int_0^z t^a f(t,\zeta)\,dt, \quad a>0, \tag{4.3.3.13}$$

then

$$I_a\left[\mathcal{S}^q_{m,l,\zeta}(\alpha)\right]\subset\mathcal{S}^q_{m,l,\zeta}(\alpha^*),\tag{4.3.3.14}$$

where

$$\alpha^*=2\alpha-\zeta+2(a+2)(\zeta-\alpha)\int_0^1\frac{t^{a+1}}{t+1}dt.\tag{4.3.3.15}$$

Proof. We obtain using the hypothesis of Theorem 4.42 for the convex function $h(z,\zeta)=\frac{\zeta+(2\delta-\zeta)z}{1+z}$, $z\in U$, $\zeta\in\overline{U}$ and following the same steps like the proof of Theorem 4.41, the strong differential subordination

$$\frac{zp'_z(z,\zeta)}{a+2}+p(z,\zeta)\prec\prec h(z,\zeta),$$

with p defined by relation (4.3.3.12).

Applying Lemma 1.7, we get the strong differential subordinations

$$p(z,\zeta)\prec\prec g(z,\zeta)\prec\prec h(z,\zeta),$$

equivalently with

$$\left(\mathcal{I}^{m,l}_q F(z,\zeta)\right)'_z\prec\prec g(z,\zeta)\prec\prec h(z,\zeta),$$

where

$$\begin{aligned}g(z,\zeta)&=\frac{a+2}{z^{a+2}}\int_0^z t^{a+1}\frac{\zeta+(2\alpha-\zeta)t}{1+t}dt\\&=2\alpha-\zeta+\frac{2(a+2)(\zeta-\alpha)}{z^{a+2}}\int_0^z\frac{t^{a+1}}{t+1}dt.\end{aligned}$$

Taking account that g is a convex function with $g\left(U\times\overline{U}\right)$ symmetric to the real axis, we obtain

$$\begin{aligned}\operatorname{Re}\left(\mathcal{I}^{m,l}_q F(z,\zeta)\right)'_z&\geq\min_{|z|=1}\operatorname{Re}\,g(z,\zeta)=\operatorname{Re}\,g(1,\zeta)=\alpha^*\\&=2\alpha-\zeta+2(a+2)(\zeta-\alpha)\int_0^z\frac{t^{a+1}}{t+1}.\end{aligned}$$

□

Theorem 4.43 (Alb Lupaş and Ghanim, 2023). *Considering the convex function $g(z,\zeta)$ such that $g(0,\zeta)=1$, we define the function*

$$h(z,\zeta)=zg'_z(z,\zeta)+g(z,\zeta),\quad z\in U,\ \zeta\in\overline{U}.$$

If $f\in\mathcal{A}^*_\zeta$ verifies the strong differential subordination

$$\left(\mathcal{I}_q^{m,l}f(z,\zeta)\right)'_z\prec\prec h(z,\zeta),\quad z\in U,\ \zeta\in\overline{U}, \tag{4.3.3.16}$$

then the sharp strong differential subordination

$$\frac{\mathcal{I}_q^{m,l}f(z,\zeta)}{z}\prec\prec g(z,\zeta),\quad z\in U,\ \zeta\in\overline{U}$$

holds.

Proof. Considering

$$\begin{aligned}p(z,\zeta)&=\frac{\mathcal{I}_q^{m,l}f(z,\zeta)}{z}\\&=\frac{z+\sum_{k=2}^{\infty}\left(\frac{[k+l]_q}{[1+l]_q}\right)^m a_k(\zeta)z^k}{z}=1+p_1(\zeta)z+p_2(\zeta)z^2+\cdots,\end{aligned}$$

$z\in U$, $\zeta\in\overline{U}$, so we can write

$$zp(z,\zeta)=\mathcal{I}_q^{m,l}f(z,\zeta),\quad z\in U,\ \zeta\in\overline{U},$$

and differentiating it with respect to z, we get

$$\left(\mathcal{I}_q^{m,l}f(z,\zeta)\right)'_z=zp'_z(z,\zeta)+p(z,\zeta),\quad z\in U,\ \zeta\in\overline{U}.$$

The strong differential subordination (4.3.3.16) take the form

$$zp'_z(z,\zeta)+p(z,\zeta)\prec\prec h(z,\zeta)=zg'_z(z,\zeta)+g(z,\zeta),\quad z\in U,\ \zeta\in\overline{U},$$

and applying Lemma 1.8, we get

$$p(z,\zeta)\prec\prec g(z,\zeta),\quad z\in U,\ \zeta\in\overline{U},$$

that means

$$\frac{\mathcal{I}_q^{m,l}f(z,\zeta)}{z}\prec\prec g(z,\zeta),\quad z\in U,\ \zeta\in\overline{U}.$$

□

Theorem 4.44 (Alb Lupaş and Ghanim, 2023). *Considering the convex function $h(z,\zeta)$ with the property $h(0,\zeta)=1$, $\zeta\in\overline{U}$, for $f\in\mathcal{A}_{\zeta}^{*}$ that verifies the strong subordination*

$$\left(\mathcal{I}_q^{m,l}f(z,\zeta)\right)_z' \prec\prec h(z,\zeta), \quad z\in U,\ \zeta\in\overline{U}, \tag{4.3.3.17}$$

we get the strong differential subordination

$$\frac{\mathcal{I}_q^{m,l}f(z,\zeta)}{z}\prec\prec g(z,\zeta), \quad z\in U,\ \zeta\in\overline{U},$$

for the convex function $g(z,\zeta)=\frac{1}{z}\int_0^z h(t,\zeta)dt$, which is the best dominant.

Proof. Denote

$$p(z,\zeta)=\frac{\mathcal{I}_q^{m,l}f(z,\zeta)}{z}=1+\sum_{k=2}^{\infty}\left(\frac{[k+l]_q}{[1+l]_q}\right)^m a_k(\zeta)\,z^{k-1}, \quad z\in U,\ \zeta\in\overline{U}.$$

Differentiating the relation with respect to z, we get

$$\left(\mathcal{I}_q^{m,l}f(z,\zeta)\right)_z'=zp_z'(z,\zeta)+p(z,\zeta), \quad z\in U,\ \zeta\in\overline{U}$$

and strong differential subordination (4.3.3.17) has the following form

$$zp_z'(z,\zeta)+p(z,\zeta)\prec\prec h(z,\zeta), \quad z\in U,\ \zeta\in\overline{U}.$$

After applying Lemma 1.7, we get

$$p(z,\zeta)\prec\prec g(z,\zeta)=\frac{1}{z}\int_0^z h(t,\zeta)dt, \quad z\in U,\ \zeta\in\overline{U},$$

written as

$$\frac{\mathcal{I}_q^{m,l}f(z,\zeta)}{z}\prec\prec g(z,\zeta)=\frac{1}{z}\int_0^z h(t,\zeta)dt, \quad z\in U,\ \zeta\in\overline{U}$$

for g the best dominant. □

Theorem 4.45 (Alb Lupaş and Ghanim, 2023). *Considering a convex function $g(z,\zeta)$ such that $g(0,\zeta)=1$, we define the function $h(z,\zeta)=zg_z'(z,\zeta)+g(z,\zeta)$, $z\in U$, $\zeta\in\overline{U}$. When $f\in\mathcal{A}_{\zeta}^{*}$ verifies the strong subordination*

$$\left(\frac{z\mathcal{I}_q^{m+1,l}f(z,\zeta)}{\mathcal{I}_q^{m,l}f(z,\zeta)}\right)_z' \prec\prec h(z,\zeta), \quad z\in U,\ \zeta\in\overline{U}, \tag{4.3.3.18}$$

then we get the sharp strong differential subordination

$$\frac{\mathcal{I}_q^{m+1,l} f(z,\zeta)}{\mathcal{I}_q^{m,l} f(z,\zeta)} \prec\prec g(z,\zeta), \quad z \in U,\ \zeta \in \overline{U}.$$

Proof. Denote

$$p(z,\zeta) = \frac{\mathcal{I}_q^{m+1,l} f(z,\zeta)}{\mathcal{I}_q^{m,l} f(z,\zeta)} = \frac{z + \sum_{k=2}^{\infty} \left(\frac{[k+l]_q}{[1+l]_q}\right)^{m+1} a_k(\zeta) z^k}{z + \sum_{k=2}^{\infty} \left(\frac{[k+l]_q}{[1+l]_q}\right)^{m} a_k(\zeta) z^k}.$$

Differentiating it with respect to z, we get $p'_z(z,\zeta) = \frac{\left(\mathcal{I}_q^{m+1,l} f(z,\zeta)\right)'_z}{\mathcal{I}_q^{m,l} f(z,\zeta)} - p(z,\zeta) \frac{\left(\mathcal{I}_q^{m,l} f(z,\zeta)\right)'_z}{\mathcal{I}_q^{m,l} f(z,\zeta)}$ written as

$$zp'_z(z,\zeta) + p(z,\zeta) = \left(\frac{z\mathcal{I}_q^{m+1,l} f(z,\zeta)}{\mathcal{I}_q^{m,l} f(z,\zeta)}\right)'_z.$$

Strong differential subordination (4.3.3.18) has the following form, for $z \in U$, $\zeta \in \overline{U}$,

$$zp'_z(z,\zeta) + p(z,\zeta) \prec\prec h(z,\zeta) = zg'_z(z,\zeta) + g(z,\zeta),$$

and applying Lemma 1.8, we get the strong differential subordination, for $z \in U$, $\zeta \in \overline{U}$,

$$p(z,\zeta) \prec\prec g(z,\zeta),$$

written as

$$\frac{\mathcal{I}_q^{m+1,l} f(z,\zeta)}{\mathcal{I}_q^{m,l} f(z,\zeta)} \prec\prec g(z,\zeta).$$

□

Next, strong differential superordinations are investigated concerning the extended q-analogue of the multiplier transformation $\mathcal{I}_q^{m,l}$ and its derivatives of first and second order with respect to z. The best subordinant is given for each of the investigated strong differential superordinations.

Theorem 4.46 (Alb Lupaş and Ghanim, 2023). *Let us consider* $f \in \mathcal{A}_\zeta^*$ *and a convex function* $h(z,\zeta)$ *in* $U \times \overline{U}$ *such that* $h(0,\zeta) = 1$. *Denote by* $F(z,\zeta) = \frac{a+2}{z^{a+1}} \int_0^z t^a f(t,\zeta)\,dt$, $z \in U$, $\zeta \in \overline{U}$, Re $a > -2$, *and assume* *that* $\left(\mathcal{I}_q^{m,l} f(z,\zeta)\right)_z'$ *is a univalent function in* $U \times \overline{U}$, $\left(\mathcal{I}_q^{m,l} F(z,\zeta)\right)_z' \in Q^* \cap \mathcal{H}^*[1,1,\zeta]$. *If the strong differential superordination*

$$h(z,\zeta) \prec\prec \left(\mathcal{I}_q^{m,l} f(z,\zeta)\right)_z', \quad z \in U, \zeta \in \overline{U}, \tag{4.3.3.19}$$

holds, then

$$g(z,\zeta) \prec\prec \left(\mathcal{I}_q^{m,l} F(z,\zeta)\right)_z', \quad z \in U, \zeta \in \overline{U},$$

with the convex function $g(z,\zeta) = \frac{a+2}{z^{a+2}} \int_0^z h(t,\zeta) t^{a+1} dt$ *the best subordinant.*

Proof. Using the relation $z^{a+1} F(z,\zeta) = (a+2) \int_0^z t^a f(t,\zeta)\, dt$ from Theorem 4.41 and differentiating it with respect to z, we can write $zF_z'(z,\zeta) + (a+1) F(z,\zeta) = (a+2) f(z,\zeta)$ in the following form $z\left(\mathcal{I}_q^{m,l} F(z,\zeta)\right)_z' + (a+1) \mathcal{I}_q^{m,l} F(z,\zeta) = (a+2) \mathcal{I}_q^{m,l} f(z,\zeta)$, $z \in U, \zeta \in \overline{U}$, which after differentiating it again with respect to z, has the form

$$\frac{z\left(\mathcal{I}_q^{m,l} F(z,\zeta)\right)_{z^2}''}{a+2} + \left(\mathcal{I}_q^{m,l} F(z,\zeta)\right)_z' = \left(\mathcal{I}_q^{m,l} f(z,\zeta)\right)_z', \quad z \in U, \ \zeta \in \overline{U}.$$

Using the last relation, the strong superordination (4.3.3.19) can be written

$$h(z,\zeta) \prec\prec \frac{z\left(\mathcal{I}_q^{m,l} F(z,\zeta)\right)_{z^2}''}{a+2} + \left(\mathcal{I}_q^{m,l} F(z,\zeta)\right)_z'. \tag{4.3.3.20}$$

Define

$$p(z,\zeta) = \left(\mathcal{I}_q^{m,l} F(z,\zeta)\right)_z', \quad z \in U, \zeta \in \overline{U}, \tag{4.3.3.21}$$

and replacing (4.3.3.21) in (4.3.3.20), we have $h(z,\zeta) \prec\prec \frac{zp_z'(z,\zeta)}{a+2} + p(z,\zeta)$, $z \in U$, $\zeta \in \overline{U}$. Applying Lemma 1.10 considering $n = 1$ and $\alpha = a+2$, it yields $g(z,\zeta) \prec\prec p(z,\zeta)$, equivalently with $g(z,\zeta) \prec\prec \left(\mathcal{I}_q^{m,l} F(z,\zeta)\right)_z'$, $z \in U$, $\zeta \in \overline{U}$, with the best subordinant $g(z,\zeta) = \frac{a+2}{z^{a+2}} \int_0^z h(t,\zeta) t^{a+1} dt$ convex function. □

Theorem 4.47 (Alb Lupaş and Ghanim, 2023). *Considering a convex function* $g(z,\zeta)$, *we define the function* $h(z,\zeta) = \frac{zg_z'(z,\zeta)}{a+2} + g(z,\zeta)$, *with* Re $a > -2$, $z \in U$, $\zeta \in \overline{U}$. *For* $f \in \mathcal{A}_\zeta^*$, *denote* $F(z,\zeta) = \frac{a+2}{z^{a+1}} \int_0^z t^a f(t,\zeta)\, dt$, $z \in U$, $\zeta \in \overline{U}$ *and assume that* $\left(\mathcal{I}_q^{m,l} f(z,\zeta)\right)_z'$ *is* *univalent in* $U \times \overline{U}$ *and* $\left(\mathcal{I}_q^{m,l} F(z,\zeta)\right)'_z \in Q^* \cap \mathcal{H}^*[1,1,\zeta]$. *When the strong differential superordination*

$$h(z,\zeta) \prec\prec \left(\mathcal{I}_q^{m,l} f(z,\zeta)\right)'_z, \quad z \in U, \ \zeta \in \overline{U}, \tag{4.3.3.22}$$

holds, then

$$g(z,\zeta) \prec\prec \left(\mathcal{I}_q^{m,l} F(z,\zeta)\right)'_z, \quad z \in U, \ \zeta \in \overline{U},$$

for $g(z,\zeta) = \frac{a+2}{z^{a+2}} \int_0^z h(t,\zeta) t^{a+1} dt$ *the best subordinant.*

Proof. Considering $p(z,\zeta) = \left(\mathcal{I}_q^{m,l} F(z,\zeta)\right)'_z$, $z \in U, \zeta \in \overline{U}$, following the proof of Theorem 4.46, we can write the strong differential superordination (4.3.3.22) in the following form

$$h(z,\zeta) = \frac{z g'_z(z,\zeta)}{a+2} + g(z,\zeta) \prec\prec \frac{z p'_z(z,\zeta)}{a+2} + p(z,\zeta), \quad z \in U, \ \zeta \in \overline{U}.$$

Applying Lemma 1.11 for $\alpha = a+2$ and $n = 1$, we obtain the strong differential superordination $g(z,\zeta) \prec\prec p(z,\zeta) = \left(\mathcal{I}_q^{m,l} F(z,\zeta)\right)'_z$, $z \in U$, $\zeta \in \overline{U}$, having $g(z,\zeta) = \frac{a+2}{z^{a+2}} \int_0^z h(t,\zeta) t^{a+1} dt$ the best subordinant. □

Theorem 4.48 (Alb Lupaş and Ghanim, 2023). *For* $f \in \mathcal{A}_\zeta^*$ *denote* $F(z,\zeta) = \frac{a+2}{z^{a+1}} \int_0^z t^a f(t,\zeta)\, dt$, $z \in U$, $\zeta \in \overline{U}$, *and* $h(z,\zeta) = \frac{\zeta + (2\alpha - \zeta) z}{1+z}$, *where* $\operatorname{Re} a > -2$, $\alpha \in [0,1)$. *Assume that* $\left(\mathcal{I}_q^{m,l} f(z,\zeta)\right)'_z$ *is univalent in* $U \times \overline{U}$, $\left(\mathcal{I}_q^{m,l} F(z,\zeta)\right)'_z \in Q^* \cap \mathcal{H}^*[1,1,\zeta]$ *and the strong differential superordination*

$$h(z,\zeta) \prec\prec \left(\mathcal{I}_q^{m,l} f(z,\zeta)\right)'_z, \quad z \in U, \ \zeta \in \overline{U}, \tag{4.3.3.23}$$

is satisfied, then the strong differential superordination

$$g(z,\zeta) \prec\prec \left(\mathcal{I}_q^{m,l} F(z,\zeta)\right)'_z, \quad z \in U, \ \zeta \in \overline{U},$$

is satisfied for the convex function $g(z,\zeta) = 2\alpha - \zeta + \frac{2(a+2)(\zeta - \alpha)}{z^{a+2}} \int_0^z \frac{t^{a+1}}{t+1} dt$, $z \in U$, $\zeta \in \overline{U}$ *as the best subordinant.*

Proof. Denoting $p(z,\zeta) = \left(\mathcal{I}_q^{m,l} F(z,\zeta)\right)'_z$, following the proof of Theorem 4.46, the strong superordination (4.3.3.23) can be written as $h(z,\zeta) = \frac{\zeta + (2\alpha - \zeta) z}{1+z} \prec\prec \frac{z p'_z(z,\zeta)}{a+2} + p(z,\zeta)$, $z \in U$, $\zeta \in \overline{U}$.

Applying Lemma 1.10, we get the strong differential superordination $g(z,\zeta) \prec\prec p(z,\zeta)$, with $g(z,\zeta) = \frac{a+2}{z^{a+2}} \int_0^z \frac{\zeta+(2\alpha-\zeta)t}{1+t} t^{a+1} dt = 2\alpha - \zeta + \frac{2(a+2)(\zeta-\alpha)}{z^{a+2}} \int_0^z \frac{t^{a+1}}{t+1} dt \prec\prec \left(\mathcal{I}_q^{m,l} F(z,\zeta)\right)_z'$, $z \in U$, $\zeta \in \overline{U}$, and g is the best subordinant and it is convex. □

Theorem 4.49 (Alb Lupaş and Ghanim, 2023). *For $f \in \mathcal{A}_\zeta^*$, consider a convex function $h(z,\zeta)$ such that $h(0,\zeta) = 1$ and assume that $\left(\mathcal{I}_q^{m,l} f(z,\zeta)\right)_z'$ is univalent and $\frac{\mathcal{I}_q^{m,l} f(z,\zeta)}{z} \in Q^* \cap \mathcal{H}^*[1,1,\zeta]$. When the strong superordination*

$$h(z,\zeta) \prec\prec \left(\mathcal{I}_q^{m,l} f(z,\zeta)\right)_z', \quad z \in U, \ \zeta \in \overline{U}, \tag{4.3.3.24}$$

holds, then the following strong differential superordination

$$g(z,\zeta) \prec\prec \frac{\mathcal{I}_q^{m,l} f(z,\zeta)}{z}, \quad z \in U, \ \zeta \in \overline{U},$$

is satisfied, for the convex function $g(z,\zeta) = \frac{1}{z}\int_0^z h(t,\zeta)dt$ the best subordinant.

Proof. Denoting $p(z,\zeta) = \frac{\mathcal{I}_q^{m,l} f(z,\zeta)}{z} = \frac{z+\sum_{k=2}^{\infty}\left(\frac{[k+l]_q}{[1+l]_q}\right)^m a_k(\zeta) z^k}{z} \in \mathcal{H}^*[1,1,\zeta]$, $z \in U$, $\zeta \in \overline{U}$, we can write $\mathcal{I}_q^{m,l} f(z,\zeta) = zp(z,\zeta)$, and differentiating it with respect to z, we have $\left(\mathcal{I}_q^{m,l} f(z,\zeta)\right)_z' = zp_z'(z,\zeta) + p(z,\zeta)$, $z \in U$, $\zeta \in \overline{U}$.

With this notation, the strong differential superordination (4.3.3.24) becomes $h(z,\zeta) \prec\prec zp_z'(z,\zeta) + p(z,\zeta)$, $z \in U$, $\zeta \in \overline{U}$, and applying Lemma 1.10, we get $g(z,\zeta) \prec\prec p(z,\zeta) = \frac{\mathcal{I}_q^{m,l} f(z,\zeta)}{z}$, $z \in U$, $\zeta \in \overline{U}$, for $g(z,\zeta) = \frac{1}{z}\int_0^z h(t,\zeta)dt$ the best subordinant and convex. □

Theorem 4.50 (Alb Lupaş and Ghanim, 2023). *Considering a convex function $g(z,\zeta)$ in $U \times \overline{U}$, we define the function $h(z,\zeta) = zg_z'(z,\zeta) + g(z,\zeta)$. Assume $\left(\mathcal{I}_q^{m,l} f(z,\zeta)\right)_z'$ is univalent, $\frac{\mathcal{I}_q^{m,l} f(z,\zeta)}{z} \in Q^* \cap \mathcal{H}^*[1,1,\zeta]$ for $f \in \mathcal{A}_\zeta^*$ and the strong superordination*

$$h(z,\zeta) = zg_z'(z,\zeta) + g(z,\zeta) \prec\prec \left(\mathcal{I}_q^{m,l} f(z,\zeta)\right)_z', \quad z \in U, \ \zeta \in \overline{U}, \tag{4.3.3.25}$$

is satisfied, then the strong differential supeordination

$$g(z,\zeta) \prec\prec \frac{\mathcal{I}_q^{m,l} f(z,\zeta)}{z}, \quad z \in U, \ \zeta \in \overline{U},$$

is satisfied for $g(z,\zeta) = \frac{1}{z}\int_0^z h(t,\zeta)dt$ the best subordinant.

Proof. Taking account the proof of Theorem 4.49 for $p(z,\zeta)=\frac{\mathcal{I}_q^{m,l}f(z,\zeta)}{z}$, the strong superordination (4.3.3.25) can be written in the following form $zg_z'(z,\zeta)+g(z,\zeta)\prec\prec zp_z'(z,\zeta)+p(z,\zeta)$, $z\in U$, $\zeta\in\overline{U}$.

Applying Lemma 1.11, we get the strong differential superordination $g(z,\zeta)\prec\prec p(z,\zeta)$, equivalently with $g(z,\zeta)=\frac{1}{z}\int_0^z h(t,\zeta)dt\prec\prec\frac{\mathcal{I}_q^{m,l}f(z,\zeta)}{z}$, $z\in U$, $\zeta\in\overline{U}$, for g the best subordinant. □

Theorem 4.51 (Alb Lupaş and Ghanim, 2023). *Let $h(z,\zeta)=\frac{\zeta+(2\alpha-\zeta)z}{1+z}$ with $0\le\alpha<1$, $z\in U$, $\zeta\in\overline{U}$. For $f\in\mathcal{A}_\zeta^*$, suppose that $\left(\mathcal{I}_q^{m,l}f(z,\zeta)\right)_z'$ is univalent and $\frac{\mathcal{I}_q^{m,l}f(z,\zeta)}{z}\in Q^*\cap\mathcal{H}^*[1,1,\zeta]$. If the strong differential superordination*

$$h(z,\zeta)\prec\prec\left(\mathcal{I}_q^{m,l}f(z,\zeta)\right)',\quad z\in U,\ \zeta\in\overline{U},\tag{4.3.3.26}$$

holds, then we get the following strong differential superordination

$$g(z,\zeta)\prec\prec\frac{\mathcal{I}_q^{m,l}f(z,\zeta)}{z},\quad z\in U,\ \zeta\in\overline{U},$$

where the best subordinant is the convex function $g(z,\zeta)=2\alpha-\zeta+2(\zeta-\alpha)\frac{\ln(1+z)}{z}$, $z\in U$, $\zeta\in\overline{U}$.

Proof. Following the proof of Theorem 4.49 for $p(z,\zeta)=\frac{\mathcal{I}_q^{m,l}f(z,\zeta)}{z}$, the strong superordination (4.3.3.26) takes the form $h(z,\zeta)=\frac{\zeta+(2\alpha-\zeta)z}{1+z}\prec\prec zp_z'(z,\zeta)+p(z,\zeta)$, $z\in U$, $\zeta\in\overline{U}$.

Applying Lemma 1.10, we get the following strong differential superordination $g(z,\zeta)\prec\prec p(z,\zeta)$, which can be written as $g(z,\zeta)=\frac{1}{z}\int_0^z\frac{\zeta+(2\alpha-\zeta)t}{1+t}dt=2\alpha-\zeta+\frac{2(\zeta-\alpha)}{z}\ln(z+1)\prec\prec\frac{\mathcal{I}_q^{m,l}f(z,\zeta)}{z}$, $z\in U$, $\zeta\in\overline{U}$. The function g is the best subordinant and it is convex. □

Theorem 4.52 (Alb Lupaş and Ghanim, 2023). *Considering a convex function $h(z,\zeta)$ such that $h(0,\zeta)=1$, for $f\in\mathcal{A}_\zeta^*$ suppose that $\left(\frac{z\mathcal{I}_q^{m+1,l}f(z,\zeta)}{\mathcal{I}_q^{m,l}f(z,\zeta)}\right)_z'$ is univalent and $\frac{\mathcal{I}_q^{m+1,l}f(z,\zeta)}{\mathcal{I}_q^{m,l}f(z,\zeta)}\in Q^*\cap\mathcal{H}^*[1,1,\zeta]$. If the strong differential superordination*

$$h(z,\zeta)\prec\prec\left(\frac{z\mathcal{I}_q^{m+1,l}f(z,\zeta)}{\mathcal{I}_q^{m,l}f(z,\zeta)}\right)_z',\quad z\in U,\ \zeta\in\overline{U},\tag{4.3.3.27}$$

holds, then we get the following strong differential superordination

$$g(z,\zeta) \prec\prec \frac{\mathcal{I}_q^{m+1,l} f(z,\zeta)}{\mathcal{I}_q^{m,l} f(z,\zeta)}, \quad z \in U, \ \zeta \in \overline{U},$$

with the best subordinant is the convex function $g(z,\zeta) = \frac{1}{z}\int_0^z h(t,\zeta)dt$.

Proof. Let $p(z,\zeta) = \frac{\mathcal{I}_q^{m+1,l} f(z,\zeta)}{\mathcal{I}_q^{m,l} f(z,\zeta)}$, after differentiating it with respect to z, we can write $p'_z(z,\zeta) = \frac{\left(\mathcal{I}_q^{m+1,l} f(z,\zeta)\right)'_z}{\mathcal{I}_q^{m,l} f(z,\zeta)} - p(z,\zeta)\frac{\left(\mathcal{I}_q^{m,l} f(z,\zeta)\right)'_z}{\mathcal{I}_q^{m,l} f(z,\zeta)}$ in the form $zp'_z(z,\zeta) + p(z,\zeta) = \left(\frac{z\mathcal{I}_q^{m+1,l} f(z,\zeta)}{\mathcal{I}_q^{m,l} f(z,\zeta)}\right)'_z$.

Strong differential superordination (4.3.3.27) for $z \in U, \zeta \in \overline{U}$, becomes $h(z,\zeta) \prec\prec zp'_z(z,\zeta) + p(z,\zeta)$.

Applying Lemma 1.10, we obtain the following strong differential superordination $g(z,\zeta) \prec\prec p(z,\zeta) = \frac{\mathcal{I}_q^{m+1,l} f(z,\zeta)}{\mathcal{I}_q^{m,l} f(z,\zeta)}$, $z \in U$, $\zeta \in \overline{U}$, for the best subordinant $g(z,\zeta) = \frac{1}{z}\int_0^z h(t,\zeta)dt$ convex. □

Theorem 4.53 (Alb Lupaş and Ghanim, 2023). *Consider a convex function* $g(z,\zeta)$ *and* $h(z,\zeta) = zg'_z(z,\zeta) + g(z,\zeta)$. *For* $f \in \mathcal{A}^*_\zeta$, *assume that* $\left(\frac{z\mathcal{I}_q^{m+1,l} f(z,\zeta)}{\mathcal{I}_q^{m,l} f(z,\zeta)}\right)'_z$ *is univalent and* $\frac{\mathcal{I}_q^{m+1,l} f(z,\zeta)}{\mathcal{I}_q^{m,l} f(z,\zeta)} \in Q^* \cap \mathcal{H}^*[1,1,\zeta]$. *If the strong differential superordination*

$$h(z,\zeta) = zg'_z(z,\zeta) + g(z,\zeta) \prec\prec \left(\frac{z\mathcal{I}_q^{m+1,l} f(z,\zeta)}{\mathcal{I}_q^{m,l} f(z,\zeta)}\right)'_z, \quad z \in U, \ \zeta \in \overline{U}, \tag{4.3.3.28}$$

states, then we get the strong differential superordination

$$g(z,\zeta) \prec\prec \frac{\mathcal{I}_q^{m+1,l} f(z,\zeta)}{\mathcal{I}_q^{m,l} f(z,\zeta)}, \quad z \in U, \ \zeta \in \overline{U},$$

and the best subordinant is $g(z,\zeta) = \frac{1}{z}\int_0^z h(t,\zeta)dt$.

Proof. Following the proof of Theorem 4.52 for $p(z,\zeta) = \frac{\mathcal{I}_q^{m+1,l} f(z,\zeta)}{\mathcal{I}_q^{m,l} f(z,\zeta)}$, the strong superordination (4.3.3.28) has the form $h(z,\zeta) = zg'_z(z,\zeta) + g(z,\zeta) \prec\prec zp'_z(z,\zeta) + p(z,\zeta)$, $z \in U$, $\zeta \in \overline{U}$.

Applying Lemma 1.11, it yields $g(z,\zeta) \prec\prec p(z,\zeta)$, equivalently with $g(z,\zeta) = \frac{1}{z}\int_0^z h(t,\zeta)dt \prec\prec \frac{\mathcal{I}_q^{m+1,l} f(z,\zeta)}{\mathcal{I}_q^{m,l} f(z,\zeta)}$, $z \in U$, $\zeta \in \overline{U}$, and the best subordinant is g. □

Theorem 4.54 (Alb Lupaş and Ghanim, 2023). *Consider* $h(z,\zeta) = \frac{\zeta+(2\alpha-\zeta)z}{1+z}$, *with* $0 \leq \alpha < 1$. *For* $f \in \mathcal{A}_{\zeta}^{*}$, *assume that* $\left(\frac{z\mathcal{I}_q^{m+1,l}f(z,\zeta)}{\mathcal{I}_q^{m,l}f(z,\zeta)}\right)_z'$ *is univalent and* $\frac{\mathcal{I}_q^{m+1,l}f(z,\zeta)}{\mathcal{I}_q^{m,l}f(z,\zeta)} \in Q^* \cap \mathcal{H}^*[1,1,\zeta]$. *If the strong differential superordination*

$$h(z,\zeta) \prec\prec \left(\frac{z\mathcal{I}_q^{m+1,l}f(z,\zeta)}{\mathcal{I}_q^{m,l}f(z,\zeta)}\right)_z', \quad z \in U,\ \zeta \in \overline{U}, \tag{4.3.3.29}$$

holds, then the strong differential superordination

$$g(z,\zeta) \prec\prec \frac{\mathcal{I}_q^{m+1,l}f(z,\zeta)}{\mathcal{I}_q^{m,l}f(z,\zeta)}, \quad z \in U,\ \zeta \in \overline{U},$$

holds, and the best subordinant is the convex function $g(z) = 2\alpha - \zeta + 2(\zeta-\alpha)\frac{\ln(1+z)}{z}$, $z \in U$, $\zeta \in \overline{U}$.

Proof. Using the notation $p(z,\zeta) = \frac{\mathcal{I}_q^{m+1,l}f(z,\zeta)}{\mathcal{I}_q^{m,l}f(z,\zeta)}$, the strong differential superordination (4.3.3.29) can be written $h(z,\zeta) = \frac{\zeta 1+(2\alpha-\zeta)z}{1+z} \prec\prec zp_z'(z,\zeta) + p(z,\zeta)$, $z \in U$, $\zeta \in \overline{U}$.

Applying Lemma 1.10, we get the strong differential superordination $g(z,\zeta) \prec\prec p(z,\zeta)$, equivalently with $g(z,\zeta) = \frac{1}{z}\int_0^z \frac{\zeta+(2\alpha-\zeta)t}{1+t}dt = 2\alpha - \zeta 1 + 2(\zeta 1-\alpha)\frac{1}{z}\ln(z+1) \prec\prec \frac{\mathcal{I}_q^{m+1,l}f(z,\zeta)}{\mathcal{I}_q^{m,l}f(z,\zeta)}$, $z \in U$, $\zeta \in \overline{U}$.

The best subordinant is the convex function g. □

Chapter 5

Recent Developments in the Fuzzy Differential Subordinations and Superordinations Theory

In this chapter, we expose some of the results obtained by the author regarding few operators presented in the previous chapter using the fuzzy differential subordinations and superordinations theory.

5.1 Applications of Differential Operators in the Theory of Fuzzy Differential Subordinations

In this section, we present several fuzzy differential subordination results involving few of the defined operators presented in Chapter 2.

5.1.1 *Some Fuzzy Differential Subordinations Using Ruscheweyh Derivative and Sălăgean Operator*

In the paper Alb Lupaş (2016a), we established some fuzzy differential subordinations regarding the differential operator defined in Definition 2.5 by $L_\alpha^m : \mathcal{A}_n \to \mathcal{A}_n$,

$$L_\alpha^m f(z) = (1-\alpha) R^m f(z) + \alpha S^m f(z),$$

where R^m is the Ruscheweyh derivative, S^m is the Sălăgean operator and $\mathcal{A}_n = \{f \in \mathcal{H}(U),\ f(z) = z + a_{n+1}z^{n+1} + \cdots,\ z \in U\}$ is the class of normalized analytic functions.

If $f \in \mathcal{A}_n$, $f(z) = z + \sum_{j=n+1}^{\infty} a_j z^j$, then $L_\alpha^m f(z) = z + \sum_{j=n+1}^{\infty} \left(\alpha j^m + (1-\alpha) C_{m+j-1}^m\right) a_j z^j$, $z \in U$.

Theorem 5.1 (Alb Lupaş, 2016a). *For a convex function g, with $g(0) = 1$, consider $h(z) = g(z) + \frac{nz}{\delta} g'(z), z \in U$. The fuzzy differential subordination*

$$F_{L_\alpha^m f(U)}\left(\left(\frac{L_\alpha^m f(z)}{z}\right)^{\delta-1}\left(L_\alpha^m f(z)\right)'\right) \leq F_{h(U)} h(z), \quad z \in U, \qquad (5.1.1.1)$$

is satisfied, when $\alpha, \delta \geq 0$, $m, n \in \mathbb{N}$, $f \in \mathcal{A}_n$, and implies the following fuzzy sharp differential subordination

$$F_{L_\alpha^m f(U)}\left(\frac{L_\alpha^m f(z)}{z}\right)^{\delta} \leq F_{g(U)} g(z), \quad z \in U.$$

Proof. Differentiating $p(z) = \left(\frac{L_\alpha^m f(z)}{z}\right)^{\delta}$, $z \in U$, we get $\left(\frac{L_\alpha^m f(z)}{z}\right)^{\delta-1} (L_\alpha^m f(z))' = p(z) + \frac{1}{\delta} z p'(z), z \in U$. Evidently, $p \in \mathcal{H}[1, n\delta]$, and (5.1.1.1) becomes $F_{p(U)}\left(p(z) + \frac{1}{\delta} z p'(z)\right) \leq F_{h(U)} h(z) = F_{g(U)}\left(g(z) + \frac{z}{\delta} g'(z)\right)$, $z \in U$.

By using Lemma 1.15, we have

$$F_{L_\alpha^m f(U)}\left(\frac{L_\alpha^m f(z)}{z}\right)^{\delta} \leq F_{g(U)} g(z), \quad z \in U.$$

□

Theorem 5.2 (Alb Lupaş, 2016a). *Consider an holomorphic function h which satisfies the inequality* Re $\left(1 + \frac{zh''(z)}{h'(z)}\right) > -\frac{1}{2}, z \in U$, *and $h(0) = 1$. When the fuzzy differential subordination*

$$F_{L_\alpha^m f(U)}\left(\left(\frac{L_\alpha^m f(z)}{z}\right)^{\delta-1}\left(L_\alpha^m f(z)\right)'\right) \leq F_{h(U)} h(z), \quad z \in U, \qquad (5.1.1.2)$$

holds, with $\alpha, \delta \geq 0$, $m, n \in \mathbb{N}$, $f \in \mathcal{A}_n$, then

$$F_{L_\alpha^m f(U)}\left(\frac{L_\alpha^m f(z)}{z}\right)^{\delta} \leq F_{q(U)} q(z), \quad z \in U,$$

and $q(z) = \frac{\delta}{nz^{\frac{\delta}{n}}} \int_0^z h(t) t^{\frac{\delta}{n}-1} dt$.

Proof. Since Re $\left(1 + \frac{zh''(z)}{h'(z)}\right) > -\frac{1}{2}, z \in U$, from Lemma 1.13, we deduce that $q(z) = \frac{\delta}{nz^{\frac{\delta}{n}}} \int_0^z h(t) t^{\frac{\delta}{n}-1} dt$ is a convex function and verifies the differential equation associated to the fuzzy differential subordination (5.1.1.2) $q(z) + \frac{1}{\delta} z q'(z) = h(z)$, therefore, it is the fuzzy best dominant.

Consider $p(z) = \left(\frac{L_\alpha^m f(z)}{z}\right)^\delta$, $z \in U$, then $p \in \mathcal{H}[1, n\delta]$. Differentiating, we obtain $\left(\frac{L_\alpha^m f(z)}{z}\right)^{\delta-1} (L_\alpha^m f(z))' = p(z) + \frac{1}{\delta} zp'(z)$, $z \in U$, and the fuzzy differential subordination (5.1.1.2) becomes $F_{p(U)}\left(p(z) + \frac{1}{\delta}zp'(z)\right) \leq F_{h(U)}h(z), z \in U$. applying Lemma 1.14, we get

$$F_{L_\alpha^m f(U)}\left(\frac{L_\alpha^m f(z)}{z}\right)^\delta \leq F_{q(U)}q(z)$$

and $q(z) = \frac{\delta}{nz^{\frac{\delta}{n}}} \int_0^z h(t)t^{\frac{\delta}{n}-1}dt$, $z \in U$, is the best dominant. □

Theorem 5.3 (Alb Lupaş, 2016a). *For a convex function g, with $g(0) = 1$, define the function $h(z) = g(z) + \frac{nz}{\gamma}g'(z)$, $z \in U$. If $\alpha \geq 0$, $\gamma \in \mathbb{C}\backslash\{0\}$, $m, n \in \mathbb{N}$, $f \in \mathcal{A}_n$ and the fuzzy differential subordination*

$$F_{L_\alpha^m f(U)}\left(\frac{z^2}{\gamma}\frac{L_\alpha^m f(z)}{\left(L_\alpha^{m+1}f(z)\right)^2}\left[\frac{(L_\alpha^m f(z))'}{L_\alpha^m f(z)} - 2\frac{\left(L_\alpha^{m+1}f(z)\right)'}{L_\alpha^{m+1}f(z)}\right]\right.$$
$$\left.+\frac{(\gamma+1)z}{\gamma}\frac{L_\alpha^m f(z)}{\left(L_\alpha^{m+1}f(z)\right)^2}\right) \leq F_{h(U)}h(z), \quad z \in U, \qquad (5.1.1.3)$$

holds, then

$$F_{L_\alpha^m f(U)}\left(z\frac{L_\alpha^m f(z)}{\left(L_\alpha^{m+1}f(z)\right)^2}\right) \leq F_{g(U)}g(z), \quad z \in U,$$

and this result is sharp.

Proof. Differentiating $p(z) = z\frac{L_\alpha^m f(z)}{\left(L_\alpha^{m+1}f(z)\right)^2}$, we get $p(z) + \frac{z}{\gamma}p'(z) = \frac{(\gamma+1)z}{\gamma}$ $\frac{L_\alpha^m f(z)}{\left(L_\alpha^{m+1}f(z)\right)^2} + \frac{z^2}{\gamma}\frac{L_\alpha^m f(z)}{\left(L_\alpha^{m+1}f(z)\right)^2}\left[\frac{(L_\alpha^m f(z))'}{L_\alpha^m f(z)} - 2\frac{\left(L_\alpha^{m+1}f(z)\right)'}{L_\alpha^{m+1}f(z)}\right]$.

The fuzzy differential subordination (5.1.1.3) becomes $F_{p(U)}$ $\left(p(z) + \frac{z}{\gamma}p'(z)\right) \leq F_{h(U)}h(z)F_{g(U)}\left(g(z) + \frac{z}{\gamma}g'(z)\right), z \in U$, and by using Lemma 1.15, we derive

$$F_{L_\alpha^m f(U)}\left(z\frac{L_\alpha^m f(z)}{\left(L_\alpha^{m+1}f(z)\right)^2}\right) \leq F_{g(U)}g(z), \quad z \in U.$$

□

Theorem 5.4 (Alb Lupaş, 2016a). *Let h be a holomorphic function with properties* $\operatorname{Re}\left(1+\frac{zh''(z)}{h'(z)}\right) > -\frac{1}{2}$, $z \in U$, *and* $h(0) = 1$. *The fuzzy differential subordination*

$$F_{L_\alpha^m f(U)}\left(\frac{z^2}{\gamma}\frac{L_\alpha^m f(z)}{\left(L_\alpha^{m+1}f(z)\right)^2}\left[\frac{\left(L_\alpha^m f(z)\right)'}{L_\alpha^m f(z)} - 2\frac{\left(L_\alpha^{m+1}f(z)\right)'}{L_\alpha^{m+1}f(z)}\right] + \frac{(\gamma+1)z}{\gamma}\frac{L_\alpha^m f(z)}{\left(L_\alpha^{m+1}f(z)\right)^2}\right) \le F_{h(U)}h(z), \quad z \in U, \qquad (5.1.1.4)$$

implies the following fuzzy differential subordination

$$F_{L_\alpha^m f(U)}\left(z\frac{L_\alpha^m f(z)}{\left(L_\alpha^{m+1}f(z)\right)^2}\right) \le F_{q(U)}q(z), \quad z \in U,$$

where $\alpha \ge 0$, $\gamma \in \mathbb{C}\backslash\{0\}$, $m, n \in \mathbb{N}$, $f \in \mathcal{A}_n$, *and* $q(z) = \frac{\gamma}{nz^{\frac{\gamma}{n}}}\int_0^z h(t)t^{\frac{\gamma}{n}-1}dt$.

Proof. Since $\operatorname{Re}\left(1+\frac{zh''(z)}{h'(z)}\right) > -\frac{1}{2}$, $z \in U$, from Lemma 1.13, we deduce that $q(z) = \frac{\gamma}{nz^{\frac{\gamma}{n}}}\int_0^z h(t)t^{\frac{\gamma}{n}-1}dt$ is a convex function and verifies the differential equation associated to the fuzzy differential subordination (5.1.1.4) $q(z) + \frac{1}{\gamma}zq'(z) = h(z)$, therefore, it is the fuzzy best dominant.

Considering $p(z) = z\frac{L_\alpha^m f(z)}{\left(L_\alpha^{m+1}f(z)\right)^2}$, $z \in U$, $p \in \mathcal{H}[1,n]$. Since $p(z) + \frac{z}{\gamma}p'(z) = \frac{(\gamma+1)z}{\gamma}\frac{L_\alpha^m f(z)}{\left(L_\alpha^{m+1}f(z)\right)^2} + \frac{z^2}{\gamma}\frac{L_\alpha^m f(z)}{\left(L_\alpha^{m+1}f(z)\right)^2}\left[\frac{\left(L_\alpha^m f(z)\right)'}{L_\alpha^m f(z)} - 2\frac{\left(L_\alpha^{m+1}f(z)\right)'}{L_\alpha^{m+1}f(z)}\right]$, $z \in U$, from (5.1.1.4), we have $F_{p(U)}\left(p(z) + \frac{z}{\gamma}p'(z)\right) \le F_{h(U)}h(z)$, $z \in U$, and from Lemma 1.14, we obtain

$$F_{L_\alpha^m f(U)}\left(z\frac{L_\alpha^m f(z)}{\left(L_\alpha^{m+1}f(z)\right)^2}\right) \le F_{q(U)}q(z), \quad z \in U,$$

and the best dominant is $q(z) = \frac{\gamma}{nz^{\frac{\gamma}{n}}}\int_0^z h(t)t^{\frac{\gamma}{n}-1}dt$. □

Theorem 5.5 (Alb Lupaş, 2016a). *For a convex function g, with* $g(0) = 0$ *define* $h(z) = g(z) + \frac{nz}{\gamma}g'(z)$, $z \in U$. *If* $\alpha \ge 0$, $\gamma \in \mathbb{C}\backslash\{0\}$, $m, n \in \mathbb{N}$, $f \in \mathcal{A}_n$ *and the fuzzy differential subordination*

$$F_{L_\alpha^m f(U)}\left(\frac{z^3}{\gamma}\left[\frac{\left(L_\alpha^m f(z)\right)''}{L_\alpha^m f(z)} - \left(\frac{\left(L_\alpha^m f(z)\right)'}{L_\alpha^m f(z)}\right)^2\right] + \frac{(\gamma+2)z^2}{\gamma}\frac{\left(L_\alpha^m f(z)\right)'}{L_\alpha^m f(z)}\right) \le F_{h(U)}h(z), \quad z \in U,$$

holds, then the following result is sharp

$$F_{L_\alpha^m f(U)}\left(z^2\frac{(L_\alpha^m f(z))'}{L_\alpha^m f(z)}\right) \le F_{g(U)}g(z), \quad z\in U.$$

Proof. Considering $p(z) = z^2\frac{(L_\alpha^m f(z))'}{L_\alpha^m f(z)}$, we deduce that $p \in \mathcal{H}[0,n]$ and differentiating it, we obtain

$$\begin{aligned} p(z)+\frac{z}{\gamma}p'(z) &= \frac{(\gamma+2)\,z^2}{\gamma}\frac{(L_\alpha^m f(z))'}{L_\alpha^m f(z)} \\ &\quad + \frac{z^3}{\gamma}\left[\frac{(L_\alpha^m f(z))''}{L_\alpha^m f(z)} - \left(\frac{(L_\alpha^m f(z))'}{L_\alpha^m f(z)}\right)^2\right], \quad z\in U. \end{aligned}$$

The fuzzy differential subordination becomes $F_{p(U)}\left(p(z)+\frac{1}{\gamma}zp'(z)\right) \le F_{h(U)}h(z) = F_{g(U)}\left(g(z)+\frac{z}{\gamma}g'(z)\right)$ and by using Lemma 1.15, we deduce

$$F_{L_\alpha^m f(U)}\left(z^2\frac{(L_\alpha^m f(z))'}{L_\alpha^m f(z)}\right) \le F_{g(U)}g(z), \quad z\in U,$$

and this result is sharp. □

Theorem 5.6 (Alb Lupaş, 2016a). *Consider h a holomorphic function with* Re $\left(1+\frac{zh''(z)}{h'(z)}\right) > -\frac{1}{2}$, $z\in U$, *and* $h(0)=0$. *If the fuzzy differential subordination*

$$\begin{aligned} F_{L_\alpha^m f(U)}\Bigg(&\frac{z^3}{\gamma}\left[\frac{(L_\alpha^m f(z))''}{L_\alpha^m f(z)} - \left(\frac{(L_\alpha^m f(z))'}{L_\alpha^m f(z)}\right)^2\right] \\ &+ \frac{(\gamma+2)\,z^2}{\gamma}\frac{(L_\alpha^m f(z))'}{L_\alpha^m f(z)}\Bigg) \le F_{h(U)}h(z), \quad z\in U, \qquad (5.1.1.5) \end{aligned}$$

holds, for $\alpha\ge 0$, $\gamma\in\mathbb{C}\backslash\{0\}$ *with* Re $\gamma \ge 0$, $m,n\in\mathbb{N}$, $f\in\mathcal{A}_n$, *then*

$$F_{L_\alpha^m f(U)}\left(z^2\frac{(L_\alpha^m f(z))'}{L_\alpha^m f(z)}\right) \le F_{q(U)}q(z), \quad z\in U,$$

where $q(z) = \frac{\gamma}{nz^{\frac{\gamma}{n}}}\int_0^z h(t)t^{\frac{\gamma}{n}-1}dt$.

Proof. Since $\operatorname{Re}\left(1+\frac{zh''(z)}{h'(z)}\right) > -\frac{1}{2}, z \in U$, from Lemma 1.13, we deduce that $q(z) = \frac{\gamma}{nz^{\frac{\gamma}{n}}}\int_0^z h(t)t^{\frac{\gamma}{n}-1}dt$ is a convex function and verifies the differential equation associated to the fuzzy differential subordination (5.1.1.5) $q(z)+\frac{1}{\gamma}zq'(z) = h(z)$, therefore it is the fuzzy best dominant.

Considering $p(z) = z^2\frac{(L_\alpha^m f(z))'}{L_\alpha^m f(z)}$, $z \in U$, $p \in \mathcal{H}[0,n]$. Differentiating it, we obtain $p(z)+\frac{z}{\gamma}p'(z) = \frac{(\gamma+2)z^2}{\gamma}\frac{(L_\alpha^m f(z))'}{L_\alpha^m f(z)} + \frac{z^3}{\gamma}\left[\frac{(L_\alpha^m f(z))''}{L_\alpha^m f(z)} - \left(\frac{(L_\alpha^m f(z))'}{L_\alpha^m f(z)}\right)^2\right], z \in U$, and (5.1.1.5) becomes $F_{p(U)}\left(p(z)+\frac{1}{\gamma}zp'(z)\right) \leq F_{h(U)}h(z), z \in U$.

Applying Lemma 1.14, we get

$$F_{L_\alpha^m f(U)}\left(z^2\frac{(L_\alpha^m f(z))'}{L_\alpha^m f(z)}\right) \leq F_{q(U)}q(z), \quad z \in U,$$

and the best dominant is $q(z) = \frac{\gamma}{nz^{\frac{\gamma}{n}}}\int_0^z h(t)t^{\frac{\gamma}{n}-1}dt$. □

Theorem 5.7 (Alb Lupaş, 2016a). *Let $h(z) = g(z)+zg'(z), z \in U$, where g is a convex function such that $g(0) = 1$. When the fuzzy differential subordination*

$$F_{L_\alpha^m f(U)}\left(1-\frac{L_\alpha^m f(z)\cdot(L_\alpha^m f(z))''}{\left[(L_\alpha^m f(z))'\right]^2}\right) \leq F_{h(U)}h(z), \quad z \in U$$

holds, for $\alpha \geq 0$, $m,n \in \mathbb{N}$, $f \in \mathcal{A}_n$, then the following result is sharp

$$F_{L_\alpha^m f(U)}\left(\frac{L_\alpha^m f(z)}{z\,(L_\alpha^m f(z))'}\right) \leq F_{g(U)}g(z), \quad z \in U.$$

Proof. Let $p(z) = \frac{L_\alpha^m f(z)}{z(L_\alpha^m f(z))'}$. We deduce that $p \in \mathcal{H}[1,n]$ and differentiating it, we get $1-\frac{L_\alpha^m f(z)\cdot(L_\alpha^m f(z))''}{[(L_\alpha^m f(z))']^2} = p(z)+zp'(z), z \in U$.

The fuzzy differential subordination becomes $F_{p(U)}\left(p(z)+zp'(z)\right) \leq F_{h(U)}h(z) = F_{g(U)}\left(g(z)+zg'(z)\right)$ and we apply Lemma 1.15 to deduce the following sharp result

$$F_{L_\alpha^m f(U)}\left(\frac{L_\alpha^m f(z)}{z\,(L_\alpha^m f(z))'}\right) \leq F_{g(U)}g(z), \quad z \in U.$$

□

Theorem 5.8 (Alb Lupaş, 2016a). *Let h be a holomorphic function with $h(0) = 1$ and* $\mathrm{Re}\left(1 + \frac{zh''(z)}{h'(z)}\right) > -\frac{1}{2}, z \in U$. *The fuzzy differential subordination*

$$F_{L_\alpha^m f(U)}\left(1 - \frac{L_\alpha^m f(z) \cdot (L_\alpha^m f(z))''}{\left[(L_\alpha^m f(z))'\right]^2}\right) \leq F_{h(U)}h(z), \quad z \in U, \qquad (5.1.1.6)$$

induces

$$F_{L_\alpha^m f(U)}\left(\frac{L_\alpha^m f(z)}{z\left(L_\alpha^m f(z)\right)'}\right) \leq F_{q(U)}q(z), \quad z \in U,$$

where $q(z) = \frac{1}{nz^{\frac{1}{n}}} \int_0^z h(t)t^{\frac{1}{n}-1}dt$, *for* $\alpha \geq 0$, $m, n \in \mathbb{N}$, $f \in \mathcal{A}_n$.

Proof. Since $\mathrm{Re}\left(1 + \frac{zh''(z)}{h'(z)}\right) > -\frac{1}{2}, z \in U$, from Lemma 1.13, we deduce that $\frac{1}{nz^{\frac{1}{n}}} \int_0^z h(t)t^{\frac{1}{n}-1}dt$ is a convex function and verifies the differential equation associated to the fuzzy differential subordination (5.1.1.6) $q(z) + zq'(z) = h(z)$, therefore, it is the fuzzy best dominant.

Consider $p(z) = \frac{L_\alpha^m f(z)}{z(L_\alpha^m f(z))'}$, $z \in U$, $p \in \mathcal{H}[1, n]$. Differentiating it, we obtain $1 - \frac{L_\alpha^m f(z) \cdot (L_\alpha^m f(z))''}{[(L_\alpha^m f(z))']^2} = p(z) + zp'(z), z \in U$, and the fuzzy differential subordination (5.1.1.6) is written $F_{p(U)}\left(p(z) + zp'(z)\right) \leq F_{h(U)}h(z)$, $z \in U$.

By using Lemma 1.14, we get

$$F_{L_\alpha^m f(U)}\left(\frac{L_\alpha^m f(z)}{z\left(L_\alpha^m f(z)\right)'}\right) \leq F_{q(U)}q(z), \quad z \in U,$$

and the best dominant is $q(z) = \frac{1}{nz^{\frac{1}{n}}} \int_0^z h(t)t^{\frac{1}{n}-1}dt$. □

Example 5.1 (Alb Lupaş, 2016a). Let $h(z) = \frac{1-z}{1+z}$ a convex function in U with $h(0) = 1$ and $\mathrm{Re}\left(\frac{zh''(z)}{h'(z)} + 1\right) > -\frac{1}{2}$.

Let $f(z) = z - z^2$, $z \in U$. For $n = 1$, $m = 1$, $\alpha = 2$, we obtain $L_2^1 f(z) = zf'(z) = z - 2z^2$, $z \in U$.

Then $\left(L_2^1 f(z)\right)' = 1 - 4z$ and $\left(L_2^1 f(z)\right)'' = -4$, $\frac{L_2^1 f(z)}{z\left(L_2^1 f(z)\right)'} = \frac{z-2z^2}{z(1-4z)} = \frac{1-2z}{1-4z}$, $1 - \frac{L_2^1 f(z) \cdot \left(L_2^1 f(z)\right)''}{\left[\left(L_2^1 f(z)\right)'\right]^2} = 1 - \frac{\left(z-2z^2\right) \cdot (-4)}{(1-4z)^2} = \frac{8z^2-4z+1}{(1-4z)^2}$.

We have $q(z) = \frac{1}{z} \int_0^z \frac{1-t}{1+t}dt = -1 + \frac{2\ln(1+z)}{z}$.

Using Theorem 5.8, we obtain $F_U\left(\frac{8z^2-4z+1}{(1-4z)^2}\right) \leq F_U\left(\frac{1-z}{1+z}\right)$, $z \in U$, induce $F_U\left(\frac{1-2z}{1-4z}\right) \leq F_U\left(-1 + \frac{2\ln(1+z)}{z}\right)$, $z \in U$.

Theorem 5.9 (Alb Lupaş, 2016a). *Consider* $h(z) = g(z)+zg'(z), z \in U$, *where* g *is a convex function with* $g(0) = 1$. *When the fuzzy differential subordination*

$$F_{L_\alpha^m f(U)}\left(\left[(L_\alpha^m f(z))'\right]^2 + L_\alpha^m f(z) \cdot (L_\alpha^m f(z))''\right) \leq F_{h(U)}h(z), \quad z \in U$$

holds, for $\alpha \geq 0$, $m, n \in \mathbb{N}$, $f \in \mathcal{A}_n$, *then the following result is sharp*

$$F_{L_\alpha^m f(U)}\left(\frac{L_\alpha^m f(z) \cdot (L_\alpha^m f(z))'}{z}\right) \leq F_{g(U)}g(z), \quad z \in U.$$

Proof. Let $p(z) = \frac{L_\alpha^m f(z) \cdot (L_\alpha^m f(z))'}{z}$. We deduce that $p \in \mathcal{H}[1, n]$.

Differentiating it, we obtain $\left[(L_\alpha^m f(z))'\right]^2 + L_\alpha^m f(z) \cdot (L_\alpha^m f(z))'' = p(z) + zp'(z), z \in U$, and the fuzzy differential subordination becomes $F_{p(U)}(p(z) + zp'(z)) \leq F_{h(U)}h(z) = F_{g(U)}(g(z) + zg'(z))$.

By using Lemma 1.15, we obtain the following result sharp

$$F_{L_\alpha^m f(U)}\left(\frac{L_\alpha^m f(z) \cdot (L_\alpha^m f(z))'}{z}\right) \leq F_{g(U)}g(z), \quad z \in U.$$

□

Theorem 5.10 (Alb Lupaş, 2016a). *For a holomorphic function* h *which satisfies the inequality* $\mathrm{Re}\left(1 + \frac{zh''(z)}{h'(z)}\right) > -\frac{1}{2}, z \in U$, *and* $h(0) = 1$, *the fuzzy differential subordination*

$$F_{L_\alpha^m f(U)}\left(\left[(L_\alpha^m f(z))'\right]^2 + L_\alpha^m f(z) \cdot (L_\alpha^m f(z))''\right) \leq F_{h(U)}h(z), \quad z \in U, \tag{5.1.1.7}$$

induce

$$F_{L_\alpha^m f(U)}\left(\frac{L_\alpha^m f(z) \cdot (L_\alpha^m f(z))'}{z}\right) \leq F_{q(U)}q(z), \quad z \in U,$$

where $q(z) = \frac{1}{nz^{\frac{1}{n}}} \int_0^z h(t)t^{\frac{1}{n}-1}dt$, *and* $\alpha \geq 0$, $m, n \in \mathbb{N}$, $f \in \mathcal{A}_n$

Proof. Since $\mathrm{Re}\left(1 + \frac{zh''(z)}{h'(z)}\right) > -\frac{1}{2}, z \in U$, from Lemma 1.13, we deduce that $q(z) = \frac{1}{nz^{\frac{1}{n}}} \int_0^z h(t)t^{\frac{1}{n}-1}dt$ is a convex function and verifies the differential equation associated to the fuzzy differential subordination (5.1.1.7) $q(z) + zq'(z) = h(z)$, therefore, it is the fuzzy best dominant.

Let $p(z) = \frac{L_\alpha^m f(z) \cdot (L_\alpha^m f(z))'}{z}$, $z \in U$, evidently $p \in \mathcal{H}[1, n]$.

We have $p(z) + zp'(z) = \left[(L_\alpha^m f(z))'\right]^2 + L_\alpha^m f(z) \cdot (L_\alpha^m f(z))''''$, $z \in U$, and (5.1.1.7) means $F_{p(U)}(p(z) + zp'(z)) \leq F_{h(U)}h(z)$, $z \in U$.

Applying Lemma 1.14, we get

$$F_{L_\alpha^m f(U)}\left(\frac{L_\alpha^m f(z)\cdot\left(L_\alpha^m f(z)\right)'}{z}\right)\le F_{q(U)}q(z),\quad z\in U,$$

and the best dominant is $q(z)=\frac{1}{nz^{\frac{1}{n}}}\int_0^z h(t)t^{\frac{1}{n}-1}dt$. □

Example 5.2 (Alb Lupaş, 2016a). Let $h(z)=\frac{1-z}{1+z}$ a convex function in U with $h(0)=1$ and $\operatorname{Re}\left(\frac{zh''(z)}{h'(z)}+1\right)>-\frac{1}{2}$.

Consider $f(z)=z-z^2$, $z\in U$. For $n=1$, $m=1$, $\alpha=2$, we obtain $L_2^1f(z)=z-2z^2$, $z\in U$.

Then $\left(L_2^1f(z)\right)'=1-4z$, $\left(L_2^1f(z)\right)''=-4$, $\frac{L_2^1f(z)\cdot\left(L_2^1f(z)\right)'}{z}=\frac{\left(z-2z^2\right)(1-4z)}{z}=8z^2-6z+1$, $\left[\left(L_2^1f(z)\right)'\right]^2+L_2^1f(z)\cdot\left(L_2^1f(z)\right)''=(1-4z)^2+\left(z-2z^2\right)\cdot(-4)=24z^2-12z+1$.

We have $q(z)=\frac{1}{z}\int_0^z\frac{1-t}{1+t}dt=-1+\frac{2\ln(1+z)}{z}$.

From Theorem 5.10, we obtain $F_U\left(24z^2-12z+1\right)\le F_U\left(\frac{1-z}{1+z}\right)$, $z\in U$, induce $F_U\left(8z^2-6z+1\right)\le F_U\left(-1+\frac{2\ln(1+z)}{z}\right)$, $z\in U$.

Theorem 5.11 (Alb Lupaş, 2016a). *Consider $h(z)=g(z)+\frac{z}{1-\delta}g'(z)$, $z\in U$, where g is a convex function with $g(0)=1$. If the fuzzy differential subordination*

$$F_{L_\alpha^m f(U)}\left(\left(\frac{z}{L_\alpha^m f(z)}\right)^\delta\frac{L_\alpha^{m+1}f(z)}{1-\delta}\left(\frac{\left(L_\alpha^{m+1}f(z)\right)'}{L_\alpha^{m+1}f(z)}-\delta\frac{\left(L_\alpha^m f(z)\right)'}{L_\alpha^m f(z)}\right)\right)$$
$$\le F_{h(U)}h(z),\quad z\in U,$$

holds, for $\alpha\ge 0$, $\delta\in(0,1)$, $m,n\in\mathbb{N}$, $f\in\mathcal{A}_n$, then the following result is sharp

$$F_{L_\alpha^m f(U)}\left(\frac{L_\alpha^{m+1}f(z)}{z}\cdot\left(\frac{z}{L_\alpha^m f(z)}\right)^\delta\right)\le F_{g(U)}g(z),\quad z\in U.$$

Proof. Let $p(z)=\frac{L_\alpha^{m+1}f(z)}{z}\cdot\left(\frac{z}{L_\alpha^m f(z)}\right)^\delta$. We deduce that $p\in\mathcal{H}[1,n]$ and differentiating the function p, we obtain $\left(\frac{z}{L_\alpha^m f(z)}\right)^\delta\frac{L_\alpha^{m+1}f(z)}{1-\delta}\left(\frac{\left(L_\alpha^{m+1}f(z)\right)'}{L_\alpha^{m+1}f(z)}-\delta\frac{\left(L_\alpha^m f(z)\right)'}{L_\alpha^m f(z)}\right)=p(z)+\frac{1}{1-\delta}zp'(z)$, $z\in U$.

The fuzzy differential subordination means $F_{p(U)}\left(p(z)+\frac{1}{1-\delta}zp'(z)\right) \leq F_{h(U)}h(z) = F_{g(U)}\left(g(z)+\frac{z}{1-\delta}g'(z)\right)$ and by using Lemma 1.15, we obtain the following sharp result

$$F_{L_\alpha^m f(U)}\left(\frac{L_\alpha^{m+1}f(z)}{z}\cdot\left(\frac{z}{L_\alpha^m f(z)}\right)^\delta\right) \leq F_{g(U)}g(z), \quad z \in U.$$

□

Theorem 5.12 (Alb Lupaş, 2016a). *Let h be an holomorphic function with $h(0) = 1$ and $\operatorname{Re}\left(1+\frac{zh''(z)}{h'(z)}\right) > -\frac{1}{2}, z \in U$. The fuzzy differential subordination*

$$F_{L_\alpha^m f(U)}\left(\left(\frac{z}{L_\alpha^m f(z)}\right)^\delta \frac{L_\alpha^{m+1}f(z)}{1-\delta}\left(\frac{\left(L_\alpha^{m+1}f(z)\right)'}{L_\alpha^{m+1}f(z)} - \delta\frac{\left(L_\alpha^m f(z)\right)'}{L_\alpha^m f(z)}\right)\right)$$
$$\leq F_{h(U)}h(z), \quad z \in U, \tag{5.1.1.8}$$

induce

$$F_{L_\alpha^m f(U)}\left(\frac{L_\alpha^{m+1}f(z)}{z}\cdot\left(\frac{z}{L_\alpha^m f(z)}\right)^\delta\right) \leq F_{q(U)}q(z), \quad z \in U,$$

where $q(z) = \frac{1-\delta}{nz^{\frac{1-\delta}{n}}}\int_0^z h(t)t^{\frac{1-\delta}{n}-1}dt$, for $\alpha, \lambda, l \geq 0$, $\delta \in (0,1)$, $m, n \in \mathbb{N}$, $f \in \mathcal{A}_n$.

Proof. Since $\operatorname{Re}\left(1+\frac{zh''(z)}{h'(z)}\right) > -\frac{1}{2}, z \in U$, from Lemma 1.13, we deduce that $q(z) = \frac{1-\delta}{nz^{\frac{1-\delta}{n}}}\int_0^z h(t)t^{\frac{1-\delta}{n}-1}dt$ is a convex function and verifies the differential equation associated to the fuzzy differential subordination (5.1.1.8) $q(z)+\frac{1}{1-\delta}zq'(z) = h(z)$, therefore, it is the fuzzy best dominant.

Consider $p(z) = \frac{L_\alpha^{m+1}f(z)}{z}\cdot\left(\frac{z}{L_\alpha^m f(z)}\right)^\delta$, $z \in U$, $p \in \mathcal{H}[1,n]$.

Differentiating it, we get $\left(\frac{z}{L_\alpha^m f(z)}\right)^\delta \frac{L_\alpha^{m+1}f(z)}{1-\delta}\left(\frac{(L_\alpha^{m+1}f(z))'}{L_\alpha^{m+1}f(z)} - \delta\frac{(L_\alpha^m f(z))'}{L_\alpha^m f(z)}\right) = p(z)+\frac{1}{1-\delta}zp'(z), z \in U$, and (5.1.1.8) becomes $F_{p(U)}\left(p(z)+\frac{1}{1-\delta}zp'(z)\right) \leq F_{h(U)}h(z)$, $z \in U$.

From Lemma 1.14, we get

$$F_{L_\alpha^m f(U)}\left(\frac{L_\alpha^{m+1}f(z)}{z}\cdot\left(\frac{z}{L_\alpha^m f(z)}\right)^\delta\right) \leq F_{q(U)}q(z), \quad z \in U,$$

and the best dominant is $q(z) = \frac{1-\delta}{nz^{\frac{1-\delta}{n}}}\int_0^z h(t)t^{\frac{1-\delta}{n}-1}dt$. □

5.1.2 *A Note on Special Fuzzy Differential Subordinations Regarding an Operator Defined by Multiplier Transformation and Ruscheweyh Derivative*

In the paper Alb Lupaş (2018a), we established some fuzzy differential subordinations regarding the differential operator from Definition 2.7, $RI^{\alpha}_{m,\lambda,l} : \mathcal{A}_n \to \mathcal{A}_n$,

$$RI^{\alpha}_{m,\lambda,l} f(z) = (1-\alpha)R^m f(z) + \alpha I\,(m,\lambda,l)\, f(z),$$

defined in Alb Lupaş (2010d), where R^m is the Ruscheweyh derivative, $I\,(m,\lambda,l)$ is the multiplier transformation and $\mathcal{A}_n = \{f \in \mathcal{H}(U),\ f(z) = z + a_{n+1}z^{n+1} + \cdots,\ z \in U\}$ is the class of normalized analytic functions. We introduce a fuzzy subclass $RI^{\delta}_{\mathcal{F}}\,(\alpha,m,\lambda,l)$ and by using the fuzzy differential subordinations, we derive some properties of this subclass. Also, several fuzzy differential subordinations are established regarding the studied differential operator.

If $f \in \mathcal{A}_n$, $f(z) = z + \sum_{j=n+1}^{\infty} a_j z^j$, then $RI^{\alpha}_{m,\lambda,l} f(z) = z + \sum_{j=n+1}^{\infty} \left\{ \alpha \left(\frac{1+\lambda(j-1)+l}{l+1} \right)^m + (1-\alpha)\frac{(n+j-1)!}{n!(j-1)!} \right\} a_j z^j, z \in U$.

Using the operator $RI^{\alpha}_{m,\lambda,l}$, we define the subclass $RI^{\delta}_{\mathcal{F}}\,(\alpha,m,\lambda,l)$ and we study fuzzy subordinations.

Definition 5.1 (Alb Lupaş, 2018a). The subclass $RI^{\delta}_{\mathcal{F}}\,(\alpha,m,\lambda,l)$ contains all the functions $f \in \mathcal{A}_n$ which satisfy the inequality

$$F_{(RI^{\alpha}_{m,\lambda,l} f)'(U)} \left(RI^{\alpha}_{m,\lambda,l} f\,(z) \right)' > \delta, \quad z \in U,$$

where $\delta \in (0,1]$, $\alpha,\lambda,l \geq 0$ and $m,n \in \mathbb{N}$.

Theorem 5.13 (Alb Lupaş, 2018a). *$RI^{\delta}_{\mathcal{F}}\,(\alpha,m,\lambda,l)$ is a convex set.*

Proof. Let $f_1,\ f_2 \in RI^{\delta}_{\mathcal{F}}\,(\alpha,m,\lambda,l)$, $f_k(z) = z + \sum_{j=n+1}^{\infty} a_{jk} z^j$, $k = 1,2$, $z \in U$. We show that the function $h(z) = \eta_1 f_1(z) + \eta_2 f_2(z)$ is in the subclass $RI^{\delta}_{\mathcal{F}}\,(\alpha,m,\lambda,l)$, where η_1 and η_2 are nonnegative such that $\eta_1 + \eta_2 = 1$.

Differentiating, we obtain $h'(z) = (\mu_1 f_1 + \mu_2 f_2)'(z) = \mu_1 f_1'(z) + \mu_2 f_2'(z)$, $z \in U$, and $\left(RI^{\alpha}_{m,\lambda,l} h(z) \right)' = \left(RI^{\alpha}_{m,\lambda,l} \left(\mu_1 f_1 + \mu_2 f_2 \right)(z) \right)' = \mu_1 \left(RI^{\alpha}_{m,\lambda,l} f_1(z) \right)' + \mu_2 \left(RI^{\alpha}_{m,\lambda,l} f_2(z) \right)'$, so we have also

$$F_{(RI^{\alpha}_{m,\lambda,l}h)'(U)}\left(RI^{\alpha}_{m,\lambda,l}h(z)\right)'$$

$$= F_{(RI^{\alpha}_{m,\lambda,l}(\mu_1 f_1+\mu_2 f_2))'(U)}\left(RI^{\alpha}_{m,\lambda,l}\left(\mu_1 f_1+\mu_2 f_2\right)(z)\right)'$$

$$= F_{(RI^{\alpha}_{m,\lambda,l}(\mu_1 f_1+\mu_2 f_2))'(U)}\left(\mu_1\left(RI^{\alpha}_{m,\lambda,l}f_1(z)\right)'+\mu_2\left(RI^{\alpha}_{m,\lambda,l}f_2(z)\right)'\right)$$

$$= \frac{F_{(\mu_1 RI^{\alpha}_{m,\lambda,l}f_1)'(U)}\left(\mu_1\left(RI^{\alpha}_{m,\lambda,l}f_1(z)\right)'\right)+F_{(\mu_2 RI^{\alpha}_{m,\lambda,l}f_2)'(U)}\left(\mu_2\left(RI^{\alpha}_{m,\lambda,l}f_2(z)\right)'\right)}{2}$$

$$= \frac{F_{(RI^{\alpha}_{m,\lambda,l}f_1)'(U)}\left(RI^{\alpha}_{m,\lambda,l}f_1(z)\right)'+F_{(RI^{\alpha}_{m,\lambda,l}f_2)'(U)}\left(RI^{\alpha}_{m,\lambda,l}f_2(z)\right)'}{2}.$$

Since $f_1, f_2 \in RI^{\delta}_{\mathcal{F}}(\alpha, m, \lambda, l)$, we have $\delta < F_{(RI^{\alpha}_{m,\lambda,l}f_1)'(U)} \left(RI^{\alpha}_{m,\lambda,l}f_1(z)\right)' \leq 1$ and $\delta < F_{(RI^{\alpha}_{m,\lambda,l}f_2)'(U)}\left(RI^{\alpha}_{m,\lambda,l}f_2(z)\right)' \leq 1$, $z \in U$.

Therefore,

$$\delta < \frac{F_{(RI^{\alpha}_{m,\lambda,l}f_1)'(U)}\left(RI^{\alpha}_{m,\lambda,l}f_1(z)\right)'+F_{(RI^{\alpha}_{m,\lambda,l}f_2)'(U)}\left(RI^{\alpha}_{m,\lambda,l}f_2(z)\right)'}{2} \leq 1$$

and we obtain that $\delta < F_{(RI^{\alpha}_{m,\lambda,l}h)'(U)}\left(RI^{\alpha}_{m,\lambda,l}h(z)\right)' \leq 1$, which means that $h \in RI^{\delta}_{\mathcal{F}}(\alpha, m, \lambda, l)$ and $RI^{\delta}_{\mathcal{F}}(\alpha, m, \lambda, l)$ is a convex set. □

Theorem 5.14 (Alb Lupaş, 2018a). *Consider g a convex function in U and $h(z) = g(z) + \frac{1}{c+2}zg'(z)$, with $z \in U$, $c > 0$. When $f \in RI^{\delta}_{\mathcal{F}}(\alpha, m, \lambda, l)$ and $G(z) = \frac{c+2}{z^{c+1}}\int_0^z t^c f(t)\,dt$, $z \in U$, then*

$$F_{(RI^{\alpha}_{m,\lambda,l}f)'(U)}\left(RI^{\alpha}_{m,\lambda,l}f(z)\right)' \leq F_{h(U)}h(z), \quad z \in U, \tag{5.1.2.1}$$

implies $F_{(RI^{\alpha}_{m,\lambda,l}G)'(U)}\left(RI^{\alpha}_{m,\lambda,l}G(z)\right)' \leq F_{g(U)}g(z)$, $z \in U$, and this result is sharp.

Proof. Differentiating with respect to z, the following relation $z^{c+1}G(z) = (c+2)\int_0^z t^c f(t)\,dt$. We obtain $(c+1)G(z) + zG'(z) = (c+2)f(z)$ and $(c+1)RI^{\alpha}_{m,\lambda,l}G(z) + z\left(RI^{\alpha}_{m,\lambda,l}G(z)\right)' = (c+2)RI^{\alpha}_{m,\lambda,l}f(z)$, $z \in U$.

Differentiating again, we get $\left(RI^{\alpha}_{m,\lambda,l}G(z)\right)' + \frac{1}{c+2}z\left(RI^{\alpha}_{m,\lambda,l}G(z)\right)'' = \left(RI^{\alpha}_{m,\lambda,l}f(z)\right)'$, $z \in U$, and the fuzzy differential subordination (5.1.2.1) becomes

$$F_{RI^{\alpha}_{m,\lambda,l}G(U)}\left(\left(RI^{\alpha}_{m,\lambda,l}G(z)\right)' + \frac{1}{c+2}z\left(RI^{\alpha}_{m,\lambda,l}G(z)\right)''\right)$$
$$\leq F_{g(U)}\left(g(z) + \frac{1}{c+2}zg'(z)\right). \tag{5.1.2.2}$$

Consider

$$p(z) = \left(RI^{\alpha}_{m,\lambda,l}G(z)\right)', \quad z \in U, \tag{5.1.2.3}$$

evidently $p \in \mathcal{H}[1,n]$ and from relation (5.1.2.2), we get

$$F_{p(U)}\left(p(z) + \frac{1}{c+2}zp'(z)\right) \leq F_{g(U)}\left(g(z) + \frac{1}{c+2}zg'(z)\right), \quad z \in U.$$

By using Lemma 1.15, we obtain $F_{\left(RI^{\alpha}_{m,\lambda,l}G\right)'(U)}\left(RI^{\alpha}_{m,\lambda,l}G(z)\right)' \leq F_{g(U)}g(z)$, $z \in U$, and g is the fuzzy best dominant. □

Theorem 5.15 (Alb Lupaş, 2018a). *Let* $h(z) = \frac{1+(2\beta-1)z}{1+z}$, $\beta \in [0,1)$ *and* $c > 0$. *If* $\lambda, l \geq 0$, $m, n \in \mathbb{N}$ *and* $I_c(f)(z) = \frac{c+2}{z^{c+1}}\int_0^z t^c f(t)\,dt$, $z \in U$, *then*

$$I_c\left[RI^{\beta}_{\mathcal{F}}(\alpha,m,\lambda,l)\right] \subset RI^{\beta^*}_{\mathcal{F}}(\alpha,m,\lambda,l), \tag{5.1.2.4}$$

where $\beta^* = 2\beta - 1 + \frac{(c+2)(2-2\beta)}{n}\int_0^1 \frac{t^{\frac{c+2}{n}-1}}{t+1}dt$.

Proof. A similar proof with Theorem 5.14 for the convex function h get us $F_{p(U)}\left(p(z) + \frac{1}{c+2}zp'(z)\right) \leq f_{h(U)}h(z)$, with $p(z)$ defined in (5.1.2.3).

Applying Lemma 1.14, we obtain $F_{\left(RI^{\alpha}_{m,\lambda,l}G\right)'(U)}\left(RI^{\alpha}_{m,\lambda,l}G(z)\right)' \leq F_{g(U)}g(z) \leq F_{h(U)}h(z)$, where $g(z) = \frac{c+2}{nz^{\frac{c+2}{n}}}\int_0^z t^{\frac{c+2}{n}-1}\frac{1+(2\beta-1)t}{1+t}dt = 2\beta - 1 + \frac{(c+2)(2-2\beta)}{nz^{\frac{c+2}{n}}}\int_0^z \frac{t^{\frac{c+2}{n}-1}}{t+1}dt$. Since g is convex and $g(U)$ is symmetric with respect to the real axis, we have

$$F_{I(m,\lambda,l)G(U)}\left(I(m,\lambda,l)G(z)\right)' \geq \min_{|z|=1} F_{g(U)}g(z) = F_{g(U)}g(1) \tag{5.1.2.5}$$

and $\beta^* = g(1) = 2\beta - 1 + \frac{(c+2)(2-2\beta)}{n}\int_0^1 \frac{t^{\frac{c+2}{n}-1}}{t+1}dt$. The inclusion (5.1.2.4) follows from (5.1.2.5). □

Theorem 5.16 (Alb Lupaş, 2018a). *Consider g a convex function with $g(0)=1$ and $h(z)=g(z)+zg'(z), z\in U$. For $\lambda, l\geq 0$, $m,n\in\mathbb{N}$, $f\in\mathcal{A}_n$ if the fuzzy differential subordination*

$$F_{\left(RI^{\alpha}_{m,\lambda,l}f\right)'(U)}\left(RI^{\alpha}_{m,\lambda,l}f(z)\right)'\leq F_{h(U)}h(z),\quad z\in U, \tag{5.1.2.6}$$

holds, then $F_{RI^{\alpha}_{m,\lambda,l}f(U)}\frac{RI^{\alpha}_{m,\lambda,l}f(z)}{z}\leq F_{g(U)}g(z)$, $z\in U$, and this result is sharp.

Proof. For $p(z)=\frac{RI^{\alpha}_{m,\lambda,l}f(z)}{z}$, $z\in U$, we have $p\in\mathcal{H}[1,n]$ and differentiating the relation $RI^{\alpha}_{m,\lambda,l}f(z)=zp(z), z\in U$, we obtain $\left(RI^{\alpha}_{m,\lambda,l}f(z)\right)'=p(z)+zp'(z), z\in U$.

The fuzzy differential subordination (5.1.2.6) becomes $F_{p(U)}(p(z)+zp'(z))\leq F_{h(U)}h(z)=F_{g(U)}\left(g(z)+zg'(z)\right), z\in U$, and applying Lemma 1.15, we deduce $F_{\left(RI^{\alpha}_{m,\lambda,l}f\right)'(U)}\frac{RI^{\alpha}_{m,\lambda,l}f(z)}{z}\leq F_{g(U)}g(z)$, $z\in U$. □

Theorem 5.17 (Alb Lupaş, 2018a). *Let h be a holomorphic function with* Re $\left(1+\frac{zh''(z)}{h'(z)}\right)>-\frac{1}{2}, z\in U$, *and $h(0)=1$. When the fuzzy differential subordination*

$$F_{\left(RI^{\alpha}_{m,\lambda,l}f\right)'(U)}\left(RI^{\alpha}_{m,\lambda,l}f(z)\right)'\leq F_{h(U)}h(z),\quad z\in U, \tag{5.1.2.7}$$

holds, where $\lambda, l\geq 0$, $m,n\in\mathbb{N}$, $f\in\mathcal{A}_n$, then $F_{RI^{\alpha}_{m,\lambda,l}f(U)}\frac{RI^{\alpha}_{m,\lambda,l}f(z)}{z}\leq F_{q(U)}q(z)$, $z\in U$. The function $q(z)=\frac{1}{nz^{\frac{1}{n}}}\int_0^z h(t)t^{\frac{1}{n}-1}dt$ is convex and it is the fuzzy best dominant.

Proof. Since Re $\left(1+\frac{zh''(z)}{h'(z)}\right)>-\frac{1}{2}, z\in U$, from Lemma 1.13, we deduce that $q(z)=\frac{1}{nz^{\frac{1}{n}}}\int_0^z h(t)t^{\frac{1}{n}-1}dt$ is a convex function and verifies the differential equation associated to the fuzzy differential subordination (5.1.2.7) $q(z)+zq'(z)=h(z)$, therefore, it is the fuzzy best dominant.

Keeping $p(z)=\frac{RI^{\alpha}_{m,\lambda,l}f(z)}{z}$, $p\in\mathcal{H}[1,n]$ and differentiating it, we obtain $\left(RI^{\alpha}_{m,\lambda,l}f(z)\right)'=p(z)+zp'(z), z\in U$ and (5.1.2.7) becomes $F_{p(U)}\left(p(z)+zp'(z)\right)\leq F_{h(U)}h(z)$, $z\in U$.

By using Lemma 1.15, we deduce $F_{RI^{\alpha}_{m,\lambda,l}f(U)}\frac{RI^{\alpha}_{m,\lambda,l}f(z)}{z}\leq F_{q(U)}q(z)$, $z\in U$. □

Example 5.3 (Alb Lupaş, 2018a). Let $h(z) = \frac{1-z}{1+z}$ with $h(0) = 1$, $h'(z) = \frac{-2}{(1+z)^2}$ and $h''(z) = \frac{4}{(1+z)^3}$.

Since $\operatorname{Re}\left(\frac{zh''(z)}{h'(z)} + 1\right) = \operatorname{Re}\left(\frac{1-z}{1+z}\right) = \operatorname{Re}\left(\frac{1-\rho\cos\theta - i\rho\sin\theta}{1+\rho\cos\theta + i\rho\sin\theta}\right) = \frac{1-\rho^2}{1+2\rho\cos\theta+\rho^2} > 0 > -\frac{1}{2}$, the function h is convex in U.

Let $f(z) = z - z^2$, $z \in U$. For $n = 1$, $m = 1$, $l = 1$, $\lambda = 2$, $\alpha = \frac{2}{3}$, we obtain $RI^{\frac{2}{3}}_{1,2,1}f(z) = \frac{1}{3}R^1 f(z) + \frac{2}{3}I(1,2,1)f(z) = \frac{1}{3}zf'(z) + \frac{2}{3}zf'(z) = z - 2z^2$. Then $\left(RI^{\frac{2}{3}}_{1,2,1}f(z)\right)' = 1 - 4z$ and $\frac{RI^{\frac{2}{3}}_{1,2,1}f(z)}{z} = 1 - 2z$. Function $q(z) = \frac{1}{z}\int_0^z \frac{1-t}{1+t}dt = -1 + \frac{2\ln(1+z)}{z}$.

From Theorem 5.17, we have $F_U(1-4z) \leq F_U\left(\frac{1-z}{1+z}\right), z \in U$, imply $F_U(1-2z) \leq F_U\left(-1 + \frac{2\ln(1+z)}{z}\right)$, $z \in U$.

Theorem 5.18 (Alb Lupaş, 2018a). *Let g be a convex function with $g(0) = 1$ and $h(z) = g(z) + zg'(z)$, $z \in U$. When the fuzzy differential subordination*

$$F_{RI^{\alpha}_{m,\lambda,l}f(U)}\left(\frac{zRI^{\alpha}_{m+1,\lambda,l}f(z)}{RI^{\alpha}_{m,\lambda,l}f(z)}\right)' \leq F_{h(U)}h(z), \quad z \in U \tag{5.1.2.8}$$

holds, for $\lambda, l \geq 0$, $m, n \in \mathbb{N}$, $f \in \mathcal{A}_n$, then $F_{RI^{\alpha}_{m,\lambda,l}f(U)}\frac{RI^{\alpha}_{m+1,\lambda,l}f(z)}{RI^{\alpha}_{m,\lambda,l}f(z)} \leq F_{g(U)}g(z)$, $z \in U$, and this result is sharp.

Proof. Differentiating $p(z) = \frac{RI^{\alpha}_{m+1,\lambda,l}f(z)}{RI^{\alpha}_{m,\lambda,l}f(z)}$, we get $p'(z) = \frac{\left(RI^{\alpha}_{m+1,\lambda,l}f(z)\right)'}{RI^{\alpha}_{m,\lambda,l}f(z)} - p(z)\cdot\frac{\left(RI^{\alpha}_{m+1,\lambda,l}f(z)\right)'}{RI^{\alpha}_{m,\lambda,l}f(z)}$ and $p(z) + z\cdot p'(z) = \left(\frac{zRI^{\alpha}_{m+1,\lambda,l}f(z)}{RI^{\alpha}_{m,\lambda,l}f(z)}\right)'$.

In this conditions, relation (5.1.2.8) becomes $F_{p(U)}(p(z) + zp'(z)) \leq F_{h(U)}h(z) = F_{g(U)}(g(z) + zg'(z))$, $z \in U$, and applying Lemma 1.15, we obtain $F_{RI^{\alpha}_{m,\lambda,l}f(U)}\frac{RI^{\alpha}_{m+1,\lambda,l}f(z)}{RI^{\alpha}_{m,\lambda,l}f(z)} \leq F_{g(U)}g(z), z \in U$. □

Theorem 5.19 (Alb Lupaş, 2018a). *For a convex function g, with $g(0) = 1$, consider $h(z) = g(z) + \frac{z}{\delta}g'(z), z \in U$. The fuzzy differential subordination*

$$F_{RI^{\alpha}_{m,\lambda,l}f(U)}\left(\left(\frac{RI^{\alpha}_{m,\lambda,l}f(z)}{z}\right)^{\delta-1}\left(RI^{\alpha}_{m,\lambda,l}f(z)\right)'\right) \leq F_{h(U)}h(z), \quad z \in U, \tag{5.1.2.9}$$

is satisfied, when $\alpha, \lambda, l, \delta \geq 0$, $m, n \in \mathbb{N}$, $f \in \mathcal{A}_n$, *and implies the following fuzzy sharp differential subordination* $F_{RI^{\alpha}_{m,\lambda,l}f(U)}\left(\frac{RI^{\alpha}_{m,\lambda,l}f(z)}{z}\right)^{\delta} \leq F_{g(U)}g(z)$, $z \in U$.

Proof. Differentiating $p(z) = \left(\frac{RI^{\alpha}_{m,\lambda,l}f(z)}{z}\right)^{\delta}$, $z \in U$, we get $\left(\frac{RI^{\alpha}_{m,\lambda,l}f(z)}{z}\right)^{\delta-1}\left(RI^{\alpha}_{m,\lambda,l}f(z)\right)' = p(z) + \frac{1}{\delta}zp'(z), z \in U$. Evidently, $p \in \mathcal{H}[1,n]$, and (5.1.2.9) becomes $F_{p(U)}\left(p(z) + \frac{1}{\delta}zp'(z)\right) \leq F_{h(U)}h(z) = F_{g(U)}\left(g(z) + \frac{z}{\delta}g'(z)\right)$, $z \in U$.

By using Lemma 1.15, we have $F_{RI^{\alpha}_{m,\lambda,l}f(U)}\left(\frac{RI^{\alpha}_{m,\lambda,l}f(z)}{z}\right)^{\delta} \leq F_{g(U)}g(z), z \in U$. □

Theorem 5.20 (Alb Lupaş, 2018a). *Consider an holomorphic function h which satisfies the inequality* $\mathrm{Re}\left(1 + \frac{zh''(z)}{h'(z)}\right) > -\frac{1}{2}, z \in U$, *and* $h(0) = 1$. *When the fuzzy differential subordination*

$$F_{RI^{\alpha}_{m,\lambda,l}f(U)}\left(\left(\frac{RI^{\alpha}_{m,\lambda,l}f(z)}{z}\right)^{\delta-1}\left(RI^{\alpha}_{m,\lambda,l}f(z)\right)'\right) \leq F_{h(U)}h(z), \quad z \in U, \tag{5.1.2.10}$$

holds, with $\alpha, \lambda, l, \delta \geq 0$, $m, n \in \mathbb{N}$, $f \in \mathcal{A}_n$, *then*

$$F_{RI^{\alpha}_{m,\lambda,l}f(U)}\left(\frac{RI^{\alpha}_{m,\lambda,l}f(z)}{z}\right)^{\delta} \leq F_{q(U)}q(z), \quad z \in U,$$

and $q(z) = \frac{\delta}{nz^{\frac{\delta}{n}}}\int_0^z h(t)t^{\frac{\delta}{n}-1}dt$.

Proof. Since $\mathrm{Re}\left(1 + \frac{zh''(z)}{h'(z)}\right) > -\frac{1}{2}, z \in U$, from Lemma 1.13, we deduce that $q(z) = \frac{\delta}{nz^{\frac{\delta}{n}}}\int_0^z h(t)t^{\frac{\delta}{n}-1}dt$ is a convex function and verifies the differential equation associated to the fuzzy differential subordination (5.1.2.10) $q(z) + \frac{1}{\delta}zq'(z) = h(z)$, therefore, it is the fuzzy best dominant.

Consider $p(z) = \left(\frac{RI^{\alpha}_{m,\lambda,l}f(z)}{z}\right)^{\delta}$, $z \in U$, then $p \in \mathcal{H}[1,n]$. Differentiating, we obtain $\left(\frac{RI^{\alpha}_{m,\lambda,l}f(z)}{z}\right)^{\delta-1}\left(RI^{\alpha}_{m,\lambda,l}f(z)\right)' = p(z) + \frac{1}{\delta}zp'(z), z \in U$, and the fuzzy differential subordination (5.1.2.10) becomes

$$F_{p(U)}\left(p(z) + \frac{1}{\delta}zp'(z)\right) \leq F_{h(U)}h(z), \quad z \in U.$$

Applying Lemma 1.14, we get $F_{RI^{\alpha}_{m,\lambda,l}f(U)}\left(\frac{RI^{\alpha}_{m,\lambda,l}f(z)}{z}\right)^{\delta} \leq F_{q(U)}q(z)$ and $q(z) = \frac{\delta}{nz^{\frac{\delta}{n}}}\int_0^z h(t)t^{\frac{\delta}{n}-1}dt$, $z \in U$, is the best dominant. □

Theorem 5.21 (Alb Lupaş, 2018a). *For a convex function g, with $g(0) = 1$, define the function $h(z) = g(z) + \frac{z}{\delta}g'(z)$, $z \in U$. If $\alpha, \lambda, l, \delta \geq 0$, $m, n \in \mathbb{N}$, $f \in \mathcal{A}_n$ and the fuzzy differential subordination*

$$F_{RI^{\alpha}_{m,\lambda,l}f(U)}\left(\frac{z^2}{\delta}\frac{RI^{\alpha}_{m,\lambda,l}f(z)}{\left(RI^{\alpha}_{m+1,\lambda,l}f(z)\right)^2}\left[\frac{\left(RI^{\alpha}_{m,\lambda,l}f(z)\right)'}{RI^{\alpha}_{m,\lambda,l}f(z)} - 2\frac{\left(RI^{\alpha}_{m+1,\lambda,l}f(z)\right)'}{RI^{\alpha}_{m+1,\lambda,l}f(z)}\right]\right.$$
$$\left.+z\frac{\delta+1}{\delta}\frac{RI^{\alpha}_{m,\lambda,l}f(z)}{\left(RI^{\alpha}_{m+1,\lambda,l}f(z)\right)^2}\right) \leq F_{h(U)}h(z), \quad z \in U, \tag{5.1.2.11}$$

holds, then $F_{RI^{\alpha}_{m,\lambda,l}f(U)}\left(z\frac{RI^{\alpha}_{m,\lambda,l}f(z)}{\left(RI^{\alpha}_{m+1,\lambda,l}f(z)\right)^2}\right) \leq F_{g(U)}g(z)$, $z \in U$, and this result is sharp.

Proof. Differentiating $p(z) = z\frac{RI^{\alpha}_{m,\lambda,l}f(z)}{\left(RI^{\alpha}_{m+1,\lambda,l}f(z)\right)^2}$, we get $p(z) + \frac{z}{\delta}p'(z) = z\frac{\delta+1}{\delta}\frac{RI^{\alpha}_{m,\lambda,l}f(z)}{\left(RI^{\alpha}_{m+1,\lambda,l}f(z)\right)^2} + \frac{z^2}{\delta}\frac{RI^{\alpha}_{m,\lambda,l}f(z)}{\left(RI^{\alpha}_{m+1,\lambda,l}f(z)\right)^2}\left[\frac{\left(RI^{\alpha}_{m,\lambda,l}f(z)\right)'}{RI^{\alpha}_{m,\lambda,l}f(z)} - 2\frac{\left(RI^{\alpha}_{m+1,\lambda,l}f(z)\right)'}{RI^{\alpha}_{m+1,\lambda,l}f(z)}\right]$.

The fuzzy differential subordination (5.1.2.11) becomes $F_{p(U)}\left(p(z) + \frac{z}{\delta}p'(z)\right) \leq F_{h(U)}h(z)F_{g(U)}\left(g(z) + \frac{z}{\delta}g'(z)\right), z \in U$, and by using Lemma 1.15, we derive

$$F_{RI^{\alpha}_{m,\lambda,l}f(U)}\left(z\frac{RI^{\alpha}_{m,\lambda,l}f(z)}{\left(RI^{\alpha}_{m+1,\lambda,l}f(z)\right)^2}\right) \leq F_{g(U)}g(z), \quad z \in U.$$

□

Theorem 5.22 (Alb Lupaş, 2018a). *Let h be a holomorphic function with properties $\operatorname{Re}\left(1 + \frac{zh''(z)}{h'(z)}\right) > -\frac{1}{2}, z \in U$, and $h(0) = 1$. The fuzzy differential subordination*

$$F_{RI^{\alpha}_{m,\lambda,l}f(U)}\left(\frac{z^2}{\delta}\frac{RI^{\alpha}_{m,\lambda,l}f(z)}{\left(RI^{\alpha}_{m+1,\lambda,l}f(z)\right)^2}\left[\frac{\left(RI^{\alpha}_{m,\lambda,l}f(z)\right)'}{RI^{\alpha}_{m,\lambda,l}f(z)} - 2\frac{\left(RI^{\alpha}_{m+1,\lambda,l}f(z)\right)'}{RI^{\alpha}_{m+1,\lambda,l}f(z)}\right]\right.$$
$$\left.+\, z\frac{\delta+1}{\delta}\frac{RI^{\alpha}_{m,\lambda,l}f(z)}{\left(RI^{\alpha}_{m+1,\lambda,l}f(z)\right)^2}\right) \leq F_{h(U)}h(z), \quad z \in U, \tag{5.1.2.12}$$

implies the following fuzzy differential subordination $F_{RI^{\alpha}_{m,\lambda,l}f(U)}\left(z\frac{RI^{\alpha}_{m,\lambda,l}f(z)}{\left(RI^{\alpha}_{m+1,\lambda,l}f(z)\right)^2}\right) \leq F_{q(U)}q(z)$, $z \in U$, *where* $\alpha, \lambda, l, \delta \geq 0$, $m, n \in \mathbb{N}$, $f \in \mathcal{A}_n$, *and* $q(z) = \frac{\delta}{nz^{\frac{\delta}{n}}} \int_0^z h(t)t^{\frac{\delta}{n}-1}dt$.

Proof. Since Re $\left(1 + \frac{zh''(z)}{h'(z)}\right) > -\frac{1}{2}$, $z \in U$, from Lemma 1.13, we deduce that $q(z) = \frac{\delta}{nz^{\frac{\delta}{n}}} \int_0^z h(t)t^{\frac{\delta}{n}-1}dt$ is a convex function and verifies the differential equation associated to the fuzzy differential subordination (5.1.2.12) $q(z) + \frac{1}{\delta}zq'(z) = h(z)$, therefore, it is the fuzzy best dominant.

Considering $p(z) = z\frac{RI^{\alpha}_{m,\lambda,l}f(z)}{\left(RI^{\alpha}_{m+1,\lambda,l}f(z)\right)^2}$, $z \in U$, $p \in \mathcal{H}[1,n]$. Since $p(z) + \frac{z}{\delta}p'(z) = z\frac{\delta+1}{\delta}\frac{RI^{\alpha}_{m,\lambda,l}f(z)}{\left(RI^{\alpha}_{m+1,\lambda,l}f(z)\right)^2} + \frac{z^2}{\delta}\frac{RI^{\alpha}_{m,\lambda,l}f(z)}{\left(RI^{\alpha}_{m+1,\lambda,l}f(z)\right)^2}$ $\left[\frac{\left(RI^{\alpha}_{m,\lambda,l}f(z)\right)'}{RI^{\alpha}_{m,\lambda,l}f(z)} - 2\frac{\left(RI^{\alpha}_{m+1,\lambda,l}f(z)\right)'}{RI^{\alpha}_{m+1,\lambda,l}f(z)}\right]$, $z \in U$, from (5.1.2.12), we have $F_{p(U)}\left(p(z) + \frac{z}{\delta}p'(z)\right) \leq F_{h(U)}h(z)$, $z \in U$, and from Lemma 1.14, we obtain $F_{RI^{\alpha}_{m,\lambda,l}f(U)}\left(z\frac{RI^{\alpha}_{m,\lambda,l}f(z)}{\left(RI^{\alpha}_{m+1,\lambda,l}f(z)\right)^2}\right) \leq F_{q(U)}q(z)$, $z \in U$, and the best dominant is $q(z) = \frac{\delta}{nz^{\frac{\delta}{n}}} \int_0^z h(t)t^{\frac{\delta}{n}-1}dt$. □

Theorem 5.23 (Alb Lupaş, 2018a). *For a convex function* g, *with* $g(0)=0$, *define* $h(z) = g(z) + \frac{z}{\delta}g'(z), z \in U$. *If* $\alpha, \lambda, l, \delta \geq 0$, $m, n \in \mathbb{N}$, $f \in \mathcal{A}_n$ *and the fuzzy differential subordination*

$$F_{RI^{\alpha}_{m,\lambda,l}f(U)}\left(z^2\frac{\delta+2}{\delta}\frac{\left(RI^{\alpha}_{m,\lambda,l}f(z)\right)'}{RI^{\alpha}_{m,\lambda,l}f(z)} + \frac{z^3}{\delta}\left[\frac{\left(RI^{\alpha}_{m,\lambda,l}f(z)\right)''}{RI^{\alpha}_{m,\lambda,l}f(z)} - \left(\frac{\left(RI^{\alpha}_{m,\lambda,l}f(z)\right)'}{RI^{\alpha}_{m,\lambda,l}f(z)}\right)^2\right]\right) \leq F_{h(U)}h(z), \quad z \in U, \tag{5.1.2.13}$$

holds, then the following result is sharp $F_{RI^{\alpha}_{m,\lambda,l}f(U)}\left(z^2\frac{\left(RI^{\alpha}_{m,\lambda,l}f(z)\right)'}{RI^{\alpha}_{m,\lambda,l}f(z)}\right) \leq F_{g(U)}g(z)$, $z \in U$.

Proof. Considering $p(z)=z^2\frac{\left(RI^{\alpha}_{m,\lambda,l}f(z)\right)'}{RI^{\alpha}_{m,\lambda,l}f(z)}$, we deduce that $p\in\mathcal{H}[0,1]$ and differentiating it, we obtain

$$p(z)+\frac{z}{\delta}p'(z)=z^2\frac{\delta+2}{\delta}\frac{\left(RI^{\alpha}_{m,\lambda,l}f(z)\right)'}{RI^{\alpha}_{m,\lambda,l}f(z)}$$

$$+\frac{z^3}{\delta}\left[\frac{\left(RI^{\alpha}_{m,\lambda,l}f(z)\right)''}{RI^{\alpha}_{m,\lambda,l}f(z)}-\left(\frac{\left(RI^{\alpha}_{m,\lambda,l}f(z)\right)'}{RI^{\alpha}_{m,\lambda,l}f(z)}\right)^2\right],\quad z\in U.$$

The fuzzy differential subordination becomes $F_{p(U)}\left(p(z)+\frac{1}{\delta}zp'(z)\right)\le F_{h(U)}h(z)=F_{g(U)}\left(g(z)+\frac{z}{\delta}g'(z)\right)$ and by using Lemma 1.15, we deduce $F_{RI^{\alpha}_{m,\lambda,l}f(U)}\left(z^2\frac{\left(RI^{\alpha}_{m,\lambda,l}f(z)\right)'}{RI^{\alpha}_{m,\lambda,l}f(z)}\right)\le F_{g(U)}g(z)$, $z\in U$, and this result is sharp. □

Theorem 5.24 (Alb Lupaş, 2018a). *Consider h a holomorphic function with* Re $\left(1+\frac{zh''(z)}{h'(z)}\right)>-\frac{1}{2}$, $z\in U$, *and* $h(0)=0$. *If the fuzzy differential subordination*

$$F_{RI^{\alpha}_{m,\lambda,l}f(U)}\left(z^2\frac{\delta+2}{\delta}\frac{\left(RI^{\alpha}_{m,\lambda,l}f(z)\right)'}{RI^{\alpha}_{m,\lambda,l}f(z)}+\frac{z^3}{\delta}\left[\frac{\left(RI^{\alpha}_{m,\lambda,l}f(z)\right)''}{RI^{\alpha}_{m,\lambda,l}f(z)}-\left(\frac{\left(RI^{\alpha}_{m,\lambda,l}f(z)\right)'}{RI^{\alpha}_{m,\lambda,l}f(z)}\right)^2\right]\right)\le F_{h(U)}h(z),\quad z\in U,\tag{5.1.2.14}$$

holds, for $\alpha,\lambda,l,\delta\ge 0$, $m,n\in\mathbb{N}$, $f\in\mathcal{A}_n$, *then*

$$F_{RI^{\alpha}_{m,\lambda,l}f(U)}\left(z^2\frac{\left(RI^{\alpha}_{m,\lambda,l}f(z)\right)'}{RI^{\alpha}_{m,\lambda,l}f(z)}\right)\le F_{q(U)}q(z),\quad z\in U,$$

where $q(z)=\frac{\delta}{nz^{\frac{\delta}{n}}}\int_0^z h(t)t^{\frac{\delta}{n}-1}dt$.

Proof. Since Re $\left(1+\frac{zh''(z)}{h'(z)}\right)>-\frac{1}{2}$, $z\in U$, from Lemma 1.13, we deduce that $q(z)=\frac{\delta}{nz^{\frac{\delta}{n}}}\int_0^z h(t)t^{\frac{\delta}{n}-1}dt$ is a convex function and verifies the differential equation associated to the fuzzy differential subordination (5.1.2.14) $q(z)+\frac{1}{\delta}zq'(z)=h(z)$, therefore, it is the fuzzy best dominant.

Considering $p(z) = z^2 \frac{\left(RI^{\alpha}_{m,\lambda,l}f(z)\right)'}{RI^{\alpha}_{m,\lambda,l}f(z)}$, $z \in U$, $p \in \mathcal{H}[0,n]$. Differentiating it, we obtain $p(z) + \frac{z}{\delta}p'(z) = z^2 \frac{\delta+2}{\delta} \frac{\left(RI^{\alpha}_{m,\lambda,l}f(z)\right)'}{RI^{\alpha}_{m,\lambda,l}f(z)} + \frac{z^3}{\delta}\left[\frac{\left(RI^{\alpha}_{m,\lambda,l}f(z)\right)''}{RI^{\alpha}_{m,\lambda,l}f(z)} - \left(\frac{\left(RI^{\alpha}_{m,\lambda,l}f(z)\right)'}{RI^{\alpha}_{m,\lambda,l}f(z)}\right)^2\right]$, $z \in U$, and (5.1.2.14) becomes $F_{p(U)}\left(p(z) + \frac{1}{\delta}zp'(z)\right) \leq F_{h(U)}h(z)$, $z \in U$.

Applying Lemma 1.14, we get $F_{RI^{\alpha}_{m,\lambda,l}f(U)}\left(z^2 \frac{\left(RI^{\alpha}_{m,\lambda,l}f(z)\right)'}{RI^{\alpha}_{m,\lambda,l}f(z)}\right) \leq F_{q(U)}q(z)$, $z \in U$, and the best dominant is $q(z) = \frac{\delta}{nz^{\frac{\delta}{n}}} \int_0^z h(t)t^{\frac{\delta}{n}-1}dt$. □

Theorem 5.25 (Alb Lupaş, 2018a). *Let $h(z) = g(z) + zg'(z), z \in U$, where g is a convex function such that $g(0) = 1$. When the fuzzy differential subordination*

$$F_{RI^{\alpha}_{m,\lambda,l}f(U)}\left(1 - \frac{RI^{\alpha}_{m,\lambda,l}f(z) \cdot \left(RI^{\alpha}_{m,\lambda,l}f(z)\right)''}{\left[\left(RI^{\alpha}_{m,\lambda,l}f(z)\right)'\right]^2}\right) \leq F_{h(U)}h(z), \quad z \in U \tag{5.1.2.15}$$

holds, for $\alpha, \lambda, l \geq 0$, $m, n \in \mathbb{N}$, $f \in \mathcal{A}_n$, then the following result is sharp

$$F_{RI^{\alpha}_{m,\lambda,l}f(U)}\left(\frac{RI^{\alpha}_{m,\lambda,l}f(z)}{z\left(RI^{\alpha}_{m,\lambda,l}f(z)\right)'}\right) \leq F_{g(U)}g(z), \quad z \in U.$$

Proof. Let $p(z) = \frac{RI^{\alpha}_{m,\lambda,l}f(z)}{z\left(RI^{\alpha}_{m,\lambda,l}f(z)\right)'}$. We deduce that $p \in \mathcal{H}[1,n]$ and differentiating it, we get $1 - \frac{RI^{\alpha}_{m,\lambda,l}f(z) \cdot \left(RI^{\alpha}_{m,\lambda,l}f(z)\right)''}{\left[\left(RI^{\alpha}_{m,\lambda,l}f(z)\right)'\right]^2} = p(z) + zp'(z), z \in U$.

The fuzzy differential subordination becomes $F_{p(U)}\left(p(z) + zp'(z)\right) \leq F_{h(U)}h(z) = F_{g(U)}\left(g(z) + zg'(z)\right)$ and we apply Lemma 1.15 to deduce the following sharp result $_{RI^{\alpha}_{m,\lambda,l}f(U)}\left(\frac{RI^{\alpha}_{m,\lambda,l}f(z)}{z\left(RI^{\alpha}_{m,\lambda,l}f(z)\right)'}\right) \leq F_{g(U)}g(z)$, $z \in U$. □

Theorem 5.26 (Alb Lupaş, 2018a). *Let h be a holomorphic function with $h(0) = 1$ and $\operatorname{Re}\left(1 + \frac{zh''(z)}{h'(z)}\right) > -\frac{1}{2}, z \in U$. The fuzzy differential* *subordination*

$$F_{RI^{\alpha}_{m,\lambda,l}f(U)}\left(1-\frac{RI^{\alpha}_{m,\lambda,l}f(z)\cdot\left(RI^{\alpha}_{m,\lambda,l}f(z)\right)''}{\left[\left(RI^{\alpha}_{m,\lambda,l}f(z)\right)'\right]^2}\right)\leq F_{h(U)}h(z),\quad z\in U, \tag{5.1.2.16}$$

induce $F_{RI^{\alpha}_{m,\lambda,l}f(U)}\left(\frac{RI^{\alpha}_{m,\lambda,l}f(z)}{z\left(RI^{\alpha}_{m,\lambda,l}f(z)\right)'}\right)\leq F_{q(U)}q(z)$, $z\in U$, *where* $q(z)=\frac{1}{nz^{\frac{1}{n}}}\int_0^z h(t)t^{\frac{1}{n}-1}dt$, *for* $\alpha,\lambda,l\geq 0$, $m,n\in\mathbb{N}$, $f\in\mathcal{A}_n$.

Proof. Since $\operatorname{Re}\left(1+\frac{zh''(z)}{h'(z)}\right)>-\frac{1}{2}$, $z\in U$, from Lemma 1.13, we deduce that $\frac{1}{nz^{\frac{1}{n}}}\int_0^z h(t)t^{\frac{1}{n}-1}dt$ is a convex function and verifies the differential equation associated to the fuzzy differential subordination (5.1.2.16) $q(z)+zq'(z)=h(z)$, therefore, it is the fuzzy best dominant.

Consider $p(z)=\frac{RI^{\alpha}_{m,\lambda,l}f(z)}{z\left(RI^{\alpha}_{m,\lambda,l}f(z)\right)'}$, $z\in U$, $p\in\mathcal{H}[1,n]$. Differentiating it, we obtain $1-\frac{RI^{\alpha}_{m,\lambda,l}f(z)\cdot\left(RI^{\alpha}_{m,\lambda,l}f(z)\right)''}{\left[\left(RI^{\alpha}_{m,\lambda,l}f(z)\right)'\right]^2}=p(z)+zp'(z)$, $z\in U$, and the fuzzy differential subordination (5.1.2.16) is written $F_{p(U)}\left(p(z)+zp'(z)\right)\leq F_{h(U)}h(z)$, $z\in U$.

By using Lemma 1.14, we get $F_{RI^{\alpha}_{m,\lambda,l}f(U)}\left(\frac{RI^{\alpha}_{m,\lambda,l}f(z)}{z\left(RI^{\alpha}_{m,\lambda,l}f(z)\right)'}\right)\leq F_{q(U)}q(z)$, $z\in U$, and the best dominant is $q(z)=\frac{1}{nz^{\frac{1}{n}}}\int_0^z h(t)t^{\frac{1}{n}-1}dt$. □

Example 5.4 (Alb Lupaş, 2018a). Let $h(z)=\frac{1-z}{1+z}$ a convex function in U with $h(0)=1$ and $\operatorname{Re}\left(\frac{zh''(z)}{h'(z)}+1\right)>-\frac{1}{2}$.

Let $f(z)=z-z^2$, $z\in U$. For $n=1$, $m=1$, $l=1$, $\lambda=2$, $\alpha=\frac{2}{3}$, we obtain $RI^{\frac{2}{3}}_{1,2,1}f(z)=z-2z^2$, $z\in U$.

Then $\left(RI^{\frac{2}{3}}_{1,2,1}f(z)\right)'=1-4z$ and $\left(RI^{\frac{2}{3}}_{1,2,1}f(z)\right)''=-4$, $\frac{RI^{\frac{2}{3}}_{1,2,1}f(z)}{z\left(RI^{\frac{2}{3}}_{1,2,1}f(z)\right)'}=\frac{z-2z^2}{z(1-4z)}=\frac{1-2z}{1-4z}$, $1-\frac{RI^{\frac{2}{3}}_{1,2,1}f(z)\cdot\left(RI^{\frac{2}{3}}_{1,2,1}f(z)\right)''}{\left[\left(RI^{\frac{2}{3}}_{1,2,1}f(z)\right)'\right]^2}=1-\frac{\left(z-2z^2\right)\cdot(-4)}{(1-4z)^2}=\frac{8z^2-4z+1}{(1-4z)^2}$. We have $q(z)=\frac{1}{z}\int_0^z\frac{1-t}{1+t}dt=-1+\frac{2\ln(1+z)}{z}$.

Using Theorem 5.26, we obtain $F_U\left(\frac{8z^2-4z+1}{(1-4z)^2}\right) \leq F_U\left(\frac{1-z}{1+z}\right)$, $z \in U$, induce $F_U\left(\frac{1-2z}{1-4z}\right) \leq F_U\left(-1+\frac{2\ln(1+z)}{z}\right)$, $z \in U$.

Theorem 5.27 (Alb Lupaş, 2018a). *Consider* $h(z) = g(z) + zg'(z)$, $z \in U$, *where* g *is a convex function with* $g(0) = 1$. *When the fuzzy differential subordination*

$$F_{RI^{\alpha}_{m,\lambda,l}f(U)}\left(\left[\left(RI^{\alpha}_{m,\lambda,l}f(z)\right)'\right]^2 + RI^{\alpha}_{m,\lambda,l}f(z)\cdot\left(RI^{\alpha}_{m,\lambda,l}f(z)\right)''\right) \leq F_{h(U)}h(z), \quad z \in U \tag{5.1.2.17}$$

holds, for $\alpha, \lambda, l \geq 0$, $m, n \in \mathbb{N}$, $f \in \mathcal{A}_n$, *then the following result is sharp*

$$F_{RI^{\alpha}_{m,\lambda,l}f(U)}\left(\frac{RI^{\alpha}_{m,\lambda,l}f(z)\cdot\left(RI^{\alpha}_{m,\lambda,l}f(z)\right)'}{z}\right) \leq F_{g(U)}g(z), \quad z \in U.$$

Proof. Let $p(z) = \frac{RI^{\alpha}_{m,\lambda,l}f(z)\cdot\left(RI^{\alpha}_{m,\lambda,l}f(z)\right)'}{z}$. We deduce that $p \in \mathcal{H}[1, n]$.

Differentiating it, we obtain $\left[\left(RI^{\alpha}_{m,\lambda,l}f(z)\right)'\right]^2 + RI^{\alpha}_{m,\lambda,l}f(z)\cdot \left(RI^{\alpha}_{m,\lambda,l}f(z)\right)'' = p(z) + zp'(z)$, $z \in U$, and the fuzzy differential subordination becomes $F_{p(U)}(p(z) + zp'(z)) \leq F_{h(U)}h(z) = F_{g(U)}(g(z) + zg'(z))$.

By using Lemma 1.15, we obtain the following result sharp $F_{RI^{\alpha}_{m,\lambda,l}f(U)}\left(\frac{RI^{\alpha}_{m,\lambda,l}f(z)\cdot\left(RI^{\alpha}_{m,\lambda,l}f(z)\right)'}{z}\right) \leq F_{g(U)}g(z)$, $z \in U$. □

Theorem 5.28 (Alb Lupaş, 2018a). *For a holomorphic function* h *which satisfies the inequality* $\text{Re}\left(1+\frac{zh''(z)}{h'(z)}\right) > -\frac{1}{2}$, $z \in U$, *and* $h(0) = 1$, *the fuzzy differential subordination*

$$F_{RI^{\alpha}_{m,\lambda,l}f(U)}\left(\left[\left(RI^{\alpha}_{m,\lambda,l}f(z)\right)'\right]^2 + RI^{\alpha}_{m,\lambda,l}f(z)\cdot\left(RI^{\alpha}_{m,\lambda,l}f(z)\right)''\right) \leq F_{h(U)}h(z), \quad z \in U, \tag{5.1.2.18}$$

induce $F_{RI^{\alpha}_{m,\lambda,l}f(U)}\left(\frac{RI^{\alpha}_{m,\lambda,l}f(z)\cdot\left(RI^{\alpha}_{m,\lambda,l}f(z)\right)'}{z}\right) \leq F_{q(U)}q(z)$, $z \in U$, *where* $q(z) = \frac{1}{nz^{\frac{1}{n}}}\int_0^z h(t)t^{\frac{1}{n}-1}dt$, *and* $\alpha, \lambda, l \geq 0$, $m, n \in \mathbb{N}$, $f \in \mathcal{A}_n$

Proof. Since Re $\left(1 + \frac{zh''(z)}{h'(z)}\right) > -\frac{1}{2}$, $z \in U$, from Lemma 1.13, we deduce that $q(z) = \frac{1}{nz^{\frac{1}{n}}}\int_0^z h(t)t^{\frac{1}{n}-1}dt$ is a convex function and verifies the differential equation associated to the fuzzy differential subordination (5.1.2.18) $q(z) + zq'(z) = h(z)$, therefore, it is the fuzzy best dominant.

Let $p(z) = \frac{RI^{\alpha}_{m,\lambda,l}f(z)\cdot\left(RI^{\alpha}_{m,\lambda,l}f(z)\right)'}{z}$, $z \in U$, evidently $p \in \mathcal{H}[1, n]$.

We have $p(z) + zp'(z) = \left[\left(RI^{\alpha}_{m,\lambda,l}f(z)\right)'\right]^2 + RI^{\alpha}_{m,\lambda,l}f(z)\cdot \left(RI^{\alpha}_{m,\lambda,l}f(z)\right)''$, $z \in U$, and (5.1.2.18) means $F_{p(U)}\left(p(z) + zp'(z)\right) \leq F_{h(U)}h(z)$, $z \in U$.

Applying Lemma 1.14, we get $F_{RI^{\alpha}_{m,\lambda,l}f(U)}\left(\frac{RI^{\alpha}_{m,\lambda,l}f(z)\cdot\left(RI^{\alpha}_{m,\lambda,l}f(z)\right)'}{z}\right) \leq F_{q(U)}q(z)$, $z \in U$, and the best dominant is $q(z) = \frac{1}{nz^{\frac{1}{n}}}\int_0^z h(t)t^{\frac{1}{n}-1}dt$. $\square$

Example 5.5 (Alb Lupaş, 2018a). Let $h(z) = \frac{1-z}{1+z}$ a convex function in U with $h(0) = 1$ and Re $\left(\frac{zh''(z)}{h'(z)} + 1\right) > -\frac{1}{2}$.

Consider $f(z) = z - z^2$, $z \in U$. For $n = 1$, $m = 1$, $l = 1$, $\lambda = 2$, $\alpha = \frac{2}{3}$, we obtain $RI^{\frac{2}{3}}_{1,2,1}f(z) = z - 2z^2$, $z \in U$.

Then $\left(RI^{\frac{2}{3}}_{1,2,1}f(z)\right)' = 1 - 4z$, $\left(RI^{\frac{2}{3}}_{1,2,1}f(z)\right)'' = -4$, $\frac{RI^{\frac{2}{3}}_{1,2,1}f(z)\cdot\left(RI^{\frac{2}{3}}_{1,2,1}f(z)\right)'}{z} = \frac{(z-2z^2)(1-4z)}{z} = 8z^2 - 6z + 1$, $\left[\left(RI^{\frac{2}{3}}_{1,2,1}f(z)\right)'\right]^2 + RI^{\frac{2}{3}}_{1,2,1}f(z)\cdot\left(RI^{\frac{2}{3}}_{1,2,1}f(z)\right)'' = (1-4z)^2 + \left(z - 2z^2\right)\cdot(-4) = 24z^2 - 12z + 1$.

We have $q(z) = \frac{1}{z}\int_0^z \frac{1-t}{1+t}dt = -1 + \frac{2\ln(1+z)}{z}$.

From Theorem 5.28, we obtain $F_U\left(24z^2 - 12z + 1\right) \leq F_U\left(\frac{1-z}{1+z}\right)$, $z \in U$, induce $F_U\left(8z^2 - 6z + 1\right) \leq F_U\left(-1 + \frac{2\ln(1+z)}{z}\right)$, $z \in U$.

Theorem 5.29 (Alb Lupaş, 2018a). *Consider* $h(z) = g(z) + \frac{z}{1-\delta}g'(z)$, $z \in U$, *where* g *is a convex function with* $g(0) = 1$. *If the fuzzy differential* *subordination*

$$F_{RI^{\alpha}_{m,\lambda,l}f(U)}\left(\left(\frac{z}{RI^{\alpha}_{m,\lambda,l}f(z)}\right)^{\delta}\frac{RI^{\alpha}_{m+1,\lambda,l}f(z)}{1-\delta}\left(\frac{\left(RI^{\alpha}_{m+1,\lambda,l}f(z)\right)'}{RI^{\alpha}_{m+1,\lambda,l}f(z)}-\delta\frac{\left(RI^{\alpha}_{m,\lambda,l}f(z)\right)'}{RI^{\alpha}_{m,\lambda,l}f(z)}\right)\right)\leq F_{h(U)}h(z), \quad (5.1.2.19)$$

$z\in U$, holds, for $\alpha,\lambda,l\geq 0$, $\delta\in(0,1)$, $m,n\in\mathbb{N}$, $f\in\mathcal{A}_n$, then the following result is sharp

$$F_{RI^{\alpha}_{m,\lambda,l}f(U)}\left(\frac{RI^{\alpha}_{m+1,\lambda,l}f(z)}{z}\cdot\left(\frac{z}{RI^{\alpha}_{m,\lambda,l}f(z)}\right)^{\delta}\right)\leq F_{g(U)}g(z), \quad z\in U.$$

Proof. Let $p(z)=\frac{RI^{\alpha}_{m+1,\lambda,l}f(z)}{z}\cdot\left(\frac{z}{RI^{\alpha}_{m,\lambda,l}f(z)}\right)^{\delta}$. We deduce that $p\in\mathcal{H}[1,n]$ and differentiating the function p, we obtain $\left(\frac{z}{RI^{\alpha}_{m,\lambda,l}f(z)}\right)^{\delta}\frac{RI^{\alpha}_{m+1,\lambda,l}f(z)}{1-\delta}\left(\frac{\left(RI^{\alpha}_{m+1,\lambda,l}f(z)\right)'}{RI^{\alpha}_{m+1,\lambda,l}f(z)}-\delta\frac{\left(RI^{\alpha}_{m,\lambda,l}f(z)\right)'}{RI^{\alpha}_{m,\lambda,l}f(z)}\right)=p(z)+\frac{1}{1-\delta}zp'(z)$, $z\in U$.

The fuzzy differential subordination means

$$F_{p(U)}\left(p(z)+\frac{1}{1-\delta}zp'(z)\right)\leq F_{h(U)}h(z)=F_{g(U)}\left(g(z)+\frac{z}{1-\delta}g'(z)\right)$$

and by using Lemma 1.15, we obtain the following sharp result

$$F_{RI^{\alpha}_{m,\lambda,l}f(U)}\left(\frac{RI^{\alpha}_{m+1,\lambda,l}f(z)}{z}\cdot\left(\frac{z}{RI^{\alpha}_{m,\lambda,l}f(z)}\right)^{\delta}\right)\leq F_{g(U)}g(z), \quad z\in U.$$

□

Theorem 5.30 (Alb Lupaş, 2018a). *Let h be an holomorphic function with $h(0)=1$ and* Re $\left(1+\frac{zh''(z)}{h'(z)}\right)>-\frac{1}{2}, z\in U$. *The fuzzy differential subordination*

$$F_{RI^{\alpha}_{m,\lambda,l}f(U)}\left(\left(\frac{z}{RI^{\alpha}_{m,\lambda,l}f(z)}\right)^{\delta}\frac{RI^{\alpha}_{m+1,\lambda,l}f(z)}{1-\delta}\left(\frac{\left(RI^{\alpha}_{m+1,\lambda,l}f(z)\right)'}{RI^{\alpha}_{m+1,\lambda,l}f(z)}-\delta\frac{\left(RI^{\alpha}_{m,\lambda,l}f(z)\right)'}{RI^{\alpha}_{m,\lambda,l}f(z)}\right)\right)\leq F_{h(U)}h(z), \quad (5.1.2.20)$$

$z \in U$, *induce* $F_{RI^{\alpha}_{m,\lambda,l}f(U)}\left(\frac{RI^{\alpha}_{m+1,\lambda,l}f(z)}{z} \cdot \left(\frac{z}{RI^{\alpha}_{m,\lambda,l}f(z)}\right)^{\delta}\right) \leq F_{q(U)}q(z)$, $z \in U$, *where* $q(z) = \frac{1-\delta}{nz^{\frac{1-\delta}{n}}} \int_0^z h(t)t^{\frac{1-\delta}{n}-1}dt$, *for* $\alpha, \lambda, l \geq 0$, $\delta \in (0,1)$, $m, n \in \mathbb{N}$, $f \in \mathcal{A}_n$.

Proof. Since $\operatorname{Re}\left(1 + \frac{zh''(z)}{h'(z)}\right) > -\frac{1}{2}$, $z \in U$, from Lemma 1.13, we deduce that $q(z) = \frac{1-\delta}{nz^{\frac{1-\delta}{n}}} \int_0^z h(t)t^{\frac{1-\delta}{n}-1}dtt$ is a convex function and verifies the differential equation associated to the fuzzy differential subordination (5.1.2.20) $q(z) + \frac{1}{1-\delta}zq'(z) = h(z)$, therefore, it is the fuzzy best dominant.

Consider $p(z) = \frac{RI^{\alpha}_{m+1,\lambda,l}f(z)}{z} \cdot \left(\frac{z}{RI^{\alpha}_{m,\lambda,l}f(z)}\right)^{\delta}$, $z \in U$, $p \in \mathcal{H}[1,n]$.

Differentiating it, we get $\left(\frac{z}{RI^{\alpha}_{m,\lambda,l}f(z)}\right)^{\delta} \frac{RI^{\alpha}_{m+1,\lambda,l}f(z)}{1-\delta} \left(\frac{\left(RI^{\alpha}_{m+1,\lambda,l}f(z)\right)'}{RI^{\alpha}_{m+1,\lambda,l}f(z)} - \delta\frac{\left(RI^{\alpha}_{m,\lambda,l}f(z)\right)'}{RI^{\alpha}_{m,\lambda,l}f(z)}\right) = p(z) + \frac{1}{1-\delta}zp'(z)$, $z \in U$, and (5.1.2.20) becomes $F_{p(U)}\left(p(z) + \frac{1}{1-\delta}zp'(z)\right) \leq F_{h(U)}h(z)$, $z \in U$.

From Lemma 1.14, we get $F_{RI^{\alpha}_{m,\lambda,l}f(U)}\left(\frac{RI^{\alpha}_{m+1,\lambda,l}f(z)}{z} \cdot \left(\frac{z}{RI^{\alpha}_{m,\lambda,l}f(z)}\right)^{\delta}\right) \leq F_{q(U)}q(z)$, $z \in U$, and the best dominant is $q(z) = \frac{1-\delta}{nz^{\frac{1-\delta}{n}}} \int_0^z h(t)t^{\frac{1-\delta}{n}-1}dt$. □

5.2 Applications of Differential Operators in the Theory of Fuzzy Differential Subordinations and Superordinations

In this section, we present several fuzzy differential subordination, superordination, respectively sandwich-type results involving few of the defined operators presented in Chapter 2.

5.2.1 *Fuzzy Differential Subordination and Superordination Results Involving Riemann–Liouville Fractional Integral of q-Hypergeometric Function*

The concepts of fuzzy differential subordination and superordination were introduced in geometric function theory as generalization of the classical notions of differential subordination and superordination. Fractional calculus is combined in the papers Alb Lupaş and Oros (2022d) and Alb Lupaş and Oros (2022e) with quantum calculus aspects for obtaining new fuzzy differential subordinations and superordinations. For the investigated fuzzy differential subordinations and superordinations, fuzzy best subordinants and fuzzy best dominants are obtained, respectively. Furthermore, interesting corollaries emerge when using particular functions, frequently involved in researches due to their geometric properties, as fuzzy best subordinant and fuzzy best dominant. The study is finalized by stating the sandwich-type results connecting the previously proved results.

Let m, n be complex numbers with $m \neq 0, -1, -2, \ldots$ and $\lambda > 0$, $0 < q < 1$. The Riemann–Liouville fractional integral of q-hypergeometric function (Alb Lupaş and Oros, 2022c) exposed in Definition 2.20 is

$$\begin{aligned} D_z^{-\lambda}\phi(m,n;q,z) &= \frac{1}{\Gamma(\lambda)}\int_0^z \frac{\phi(m,n;q,t)}{(z-t)^{1-\lambda}}dt \\ &= \frac{1}{\Gamma(\lambda)}\sum_{k=0}^{\infty}\frac{(m,q)_k}{(n,q)_k(q,q)_k}\int_0^z \frac{t^k}{(z-t)^{1-\lambda}}dt. \end{aligned} \tag{5.2.1.1}$$

After a simple calculation, the Riemann–Liouville fractional integral of q-hypergeometric function has the following form

$$D_z^{-\lambda}\phi(m,n;q,z) = \sum_{k=0}^{\infty}\frac{(m,q)_k}{(n,q)_k(q,q)_k(k+1)_\lambda}z^{\lambda+k}. \tag{5.2.1.2}$$

We note that $D_z^{-\lambda}\phi(m,n;q,z) \in \mathcal{H}[0,\lambda]$.

The operator seen in (5.2.1.1) will be used for obtaining new fuzzy differential subordinations and superordinations for which the best fuzzy dominants and the best fuzzy subordinants are established in the theorems proved in the following. Furthermore, using particular functions well-known due to their geometric properties as fuzzy best dominant and fuzzy best subordinant, corollaries are established following each theorem. At the end of the study, the results obtained using the two dual theories are combined into sandwich-type results familiar to geometric function theory.

The first theorem and its corollaries are related to fuzzy differential subordination.

Theorem 5.31 (Alb Lupaş and Oros, 2022d). *Consider g an analytic and univalent function in U such that $g(z) \neq 0, z \in U$, and $\left(\frac{D_z^{-\lambda}\phi(m,n;q,z)}{z}\right)^\alpha \in \mathcal{H}(U)$, where m, n are complex numbers with $n \neq 0, -1, -2, \ldots$ and $\alpha, \lambda > 0$, $0 < q < 1$. Assuming $\frac{zg'(z)}{g(z)}$ is starlike univalent in U and*

$$\operatorname{Re}\left(1 + \frac{\psi}{\delta}g(z) + \frac{2\varepsilon}{\delta}(g(z))^2 - \frac{zg'(z)}{g(z)} + \frac{zg''(z)}{g'(z)}\right) > 0, \tag{5.2.1.3}$$

for $\beta, \psi, \varepsilon, \delta \in \mathbb{C}$, $\delta \neq 0$, $z \in U$ *and*

$$\Psi_\lambda^{m,n,q}(\alpha, \beta, \psi, \varepsilon, \delta; z) := \beta + \psi\left[\frac{D_z^{-\lambda}\phi(m,n;q,z)}{z}\right]^\alpha$$

$$+ \varepsilon\left[\frac{D_z^{-\lambda}\phi(m,n;q,z)}{z}\right]^{2\alpha} + \delta\alpha\left[\frac{z\left(D_z^{-\lambda}\phi(m,n;q,z)\right)'}{D_z^{-\lambda}\phi(m,n;q,z)} - 1\right]. \quad (5.2.1.4)$$

If the fuzzy differential subordination is satisfied by g

$$F_{\Psi_\lambda^{m,n,q}(U)}\Psi_\lambda^{m,n,q}(\alpha, \beta, \psi, \varepsilon, \delta; z)$$
$$\leq F_{q(U)}\left(\beta + \psi g(z) + \varepsilon(g(z))^2 + \delta\frac{zg'(z)}{g(z)}\right), \quad (5.2.1.5)$$

for $\beta, \psi, \varepsilon, \delta \in \mathbb{C}$, $\delta \neq 0$, *then*

$$F_{D_z^{-\lambda}\phi(U)}\left(\frac{D_z^{-\lambda}\phi(m,n;q,z)}{z}\right)^\alpha \leq F_{g(U)}g(z), \quad z \in U, \quad (5.2.1.6)$$

and the fuzzy best dominant is g.

Proof. Define $p(z) := \left(\frac{D_z^{-\lambda}\phi(m,n;q,z)}{z}\right)^\alpha$, $z \in U$, $z \neq 0$. Differentiating it, we obtain

$$p'(z) = \alpha\left(\frac{D_z^{-\lambda}\phi(m,n;q,z)}{z}\right)^{\alpha-1}\left[\frac{(D_z^{-\lambda}\phi(m,n;q,z))'}{z} - \frac{D_z^{-\lambda}\phi(m,n;q,z)}{z^2}\right]$$
$$= \alpha\left(\frac{D_z^{-\lambda}\phi(m,n;q,z)}{z}\right)^{\alpha-1}\frac{(D_z^{-\lambda}\phi(m,n;q,z))'}{z} - \frac{\alpha}{z}p(z).$$

Then $\frac{zp'(z)}{p(z)} = \alpha\left[\frac{z\left(D_z^{-\lambda}\phi(m,n;q,z)\right)'}{D_z^{-\lambda}\phi(m,n;q,z)} - 1\right]$.

Set $\varphi(u) := \beta + \psi u + \varepsilon u^2$ and $\gamma(u) := \frac{\delta}{u}$, we can show easily that φ is analytic in $\mathbb{C}$, γ is analytic in $\mathbb{C}\backslash\{0\}$ and $\gamma(u) \neq 0$, $u \in \mathbb{C}\backslash\{0\}$.

Also, we set $G(z) = zg'(z)\gamma(g(z)) = \delta\frac{zg'(z)}{g(z)}$, which is starlike univalent in U, and $h(z) = G(z) + \varphi(g(z)) = \beta + \psi g(z) + \varepsilon(g(z))^2 + \delta\frac{zg'(z)}{g(z)}$.

Differentiating it, we obtain $h'(z) = \delta + g'(z) + 2\varepsilon g(z)g'(z) + \delta\frac{(g'(z)+zg''(z))g(z) - z(g'(z))^2}{(g(z))^2}$ and $\frac{zh'(z)}{G(z)} = \frac{zh'(z)}{\delta\frac{zg'(z)}{g(z)}} = 1 + \frac{\psi}{\delta}g(z) + \frac{2\varepsilon}{\delta}(g(z))^2 - \frac{zg'(z)}{g(z)} + \frac{zg''(z)}{g'(z)}$.

It yields that

$$\text{Re}\left(\frac{zh'(z)}{G(z)}\right) = \text{Re}\left(1 + \frac{\psi}{\delta}g(z) + \frac{2\varepsilon}{\delta}(g(z))^2 - \frac{zg'(z)}{g(z)} + \frac{zg''(z)}{g'(z)}\right) > 0.$$

We get $\beta + \psi p(z) + \varepsilon (p(z))^2 + \delta \frac{zp'(z)}{p(z)} = \beta + \psi \left[\frac{D_z^{-\lambda}\phi(m,n;q,z)}{z}\right]^{\alpha} + \varepsilon \left[\frac{D_z^{-\lambda}\phi(m,n;q,z)}{z}\right]^{2\alpha} + \alpha\delta \left[\frac{z\left(D_z^{-\lambda}\phi(m,n;q,z)\right)'}{D_z^{-\lambda}\phi(m,n;q,z)} - 1\right]$.

By relation (5.2.1.5), we obtain $F_{p(U)}(\beta+\psi p(z)+\varepsilon(p(z))^2+\delta\frac{zp'(z)}{p(z)}) \leq F_{g(U)}\left(\beta + \psi g(z) + \varepsilon (g(z))^2 + \delta \frac{zg'(z)}{g(z)}\right)$.

Using Lemma 1.3, we get $F_{p(U)}p(z) \leq F_{g(U)}g(z)$, $z \in U$, i.e. $F_{D_z^{-\lambda}\phi(U)}\left(\frac{D_z^{-\lambda}\phi(m,n;q,z)}{z}\right)^{\alpha} \leq F_{g(U)}g(z)$, $z \in U$ and the fuzzy best dominant is g. □

Example 5.6 (Alb Lupaş and Oros, 2022d). Let $\phi(3,2,q,z) = 1 + \frac{2}{1-q}z + \frac{2(1-3q)}{1-2q}z^2$ for $m = 3$, $n = 2$, $z \in U$.

Then $D_z^{-\lambda}\phi(3,2,q,z) = \frac{1}{\Gamma(\lambda)}\int_0^z \frac{\phi(3,2,q,t)}{(z-t)^{1-\lambda}}dt = \frac{1}{\Gamma(\lambda)}\int_0^z \left(1 + \frac{2}{1-q}z + \frac{2(1-3q)}{1-2q}z^2\right)(z-t)^{\lambda-1}\,dt$ and after changing the variable $z-t=u$ and making a simple calculus, we get $D_z^{-\lambda}\phi(3,2,q,z) = \frac{z^{\lambda-1}}{\Gamma(\lambda)}$, $z \in U$.

Consider $g(z) = \frac{1-z}{1+z}$, $z \in U$ and differentiating it, we obtain $g'(z) = \frac{-2}{(1+z)^2}$ and $g''(z) = \frac{4}{(1+z)^3}$, $z \in U$.

Choosing $\alpha = 1$, $\beta = 1$, $\delta = 1$, $\psi = 1$, $\varepsilon = 1$, we obtain after a long calculus that $\Psi_{\lambda}^{3,2,q}(1,1,1,1,1;z) = \lambda - 1 + \frac{z^{\lambda-2}}{\Gamma(\lambda)} + \left(\frac{z^{\lambda-2}}{\Gamma(\lambda)}\right)^2$ and $\mathrm{Re}\left(1 + \frac{\psi}{\delta}g(z) + \frac{2\varepsilon}{\delta}(g(z))^2 - \frac{zg'(z)}{g(z)} + \frac{zg''(z)}{g'(z)}\right) = \mathrm{Re}\left(1 + \frac{1-z}{1+z} + 2\left(\frac{1-z}{1+z}\right)^2 + \frac{2z}{1-z^2} - \frac{2}{1+z}\right) = \mathrm{Re}\frac{2-4z+8z^2-2z^3}{(1+z)^2(1-z)}$.

We have also that $\beta + \psi g(z) + \varepsilon (g(z))^2 + \delta \frac{zg'(z)}{g(z)} = 1 + \frac{1-z}{1+z} + \left(\frac{1-z}{1+z}\right)^2 - \frac{2z}{1-z^2} = \frac{3-5z-z^2-z^3}{(1+z)^2(1-z)}$.

Using Theorem 5.31, when $\mathrm{Re}\frac{2-4z+8z^2-2z^3}{(1+z)^2(1-z)} > 0$ and $\frac{zg'(z)}{g(z)} = \frac{-2z}{1-z^2}$ is starlike univalent in U, we obtain $F_U\left(\lambda - 1 + \frac{z^{\lambda-2}}{\Gamma(\lambda)} + \left(\frac{z^{\lambda-2}}{\Gamma(\lambda)}\right)^2\right) \leq F_U\left(\frac{3-5z-z^2-z^3}{(1+z)^2(1-z)}\right)$, $z \in U$, induce $F_U\left(\frac{z^{\lambda-2}}{\Gamma(\lambda)}\right) \leq F_U\left(\frac{1-z}{1+z}\right)$.

Corollary 5.1 (Alb Lupaş and Oros, 2022d). *Consider m, n complex numbers with $n \neq 0, -1, -2, \ldots$; $\lambda, \alpha > 0$, $0 < q < 1$ and suppose that relation* (5.2.1.3) *holds. When*

$$F_{\Psi_{\lambda}^{m,n,q}(U)}\Psi_{\lambda}^{m,n,q}(\alpha,\beta,\psi,\varepsilon,\delta;z) \leq F_{g(U)}\left(\beta + \psi\frac{Mz+1}{Nz+1} + \varepsilon\left(\frac{Mz+1}{Nz+1}\right)^2 + \delta\frac{(M-N)z}{(Mz+1)(Nz+1)}\right),$$

with $\beta, \psi, \varepsilon, \delta \in \mathbb{C}$, $\delta \neq 0$, $-1 \leq N < M \leq 1$, *and* $\Psi_{\lambda}^{m,n,q}$ *defined by relation* (5.2.1.4), *then*

$$F_{D_z^{-\lambda}\phi(U)}\left(\frac{D_z^{-\lambda}\phi(m,n;q,z)}{z}\right)^{\alpha} \leq F_{g(U)}\left(\frac{Mz+1}{Nz+1}\right), \quad z \in U,$$

and the fuzzy best dominant is $\frac{Mz+1}{Nz+1}$.

Proof. Put $g(z) = \frac{Mz+1}{Nz+1}$, $-1 \leq N < M \leq 1$ in Theorem 5.31 and we obtain the corollary. □

Corollary 5.2 (Alb Lupaş and Oros, 2022d). *Consider* m, n *complex numbers with* $n \neq 0, -1, -2, \ldots$; $\lambda, \alpha > 0$, $0 < q < 1$ *and suppose that relation* (5.2.1.3) *holds. When*

$$\begin{aligned} &F_{\Psi_{\lambda}^{m,n,q}(U)}\Psi_{\lambda}^{m,n,q}(\alpha,\beta,\psi,\varepsilon,\delta;z) \\ &\quad \leq F_{g(U)}\left(\beta + \psi\left(\frac{z+1}{1-z}\right)^{\kappa} + \varepsilon\left(\frac{z+1}{1-z}\right)^{2\kappa} + \delta\frac{2\kappa z}{1-z^2}\right), \end{aligned}$$

with $\beta, \psi, \varepsilon, \delta \in \mathbb{C}$, $0 < \kappa \leq 1$, $\delta \neq 0$, *and* $\Psi_{\lambda}^{m,n,q}$ *defined by relation* (5.2.1.4), *then*

$$F_{D_z^{-\lambda}\phi(U)}\left(\frac{D_z^{-\lambda}\phi(m,n;q,z)}{z}\right)^{\alpha} \leq F_{g(U)}\left(\frac{z+1}{1-z}\right)^{\kappa}, \quad z \in U,$$

and the fuzzy best dominant is $\left(\frac{z+1}{1-z}\right)^{\kappa}$.

Proof. For $g(z) = \left(\frac{z+1}{1-z}\right)^{\kappa}$, $0 < \kappa \leq 1$, in Theorem 5.31, it follows the corollary. □

The next theorem and the corollaries associated involve fuzzy differential superordination aspects.

Theorem 5.32 (Alb Lupaş and Oros, 2022d). *Consider* g *an analytic and univalent function in* U *with the properties* $g(z) \neq 0$ *and* $\frac{zg'(z)}{g(z)}$ *is starlike univalent in* U. *Suppose that*

$$\operatorname{Re}\left(\frac{2\varepsilon}{\delta}(g(z))^2 + \frac{\psi}{\delta}g(z)\right) > 0, \quad \text{for } \psi, \varepsilon, \delta \in \mathbb{C},\ \delta \neq 0. \qquad (5.2.1.7)$$

For m, n complex numbers with $n \neq 0, -1, -2, \ldots$ *and* $\lambda, \alpha > 0$, $0 < q < 1$, *when* $\left(\frac{D_z^{-\lambda}\phi(m,n;q,z)}{z}\right)^{\alpha} \in \mathcal{H}[0, (\lambda-1)\alpha] \cap Q$ *and* $\Psi_{\lambda}^{m,n,q}(\alpha, \beta, \psi, \varepsilon, \delta; z)$ *defined by* (5.2.1.4) *is univalent in* U, *then*

$$F_{g(U)}\left(\beta + \psi g(z) + \varepsilon (g(z))^2 + \delta \frac{zg'(z)}{g(z)}\right) \le F_{\Psi_{\lambda}^{m,n,q}(U)} \Psi_{\lambda}^{m,n,q}(\alpha, \beta, \psi, \varepsilon, \delta; z) \tag{5.2.1.8}$$

implies

$$F_{g(U)} g(z) \le F_{D_z^{-\lambda}\phi(U)} \left(\frac{D_z^{-\lambda}\phi(m,n;q,z)}{z}\right)^{\alpha}, \quad z \in U, \tag{5.2.1.9}$$

and the fuzzy best subordinant is g.

Proof. Define $p(z) := \left(\frac{D_z^{-\lambda}\phi(m,n;q,z)}{z}\right)^{\alpha}$, $z \in U$, $z \neq 0$.

Set $\varphi(u) := \beta + \psi u + \varepsilon u^2$ and $\gamma(u) := \frac{\delta}{u}$ and it is easy to show that φ is analytic in $\mathbb{C}$, γ is analytic in $\mathbb{C}\backslash\{0\}$ with $\gamma(u) \neq 0$, $u \in \mathbb{C}\backslash\{0\}$.

We can write $\frac{\varphi'(g(z))}{\gamma(g(z))} = \frac{g'(z)[\psi + 2\varepsilon g(z)]g(z)}{\delta}$, and $\operatorname{Re}\left(\frac{\varphi'(g(z))}{\gamma(g(z))}\right) = \operatorname{Re}\left(\frac{\psi}{\delta} g(z) + \frac{2\varepsilon}{\delta}(g(z))^2\right) > 0$, for $\varepsilon, \psi, \delta \in \mathbb{C}$, $\delta \neq 0$.

We get

$$F_{g(U)}\left(\beta + \psi g(z) + \varepsilon (g(z))^2 + \delta \frac{zg'(z)}{g(z)}\right) \le F_{p(U)}\left(\beta + \psi p(z) + \varepsilon (p(z))^2 + \delta \frac{zp'(z)}{p(z)}\right).$$

Applying Lemma 1.19, we obtain

$$F_{g(U)} g(z) \le F_{D_z^{-\lambda}\phi(U)} \left(\frac{D_z^{-\lambda}\phi(m,n;q,z)}{z}\right)^{\alpha}, \quad z \in U,$$

and the fuzzy best subordinant is g. □

Example 5.7 (Alb Lupaş and Oros, 2022d). Taking the same functions as in Example 5.6, $\phi(3,2,q,z) = 1 + \frac{2}{1-q}z + \frac{2(1-3q)}{1-2q}z^2$ for $m = 3$, $n = 2$, $z \in U$, with $D_z^{-\lambda}\phi(3,2,q,z) = \frac{z^{\lambda-1}}{\Gamma(\lambda)}$, $z \in U$, and $g(z) = \frac{1-z}{1+z}$, $z \in U$ with $g'(z) = \frac{-2}{(1+z)^2}$ and $g''(z) = \frac{4}{(1+z)^3}$, $z \in U$.

Choosing $\alpha = 1$, $\beta = 1$, $\delta = 1$, $\psi = 1$, $\varepsilon = 1$, we get $\Psi_{\lambda}^{3,2,q}(1,1,1,1,1;z) = \Psi_{\lambda}^{3,2,q}(1,1,1,1,1;z) = \lambda - 1 + \frac{z^{\lambda-2}}{\Gamma(\lambda)} + \left(\frac{z^{\lambda-2}}{\Gamma(\lambda)}\right)^2$ and $\operatorname{Re}\left(\frac{2\varepsilon}{\delta}(g(z))^2 + \frac{\psi}{\delta}g(z)\right) = \operatorname{Re}\left(2\left(\frac{1-z}{1+z}\right)^2 + \frac{1-z}{1+z}\right) = \operatorname{Re}\frac{z^2-4z+3}{(1+z)^2}$.

We have also that $\beta+\psi g(z)+\varepsilon\left(g(z)\right)^2+\delta\frac{zg'(z)}{g(z)}=1+\frac{1-z}{1+z}+\left(\frac{1-z}{1+z}\right)^2-\frac{2z}{1-z^2}=\frac{3-5z-z^2-z^3}{(1+z)^2(1-z)}$.

Using Theorem 5.32, when $\mathrm{Re}\frac{z^2-4z+3}{(1+z)^2}>0$ and $\frac{zg'(z)}{g(z)}=\frac{-2z}{1-z^2}$ is starlike univalent in U, we obtain $F_U\left(\frac{3-5z-z^2-z^3}{(1+z)^2(1-z)}\right)\leq F_U\left(\lambda-1+\frac{z^{\lambda-2}}{\Gamma(\lambda)}+\left(\frac{z^{\lambda-2}}{\Gamma(\lambda)}\right)^2\right)$, $z\in U$, induce $F_U\left(\frac{1-z}{1+z}\right)\leq F_U\left(\frac{z^{\lambda-2}}{\Gamma(\lambda)}\right)$.

Corollary 5.3 (Alb Lupaş and Oros, 2022d). *For m, n complex numbers with $n\neq 0,-1,-2,\ldots$ and $\lambda,\alpha>0$, $0<q<1$, suppose that relation (5.2.1.7) holds. When $\left(\frac{D_z^{-\lambda}\phi(m,n;q,z)}{z}\right)^\alpha\in\mathcal{H}\left[0,(\lambda-1)\alpha\right]\cap Q$ and*

$$F_{g(U)}\left(\beta+\psi\frac{Mz+1}{Nz+1}+\varepsilon\left(\frac{Mz+1}{Nz+1}\right)^2+\delta\frac{(M-N)z}{(Mz+1)(Nz+1)}\right)$$
$$\leq F_{\Psi_\lambda^{m,n,q}(U)}\Psi_\lambda^{m,n,q}\left(\alpha,\beta,\psi,\varepsilon,\delta;z\right),$$

where $\beta,\psi,\varepsilon,\delta\in\mathbb{C}$, $\delta\neq 0$, $-1\leq N<M\leq 1$, and $\Psi_\lambda^{m,n,q}$ is defined by relation (5.2.1.4), then

$$F_{g(U)}\left(\frac{Mz+1}{Nz+1}\right)\leq F_{D_z^{-\lambda}\phi(U)}\left(\frac{D_z^{-\lambda}\phi\left(m,n;q,z\right)}{z}\right)^\alpha,\quad z\in U,$$

and the fuzzy best subordinant is $\frac{Mz+1}{Nz+1}$.

Proof. The corollary yields for $g(z)=\frac{Mz+1}{Nz+1}$, $-1\leq N<M\leq 1$, in Theorem 5.32. □

Corollary 5.4 (Alb Lupaş and Oros, 2022d). *For m, n complex numbers with $n\neq 0,-1,-2,\ldots$; $\lambda,\alpha>0$, $0<q<1$, suppose that relation (5.2.1.7) holds. When $\left(\frac{D_z^{-\lambda}\phi(m,n;q,z)}{z}\right)^\alpha\in\mathcal{H}\left[0,(\lambda-1)\alpha\right]\cap Q$ and*

$$F_{g(U)}\left(\beta+\psi\left(\frac{z+1}{1-z}\right)^\kappa+\varepsilon\left(\frac{z+1}{1-z}\right)^{2\kappa}+\delta\frac{2\kappa z}{1-z^2}\right)$$
$$\leq F_{\Psi_\lambda^{m,n,q}(U)}\Psi_\lambda^{m,n,q}\left(\alpha,\beta,\psi,\varepsilon,\delta;z\right),$$

where $\beta,\psi,\varepsilon,\delta\in\mathbb{C}\in\mathbb{C}$, $0<\kappa\leq 1$, $\delta\neq 0$, and $\Psi_\lambda^{m,n,q}$ is defined by relation (5.2.1.4), then

$$F_{g(U)}\left(\frac{z+1}{1-z}\right)^\kappa\leq F_{D_z^{-\lambda}\phi(U)}\left(\frac{D_z^{-\lambda}\phi\left(m,n;q,z\right)}{z}\right)^\alpha,\quad z\in U,$$

and the fuzzy best subordinant is $\left(\frac{z+1}{1-z}\right)^\kappa$.

Proof. Corollary follows by using Theorem 5.32 when $g(z) = \left(\frac{z+1}{1-z}\right)^{\kappa}$, $0 < \kappa \leq 1$. □

Combining the results obtained in Theorems 5.31 and 5.32, the following sandwich-type result can be established.

Theorem 5.33 (Alb Lupaş and Oros, 2022d). *Consider g_1, g_2 analytic and univalent functions in U with the properties $g_1(z) \neq 0$, $g_2(z) \neq 0$, for all $z \in U$, and $\frac{zg_1'(z)}{g_1(z)}$, $\frac{zg_2'(z)}{g_2(z)}$ are starlike univalent functions. Assume that g_1 satisfies relation (5.2.1.3) and g_2 satisfies relation (5.2.1.7). For m, n complex numbers with $n \neq 0, -1, -2, \ldots$ and $\lambda, \alpha > 0$, $0 < q < 1$, when $\left(\frac{D_z^{-\lambda}\phi(m,n;q,z)}{z}\right)^{\alpha} \in \mathcal{H}[0, (\lambda-1)\alpha] \cap Q$ and $\Psi_{\lambda}^{m,n,q}(\alpha, \beta, \psi, \varepsilon, \delta; z)$ defined by (5.2.1.4) is univalent in U, then*

$$F_{g_1(U)}\left(\beta + \psi g_1(z) + \varepsilon (g_1(z))^2 + \delta \frac{zg_1'(z)}{g_1(z)}\right)$$
$$\leq F_{\Psi_{\lambda}^{m,n,q}(U)}\Psi_{\lambda}^{m,n,q}(\alpha, \beta, \psi, \varepsilon, \delta; z)$$
$$\leq F_{g_2(U)}\left(\beta + \psi g_2(z) + \varepsilon (g_2(z))^2 + \delta \frac{zg_2'(z)}{g_2(z)}\right),$$

for $\beta, \psi, \varepsilon, \delta \in \mathbb{C}$, $\delta \neq 0$, implies

$$F_{g_1(U)}g_1(z) \leq F_{D_z^{-\lambda}\phi(U)}\left(\frac{D_z^{-\lambda}\phi(m,n;q,z)}{z}\right)^{\alpha} \leq F_{g_2(U)}g_2(z), \quad z \in U,$$

and the fuzzy best subordinant is g_1 and the fuzzy best dominant is g_2.

Considering $g_1(z) = \frac{M_1z+1}{N_1z+1}$ and $g_2(z) = \frac{M_2z+1}{N_2z+1}$, with $-1 \leq N_2 < N_1 < M_1 < M_2 \leq 1$, the following corollary yields.

Corollary 5.5 (Alb Lupaş and Oros, 2022d). *For m, n complex numbers with $n \neq 0, -1, -2, \ldots$; $\lambda, \alpha > 0$, $0 < q < 1$, suppose that relations (5.2.1.3) and (5.2.1.7) hold. When $\left(\frac{D_z^{-\lambda}\phi(m,n;q,z)}{z}\right)^{\alpha} \in \mathcal{H}[0, (\lambda-1)\alpha] \cap Q$ and*

$$F_{g_1(U)}\left(\beta + \psi \frac{M_1z+1}{N_1z+1} + \varepsilon \left(\frac{M_1z+1}{N_1z+1}\right)^2 + \delta \frac{(M_1-N_1)z}{(M_1z+1)(N_1z+1)}\right)$$
$$\leq F_{\Psi_{\lambda}^{m,n,q}(U)}\Psi_{\lambda}^{m,n,q}(\alpha, \beta, \psi, \varepsilon, \delta; z)$$
$$\leq F_{g_2(U)}\left(\beta + \psi \frac{M_2z+1}{N_2z+1} + \varepsilon \left(\frac{M_2z+1}{N_2z+1}\right)^2 + \delta \frac{(M_2-N_2)z}{(M_2z+1)(N_2z+1)}\right),$$

with $\beta, \psi, \varepsilon, \delta \in \mathbb{C}$, $\delta \neq 0$, $-1 \leq N_2 \leq N_1 < M_1 \leq M_2 \leq 1$, *and* $\Psi_\lambda^{m,n,q}$ *defined by relation* (5.2.1.4), *then*

$$F_{g_1(U)}\left(\frac{M_1 z+1}{N_1 z+1}\right) \leq F_{D_z^{-\lambda}\phi(U)}\left(\frac{D_z^{-\lambda}\phi(m,n;q,z)}{z}\right)^\alpha \leq F_{g_2(U)}\left(\frac{M_2 z+1}{N_2 z+1}\right),$$

the fuzzy best subordinant is $\frac{M_1 z+1}{N_1 z+1}$ *and the fuzzy best dominant is* $\frac{M_2 z+1}{N_2 z+1}$, *respectively.*

Considering $g_1(z) = \left(\frac{z+1}{1-z}\right)^{\kappa_1}$ and $g_2(z) = \left(\frac{z+1}{1-z}\right)^{\kappa_2}$, with $0 < \kappa_1, \kappa_2 \leq 1$, the following corollary yields.

Corollary 5.6 (Alb Lupaş and Oros, 2022d). *For* m, n *complex numbers with* $n \neq 0, -1, -2, \ldots$ *and* $\lambda, \alpha > 0$, $0 < q < 1$, *suppose that relations* (5.2.1.3) *and* (5.2.1.7) *hold. When* $\left(\frac{D_z^{-\lambda}\phi(m,n;q,z)}{z}\right)^\alpha \in \mathcal{H}[0,(\lambda-1)\alpha] \cap Q$ *and*

$$\begin{aligned} &F_{g_1(U)}\left(\beta+\psi\left(\frac{z+1}{1-z}\right)^{\kappa_1}+\varepsilon\left(\frac{z+1}{1-z}\right)^{2\kappa_1}+\delta\frac{2\kappa_1 z}{1-z^2}\right) \\ &\quad\leq F_{\Psi_\lambda^{m,n,q}(U)}\Psi_\lambda^{m,n,q}(\alpha,\beta,\psi,\varepsilon,\delta;z) \\ &\quad\leq F_{g_2(U)}\left(\beta+\psi\left(\frac{z+1}{1-z}\right)^{\kappa_2}+\varepsilon\left(\frac{z+1}{1-z}\right)^{2\kappa_2}+\delta\frac{2\kappa_2 z}{1-z^2}\right), \end{aligned}$$

with $\beta, \psi, \varepsilon, \delta \in \mathbb{C}$, $0 < \kappa_1, \kappa_2 \leq 1$, $\delta \neq 0$, *and* $\Psi_\lambda^{m,n,q}$ *defined by relation* (5.2.1.4), *then*

$$F_{g_1(U)}\left(\frac{z+1}{1-z}\right)^{\kappa_1} \leq F_{D_z^{-\lambda}\phi(U)}\left(\frac{D_z^{-\lambda}\phi(m,n;q,z)}{z}\right)^\alpha \leq F_{g_2(U)}\left(\frac{z+1}{1-z}\right)^{\kappa_2},$$

the fuzzy best subordinant is $\left(\frac{z+1}{1-z}\right)^{\kappa_1}$ *and the fuzzy best dominant is* $\left(\frac{z+1}{1-z}\right)^{\kappa_2}$, *respectively.*

Considering $\varphi(u) := \beta + \psi u + \varepsilon u^2$ and $\gamma(u) := \frac{\delta}{u}$ in Lemmas 1.16 and 1.19, we obtain the following results regarding fuzzy differential subordination:

Theorem 5.34 (Alb Lupaş and Oros, 2022e). *Consider* g *an analytic and univalent function in* U *such that* $g(z) \neq 0$, *for all* $z \in U$, *and* $\frac{z\left(D_z^{-\lambda}\phi(m,n;q,z)\right)'}{D_z^{-\lambda}\phi(m,n;q,z)} \in \mathcal{H}(U)$, *with* m, n *be complex numbers with*

$n \neq 0, -1, -2, \ldots$ and $\lambda > 0$, $0 < q < 1$. Assume that $\frac{zg'(z)}{g(z)}$ is starlike univalent in U,

$$\operatorname{Re}\left(\frac{\psi}{\delta}g(z) + \frac{2\varepsilon}{\delta}g^2(z) + 1 + z\frac{g''(z)}{g'(z)} - z\frac{g'(z)}{g(z)}\right) > 0, \tag{5.2.1.10}$$

for $\beta, \psi, \delta, \varepsilon \in \mathbb{C}$, $\delta \neq 0$, $z \in U$ and

$$\begin{aligned}\Psi_{\lambda}^{m,n,q}(\beta, \psi, \varepsilon, \delta; z) &:= \beta + \delta + (\psi - \delta)\frac{z\left(D_z^{-\lambda}\phi(m,n;q,z)\right)'}{D_z^{-\lambda}\phi(m,n;q,z)} \\ &+ \varepsilon\left(\frac{z\left(D_z^{-\lambda}\phi(m,n;q,z)\right)'}{D_z^{-\lambda}\phi(m,n;q,z)}\right)^2 + \delta\frac{z\left(D_z^{-\lambda}\phi(m,n;q,z)\right)''}{\left(D_z^{-\lambda}\phi(m,n;q,z)\right)'}.\end{aligned} \tag{5.2.1.11}$$

If the following fuzzy differential subordination is satisfied by g

$$\begin{aligned}&F_{\Psi_{\lambda}^{m,n,q}(U)}\Psi_{\lambda}^{m,n,q}(\beta, \psi, \varepsilon, \delta; z) \\ &\quad\leq F_{g(U)}\left(\beta + \psi g(z) + \varepsilon(g(z))^2 + \delta\frac{zg'(z)}{g(z)}\right),\end{aligned} \tag{5.2.1.12}$$

for $\beta, \psi, \delta, \varepsilon \in \mathbb{C}$, $\delta \neq 0$, then

$$F_{D_z^{-\lambda}\phi(U)}\left(\frac{z\left(D_z^{-\lambda}\phi(m,n;q,z)\right)'}{D_z^{-\lambda}\phi(m,n;q,z)}\right) \leq F_{g(U)}g(z), \tag{5.2.1.13}$$

and the fuzzy best dominant is g.

Proof. Consider $p(z) := \frac{z\left(D_z^{-\lambda}\phi(m,n;q,z)\right)'}{D_z^{-\lambda}\phi(m,n;q,z)}$, $z \in U$, $z \neq 0$. We have $p'(z) = \frac{\left(D_z^{-\lambda}\phi(m,n;q,z)\right)'}{D_z^{-\lambda}\phi(m,n;q,z)} + z\frac{\left(D_z^{-\lambda}\phi(m,n;q,z)\right)''}{D_z^{-\lambda}\phi(m,n;q,z)} - z\left(\frac{\left(D_z^{-\lambda}\phi(m,n;q,z)\right)'}{D_z^{-\lambda}\phi(m,n;q,z)}\right)^2$ and

$$\frac{zp'(z)}{p(z)} = 1 + z\frac{\left(D_z^{-\lambda}\phi(m,n;q,z)\right)''}{\left(D_z^{-\lambda}\phi(m,n;q,z)\right)'} - z\frac{\left(D_z^{-\lambda}\phi(m,n;q,z)\right)'}{D_z^{-\lambda}D_z^{-\lambda}\phi(m,n;q,z)}. \tag{5.2.1.14}$$

Let $\varphi(u) := \beta + \psi u + \varepsilon u^2$ and $\gamma(u) := \frac{\delta}{u}$, it is easy to show that φ is analytic in $\mathbb{C}$, γ is analytic in $\mathbb{C}\backslash\{0\}$ and that $\gamma(u) \neq 0$, $u \in \mathbb{C}\backslash\{0\}$.

Also, setting $G(z) = zg'(z)\gamma(g(z)) = \delta\frac{zg'(z)}{g(z)}$, which is starlike univalent in U, and $h(z) = G(z) + \varphi(g(z)) = \beta + \psi g(z) + \varepsilon(g(z))^2 + \delta\frac{zg'(z)}{g(z)}$, we obtain by differentiating it $h'(z) = \psi g'(z) + 2\varepsilon g(z)g'(z) + \delta\frac{g'(z)}{g(z)} + \delta z\frac{g''(z)}{g(z)} - \delta z\left(\frac{g'(z)}{g(z)}\right)^2$ and $\frac{zh'(z)}{G(z)} = \frac{\psi}{\delta}g(z) + \frac{2\varepsilon}{\delta}g^2(z) + 1 + z\frac{g''(z)}{g(z)} - z\frac{g'(z)}{g(z)}$.

We get that

$$\operatorname{Re}\left(\frac{zh'(z)}{G(z)}\right) = \operatorname{Re}\left(\frac{\psi}{\delta}g(z) + \frac{2\varepsilon}{\delta}g^2(z) + 1 + z\frac{g''(z)}{g(z)} - z\frac{g'(z)}{g(z)}\right) > 0.$$

Using relation (5.2.1.14), we get $\beta + \psi p(z) + \varepsilon\left(p(z)\right)^2 + \delta\frac{zp'(z)}{p(z)} = \beta + \delta + (\psi - \delta)\frac{z\left(D_z^{-\lambda}\phi(m,n;q,z)\right)'}{D_z^{-\lambda}\phi(m,n;q,z)} + \varepsilon\left(\frac{z\left(D_z^{-\lambda}\phi(m,n;q,z)\right)'}{D_z^{-\lambda}\phi(m,n;q,z)}\right)^2 + \delta\frac{z\left(D_z^{-\lambda}\phi(m,n;q,z)\right)''}{\left(D_z^{-\lambda}\phi(m,n;q,z)\right)'}$.

By using fuzzy differential subordination (5.2.1.12), we obtain

$$F_{p(U)}\left(\beta + \psi p(z) + \varepsilon\left(p(z)\right)^2 + \delta\frac{zp'(z)}{p(z)}\right)$$
$$\leq F_{g(U)}\left(\beta + \psi g(z) + \varepsilon\left(g(z)\right)^2 + \delta\frac{zg'(z)}{g(z)}\right).$$

Applying Lemma 1.16, we have $F_{p(U)}p(z) \leq F_{g(U)}g(z)$, $z \in U$, i.e. $F_{D_z^{-\lambda}\phi(U)}\left(\frac{z\left(D_z^{-\lambda}\phi(m,n;q,z)\right)'}{D_z^{-\lambda}\phi(m,n;q,z)}\right) \leq F_{g(U)}g(z)$, $z \in U$ and the fuzzy best dominant is g. □

Example 5.8 (Alb Lupaş and Oros, 2022e). Let $\phi(3,2,q,z) = 1 + \frac{2}{1-q}z + \frac{2(1-3q)}{1-2q}z^2$ for $m = 3$, $n = 2$, $z \in U$.

Then $D_z^{-\lambda}\phi(3,2,q,z) = \frac{1}{\Gamma(\lambda)}\int_0^z \frac{\phi(3,2,q,t)}{(z-t)^{1-\lambda}}dt = \frac{1}{\Gamma(\lambda)}\int_0^z\left(1 + \frac{2}{1-q}z + \frac{2(1-3q)}{1-2q}z^2\right)(z-t)^{\lambda-1}\,dt$ and after changing the variable $z - t = u$ and making a simple calculus, we get $D_z^{-\lambda}\phi(3,2,q,z) = \frac{z^{\lambda-1}}{\Gamma(\lambda)}$, $z \in U$.

Consider $g(z) = \frac{1-z}{1+z}$, $z \in U$ and differentiating it, we obtain $g'(z) = \frac{-2}{(1+z)^2}$ and $g''(z) = \frac{4}{(1+z)^3}$, $z \in U$.

Choosing $\beta = 0$, $\delta = 1$, $\psi = 1$, $\varepsilon = 1$, we obtain after a long calculus that $\Psi_\lambda^{3,2,q}(0,1,1,1;z) = \lambda^2 - \lambda$.

Using Theorem 5.34, when $\operatorname{Re}\frac{2z^3+6z^2-6z+4}{(1+z)^2(1-z)} > 0$ and $\frac{zg'(z)}{g(z)} = \frac{-2z}{1-z^2}$ is starlike univalent in U, we obtain $F_U\left(\lambda^2 - \lambda\right) \leq F_U\left(\frac{-6z+2}{(1+z)^2(1-z)}\right)$, $z \in U$, induce $F_U(\lambda - 1) \leq F_U\left(\frac{1-z}{1+z}\right)$.

Corollary 5.7 (Alb Lupaş and Oros, 2022e). *Consider m, n complex numbers with $n \neq 0, -1, -2, \ldots$; $\lambda > 0$, $0 < q < 1$ and suppose that relation (5.2.1.10) holds. When*

$$F_{\Psi_{\lambda}^{m,n,q}(U)} \Psi_{\lambda}^{m,n,q} (\beta, \psi, \varepsilon, \delta; z) \leq F_{g(U)} \left(\beta + \psi \frac{Mz+1}{Nz+1} + \varepsilon \left(\frac{Mz+1}{Nz+1} \right)^2 + \frac{\delta (M-N) z}{(Mz+1)(Nz+1)} \right),$$

with $\beta, \psi, \delta, \varepsilon \in \mathbb{C}$, $\delta \neq 0$, $-1 \leq N < M \leq 1$, and $\Psi_{\lambda}^{m,n,q}$ defined by relation (5.2.1.11), then

$$F_{D_z^{-\lambda} \phi(U)} \left(\frac{z \left(D_z^{-\lambda} \phi(m, n; q, z) \right)'}{D_z^{-\lambda} \phi(m, n; q, z)} \right) \leq F_{g(U)} \left(\frac{Mz+1}{Nz+1} \right),$$

and the fuzzy best dominant is $\frac{Mz+1}{Nz+1}$.

Proof. Put $g(z) = \frac{Mz+1}{Nz+1}$, $-1 \leq N < M \leq 1$ in Theorem 5.34 and we obtain the corollary. □

Corollary 5.8 (Alb Lupaş and Oros, 2022e). *Consider m, n complex numbers with $n \neq 0, -1, -2, \ldots$; $\lambda > 0$, $0 < q < 1$ and suppose that relation (5.2.1.10) holds. When*

$$F_{\Psi_{\lambda}^{m,n,q}(U)} \Psi_{\lambda}^{m,n,q} (\beta, \psi, \varepsilon, \delta; z) \leq F_{g(U)} \left(\beta + \psi \left(\frac{z+1}{1-z} \right)^{\alpha} + \varepsilon \left(\frac{z+1}{1-z} \right)^{2\alpha} + \frac{2\alpha\delta z}{1-z^2} \right),$$

with $\beta, \psi, \delta, \varepsilon \in \mathbb{C}$, $0 < \alpha \leq 1$, $\delta \neq 0$, and $\Psi_{\lambda}^{m,n,q}$ defined by relation (5.2.1.11), then

$$F_{D_z^{-\lambda} \phi(U)} \left(\frac{z \left(D_z^{-\lambda} \phi(m, n; q, z) \right)'}{D_z^{-\lambda} \phi(m, n; q, z)} \right) \leq F_{g(U)} \left(\frac{z+1}{1-z} \right)^{\alpha},$$

and the fuzzy best dominant is $\left(\frac{z+1}{1-z} \right)^{\alpha}$.

Proof. For $g(z) = \left(\frac{z+1}{1-z} \right)^{\alpha}$, $0 < \alpha \leq 1$, corollary follows by using Theorem 5.34. □

The next theorem and the corollaries associated involve fuzzy differential superordination aspects.

Theorem 5.35 (Alb Lupaş and Oros, 2022e). *Consider g an analytic and univalent function in U with the properties $g(z) \neq 0$ and $\frac{zg'(z)}{g(z)}$ is starlike univalent in U. Suppose that*

$$\operatorname{Re}\left(\frac{\psi}{\delta}g(z)g'(z) + \frac{2\varepsilon}{\delta}g^2(z)g'(z)\right) > 0, \text{ for } \psi, \delta, \varepsilon \in \mathbb{C},\ \delta \neq 0. \quad (5.2.1.15)$$

For m, n complex numbers with $n \neq 0, -1, -2, \ldots$; $\lambda > 0$, $0 < q < 1$, when $\frac{z\left(D_z^{-\lambda}\phi(m,n;q,z)\right)'}{D_z^{-\lambda}\phi(m,n;q,z)} \in \mathcal{H}\left[g(0), 1\right] \cap Q$ and $\Psi_\lambda^{m,n,q}(\beta, \psi, \varepsilon, \delta; z)$ defined by relation (5.2.1.11) is univalent in U, then

$$F_{g(U)}\left(\beta + \psi g(z) + \varepsilon\left(g(z)\right)^2 + \frac{\delta z g'(z)}{g(z)}\right) \leq F_{\Psi_\lambda^{m,n,q}(U)}\Psi_\lambda^{m,n,q}(\beta, \psi, \varepsilon, \delta; z) \quad (5.2.1.16)$$

implies

$$F_{g(U)}g(z) \leq F_{D_z^{-\lambda}\phi(U)}\left(\frac{z\left(D_z^{-\lambda}\phi\left(m,n;q,z\right)\right)'}{D_z^{-\lambda}\phi\left(m,n;q,z\right)}\right), \quad z \in U, \quad (5.2.1.17)$$

and the fuzzy best subordinant is g.

Proof. Define $p(z) := \frac{z\left(D_z^{-\lambda}\phi(m,n;q,z)\right)'}{D_z^{-\lambda}\phi(m,n;q,z)}$, $z \in U$, $z \neq 0$.

Let $\varphi(u) := \beta + \psi u + \varepsilon u^2$ and $\gamma(u) := \frac{\delta}{u}$ and it is easy to show that φ is analytic in $\mathbb{C}$, γ is analytic in $\mathbb{C}\backslash\{0\}$ with $\gamma(u) \neq 0$, $u \in \mathbb{C}\backslash\{0\}$.

We can write $\frac{\varphi'(g(z))}{\gamma(g(z))} = \frac{g'(z)g(z)[\psi + 2\varepsilon g(z)]}{\tau}$, and $\operatorname{Re}\left(\frac{\theta'(g(z))}{\eta(g(z))}\right) = \operatorname{Re}\left(\frac{\psi}{\delta}g(z)g'(z) + \frac{2\varepsilon}{\delta}g^2(z)g'(z)\right) > 0$, for $\psi, \delta, \varepsilon \in \mathbb{C}$, $\delta \neq 0$.

By using relations (5.2.1.14) and (5.2.1.16), we get

$$F_{g(U)}\left(\beta + \psi g(z) + \varepsilon\left(g(z)\right)^2 + \frac{\delta z g'(z)}{g(z)}\right)$$
$$\leq F_{p(U)}\left(\beta + \psi p(z) + \varepsilon\left(p(z)\right)^2 + \frac{\delta z p'(z)}{p(z)}\right).$$

Applying Lemma 1.19, we obtain

$$F_{g(U)}g(z) \leq F_{D_z^{-\lambda}\phi(U)}\left(\frac{z\left(D_z^{-\lambda}\phi\left(m,n;q,z\right)\right)'}{D_z^{-\lambda}\phi\left(m,n;q,z\right)}\right), \quad z \in U,$$

and the fuzzy best subordinant is g. □

Example 5.9 (Alb Lupaş and Oros, 2022e). Taking the same functions as in Example 5.8, $\phi(3,2,q,z) = 1 + \frac{2}{1-q}z + \frac{2(1-3q)}{1-2q}z^2$ for $m = 3$, $n = 2$, $z \in U$, with $D_z^{-\lambda}\phi(3,2,q,z) = \frac{z^{\lambda-1}}{\Gamma(\lambda)}$, $z \in U$, and $g(z) = \frac{1-z}{1+z}$, $z \in U$ with $g'(z) = \frac{-2}{(1+z)^2}$ and $g''(z) = \frac{4}{(1+z)^3}$, $z \in U$.

Choosing $\beta = 0$, $\delta = 1$, $\psi = 1$, $\varepsilon = 1$, we get $\Psi_\lambda^{3,2,q}(0,1,1,1;z) = \lambda^2 - \lambda$.

Using Theorem 5.35, when $\text{Re}\frac{-6z^2+8z-6}{(1+z)^4} > 0$ and $\frac{zg'(z)}{g(z)} = \frac{-2z}{1-z^2}$ is starlike univalent in U, we obtain $F_U\left(\frac{-6z+2}{(1+z)^2(1-z)}\right) \le F_U(\lambda^2 - \lambda)$, $z \in U$, induce $F_U\left(\frac{1-z}{1+z}\right) \le F_U(\lambda - 1)$.

Corollary 5.9 (Alb Lupaş and Oros, 2022e). *For m, n complex numbers with $n \ne 0, -1, -2, \ldots$; $\lambda > 0$, $0 < q < 1$, suppose that relation* (5.2.1.15) *holds. When* $\frac{z\left(D_z^{-\lambda}\phi(m,n;q,z)\right)'}{D_z^{-\lambda}\phi(m,n;q,z)} \in \mathcal{H}[g(0),1] \cap Q$ *and*

$$F_{g(U)}\left(\beta + \psi\frac{Mz+1}{Nz+1} + \varepsilon\left(\frac{Mz+1}{Nz+1}\right)^2 + \frac{\delta(M-N)z}{(Mz+1)(Nz+1)}\right)$$
$$\le F_{\Psi_\lambda^{m,n,q}(U)}\Psi_\lambda^{m,n,q}(\beta,\psi,\varepsilon,\delta;z),$$

where $\beta, \psi, \delta, \varepsilon \in \mathbb{C}$, $\delta \ne 0$, $-1 \le N < M \le 1$, and $\Psi_\lambda^{m,n,q}$ defined by relation (5.2.1.11), *then*

$$F_{g(U)}\left(\frac{Mz+1}{Nz+1}\right) \le F_{D_z^{-\lambda}\phi(U)}\left(\frac{z\left(D_z^{-\lambda}\phi(m,n;q,z)\right)'}{D_z^{-\lambda}\phi(m,n;q,z)}\right),$$

and the fuzzy best subordinant is $\frac{Mz+1}{Nz+1}$.

Proof. Considering $g(z) = \frac{Mz+1}{Nz+1}$, $-1 \le N < M \le 1$ in Theorem 5.35, we have the corollary. □

Corollary 5.10 (Alb Lupaş and Oros, 2022e). *For m, n complex numbers with $n \ne 0, -1, -2, \ldots$; $\lambda > 0$, $0 < q < 1$, suppose that relation* (5.2.1.15) *holds. When* $\frac{z\left(D_z^{-\lambda}\phi(m,n;q,z)\right)'}{D_z^{-\lambda}\phi(m,n;q,z)} \in \mathcal{H}[g(0),1] \cap Q$ *and*

$$F_{g(U)}\left(\beta + \psi\left(\frac{z+1}{1-z}\right)^\alpha + \varepsilon\left(\frac{z+1}{1-z}\right)^{2\alpha} + \frac{2\alpha\delta z}{1-z^2}\right)$$
$$\le F_{\Psi_\lambda^{m,n,q}(U)}\Psi_\lambda^{m,n,q}(\beta,\psi,\varepsilon,\delta;z),$$

where $\beta, \psi, \delta, \varepsilon \in \mathbb{C}$, $\delta \neq 0$, $0 < \alpha \leq 1$, *and* $\Psi_{\lambda}^{m,n,q}$ *defined by relation* (5.2.1.11), *then*

$$F_{g(U)}\left(\frac{z+1}{1-z}\right)^{\alpha} \leq F_{D_z^{-\lambda}\phi(U)} \frac{z\left(D_z^{-\lambda}\phi(m,n;q,z)\right)'}{D_z^{-\lambda}\phi(m,n;q,z)},$$

and the fuzzy best subordinant is $\left(\frac{z+1}{1-z}\right)^{\alpha}$.

Proof. Letting $g(z) = \left(\frac{z+1}{1-z}\right)^{\alpha}$, $0 < \alpha \leq 1$ in Theorem 5.35, we obtain the corollary. □

Combining the results obtained in Theorems 5.34 and 5.35, the following sandwich-type result can be established.

Theorem 5.36 (Alb Lupaş and Oros, 2022e). *Consider* g_1, g_2 *analytic and univalent functions in* U *with the properties* $g_1(z) \neq 0, g_2(z) \neq 0$, *for all* $z \in U$, *and* $\frac{zg_1'(z)}{g_1(z)}$, $\frac{zg_2'(z)}{g_2(z)}$ *are starlike univalent functions. Assume that* g_1 *satisfies relation* (5.2.1.10) *and* g_2 *satisfies relation* (5.2.1.15). *For* m, n *complex numbers with* $n \neq 0, -1, -2, \ldots$; $\lambda > 0$, $0 < q < 1$, *when* $\frac{z\left(D_z^{-\lambda}\phi(m,n;q,z)\right)'}{D_z^{-\lambda}\phi(m,n;q,z)} \in \mathcal{H}[g(0), 1] \cap Q$ *and* $\Psi_{\lambda}^{m,n,q}(\beta, \psi, \varepsilon, \delta; z)$ *defined by relation* (5.2.1.11) *is univalent in* U, *then*

$$\begin{aligned} &F_{g_1(U)}\left(\beta + \psi g_1(z) + \varepsilon (g_1(z))^2 + \frac{\delta z g_1'(z)}{g_1(z)}\right) \\ &\quad \leq F_{\Psi_{\lambda}^{m,n,q}(U)} \Psi_{\lambda}^{m,n,q}(\beta, \psi, \varepsilon, \delta; z) \\ &\quad \leq F_{g_2(U)}\left(\beta + \psi g_2(z) + \varepsilon (g_2(z))^2 + \frac{\delta z g_2'(z)}{g_2(z)}\right), \end{aligned}$$

for $\beta, \psi, \delta, \varepsilon \in \mathbb{C}$, $\delta \neq 0$, *implies*

$$F_{g_1(U)} g_1(z) \leq F_{D_z^{-\lambda}\phi(U)}\left(\frac{z\left(D_z^{-\lambda}\phi(m,n;q,z)\right)'}{D_z^{-\lambda}\phi(m,n;q,z)}\right) \leq F_{g_2(U)} g_2(z),$$

and the fuzzy best subordinant is g_1 *and the fuzzy best dominant is* g_2.

Considering $g_1(z) = \frac{M_1 z+1}{N_1 z+1}$ and $g_2(z) = \frac{M_2 z+1}{N_2 z+1}$, with $-1 \leq N_2 < N_1 < M_1 < M_2 \leq 1$, we get the following corollary.

Corollary 5.11 (Alb Lupaş and Oros, 2022e). *For* m, n *complex numbers with* $n \neq 0, -1, -2, \ldots$; $\lambda > 0$, $0 < q < 1$, *suppose that relations*

(5.2.1.10) *and* (5.2.1.15) *hold. When* $\frac{z\left(D_z^{-\lambda}\phi(m,n;q,z)\right)'}{D_z^{-\lambda}\phi(m,n;q,z)} \in \mathcal{H}[g(0),1] \cap Q$ *and*

$$F_{g_1(U)}\left(\beta+\psi\frac{M_1z+1}{N_1z+1}+\varepsilon\left(\frac{M_1z+1}{N_1z+1}\right)^2+\delta\frac{(M_1-N_1)z}{(M_1z+1)(N_1z+1)}\right)$$
$$\leq F_{\Psi_\lambda^{m,n,q}(U)}\Psi_\lambda^{m,n,q}(\beta,\psi,\varepsilon,\delta;z)$$
$$\leq F_{g_2(U)}\left(\beta+\psi\frac{M_2z+1}{N_2z+1}+\varepsilon\left(\frac{M_2z+1}{N_2z+1}\right)^2+\delta\frac{(M_2-N_2)z}{(M_2z+1)(N_2z+1)}\right),$$

with $\beta,\psi,\delta,\varepsilon \in \mathbb{C}$, $\tau\delta \neq 0$, $-1 \leq N_2 \leq N_1 < M_1 \leq M_2 \leq 1$, *and* $\Psi_\lambda^{m,n,q}$ *defined by relation* (5.2.1.11), *then*

$$F_{g_1(U)}\left(\frac{M_1z+1}{N_1z+1}\right) \leq F_{D_z^{-\lambda}\phi(U)}\left(\frac{z\left(D_z^{-\lambda}\phi(m,n;q,z)\right)'}{D_z^{-\lambda}\phi(m,n;q,z)}\right)$$
$$\leq F_{g_2(U)}\left(\frac{M_2z+1}{N_2z+1}\right),$$

the fuzzy best subordinant is $\frac{M_1z+1}{N_1z+1}$ *and the fuzzy best dominant is* $\frac{M_2z+1}{N_2z+1}$, *respectively.*

Let $g_1(z)=\left(\frac{z+1}{1-z}\right)^{\alpha_1}$ and $g_2(z)=\left(\frac{z+1}{1-z}\right)^{\alpha_2}$, with $0<\alpha_1<\alpha_2\leq 1$, we get the following corollary.

Corollary 5.12 (Alb Lupaş and Oros, 2022e). *For* m, n *complex numbers with* $n \neq 0,-1,-2,\ldots$; $\lambda>0$, $0<q<1$, *suppose that relations* (5.2.1.10) *and* (5.2.1.15) *hold. When* $\frac{z\left(D_z^{-\lambda}\phi(m,n;q,z)\right)'}{D_z^{-\lambda}\phi(m,n;q,z)} \in \mathcal{H}[g(0),1] \cap Q$ *and*

$$F_{g_1(U)}\left(\beta+\psi\left(\frac{z+1}{1-z}\right)^{\alpha_1}+\varepsilon\left(\frac{z+1}{1-z}\right)^{2\alpha_1}+\frac{2\alpha_1\delta z}{1-z^2}\right)$$
$$\leq F_{\Psi_\lambda^{m,n,q}(U)}\Psi_\lambda^{m,n,q}(\beta,\psi,\varepsilon,\delta;z)$$
$$\leq F_{g_2(U)}\left(\beta+\psi\left(\frac{z+1}{1-z}\right)^{\alpha_2}+\varepsilon\left(\frac{z+1}{1-z}\right)^{2\alpha_2}+\frac{2\alpha_2\delta z}{1-z^2}\right),$$

with $\beta,\psi,\delta,\varepsilon \in \mathbb{C}$, $\delta \neq 0$, $0<\alpha_1<\alpha_2\leq 1$, *and* $\Psi_\lambda^{m,n,q}$ *defined by relation* (5.2.1.11), *then*

$$F_{g_1(U)}\left(\frac{z+1}{1-z}\right)^{\alpha_1} \leq F_{D_z^{-\lambda}\phi(U)}\frac{z\left(D_z^{-\lambda}\phi(m,n;q,z)\right)'}{D_z^{-\lambda}\phi(m,n;q,z)} \leq F_{g_2(U)}\left(\frac{z+1}{1-z}\right)^{\alpha_2},$$

the fuzzy best subordinant is $\left(\frac{z+1}{1-z}\right)^{\alpha_1}$ *and the fuzzy best dominant is* $\left(\frac{z+1}{1-z}\right)^{\alpha_2}$, *respectively.*

Considering $\varphi(u) := \beta u$ and $\gamma(u) := \delta$ in Lemmas 1.16 and 1.19, we obtain the following results regarding fuzzy differential subordination:

Theorem 5.37 (Alb Lupaş and Oros, 2022e). *Consider g a convex and univalent function in U such that $g(0) = \lambda$, $z \in U$, and $\frac{z\left(D_z^{-\lambda}\phi(m,n;q,z)\right)'}{D_z^{-\lambda}\phi(m,n;q,z)} \in \mathcal{H}(U)$, $z \in U$, where $\lambda > 0$, $0 < q < 1$, m, n are complex numbers with $n \neq 0, -1, -2, \ldots$. Suppose that*

$$\operatorname{Re}\left(\frac{\beta+\delta}{\delta} + z\frac{g''(z)}{g'(z)}\right) > 0, \tag{5.2.1.18}$$

for $\beta, \delta \in \mathbb{C}$, $\delta \neq 0$, $z \in U$, and

$$\Psi_\lambda^{m,n,q}(\beta,\delta;q,z) := (\beta+\delta)\frac{z\left(D_z^{-\lambda}\phi(m,n;q,z)\right)'}{D_z^{-\lambda}\phi(m,n;q,z)} + \delta\frac{z^2\left(D_z^{-\lambda}\phi(m,n;q,z)\right)''}{D_z^{-\lambda}\phi(m,n;q,z)} - \delta\left(\frac{z\left(D_z^{-\lambda}\phi(m,n;q,z)\right)'}{D_z^{-\lambda}\phi(m,n;q,z)}\right)^2. \tag{5.2.1.19}$$

If g verifies the fuzzy differential subordination

$$F_{\Psi_\lambda^{m,n,q}(U)}\Psi_\lambda^{m,n,q}(\beta,\delta;q,z) \leq F_{g(U)}(\beta g(z) + \delta z g'(z)), \tag{5.2.1.20}$$

for $\beta, \delta \in \mathbb{C}$, $\delta \neq 0, z \in U$, then

$$F_{D_z^{-\lambda}\phi(U)}\left(\frac{z\left(D_z^{-\lambda}\phi(m,n;q,z)\right)'}{D_z^{-\lambda}\phi(m,n;q,z)}\right) \leq F_{g(U)}g(z), \quad z \in U, \tag{5.2.1.21}$$

and the fuzzy best dominant is g.

Proof. Define again $p(z) := \frac{z\left(D_z^{-\lambda}\phi(m,n;q,z)\right)'}{D_z^{-\lambda}\phi(m,n;q,z)}$, $z \in U$, $z \neq 0$, which is analytic in U and $p(0) = \lambda$.

Differentiating it, we can write

$$p'(z) = \frac{\left(D_z^{-\lambda}\phi(m,n;q,z)\right)'}{D_z^{-\lambda}\phi(m,n;q,z)} + z\frac{\left(D_z^{-\lambda}\phi(m,n;q,z)\right)''}{D_z^{-\lambda}\phi(m,n;q,z)} - z\left(\frac{\left(D_z^{-\lambda}\phi(m,n;q,z)\right)'}{D_z^{-\lambda}\phi(m,n;q,z)}\right)^2$$

and

$$zp'(z) = \frac{z\left(D_z^{-\lambda}\phi(m,n;q,z)\right)'}{D_z^{-\lambda}\phi(m,n;q,z)} + \frac{z^2\left(D_z^{-\lambda}\phi(m,n;q,z)\right)''}{D_z^{-\lambda}\phi(m,n;q,z)} - \left(\frac{z\left(D_z^{-\lambda}\phi(m,n;q,z)\right)'}{D_z^{-\lambda}\phi(m,n;q,z)}\right)^2. \quad (5.2.1.22)$$

Set $\varphi(u) := \beta u$ and $\gamma(u) := \delta$, we can show easily that φ is analytic in $\mathbb{C}$, γ is analytic in $\mathbb{C}\backslash\{0\}$ and $\gamma(u) \neq 0$, $u \in \mathbb{C}\backslash\{0\}$.

Also, we set $G(z) = zg'(z)\gamma(g(z)) = \delta zg'(z)$, which is starlike univalent in U.

Let $h(z) = G(z) + \varphi(g(z)) = \beta g(z) + \delta zg'(z)$.

We get $\operatorname{Re}\left(\frac{zh'(z)}{G(z)}\right) = \operatorname{Re}\left(\frac{\beta+\delta}{\delta} + z\frac{g''(z)}{g'(z)}\right) > 0$.

Taking account the relation (5.2.1.22), we can write $\beta p(z) + \delta zp'(z) = (\beta+\delta)\frac{z\left(D_z^{-\lambda}\phi(m,n;q,z)\right)'}{D_z^{-\lambda}\phi(m,n;q,z)} + \delta\frac{z^2\left(D_z^{-\lambda}\phi(m,n;q,z)\right)''}{D_z^{-\lambda}\phi(m,n;q,z)} - \delta\left(\frac{z\left(D_z^{-\lambda}\phi(m,n;q,z)\right)'}{D_z^{-\lambda}\phi(m,n;q,z)}\right)^2$.

Using fuzzy differential subordination (5.2.1.20), we get $F_{p(U)}(\beta p(z) + \delta zp'(z)) \leq F_{g(U)}(\beta g(z) + \delta zg'(z))$.

Applying Lemma 1.16, we obtain $F_{p(U)}p(z) \leq F_{g(U)}g(z)$, $z \in U$, i.e. $F_{D_z^{-\lambda}\phi(U)}\left(\frac{z\left(D_z^{-\lambda}\phi(m,n;q,z)\right)'}{D_z^{-\lambda}\phi(m,n;q,z)}\right) \leq F_{g(U)}g(z)$, $z \in U$, and the fuzzy best dominant is g. □

Corollary 5.13 (Alb Lupaş and Oros, 2022e). *Consider $\lambda > 0$, $0 < q < 1$, m, n complex numbers with $n \neq 0, -1, -2, \ldots$ and $g(z) = \frac{Mz+1}{Nz+1}$, $z \in U$, $-1 \leq N < M \leq 1$. Suppose that relation* (5.2.1.18) *holds. When*

$$F_{\Psi_\lambda^{m,n,q}(U)}\Psi_\lambda^{m,n,q}(\beta,\delta;q,z) \leq F_{g(U)}\left(\beta\frac{Mz+1}{Nz+1} + \frac{\delta(M-N)z}{(Nz+1)^2}\right),$$

with $\beta, \delta \in \mathbb{C}$, $\delta \neq 0$, $-1 \leq N < M \leq 1$, and $\Psi_\lambda^{m,n,q}$ defined by relation (5.2.1.19), *then*

$$F_{D_z^{-\lambda}\phi(U)}\left(\frac{z\left(D_z^{-\lambda}\phi(m,n;q,z)\right)'}{D_z^{-\lambda}\phi(m,n;q,z)}\right) \leq F_{g(U)}\left(\frac{Mz+1}{Nz+1}\right),$$

and the fuzzy best dominant is $\frac{Mz+1}{Nz+1}$.

Proof. Consider $g(z) = \frac{Mz+1}{Nz+1}$, $-1 \leq N < M \leq 1$, in Theorem 5.37 and we obtain the corollary. □

Corollary 5.14 (Alb Lupaş and Oros, 2022e). *Consider $\lambda > 0$, $0 < q < 1$, m, n complex numbers with $n \neq 0, -1, -2, \ldots$ and $g(z) = \left(\frac{z+1}{1-z}\right)^{\alpha}$. Suppose that relation (5.2.1.18) holds. When*

$$F_{\Psi_{\lambda}^{m,n,q}(U)} \Psi_{\lambda}^{m,n,q}(\beta, \delta; q, z) \leq F_{g(U)}\left(\beta\left(\frac{z+1}{1-z}\right)^{\alpha} + \frac{2\alpha\delta z}{1-z^2}\left(\frac{z+1}{1-z}\right)^{\alpha}\right),$$

with $\beta, \delta \in \mathbb{C}$, $0 < \alpha \leq 1$, $\delta \neq 0$, and $\Psi_{\lambda}^{m,n,q}$ defined by relation (5.2.1.19), then

$$F_{D_z^{-\lambda}\phi(U)}\left(\frac{z\left(D_z^{-\lambda}\phi(m,n;q,z)\right)'}{D_z^{-\lambda}\phi(m,n;q,z)}\right) \leq F_{g(U)}\left(\frac{z+1}{1-z}\right)^{\alpha}, \quad z \in U,$$

and the fuzzy best dominant is $\left(\frac{z+1}{1-z}\right)^{\alpha}$.

Proof. Corollary follows considering in Theorem 5.37 $g(z) = \left(\frac{z+1}{1-z}\right)^{\alpha}$, $0 < \alpha \leq 1$. □

The next theorem and the corollaries associated involve fuzzy differential superordination aspects, too.

Theorem 5.38 (Alb Lupaş and Oros, 2022e). *Consider g a convex and univalent function in U such that $g(0) = \lambda$, with $0 < q < 1$, $\lambda > 0$, m, n complex numbers with $n \neq 0, -1, -2, \ldots$. Suppose that*

$$\mathrm{Re}\left(\frac{\beta}{\delta} g'(z)\right) > 0, \quad \text{for } \beta, \delta \in \mathbb{C}, \delta \neq 0. \tag{5.2.1.23}$$

When $\frac{z\left(D_z^{-\lambda}\phi(m,n;q,z)\right)'}{D_z^{-\lambda}\phi(m,n;q,z)} \in \mathcal{H}[g(0), 1] \cap Q$ and $\Psi_{\lambda}^{m,n,q}(\beta, \delta; q, z)$ defined by relation (5.2.1.19) is univalent in U, then

$$F_{g(U)}(\beta g(z) + \delta z g'(z)) \leq F_{\Psi_{\lambda}^{m,n,q}(U)} \Psi_{\lambda}^{m,n,q}(\beta, \delta; q, z) \tag{5.2.1.24}$$

implies

$$F_{g(U)} g(z) \leq F_{D_z^{-\lambda}\phi(U)} \frac{z\left(D_z^{-\lambda}\phi(m,n;q,z)\right)'}{D_z^{-\lambda}\phi(m,n;q,z)}, \quad z \in U, \tag{5.2.1.25}$$

and the fuzzy best subordinant is g.

Proof. Define also $p(z) := \frac{z\left(D_z^{-\lambda}\phi(m,n;q,z)\right)'}{D_z^{-\lambda}\phi(m,n;q,z)}$, $z \in U$, $z \neq 0$, which is analytic in U and $p(0) = \lambda$.

Let $\varphi(u) := \beta u$ and $\gamma(u) := \delta$, it is easy to show that φ is analytic in $\mathbb{C}$, γ is analytic in $\mathbb{C}\backslash\{0\}$ with $\gamma(u) \neq 0$, $u \in \mathbb{C}\backslash\{0\}$.

Since $\frac{\varphi'(g(z))}{\gamma(g(z))} = \frac{\beta}{\delta}g'(z)$, it yields that $\operatorname{Re}\left(\frac{\varphi'(g(z))}{\gamma(g(z))}\right) = \operatorname{Re}\left(\frac{\beta}{\delta}g'(z)\right) > 0$, for $\beta, \delta \in \mathbb{C}$, $\delta \neq 0$.

The fuzzy differential superordination (5.2.1.24) can be written as

$$F_{g(U)}\left(\beta g(z) + \delta z g'(z)\right) \leq F_{p(U)}\left(\beta p(z) + \delta z p'(z)\right), \quad z \in U,$$

and applying Lemma 1.19, we obtain

$$F_{g(U)}g(z) \leq F_{D_z^{-\lambda}\phi(U)}\left(\frac{z\left(D_z^{-\lambda}\phi(m,n;q,z)\right)'}{D_z^{-\lambda}\phi(m,n;q,z)}\right), \quad z \in U,$$

and the fuzzy best subordinant is g. □

Corollary 5.15 (Alb Lupaş and Oros, 2022e). *For $\lambda > 0$, $0 < q < 1$, m, n complex numbers with $n \neq 0, -1, -2, ..$ and $g(z) = \frac{Mz+1}{Nz+1}$, $-1 \leq N < M \leq 1$, $z \in U$, suppose that relation* (5.2.1.23) *holds. When $\frac{z\left(D_z^{-\lambda}\phi(m,n;q,z)\right)'}{D_z^{-\lambda}\phi(m,n;q,z)} \in \mathcal{H}[g(0),1] \cap Q$, and*

$$F_{g(U)}\left(\beta\frac{Mz+1}{Nz+1} + \frac{\delta(M-N)z}{(Nz+1)^2}\right) \leq F_{\Psi_\lambda^{m,n,q}(U)}\Psi_\lambda^{m,n,q}(\beta,\delta;q,z),$$

with $\beta, \delta \in \mathbb{C}$, $\delta \neq 0$, $-1 \leq N < M \leq 1$, and $\Psi_\lambda^{m,n,q}$ defined by relation (5.2.1.19), *then*

$$F_{g(U)}\left(\frac{Mz+1}{Nz+1}\right) \leq F_{D_z^{-\lambda}\phi(U)}\left(\frac{z\left(D_z^{-\lambda}\phi(m,n;q,z)\right)'}{D_z^{-\lambda}\phi(m,n;q,z)}\right),$$

and the fuzzy best subordinant is $\frac{Mz+1}{Nz+1}$.

Proof. Considering $g(z) = \frac{Mz+1}{Nz+1}$, $-1 \leq N < M \leq 1$, in Theorem 5.38 we obtain the corollary. □

Corollary 5.16 (Alb Lupaş and Oros, 2022e). *For $\lambda > 0$, $0 < q < 1$, m, n complex numbers with $n \neq 0, -1, -2, \ldots$ and $g(z) = \left(\frac{z+1}{1-z}\right)^\alpha$, suppose that relation* (5.2.1.23) *holds. When $\frac{z\left(D_z^{-\lambda}\phi(m,n;q,z)\right)'}{D_z^{-\lambda}\phi(m,n;q,z)} \in \mathcal{H}[g(0),1] \cap Q$ and*

$$F_{g(U)}\left(\beta\left(\frac{z+1}{1-z}\right)^\alpha + \frac{2\alpha\delta z}{1-z^2}\left(\frac{z+1}{1-z}\right)^\alpha\right) \leq F_{\Psi_\lambda^{m,n,q}(U)}\Psi_\lambda^{m,n,q}(\beta,\delta;q,z),$$

with $\beta, \delta \in \mathbb{C}$, $0 < \alpha \le 1$, $\delta \neq 0$, *and* $\Psi_\lambda^{m,n,q}$ *defined by relation* (5.2.1.19), *then*

$$F_{g(U)}\left(\frac{z+1}{1-z}\right)^\alpha \le F_{D_z^{-\lambda}\phi(U)}\left(\frac{z\left(D_z^{-\lambda}\phi(m,n;q,z)\right)'}{D_z^{-\lambda}\phi(m,n;q,z)}\right),$$

and the fuzzy best subordinant is $\left(\frac{z+1}{1-z}\right)^\alpha$.

Proof. Corollary follows considering in Theorem 5.38 $g(z) = \left(\frac{z+1}{1-z}\right)^\alpha$, $0 < \alpha \le 1$. □

Combining the results obtained in Theorems 5.37 and 5.38, the following sandwich-type result can be established.

Theorem 5.39 (Alb Lupaş and Oros, 2022e). *Consider* g_1, g_2 *convex and univalent functions in* U *with the properties* $g_1(z) \neq 0$, $g_2(z) \neq 0$, *for all* $z \in U$. *Assume that* g_1 *satisfies relation* (5.2.1.18) *and* g_2 *satisfies relation* (5.2.1.23). *For* $\lambda > 0$, $0 < q < 1$, m, n *complex numbers with* $n \neq 0, -1, -2, \ldots$, *when* $\frac{z\left(D_z^{-\lambda}\phi(m,n;q,z)\right)'}{D_z^{-\lambda}\phi(m,n;q,z)} \in \mathcal{H}[g(0), 1] \cap Q$, *and* $\Psi_\lambda^{m,n,q}(\beta, \delta; q, z)$ *defined by* (5.2.1.19) *is univalent in* U, *then*

$$\begin{aligned} F_{g_1(U)}\left(\beta g_1(z) + \delta z g_1'(z)\right) &\le F_{\Psi_\lambda^{m,n,q}(U)}\Psi_\lambda^{m,n,q}(\beta, \delta; q, z) \\ &\le F_{g_2(U)}\left(\beta g_2(z) + \delta z g_2'(z)\right), \end{aligned}$$

for $\beta, \delta \in \mathbb{C}$, $\delta \neq 0$, *implies*

$$F_{g_1(U)} g_1(z) \le F_{D_z^{-\lambda}\phi(U)}\left(\frac{z\left(D_z^{-\lambda}\phi(m,n;q,z)\right)'}{D_z^{-\lambda}\phi(m,n;q,z)}\right) \le F_{g_2(U)} g_2(z), \quad z \in U,$$

and the fuzzy best subordinant is g_1 *and the fuzzy best dominant is* g_2.

Considering $g_1(z) = \frac{M_1 z+1}{N_1 z+1}$ and $g_2(z) = \frac{M_2 z+1}{N_2 z+1}$, with $-1 \le N_2 < N_1 < M_1 < M_2 \le 1$, we get the following corollary.

Corollary 5.17 (Alb Lupaş and Oros, 2022e). *For* m, n *complex numbers with* $n \neq 0, -1, -2, \ldots$; $\lambda > 0$, $0 < q < 1$, *suppose that relations* (5.2.1.18) *and* (5.2.1.23) *hold for* $g_1(z) = \frac{M_1 z+1}{N_1 z+1}$ *and* $g_2(z) = \frac{M_2 z+1}{N_2 z+1}$, *respectively. When* $\frac{z\left(D_z^{-\lambda}\phi(m,n;q,z)\right)'}{D_z^{-\lambda}\phi(m,n;q,z)} \in \mathcal{H}\left[g\left(0\right),1\right] \cap Q$ *and*

$$F_{g_1(U)}\left(\beta\frac{M_1z+1}{N_1z+1}+\frac{\delta\left(M_1-N_1\right)z}{\left(N_1z+1\right)^2}\right) \leq F_{\Psi_\lambda^{m,n,q}(U)}\Psi_\lambda^{m,n,q}\left(\beta,\delta;q,z\right)$$

$$\leq F_{g_2(U)}\left(\beta\frac{M_2z+1}{N_2z+1}+\frac{\delta\left(M_2-N_2\right)z}{\left(N_2z+1\right)^2}\right), \quad z\in U,$$

with $\beta,\delta\in\mathbb{C}$, $\delta\neq 0$, $-1\leq N_2\leq N_1<M_1\leq M_2\leq 1$, *and* $\Psi_\lambda^{m,n,q}$ *defined by relation* (5.2.1.19), *then*

$$F_{g_1(U)}\left(\frac{M_1z+1}{N_1z+1}\right) \leq F_{D_z^{-\lambda}\phi(U)}\left(\frac{z\left(D_z^{-\lambda}\phi\left(m,n;q,z\right)\right)'}{D_z^{-\lambda}\phi\left(m,n;q,z\right)}\right)$$

$$\leq F_{g_2(U)}\left(\frac{M_2z+1}{N_2z+1}\right), \quad z\in U,$$

the fuzzy best subordinant is $\frac{M_1z+1}{N_1z+1}$ *and the fuzzy best dominant is* $\frac{M_2z+1}{N_2z+1}$, *respectively.*

Considering $g_1(z)=\left(\frac{z+1}{1-z}\right)^{\alpha_1}$ and $g_2(z)=\left(\frac{z+1}{1-z}\right)^{\alpha_2}$, with $0<\alpha_1<\alpha_2\leq 1$, we obtain the following corollary.

Corollary 5.18 (Alb Lupaş and Oros, 2022e). *For* m, n *complex numbers with* $n\neq 0,-1,-2,\ldots$; $\lambda>0$, $0<q<1$, *suppose that relations* (5.2.1.18) *and* (5.2.1.23) *hold for* $g_1(z)=\left(\frac{z+1}{1-z}\right)^{\alpha_1}$ *and* $g_2(z)=\left(\frac{z+1}{1-z}\right)^{\alpha_2}$, *respectively. When* $\frac{z\left(D_z^{-\lambda}\phi(m,n;q,z)\right)'}{D_z^{-\lambda}\phi(m,n;q,z)} \in \mathcal{H}\left[g\left(0\right),1\right] \cap Q$ *and*

$$F_{g_1(U)}\left(\beta\left(\frac{z+1}{1-z}\right)^{\alpha_1}+\frac{2\alpha_1\delta z}{1-z^2}\left(\frac{z+1}{1-z}\right)^{\alpha_1}\right) \leq F_{\Psi_\lambda^{m,n,q}(U)}\Psi_\lambda^{m,n,q}\left(\beta,\delta;q,z\right)$$

$$\leq F_{g_2(U)}\left(\beta\left(\frac{z+1}{1-z}\right)^{\alpha_2}+\frac{2\alpha_2\delta z}{1-z^2}\left(\frac{z+1}{1-z}\right)^{\alpha_2}\right), \quad z\in U,$$

with $\beta,\delta\in\mathbb{C}$, $\delta\neq 0$, $0<\alpha_1<\alpha_2\leq 1$, *and* $\Psi_\lambda^{m,n,q}$ *defined by relation* (5.2.1.19), *then*

$$F_{g_1(U)}\left(\frac{z+1}{1-z}\right)^{\alpha_1} \leq F_{D_z^{-\lambda}\phi(U)}\left(\frac{z\left(D_z^{-\lambda}\phi\left(a,b;q,z\right)\right)'}{D_z^{-\lambda}\phi\left(a,b;q,z\right)}\right)$$

$$\leq F_{g_2(U)}\left(\frac{z+1}{1-z}\right)^{\alpha_2}, \quad z\in U,$$

the fuzzy best subordinant is $\left(\frac{z+1}{1-z}\right)^{\alpha_1}$ *and the fuzzy best dominant is* $\left(\frac{z+1}{1-z}\right)^{\alpha_2}$, *respectively.*

5.2.2 *Fuzzy Differential Subordination and Superordination Results for q-Analogue of Multiplier Transformation*

In the paper Alb Lupaş *et al.* (2023), we combined the quantum calculus applications regarding the theories of differential subordination and superordination with fuzzy theory. These results are established by means of an operator defined as the q-analogue of the multiplier transformation in Alb Lupaş and Cătas (2023). Interesting fuzzy differential subordination and superordination results are derived involving the functions belonging to a new subclass of normalized analytic functions in the open unit disc U which is defined and investigated here by using this q-operator.

$\mathcal{I}_q^{m,l}$ is the q-analogue of multiplier transformation from Definition 2.12

$$\mathcal{I}_q^{m,l}f(z) := z + \sum_{k=2}^{\infty}\left(\frac{[k+l]_q}{[1+l]_q}\right)^m a_k z^k,$$

with $l > -1$, $q \in (0,1)$, m a real number and $f(z) = z + \sum_{k=2}^{\infty} a_k z^k \in \mathcal{A}$, $z \in U$.

Applying the properties of q-calculus, we get

$$z\mathcal{D}_q\left(\mathcal{I}_q^{m,l}f(z)\right) = \left(1 + \frac{[l]_q}{q^l}\right)\mathcal{I}_q^{m+1,l}f(z) - \frac{[l]_q}{q^l}\mathcal{I}_q^{m,l}f(z).$$

Using the q-analogue of the multiplier transformation $\mathcal{I}_q^{m,l}$, a new subclass of normalized analytic functions in the open unit disc U is introduced. It is proved that this new subclass is convex and using this property, fuzzy subordination results are investigated. Fuzzy differential superordinations involving the operator $\mathcal{I}_q^{m,l}$ are considered for which the best subordinants are also found.

The new subclass of normalized analytic functions in the open unit disc U is defined using the q-analogue of the multiplier transformation $\mathcal{I}_q^{m,l}$.

Definition 5.2 (Alb Lupaş *et al.*, 2023). Let $\alpha \in [0,1)$. The subclass $\mathcal{F}S_{m,l}^q(\alpha)$ consists the functions $f \in \mathcal{A}$ with property

$$F_{(\mathcal{I}_q^{m,l}f)'(U)}\left(\mathcal{I}_q^{m,l}f(z)\right)' > \alpha, \quad z \in U.$$

The first result concerning the subclass $\mathcal{F}S^q_{m,l}(\alpha)$ establishes its convexity.

Theorem 5.40 (Alb Lupaş *et al.*, 2023). *$\mathcal{F}S^q_{m,l}(\alpha)$ is a convex set.*

Proof. Consider the functions

$$f_j(z) = z + \sum_{k=2}^{\infty} a_{jk} z^k, \quad z \in U, \ j = 1, 2,$$

belonging to the subclass $\mathcal{F}S^q_{m,l}(\alpha)$. It is enough to prove that the function

$$f(z) = \lambda_1 f_1(z) + \lambda_2 f_2(z)$$

belongs to the subclass $\mathcal{F}S_m(\delta, \alpha)$, with λ_1 and λ_2 positive real numbers such that $\lambda_1 + \lambda_2 = 1$.

The function f has the following form

$$f(z) = z + \sum_{k=2}^{\infty} (\lambda_1 a_{1k} + \lambda_2 a_{2k}) z^k, \quad z \in U,$$

and

$$\mathcal{I}^{m,l}_q f(z) = z + \sum_{k=2}^{\infty} \left(\frac{[k+l]_q}{[1+l]_q} \right)^m (\lambda_1 a_{1k} + \lambda_2 a_{2k}) z^k, \quad z \in U. \quad (5.2.2.1)$$

Differentiating relation (5.2.2.1), we obtain

$$\left(\mathcal{I}^{m,l}_q f(z)\right)' = 1 + \sum_{k=2}^{\infty} \left(\frac{[k+l]_q}{[1+l]_q} \right)^m (\lambda_1 a_{1k} + \lambda_2 a_{2k}) k z^{k-1}, \quad z \in U.$$

We have $f'(z) = (\lambda_1 f_1 + \lambda_2 f_2)'(z) = \lambda_1 f_1'(z) + \lambda_2 f_2'(z)$, $z \in U$, and $\left(\mathcal{I}^{m,l}_q f(z)\right)' = \left(\mathcal{I}^{m,l}_q (\lambda_1 f_1 + \lambda_2 f_2)(z)\right)' = \lambda_1 \left(\mathcal{I}^{m,l}_q f_1(z)\right)' + \lambda_2 \left(\mathcal{I}^{m,l}_q f_2(z)\right)'$.

Applying fuzzy theory, we obtain that

$$\begin{aligned}
&\mathcal{F}_{(\mathcal{I}^{m,l}_q f)'(U)} \left(\mathcal{I}^{m,l}_q f(z)\right)' \\
&= \mathcal{F}_{(\mathcal{I}^{m,l}_q (\lambda_1 f_1 + \lambda_2 f_2))'(U)} \left(\mathcal{I}^{m,l}_q (\lambda_1 f_1 + \lambda_2 f_2)(z)\right)' \\
&= \mathcal{F}_{(\mathcal{I}^{m,l}_q (\lambda_1 f_1 + \lambda_2 f_2))'(U)} \left(\lambda_1 \left(\mathcal{I}^{m,l}_q f_1(z)\right)' + \lambda_2 \left(\mathcal{I}^{m,l}_q f_2(z)\right)'\right) \\
&= \frac{\mathcal{F}_{(\lambda_1 \mathcal{I}^{m,l}_q f_1)'(U)} \left(\lambda_1 \left(\mathcal{I}^{m,l}_q f_1(z)\right)'\right) + \mathcal{F}_{(\lambda_2 \mathcal{I}^{m,l}_q f_2)'(U)} \left(\lambda_2 \left(\mathcal{I}^{m,l}_q f_2(z)\right)'\right)}{2} \\
&= \frac{\mathcal{F}_{(\mathcal{I}^{m,l}_q f_1)'(U)} \left(\mathcal{I}^{m,l}_q f_1(z)\right)' + \mathcal{F}_{(\mathcal{I}^{m,l}_q f_2)'(U)} \left(\mathcal{I}^{m,l}_q f_2(z)\right)'}{2}.
\end{aligned}$$

Since $f_1, f_2 \in \mathcal{F}S^q_{m,l}(\alpha)$, we have $\alpha < \mathcal{F}_{(\mathcal{I}^{m,l}_q f_1)'(U)}\left(\mathcal{I}^{m,l}_q f_1(z)\right)' \leq 1$ and $\alpha < \mathcal{F}_{(\mathcal{I}^{m,l}_q f_2)'(U)}\left(\mathcal{I}^{m,l}_q f_2(z)\right)' \leq 1$, $z \in U$.

Therefore, $\alpha < \frac{\mathcal{F}_{(\mathcal{I}^{m,l}_q f_1)'(U)}\left(\mathcal{I}^{m,l}_q f_1(z)\right)' + \mathcal{F}_{(\mathcal{I}^{m,l}_q f_2)'(U)}\left(\mathcal{I}^{m,l}_q f_2(z)\right)'}{2} \leq 1$ and we obtain that $\alpha < \mathcal{F}_{(\mathcal{I}^{m,l}_q f)'(U)}\left(\mathcal{I}^{m,l}_q f(z)\right)' \leq 1$, which means that $f \in \mathcal{F}S^q_{m,l}(\alpha)$ and $\mathcal{F}S^q_{m,l}(\alpha)$ is convex. □

We next investigate a series of fuzzy differential subordinations involving the convex functions of the subclass $S^q_{m,l}(\alpha)$ and the q-analogue of the multiplier transformation $\mathcal{I}^{m,l}_q$.

Theorem 5.41 (Alb Lupaş *et al.*, 2023). *Considering g a convex function, we define the function*

$$h(z) = \frac{zg'(z)}{a+2} + g(z), \quad a > 0, \quad z \in U. \tag{5.2.2.2}$$

For $f \in \mathcal{F}S^q_{m,l}(\alpha)$, consider

$$F(z) = \frac{a+2}{z^{a+1}} \int_0^z t^a f(t)\, dt, \quad z \in U, \tag{5.2.2.3}$$

then the fuzzy differential subordination

$$\mathcal{F}_{(\mathcal{I}^{m,l}_q f)'(U)}\left(\mathcal{I}^{m,l}_q f(z)\right)' \leq \mathcal{F}_{h(U)} h(z), \quad z \in U, \tag{5.2.2.4}$$

implies the sharp fuzzy differential subordination

$$\mathcal{F}_{(\mathcal{I}^{m,l}_q F)'(U)}\left(\mathcal{I}^{m,l}_q F(z)\right)' \leq \mathcal{F}_{g(U)} g(z), \quad z \in U.$$

Proof. Relation (5.2.2.3) can be written as following

$$z^{a+1} F(z) = (a+2) \int_0^z t^a f(t)\, dt, \tag{5.2.2.5}$$

and differentiating it, we have

$$zF'(z) + (a+1) F(z) = (a+2) f(z) \tag{5.2.2.6}$$

and

$$z\left(\mathcal{I}^{m,l}_q F(z)\right)' + (a+1)\, \mathcal{I}^{m,l}_q F(z) = (a+2)\, \mathcal{I}^{m,l}_q f(z), \quad z \in U. \tag{5.2.2.7}$$

Differentiating the last relation, we get

$$\frac{z\left(\mathcal{I}_q^{m,l}F(z)\right)''}{a+2}+\left(\mathcal{I}_q^{m,l}F(z)\right)'=\left(\mathcal{I}_q^{m,l}f(z)\right)',\quad z\in U, \tag{5.2.2.8}$$

and the fuzzy subordination (5.2.2.4) can be written in the form

$$\mathcal{F}_{\left(\mathcal{I}_q^{m,l}F\right)'(U)}\left(\frac{z\left(\mathcal{I}_q^{m,l}F(z)\right)''}{a+2}+\left(\mathcal{I}_q^{m,l}F(z)\right)'\right)\leq\mathcal{F}_{g(U)}\left(\frac{zg'(z)}{a+2}+g(z)\right). \tag{5.2.2.9}$$

Denoting

$$p(z)=\left(\mathcal{I}_q^{m,l}F(z)\right)'\in\mathcal{H}\left[1,1\right], \tag{5.2.2.10}$$

fuzzy differential subordination (5.2.2.9) has the following form

$$\mathcal{F}_{p(U)}\left(\frac{zp'(z)}{a+2}+p(z)\right)\leq\mathcal{F}_{g(U)}\left(\frac{zg'(z)}{a+2}+g(z)\right),\quad z\in U.$$

Applying Lemma 1.15, we get

$$\mathcal{F}_{p(U)}p(z)\leq\mathcal{F}_{g(U)}g(z),$$

that means

$$\mathcal{F}_{\left(\mathcal{I}_q^{m,l}F\right)'(U)}\left(\mathcal{I}_q^{m,l}F(z)\right)'\leq F_{g(U)}g(z),\quad z\in U$$

for g the fuzzy best dominant. □

Theorem 5.42 (Alb Lupaş *et al.*, 2023). *Denoting*

$$I_a\left(f\right)(z)=\frac{a+2}{z^{a+1}}\int_0^z t^a f\left(t\right)dt,\quad a>0, \tag{5.2.2.11}$$

then

$$I_a\left[S_{m,l}^q\left(\alpha\right)\right]\subset S_{m,l}^q\left(\alpha^*\right), \tag{5.2.2.12}$$

where

$$\alpha^*=2\alpha-1+(a+2)\left(2-2\alpha\right)\int_0^1\frac{t^{a+1}}{t+1}dt. \tag{5.2.2.13}$$

Proof. We obtain using the hypothesis of Theorem 5.42 for the convex function $h(z) = \frac{1+(2\alpha-1)z}{1+z}$ and following the same steps like the proof of Theorem 5.41 the fuzzy differential subordination

$$\mathcal{F}_{p(U)}\left(\frac{zp'(z)}{a+2}+p(z)\right) \le F_{h(U)}h(z),$$

with p defined by relation (5.2.2.10).

Applying Lemma 1.14, we get the fuzzy differential subordinations

$$\mathcal{F}_{p(U)}p(z) \le \mathcal{F}_{g(U)}g(z) \le \mathcal{F}_{h(U)}h(z),$$

equivalently with

$$\mathcal{F}_{(\mathcal{I}_q^{m,l}F)'(U)}\left(\mathcal{I}_q^{m,l}F(z)\right)' \le \mathcal{F}_{g(U)}g(z) \le \mathcal{F}_{h(U)}h(z),$$

where

$$\begin{aligned} g(z) &= \frac{a+2}{z^{a+2}}\int_0^z t^{a+1}\frac{1+(2\alpha-1)t}{1+t}dt \\ &= 2\alpha-1+\frac{(a+2)(2-2\alpha)}{z^{a+2}}\int_0^z \frac{t^{a+1}}{t+1}dt. \end{aligned}$$

Taking account that g is a convex function with $g(U)$ symmetric to the real axis, we obtain

$$\begin{aligned} \mathcal{F}_{(\mathcal{I}_q^{m,l}F)'(U)}\left(\mathcal{I}_q^{m,l}F(z)\right)' &\ge \min_{|z|=1}\mathcal{F}_{g(U)}g(z) = \mathcal{F}_{g(U)}g(1) \\ &= \alpha^* = 2\alpha-1+(a+2)(2-2\alpha)\int_0^1 \frac{t^{a+1}}{t+1}. \end{aligned}$$

□

Theorem 5.43 (Alb Lupaş *et al.*, 2023). *Considering the convex function g such that $g(0)=1$, we define the function*

$$h(z) = zg'(z)+g(z), \quad z \in U.$$

If $f \in \mathcal{A}$ verifies the fuzzy subordination

$$\mathcal{F}_{(\mathcal{I}_q^{m,l}f)'(U)}\left(\mathcal{I}_q^{m,l}f(z)\right)' \le \mathcal{F}_{h(U)}h(z), \quad z \in U, \tag{5.2.2.14}$$

then the sharp fuzzy differential subordination

$$\mathcal{F}_{\mathcal{I}_q^{m,l}f(U)}\frac{\mathcal{I}_q^{m,l}f(z)}{z} \le F_{g(U)}g(z), \quad z \in U$$

holds.

Proof. Considering

$$p(z)=\frac{\mathcal{I}_q^{m,l}f(z)}{z}=\frac{z+\sum_{k=2}^{\infty}\left(\frac{[k+l]_q}{[1+l]_q}\right)^m a_k z^k}{z}=1+p_1z+p_2z^2+\cdots,\quad z\in U,$$

evidently $p\in\mathcal{H}[1,1]$, so we can write

$$zp(z)=\mathcal{I}_q^{m,l}f(z),\quad z\in U,$$

and differentiating, it we get

$$\left(\mathcal{I}_q^{m,l}f(z)\right)'=zp'(z)+p(z),\quad z\in U.$$

The fuzzy subordination (5.2.2.14) take the form

$$\mathcal{F}_{p(U)}\left(zp'(z)+p(z)\right)\leq\mathcal{F}_{g(U)}\left(zg'(z)+g(z)\right),\quad z\in U,$$

and applying Lemma 1.15, we get

$$\mathcal{F}_{p(U)}p(z)\leq\mathcal{F}_{g(U)}g(z),\quad z\in U,$$

that means

$$\mathcal{F}_{\mathcal{I}_q^{m,l}f(U)}\frac{\mathcal{I}_q^{m,l}f(z)}{z}\leq\mathcal{F}_{g(U)}g(z),\quad z\in U.$$

□

Theorem 5.44 (Alb Lupaş *et al.*, 2023). *Considering the convex function h with the property $h(0)=1$, for $f\in\mathcal{A}$ that verifies the fuzzy subordination*

$$\mathcal{F}_{(\mathcal{I}_q^{m,l}f)'(U)}\left(\mathcal{I}_q^{m,l}f(z)\right)'\leq\mathcal{F}_{h(U)}h(z),\quad z\in U,\tag{5.2.2.15}$$

we get the fuzzy differential subordination

$$\mathcal{F}_{\mathcal{I}_q^{m,l}f(U)}\frac{\mathcal{I}_q^{m,l}f(z)}{z}\leq\mathcal{F}_{g(U)}g(z),\quad z\in U,$$

for the convex function $g(z)=\frac{1}{z}\int_0^z h(t)dt$, which is the fuzzy best dominant.

Proof. Denote

$$p(z)=\frac{\mathcal{I}_q^{m,l}f(z)}{z}=1+\sum_{k=2}^{\infty}\left(\frac{[k+l]_q}{[1+l]_q}\right)^m a_k z^{k-1}\in\mathcal{H}[1,1],\quad z\in U.$$

Differentiating it the relation, we get

$$\left(\mathcal{I}_q^{m,l} f(z)\right)' = zp'(z) + p(z), \quad z \in U$$

and fuzzy differential subordination (5.2.2.15) has the following form

$$\mathcal{F}_{p(U)}\left(zp'(z) + p(z)\right) \leq \mathcal{F}_{h(U)} h(z), \quad z \in U.$$

After applying Lemma 1.14, we get

$$\mathcal{F}_{p(U)} p(z) \leq \mathcal{F}_{g(U)} g(z), \quad z \in U,$$

written as

$$\mathcal{F}_{\mathcal{I}_q^{m,l} f(U)} \frac{\mathcal{I}_q^{m,l} f(z)}{z} \leq F_{g(U)} g(z), \quad z \in U$$

for $g(z) = \frac{1}{z}\int_0^z h(t)dt$ the fuzzy best dominant. □

Theorem 5.45 (Alb Lupaş *et al.*, 2023). *Considering a convex function g such that $g(0) = 1$, we define the function $h(z) = zg'(z) + g(z)$, $z \in U$. When $f \in \mathcal{A}$ verifies the fuzzy subordination*

$$\mathcal{F}_{\mathcal{I}_q^{m,l} f(U)} \left(\frac{z\mathcal{I}_q^{m+1,l} f(z)}{\mathcal{I}_q^{m,l} f(z)}\right)' \leq \mathcal{F}_{h(U)} h(z), \quad z \in U, \tag{5.2.2.16}$$

then we get the sharp fuzzy differential subordination

$$\mathcal{F}_{\mathcal{I}_q^{m,l} f(U)} \frac{\mathcal{I}_q^{m+1,l} f(z)}{\mathcal{I}_q^{m,l} f(z)} \leq \mathcal{F}_{g(U)} g(z), \quad z \in U.$$

Proof. Denote

$$p(z) = \frac{\mathcal{I}_q^{m+1,l} f(z)}{\mathcal{I}_q^{m,l} f(z)} = \frac{z + \sum_{k=2}^{\infty}\left(\frac{[k+l]_q}{[1+l]_q}\right)^{m+1} a_k z^k}{z + \sum_{k=2}^{\infty}\left(\frac{[k+l]_q}{[1+l]_q}\right)^{m} a_k z^k}.$$

Differentiating it, we get $p'(z) = \frac{\left(\mathcal{I}_q^{m+1,l} f(z)\right)'}{\mathcal{I}_q^{m,l} f(z)} - p(z)\frac{\left(\mathcal{I}_q^{m,l} f(z)\right)'}{\mathcal{I}_q^{m,l} f(z)}$ written as

$$zp'(z) + p(z) = \left(\frac{z\mathcal{I}_q^{m+1,l} f(z)}{\mathcal{I}_q^{m,l} f(z)}\right)'.$$

Fuzzy differential subordination (5.2.2.16) has the following form, for $z \in U$,

$$\mathcal{F}_{p(U)}\left(zp'(z)+p(z)\right) \leq \mathcal{F}_{g(U)}\left(zg'(z)+g(z)\right),$$

and applying Lemma 1.15, we get the fuzzy differential subordination, for $z \in U$,

$$\mathcal{F}_{p(U)}p(z) \leq \mathcal{F}_{g(U)}g(z),$$

written as

$$\mathcal{F}_{\mathcal{I}_q^{m,l}f(U)}\frac{\mathcal{I}_q^{m+1,l}f(z)}{\mathcal{I}_q^{m,l}f(z)} \leq \mathcal{F}_{g(U)}g(z).$$

□

Next, fuzzy differential superordinations are investigated concerning q-analogue of the multiplier transformation $\mathcal{I}_q^{m,l}$ and its derivatives of first and second order. The fuzzy best subordinant is given for each of the investigated fuzzy differential superordinations.

Theorem 5.46 (Alb Lupaş *et al.*, 2023). *Considering $f \in \mathcal{A}$ and a convex function h in U such that $h(0)=1$, denote $F(z)=\frac{a+2}{z^{a+1}}\int_0^z t^a f(t)\,dt$, $z \in U$,* Re $a > -2$, *and assume that $\left(\mathcal{I}_q^{m,l}f(z)\right)'$ is a univalent function in U, $\left(\mathcal{I}_q^{m,l}F(z)\right)' \in Q \cap \mathcal{H}[1,1]$. If the fuzzy differential superordination*

$$\mathcal{F}_{h(U)}h(z) \leq \mathcal{F}_{\left(\mathcal{I}_q^{m,l}f\right)'(U)}\left(\mathcal{I}_q^{m,l}f(z)\right)', \quad z \in U, \tag{5.2.2.17}$$

states, then we get the fuzzy differential superordination

$$\mathcal{F}_{g(U)}g(z) \leq \mathcal{F}_{\left(\mathcal{I}_q^{m,l}F\right)'(U)}\left(\mathcal{I}_q^{m,l}F(z)\right)', \quad z \in U,$$

with the convex function $g(z) = \frac{a+2}{z^{a+2}}\int_0^z h(t)t^{a+1}dt$ the fuzzy best subordinant.

Proof. Using the relation $z^{a+1}F(z) = (a+2)\int_0^z t^a f(t)\,dt$ from Theorem 5.41 and differentiating it, we can write $zF'(z)+(a+1)F(z) = (a+2)f(z)$ in the following form $z\left(\mathcal{I}_q^{m,l}F(z)\right)' + (a+1)\mathcal{I}_q^{m,l}F(z) = (a+2)\mathcal{I}_q^{m,l}f(z)$, $z \in U$, which after differentiating it again has the form

$$\frac{z\left(\mathcal{I}_q^{m,l}F(z)\right)''}{a+2} + \left(\mathcal{I}_q^{m,l}F(z)\right)' = \left(\mathcal{I}_q^{m,l}f(z)\right)', \quad z \in U.$$

Using the last relation, the fuzzy differential superordination (5.2.2.17) can be written

$$\mathcal{F}_{h(U)}h(z) \leq \mathcal{F}_{(\mathcal{I}_q^{m,l}F)'(U)}\left(\frac{z\left(\mathcal{I}_q^{m,l}F(z)\right)''}{a+2} + \left(\mathcal{I}_q^{m,l}F(z)\right)'\right). \quad (5.2.2.18)$$

Define

$$p(z) = \left(\mathcal{I}_q^{m,l}F(z)\right)', \quad z \in U, \quad (5.2.2.19)$$

and replacing (5.2.2.19) in (5.2.2.18), we have $\mathcal{F}_{h(U)}h(z) \leq \mathcal{F}_{p(U)}\left(\frac{zp'(z)}{a+2} + p(z)\right)$, $z \in U$. Applying Lemma 1.17 considering $n = 1$ and $\alpha = a+2$, it yields $\mathcal{F}_{g(U)}g(z) \leq \mathcal{F}_{p(U)}p(z)$, equivalently with $\mathcal{F}_{g(U)}g(z) \leq \mathcal{F}_{(\mathcal{I}_q^{m,l}F)'(U)}\left(\mathcal{I}_q^{m,l}F(z)\right)'$, $z \in U$, with the fuzzy best subordinant $g(z) = \frac{a+2}{z^{a+2}}\int_0^z h(t)t^{a+1}dt$ convex function. □

Theorem 5.47 (Alb Lupaş *et al.*, 2023). *Considering a convex function g, we define the function $h(z) = \frac{zg'(z)}{a+2} + g(z)$, with* Re *$a > -2, z \in U$. For $f \in \mathcal{A}$, denote $F(z) = \frac{a+2}{z^{a+1}}\int_0^z t^a f(t)\,dt$, $z \in U$ and assume that $\left(\mathcal{I}_q^{m,l}f(z)\right)'$ is univalent in U and $\left(\mathcal{I}_q^{m,l}F(z)\right)' \in Q \cap \mathcal{H}[1,1]$. When the fuzzy differential superordination*

$$\mathcal{F}_{h(U)}h(z) \leq \mathcal{F}_{(\mathcal{I}_q^{m,l}f)'(U)}\left(\mathcal{I}_q^{m,l}f(z)\right)', \quad z \in U, \quad (5.2.2.20)$$

states, then we get the fuzzy differential superordination

$$\mathcal{F}_{g(U)}g(z) \leq \mathcal{F}_{(\mathcal{I}_q^{m,l}F)'(U)}\left(\mathcal{I}_q^{m,l}F(z)\right)', \quad z \in U,$$

for $g(z) = \frac{a+2}{z^{a+2}}\int_0^z h(t)t^{a+1}dt$ the fuzzy best subordinant.

Proof. Considering $p(z) = \left(\mathcal{I}_q^{m,l}F(z)\right)', z \in U$, following the proof of Theorem 5.46, we can write the fuzzy differential superordination (5.2.2.20) in the following form

$$\mathcal{F}_{g(U)}\left(\frac{zg'(z)}{a+2} + g(z)\right) \leq \mathcal{F}_{p(U)}\left(\frac{zp'(z)}{a+2} + p(z)\right), \quad z \in U.$$

Applying Lemma 1.18 for $\alpha = a+2$ and $n = 1$, we obtain the fuzzy differential superordination $\mathcal{F}_{g(U)}g(z) \leq \mathcal{F}_{p(U)}p(z) = \mathcal{F}_{(\mathcal{I}_q^{m,l}F)'(U)}\left(\mathcal{I}_q^{m,l}F(z)\right)'$, $z \in U$, having $g(z) = \frac{a+2}{z^{a+2}}\int_0^z h(t)t^{a+1}dt$ the fuzzy best subordinant. □

Theorem 5.48 (Alb Lupaş *et al.*, 2023). *For $f \in \mathcal{A}$, denote $F(z) = \frac{a+2}{z^{a+1}} \int_0^z t^a f(t)\,dt$, $z \in U$, and $h(z) = \frac{1+(2\alpha-1)z}{1+z}$, where* Re $a > -2$, $\alpha \in [0,1)$. *Assume that $\left(\mathcal{I}_q^{m,l} f(z)\right)'$ is univalent in U, $\left(\mathcal{I}_q^{m,l} F(z)\right)' \in Q \cap \mathcal{H}[1,1]$ and the fuzzy differential superordination*

$$\mathcal{F}_{h(U)} h(z) \leq \mathcal{F}_{\left(\mathcal{I}_q^{m,l} f\right)'(U)} \left(\mathcal{I}_q^{m,l} f(z)\right)', \quad z \in U, \tag{5.2.2.21}$$

is satisfied, then the fuzzy differential superordination

$$\mathcal{F}_{g(U)} g(z) \leq \mathcal{F}_{\left(\mathcal{I}_q^{m,l} F\right)'(U)} \left(\mathcal{I}_q^{m,l} F(z)\right)', \quad z \in U,$$

is satisfied for the convex function $g(z) = 2\alpha - 1 + \frac{(a+2)(2-2\alpha)}{z^{a+2}} \int_0^z \frac{t^{a+1}}{t+1} dt$, $z \in U$ as the fuzzy best subordinant.

Proof. Denoting $p(z) = \left(\mathcal{I}_q^{m,l} F(z)\right)'$, following the proof of Theorem 5.46, the fuzzy differential superordination (5.2.2.21) can be written as $\mathcal{F}_{h(U)} h(z) \leq \mathcal{F}_{p(U)} \left(\frac{zp'(z)}{a+2} + p(z)\right)$, $z \in U$.

Applying Lemma 1.17, we get the fuzzy differential superordination $\mathcal{F}_{g(U)} g(z) \leq \mathcal{F}_{p(U)} p(z)$, with $g(z) = \frac{a+2}{z^{a+2}} \int_0^z \frac{1+(2\alpha-1)t}{1+t} t^{a+1} dt = 2\alpha - 1 + \frac{(a+2)(2-2\alpha)}{z^{a+2}} \int_0^z \frac{t^{a+1}}{t+1} dt$ and $\mathcal{F}_{g(U)} g(z) \leq \mathcal{F}_{\left(\mathcal{I}_q^{m,l} F\right)'(U)} \left(\mathcal{I}_q^{m,l} F(z)\right)'$, $z \in U$, g is the fuzzy best subordinant and it is convex. □

Theorem 5.49 (Alb Lupaş *et al.*, 2023). *For $f \in \mathcal{A}$, consider a convex function h such that $h(0) = 1$ and assume that $\left(\mathcal{I}_q^{m,l} f(z)\right)'$ is univalent and $\frac{\mathcal{I}_q^{m,l} f(z)}{z} \in Q \cap \mathcal{H}[1,1]$. When the fuzzy differential superordination*

$$\mathcal{F}_{h(U)} h(z) \leq \mathcal{F}_{\left(\mathcal{I}_q^{m,l} f\right)'(U)} \left(\mathcal{I}_q^{m,l} f(z)\right)', \quad z \in U, \tag{5.2.2.22}$$

holds, then the following fuzzy differential superordination

$$\mathcal{F}_{g(U)} g(z) \leq \mathcal{F}_{\mathcal{I}_q^{m,l} f(U)} \frac{\mathcal{I}_q^{m,l} f(z)}{z}, \quad z \in U,$$

is satisfied, for the convex function $g(z) = \frac{1}{z} \int_0^z h(t)dt$ the fuzzy best subordinant.

Proof. Denoting $p(z) = \frac{\mathcal{I}_q^{m,l} f(z)}{z} = \frac{z + \sum_{k=2}^{\infty} \left(\frac{[k+l]_q}{[1+l]_q}\right)^m a_k z^k}{z} \in \mathcal{H}[1,1]$, $z \in U$, we can write $\mathcal{I}_q^{m,l} f(z) = zp(z)$, and differentiating it, we have $\left(\mathcal{I}_q^{m,l} f(z)\right)' = zp'(z) + p(z)$, $z \in U$.

With this notation, the fuzzy differential superordination (5.2.2.22) becomes $\mathcal{F}_{h(U)} h(z) \leq \mathcal{F}_{p(U)} (zp'(z) + p(z))$, $z \in U$, and applying Lemma 1.17, we get $\mathcal{F}_{g(U)} g(z) \leq \mathcal{F}_{p(U)} p(z) = \mathcal{F}_{\mathcal{I}_q^{m,l} f((U)} \frac{\mathcal{I}_q^{m,l} f(z)}{z}$, $z \in U$, for $g(z) = \frac{1}{z} \int_0^z h(t)dt$ the fuzzy best subordinant and convex. □

Theorem 5.50 (Alb Lupaş *et al.*, 2023). *Considering a convex function g in U, we define the function h by $h(z) = zg'(z)+g(z)$. Assume $\left(\mathcal{I}_q^{m,l} f(z)\right)'$ is univalent, $\frac{\mathcal{I}_q^{m,l} f(z)}{z} \in Q \cap \mathcal{H}[1,1]$ for $f \in \mathcal{A}$ and the fuzzy superordination*

$$\mathcal{F}_{g(U)}\left(zg'(z)+g(z)\right) \leq \mathcal{F}_{\left(\mathcal{I}_q^{m,l} f\right)'(U)}\left(\mathcal{I}_q^{m,l} f(z)\right)', \quad z \in U, \qquad (5.2.2.23)$$

is satisfied, then the fuzzy differential superordination

$$\mathcal{F}_{g(U)} g(z) \leq \mathcal{F}_{\mathcal{I}_q^{m,l} f(U)} \frac{\mathcal{I}_q^{m,l} f(z)}{z}, \quad z \in U,$$

is satisfied for $g(z) = \frac{1}{z}\int_0^z h(t)dt$ the fuzzy best subordinant.

Proof. Taking account the proof of Theorem 5.49 for $p(z) = \frac{\mathcal{I}_q^{m,l} f(z)}{z}$, the fuzzy superordination (5.2.2.23) can be written in the following form $\mathcal{F}_{g(U)}\left(zg'(z)+g(z)\right) \leq \mathcal{F}_{p(U)}\left(zp'(z)+p(z)\right)$, $z \in U$.

Applying Lemma 1.18, we get the fuzzy differential superordination $\mathcal{F}_{g(U)} g(z) \leq \mathcal{F}_{p(U)} p(z)$, equivalently with $\mathcal{F}_{g(U)} g(z) \leq \mathcal{F}_{\mathcal{I}_q^{m,l} f(U)} \frac{\mathcal{I}_q^{m,l} f(z)}{z}$, where $g(z) = \frac{1}{z}\int_0^z h(t)dt$ the fuzzy best subordinant. □

Theorem 5.51 (Alb Lupaş *et al.*, 2023). *Let $h(z) = \frac{1+(2\alpha-1)z}{1+z}$ with $0 \leq \alpha < 1$, $z \in U$. For $f \in \mathcal{A}$, suppose that $\left(\mathcal{I}_q^{m,l} f(z)\right)'$ is univalent and $\frac{\mathcal{I}_q^{m,l} f(z)}{z} \in Q \cap \mathcal{H}[1,1]$. If the fuzzy differential superordination*

$$\mathcal{F}_{h(U)} h(z) \leq \mathcal{F}_{\left(\mathcal{I}_q^{m,l} f\right)'(U)}\left(\mathcal{I}_q^{m,l} f(z)\right)', \quad z \in U, \qquad (5.2.2.24)$$

holds, then we get the following fuzzy differential superordination

$$\mathcal{F}_{g(U)} g(z) \leq \mathcal{F}_{\mathcal{I}_q^{m,l} f(U)} \frac{\mathcal{I}_q^{m,l} f(z)}{z}, \quad z \in U,$$

where the fuzzy best subordinant is the convex function $g(z) = 2\alpha - 1 + 2(1-\alpha)\frac{\ln(1+z)}{z}$, $z \in U$.

Proof. Following the proof of Theorem 5.49 for $p(z) = \frac{\mathcal{I}_q^{m,l} f(z)}{z}$, the fuzzy superordination (5.2.2.24) takes the form $\mathcal{F}_{h(U)} h(z) \leq \mathcal{F}_{p(U)}\left(zp'(z)+p(z)\right)$, $z \in U$.

Applying Lemma 1.17, we get the following fuzzy differential superordination $\mathcal{F}_{g(U)} g(z) \leq \mathcal{F}_{p(U)} p(z)$, which can be written as $\mathcal{F}_{g(U)} g(z) \leq \mathcal{F}_{\mathcal{I}_q^{m,l} f(U)} \frac{\mathcal{I}_q^{m,l} f(z)}{z}$, $z \in U$. The function $g(z) = \frac{1}{z}\int_0^z \frac{1+(2\alpha-1)t}{1+t}dt = 2\alpha - 1 + \frac{2(1-\alpha)}{z}\ln(z+1)$ is the best subordinant and it is convex. □

Theorem 5.52 (Alb Lupaş *et al.*, 2023). *Considering a convex function h such that $h(0) = 1$, for $f \in \mathcal{A}$ suppose that $\left(\frac{z\mathcal{I}_q^{m+1,l}f(z)}{\mathcal{I}_q^{m,l}f(z)}\right)'$ is univalent and $\frac{\mathcal{I}_q^{m+1,l}f(z)}{\mathcal{I}_q^{m,l}f(z)} \in Q \cap \mathcal{H}[1,1]$. If the fuzzy differential superordination*

$$\mathcal{F}_{h(U)}h(z) \leq \mathcal{F}_{\mathcal{I}_q^{m,l}f(U)}\left(\frac{z\mathcal{I}_q^{m+1,l}f(z)}{\mathcal{I}_q^{m,l}f(z)}\right)', \quad z \in U, \tag{5.2.2.25}$$

holds, then we get the following fuzzy differential superordination

$$\mathcal{F}_{g(U)}g(z) \leq \mathcal{F}_{\mathcal{I}_q^{m,l}f(U)}\frac{\mathcal{I}_q^{m+1,l}f(z)}{\mathcal{I}_q^{m,l}f(z)}, \quad z \in U,$$

with the fuzzy best subordinant is the convex function $g(z) = \frac{1}{z}\int_0^z h(t)dt$.

Proof. Let $p(z) = \frac{\mathcal{I}_q^{m+1,l}f(z)}{\mathcal{I}_q^{m,l}f(z)}$, after differentiating it, we can write $p'(z) = \frac{\left(\mathcal{I}_q^{m+1,l}f(z)\right)'}{\mathcal{I}_q^{m,l}f(z)} - p(z)\frac{\left(\mathcal{I}_q^{m,l}f(z)\right)'}{\mathcal{I}_q^{m,l}f(z)}$ in the form $zp'(z) + p(z) = \left(\frac{z\mathcal{I}_q^{m+1,l}f(z)}{\mathcal{I}_q^{m,l}f(z)}\right)'$.

Fuzzy differential superordination (5.2.2.25) for $z \in U$, becomes $\mathcal{F}_{h(U)}h(z) \leq \mathcal{F}_{p(U)}\left(zp'(z) + p(z)\right)$.

Applying Lemma 1.17, we obtain the following fuzzy differential superordination $\mathcal{F}_{g(U)}g(z) \leq \mathcal{F}_{p(U)}p(z) = \mathcal{F}_{\mathcal{I}_q^{m,l}f(U)}\frac{\mathcal{I}_q^{m+1,l}f(z)}{\mathcal{I}_q^{m,l}f(z)}$, $z \in U$, for the fuzzy best subordinant $g(z) = \frac{1}{z}\int_0^z h(t)dt$ convex. □

Theorem 5.53 (Alb Lupaş *et al.*, 2023). *Consider a convex function g and the function h defined by $h(z) = zg'(z) + g(z)$. For $f \in \mathcal{A}$, assume that $\left(\frac{z\mathcal{I}_q^{m+1,l}f(z)}{\mathcal{I}_q^{m,l}f(z)}\right)'$ is univalent and $\frac{\mathcal{I}_q^{m+1,l}f(z)}{\mathcal{I}_q^{m,l}f(z)} \in Q \cap \mathcal{H}[1,1]$. If the fuzzy differential superordination*

$$\mathcal{F}_{g(U)}h(z) = \mathcal{F}_{g(U)}\left(zg'(z) + g(z)\right) \leq \mathcal{F}_{\mathcal{I}_q^{m,l}f(U)}\left(\frac{z\mathcal{I}_q^{m+1,l}f(z)}{\mathcal{I}_q^{m,l}f(z)}\right)', \quad z \in U, \tag{5.2.2.26}$$

states, then we get the fuzzy differential superordination

$$\mathcal{F}_{g(U)}g(z) \leq \mathcal{F}_{\mathcal{I}_q^{m,l}f(U)}\frac{\mathcal{I}_q^{m+1,l}f(z)}{\mathcal{I}_q^{m,l}f(z)}, \quad z \in U,$$

and the fuzzy best subordinant is $g(z) = \frac{1}{z}\int_0^z h(t)dt$.

Proof. Following the proof of Theorem 5.52 for $p(z) = \frac{\mathcal{I}_q^{m+1,l} f(z)}{\mathcal{I}_q^{m,l} f(z)}$, the fuzzy superordination (5.2.2.26) has the form $\mathcal{F}_{g(U)} h(z) = \mathcal{F}_{g(U)} (zg'(z) + g(z)) \leq \mathcal{F}_{p(U)} (zp'(z) + p(z))$, $z \in U$.

Applying Lemma 1.18, it yields $\mathcal{F}_{g(U)} g(z) \leq \mathcal{F}_{p(U)} p(z)$, equivalently with $\mathcal{F}_{g(U)} g(z) = \mathcal{F}_{g(U)} \left(\frac{1}{z} \int_0^z h(t)dt \right) \leq \mathcal{F}_{\mathcal{I}_q^{m,l} f(U)} \frac{\mathcal{I}_q^{m+1,l} f(z)}{\mathcal{I}_q^{m,l} f(z)}$, $z \in U$, and the fuzzy best subordinant is g. □

Theorem 5.54 (Alb Lupaş *et al.*, 2023). *Consider $h(z) = \frac{1+(2\alpha-1)z}{1+z}$, with $0 \leq \alpha < 1$. For $f \in \mathcal{A}$, assume that $\left(\frac{z\mathcal{I}_q^{m+1,l} f(z)}{\mathcal{I}_q^{m,l} f(z)} \right)'$ is univalent and $\frac{\mathcal{I}_q^{m+1,l} f(z)}{\mathcal{I}_q^{m,l} f(z)} \in Q \cap \mathcal{H}[1,1]$. If the fuzzy differential superordination*

$$\mathcal{F}_{h(U)} h(z) \leq \mathcal{F}_{\mathcal{I}_q^{m,l} f(U)} \left(\frac{z\mathcal{I}_q^{m+1,l} f(z)}{\mathcal{I}_q^{m,l} f(z)} \right)', \quad z \in U, \tag{5.2.2.27}$$

holds, then the fuzzy differential superordination

$$\mathcal{F}_{g(U)} g(z) \leq \mathcal{F}_{\mathcal{I}_q^{m,l} f(U)} \frac{\mathcal{I}_q^{m+1,l} f(z)}{\mathcal{I}_q^{m,l} f(z)}, \quad z \in U,$$

holds, and the fuzzy best subordinant is the convex function $g(z) = 2\alpha - 1 + 2(1-\alpha)\frac{\ln(1+z)}{z}$, $z \in U$.

Proof. Using the notation $p(z) = \frac{\mathcal{I}_q^{m+1,l} f(z)}{\mathcal{I}_q^{m,l} f(z)}$, the fuzzy differential superordination (5.2.2.27) can be written $\mathcal{F}_{h(U)} h(z) \leq \mathcal{F}_{p(U)} (zp'(z) + p(z))$, $z \in U$.

Applying Lemma 1.17, we get the fuzzy differential superordination $\mathcal{F}_{g(U)} g(z) \leq \mathcal{F}_{p(U)} p(z)$, equivalently with $\mathcal{F}_{g(U)} g(z) \leq \mathcal{F}_{\mathcal{I}_q^{m,l} f(U)} \frac{\mathcal{I}_q^{m+1,l} f(z)}{\mathcal{I}_q^{m,l} f(z)}$, $z \in U$.

The best subordinant is the convex function $g(z) = \frac{1}{z} \int_0^z \frac{1+(2\alpha-1)t}{1+t} dt = 2\alpha - 1 + 2(1-\alpha)\frac{1}{z} \ln(z+1)$, $z \in U$. □

Bibliography

F.F. Abdulnabi, H.F. Al-Janaby, F. Ghanim, A. Alb Lupaş, Some results on third-order differential subordination and differential superordination for analytic functions using a fractional differential operator, *Mathematics*, 2023, 11, 4021. https://doi.org/10.3390/math11184021.

M. Acu and S. Owa, *Note on a Class of Starlike Functions.* RIMS, Kyoto, 2006.

A. Alb Lupaş, On special differential subordinations using Sălăgean and Ruscheweyh operators, *Mathematical Inequalities and Applications*, 2009, 12(4), 781–790.

A. Alb Lupaş, A special comprehensive class of analytic functions defined by multiplier transformation, *Journal of Computational Analysis and Applications*, 2010a, 12(2), 387–395.

A. Alb Lupaş, Certain differential subordinations using a generalized Sălăgean operator and Ruscheweyh operator II, *Fractional Calculus and Applied Analysis II*, 2010b, 13(4), 355–360.

A. Alb Lupaş, A note on a certain subclass of analytic functions defined by multiplier transformation, *Journal of Computational Analysis and Applications*, 2010c, 12(1-B), 369–373.

A. Alb Lupaş, On special differential subordinations using multiplier transformation and Ruscheweyh derivative, *ROMAI Journal*, 2010d, 6(2), 1–14.

A. Alb Lupaş, Certain differential superordinations using Sălăgean and Ruscheweyh operators, *Analele Universităţii din Oradea, Fascicola Matematica*, 2010e, 17(2), 209–216.

A. Alb Lupaş, On special differential subordinations using a generalized Sălăgean operator and Ruscheweyh derivative, *Journal of Computational Analysis and Applications*, 2011a, 13(1), 98–107.

A. Alb Lupaş, Certain differential superordinations using a generalized Sălăgean and Ruscheweyh operators, *Acta Universitatis Apulensis*, 2011b, 25, 31–40.

A. Alb Lupaş, On special differential superordinations using multiplier transformation, *Journal of Computational Analysis and Applications*, 2011c, 13(1), 121–126.

A. Alb Lupaş, On special differential superordinations using a generalized Sălăgean operator and Ruscheweyh derivative, *Computers and Mathematics with Applications*, 2011d, 61, 1048–1058.

A. Alb Lupaş, Certain special differential superordinations using multiplier transformation and Ruscheweyh derivative, *Journal of Computational Analysis and Applications*, 2011e, 13(1), 108–115.

A. Alb Lupaş, Certain strong differential subordinations using Sălăgean and Ruscheweyh operators, *Advances in Applied Mathematical Analysis*, 2011f, 6(1), 27–34.

A. Alb Lupaş, Certain strong differential subordinations using a multiplier transformation and Ruscheweyh operator, *International Journal of Open Problems in Complex Analysis*, 2011g, 3(1), 1–8.

A. Alb Lupaş, Certain differential subordinations using Sălăgean and Ruscheweyh operators, *Acta Universitatis Apulensis*, 2012a, 29, 125–129.

A. Alb Lupaş, On special strong differential subordinations using multiplier transformation, *Applied Mathematics Letters*, 2012b, 25, 624–630. https://doi.org/10.1016/j.aml.2011.09.074.

A. Alb Lupaş, On special strong differential subordinations using a generalized Sălăgean operator and Ruscheweyh derivative, *Journal of Concrete and Applicable Mathematics*, 2012c, 10(1–2), 17–23.

A. Alb Lupaş, On special strong differential subordinations using a multiplier transformation and Ruscheweyh derivative, *Journal of Advanced Mathematical Studies*, 2012d, 5(1), 59–70.

A. Alb Lupaş, On special strong differential superordinations using a generalized Sălăgean operator and Ruscheweyh derivative, *Cayley Journal of Mathematics*, 2012e, 1(1), 11–27.

A. Alb Lupaş, A note on special strong differential superordinations using a multiplier transformation and Ruscheweyh derivative, *International Journal of Open Problems Complex Analysis*, 2012f, 5(1), 33–47.

A. Alb Lupaş, Certain strong differential superordinations using Sălăgean and Ruscheweyh operators, *Acta Universitatis Apulensis*, 2012g, 30, 325–336.

A. Alb Lupaş, A note on strong differential superordinations using a generalized Sălăgean operator and Ruscheweyh operator, *Studia Universitatis Babeş-Bolyai, Mathematica*, 2012h, 57(2), 153–165.

A. Alb Lupaş, Certain strong differential superordinations using a multiplier transformation and Ruscheweyh operator, *Analele Universităţii din Oradea, Fascicola Matematica*, 2012i, 12(1), 125–136.

A. Alb Lupaş, Certain differential superordinations using a multiplier transformation and Ruscheweyh derivative, *Buletinul Academiei de Ştiinţe a Republicii Moldova. Matematica*, 2013a, 2(72)–3(73), 119–131.

A. Alb Lupaş, A note on strong differential subordinations using a generalized Sălăgean operator and Ruscheweyh operator, *Acta Universitatis Apulensis*, 2013b, 34, 105–114.

A. Alb Lupaş, A note on special fuzzy differential subordinations using generalized Salagean operator and Ruscheweyh derivative, *Journal of Computational Analysis and Applications*, 2013c, 15(8), 1476–1483.

A. Alb Lupaş, On special fuzzy differential subordinations using convolution product of Sălăgean operator and Ruscheweyh derivative, *Journal of Computational Analysis and Applications*, 2013d, 15(8), 1484–1489.

A. Alb Lupaş, A note on special strong differential superordinations using multiplier transformation, *Journal of Computational Analysis and Applications*, 2014a, 17(4), 746–751.

A. Alb Lupaş, Aspects of univalent holomorphic functions involving multiplier transformation and Ruscheweyh derivative, *Advances in Difference Equations*, 2014b, 2014, 117. https://doi.org/10.1186/1687-1847-2014-117.

A. Alb Lupaş, New classes containing Sălăgean operator and Ruscheweyh derivative, *International Journals of Open Problems in Complex Analysis*, 2014c, 6(1), 33–39.

A. Alb Lupaş, On univalence criteria for analytic functions defined by Sălăgean operator and Ruscheweyh derivative, *Acta Universitatis Apulensis*, 2014d, 39, 25–30.

A. Alb Lupaş, On univalence criteria for analytic functions defined by multiplier transformation and Ruscheweyh derivative, *International Journals of Open Problems in Computer Science and Mathematics*, 2014e, 7(1), 14–21.

A. Alb Lupaş, On special strong differential superordinations using Sălăgean and Ruscheweyh operators, *Journal of Advances in Applied & Computational Mathematics*, 2014f, 1(1), 1–7.

A. Alb Lupaş, Differential subordinations using multiplier transformation, *Advances and Applications in Mathematical Sciences*, 2015a, 14(3), 71–89.

A. Alb Lupaş, Some differential subordinations using Ruscheweyh derivative and a multiplier transformation, in *Proceedings Volume "Computational Analysis: Contributions from AMAT 2015"*, Chapter 8. Springer, New York, 2015b, pp. 97–117.

A. Alb Lupaş, Aspects of univalent holomorphic functions involving Sălăgean operator and Ruscheweyh derivative, *Journal of Concrete and Applicable Mathematics*, 2015c, 13(1–2), 51–59.

A. Alb Lupaş, Properties on a subclass of univalent functions defined by using a multiplier transformation and Ruscheweyh derivative, *Analele Stiintifice ale Universitatii Ovidius Constanta: Seria Matematica*, 2015d, 23(1), 9–24.

A. Alb Lupaş, On a certain subclass of analytic functions involving Sălăgean operator and Ruscheweyh derivative, *Journal of Computational Analysis and Applications*, 2015e, 19(2), 278–281.

A. Alb Lupaş, On a certain subclass of analytic functions involving multiplier transformation and Ruscheweyh derivative, *Analele Universitatii din Oradea. Fascicola Matematica*, 2015f, 22(2), 73–80.

A. Alb Lupaş, New classes containing multiplier transformation and Ruscheweyh derivative, *Journal of Computational Analysis and Applications*, 2015g, 19(2), 282–286.

A. Alb Lupaş, Differential sandwich theorems using a multiplier transformation and Ruscheweyh derivative, *Advances in Mathematics: Scientific Journal*, 2015h, 4(2), 195–207.

A. Alb Lupaş, Some differential sandwich theorems using a multiplier transformation and Ruscheweyh derivative, *Electronic Journal of Mathematics and Its Application (EJMIA)*, 2015i, 1(2), 76–86.

A. Alb Lupaş, On some differential sandwich theorems using a multiplier transformation and Ruscheweyh derivative, in *Proceedings of ICTAMI International Conference on Theory and Applications in Mathematics and Informatics 17th–20th of September, 2015, Alba-Iulia*. Romania Mathematics Section, 2015j, pp. 61–75.

A. Alb Lupaş, Some fuzzy differential subordinations using Ruscheweyh derivative and Sălăgean operator, in *Proceedings of the 12th International Symposium on Geometric Function Theory and Applications*, August 25–28, 2016, *Alba Iulia, Romania*, 2016a, pp. 89–98.

A. Alb Lupaş, Properties on a subclass of univalent functions defined by using Sălăgean operator and Ruscheweyh derivative, *Journal of Computational Analysis and Applications*, 2016b, 21(7), 1213–1217.

A. Alb Lupaş, About some differential sandwich theorems using a multiplier transformation and Ruscheweyh derivative, *Journal of Computational Analysis and Applications*, 2016c, 21(7), 1218–1224.

A. Alb Lupaş, About differential sandwich theorems involving a multiplier transformation and Ruscheweyh derivative, *TJMM*, 2016d, 8(1), 67–76.

A. Alb Lupaş, On special fuzzy differential subordinations using multiplier transformation, *Journal of Computational Analysis and Applications*, 2017a, 23(6), 1029–1035.

A. Alb Lupaş, Differential sandwich theorems involving a multiplier transformation and Ruscheweyh derivative, *Analele Universitatii din Oradea. Fascicola Matematica*, 2017b, 24(1), 93–102.

A. Alb Lupaş, On some differential sandwich theorems involving a multiplier transformation and Ruscheweyh derivative, *Journal of Computational Analysis and Applications*, 2017c, 23(6), 1036–1042.

A. Alb Lupaş, About differential sandwich theorems using multiplier transformation and Ruscheweyh derivative, in *Recent Trends in Pure and Applied Mathematics, TREPAM 31 July–4 August 2017, Alba Iulia*, 2017d, pp. 55–66.

A. Alb Lupaş, A note on special fuzzy differential subordinations using multiplier transformation and Ruscheweyh derivative, *Journal of Computational Analysis and Applications*, 2018a, 25(6), 1125–1131.

A. Alb Lupaş, About some differential sandwich theorems involving a multiplier transformation and Ruscheweyh derivative, *Journal of Computational Analysis and Applications*, 2018b, 25(6), 1116–1124.

A. Alb Lupaş, About a subclass of analytic functions defined by a fractional integral operator, *International Release of Mathematics and Mathematical Sciences*, 2018c, 1(2), 1–10.

A. Alb Lupaş, A subclass of analytic functions defined by a fractional integral operator, *Journal of Computational Analysis and Applications*, 2019a, 27(3), 502–505.

A. Alb Lupaş, Properties on a subclass of analytic functions defined by a fractional integral operator, *Journal of Computational Analysis and Applications*, 2019b, 27(3), 506–510.

A. Alb Lupaş, Applications of the fractional calculus in fuzzy differential subordinations and superordinations, *Mathematics*, 2021a, 9, 2601. https://doi.org/10.3390/math9202601.

A. Alb Lupaş, New applications of the fractional integral on analytic functions, *Symmetry Basel*, 2021b, 13, 423. https://doi.org/10.3390/sym13030423.

A. Alb Lupaş, Fuzzy differential sandwich theorems involving fractional integral of confluent hypergeometric function, *Symmetry*, 2021c, 13, 1992. https://doi.org/10.3390/sym13111992.

A. Alb Lupaş, Some differential sandwich theorems involving a multiplier transformation and Ruscheweyh derivative, *Advances in Mathematical Analysis and its Applications*, CRC Press, Taylor & Francis Group, A Chapman & Hall Group, 2022a, 284–302.

A. Alb Lupaş, Subordination results for a fractional integral operator, *Problems of Analysis—Issues of Analysis*, 2022b, 11(1), 29.

A. Alb Lupaş, On special fuzzy differential subordinations obtained for Riemann–Liouville fractional integral of Ruscheweyh and Sălăgean operators, *Axioms*, 2022c, 11, 428. https://doi.org/10.3390/axioms11090428.

A. Alb Lupaş, New applications of fractional integral for introducing subclasses of analytic functions, *Symmetry*, 2022d, 14, 419. https://doi.org/10.3390/sym14020419.

A. Alb Lupaş, Subordination results on the q-analogue of the Sălăgean differential operator, *Symmetry*, 2022e, 14, 1744. https://doi.org/10.3390/sym14081744.

A. Alb Lupaş, Applications of the q-Sălăgean differential operator involving multivalent functions, *Axioms*, 2022f, 11, 512. https://doi.org/10.3390/axioms11100512.

A. Alb Lupaş, New applications of fuzzy set concept in the geometric theory of analytic functions, *Axioms*, 2023a, 12, 494. https://doi.org/10.3390/axioms12050494.

A. Alb Lupaş, Fuzzy differential inequalities for convolution product of Ruscheweyh derivative and multiplier transformation, *Axioms*, 2023b, 12, 470. https://doi.org/10.3390/axioms12050470.

A. Alb Lupaş, Strong sandwich results involving Riemann-Liouville fractional integral of extended q-hypergeometric function, *Scientific Reports*, 2024 (Submitted).

A. Alb Lupaş, M. Acu, Properties of a subclass of analytic functions defined by Riemann-Liouville fractional integral applied to convolution product of multiplier transformation and Ruscheweyh derivative, *Demonstratio Mathematica*, 2023, 56, 20220249. https://doi.org/10.1515/dema-2022-024.

A. Alb Lupaş, L. Andrei, New classes containing generalized Sălăgean operator and Ruscheweyh derivative, *Acta Universitatis Apulensis*, 2014a, 38, 319–328.

A. Alb Lupaş, L. Andrei, On univalence criteria for analytic functions defined by generalized Sălăgean operator and Ruscheweyh derivative, *Analele Universitatii din Oradea. Fascicola Matematica*, 2014b, 12(2), 179–185.

A. Alb Lupaş, L. Andrei, Aspects of univalent holomorphic functions involving Ruscheweyh derivative and generalized Sălăgean operator, *Journal of Computational Analysis and Applications*, 2015a, 19(2), 272–277.

A. Alb Lupaş, L. Andrei, Properties on a subclass of univalent functions defined by using a generalized Sălăgean operator and Ruscheweyh derivative, *Journal of Concrete and Applicable Mathematics*, 2015b, 13(1–2), 60–68.

A. Alb Lupaş, L. Andrei, On a certain subclass of analytic functions involving generalized Sălăgean operator and Ruscheweyh derivative, *Journal of Applied Functional Analysis*, 2015c, 10(1–2), 89–94.

A. Alb Lupaş, D. Breaz, On special differential superordinations using Sălăgean and Ruscheweyh operators, in *Geometric Function Theory and Applications'2010 (Proceedings of International Symposium, Sofia*, 27–31 August 2010), 2010, pp. 98–103.

A. Alb Lupaş, A. Cătas, Differential subordination and superordination results for q-analogue of multiplier transformation, *Fractal and Fractional*, 2023, 7, 199. https://doi.org/10.3390/fractalfract7020199.

A. Alb Lupaş, M. Cioban, On a subclass of analytic functions defined by a fractional integral operator, in *International Conference on Mathematics, Informatics and Information Technologies*, 19–21 April, 2018, *Bălţi, Republica Moldova*, 2018, pp. 8–14.

A. Alb Lupaş, F. Ghanim, Strong differential subordination and superordination results for extended q-analogue of multiplier transformation, *Symmetry*, 2023, 15, 713. https://doi.org/10.3390/sym15030713.

A. Alb Lupaş, G. Oros, On special fuzzy differential subordinations using Sălăgean and Ruscheweyh operators, *Applied Mathematics and Computation*, 2015, 261, 119–127. https://doi.org/10.1016/j.amc.2015.03.087.

A. Alb Lupaş, G.I. Oros, On special differential subordinations using fractional integral of Salagean and Ruscheweyh operators, *Symmetry*, 2021a, 13, 1553. https://doi.org/10.3390/sym13091553.

A. Alb Lupaş, G.I. Oros, Differential subordination and superordination results using fractional integral of confluent hypergeometric function, *Symmetry Basel*, 2021b, 13, 327. https://doi.org/10.3390/sym13020327.

A. Alb Lupaş, G.I. Oros, Fractional integral of a confluent hypergeometric function applied to defining a new class of analytic functions, *Symmetry*, 2022a, 14, 427. https://doi.org/10.3390/sym14020427.

A. Alb Lupaş, G.I. Oros, Fractional calculus and confluent hypergeometric function applied in the study of subclasses of analytic functions, *Mathematics*, 2022b, 10, 705. https://doi.org/10.3390/math10050705.

A. Alb Lupaş, G.I. Oros, Sandwich type results regarding Riemann-Liouville fractional integral of q-confluent hypergeometric function, *Demonstratio Mathematica*, 2022c, 55, 1–9. https://doi.org/10.1515/dema-2022-0186.

A. Alb Lupaş, G.I. Oros, Fuzzy differential subordination and superordination results involving the q-hypergeometric function and fractional calculus aspects, *Mathematics*, 2022d, 10, 4121. https://doi.org/10.3390/math10214121.

A. Alb Lupaş, G.I. Oros, Applications of Riemann–Liouville fractional integral of q-hypergeometric function for obtaining fuzzy differential sandwich results, *Symmetry*, 2022e, 14, 2097. https://doi.org/10.3390/sym14102097.

A. Alb Lupaş, G.I. Oros, Strong differential subordinations and superordinations for Riemann–Liouville fractional integral of extended q-hypergeometric function, *Mathematics*, 2023a, 11, 4474. https://doi.org/10.3390/math11 214474.

A. Alb Lupaş, G.I. Oros, Differential sandwich theorems involving Riemann-Liouville fractional integral of qq-hypergeometric function, *AIMS Mathematics*, 2023b, 8(2), 4930–4943. https://doi.org/10.3934/math.2023246.

A. Alb Lupaş, G.I. Oros, G. Oros, On special strong differential subordinations using Sălăgean and Ruscheweyh operators, *Journal of Computational Analysis and Applications*, 2012, 14(2), 266–270.

A. Alb Lupaş, S.A. Shah, L. Iambor, Fuzzy differential subordination and superordination results for q-analogue of multiplier transformation, *AIMS Mathematics*, 2023, 8(7), 15569–15584. https://doi.org/10.3934/math.2023794.

S.S. Alhily, A. Alb Lupaş, Sandwich theorems involving fractional integrals applied to the q-analogue of the multiplier transformation, *AIMS Mathematics*, 2024, 9(3), 5850–5862. https://doi.org/10.3934/math.2024284.

F.M. Al-Oboudi, On univalent functions defined by a generalized Sălăgean operator, *Indian Journal of Mathematics and Mathematical Sciences*, 2004, 25–28, 1429–1436.

Y. Almalki, A.K. Wanas, T.G. Shaba, A. Alb Lupaş, M. Abdalla, Coefficient bounds and Fekete–Szegö inequalities for a two families of bi-univalent functions related to Gegenbauer polynomials, *Axioms*, 2023, 12, 1018. https://doi.org/10.3390/axioms12111018.

J.A. Antonino, S.S. Miller, Third-order differential inequalities and subordination in the complex plane, *Complex Variables and Elliptic Equations*, 2011, 56, 439–454.

J.A. Antonino, S. Romaguera, Strong differential subordination to Briot-Bouquet differential equations, *Journal of Differential Equations*, 1994, 114, 101–105.

W.G. Atshan, K.O. Hussain, Fuzzy differential superordination, *Theory and Applications of Mathematics and Computer Science*, 2017, 7(1), 27–38.

W.G. Atshan, I.A. Abbas, S. Yalcin, New concept on fourth-order differential subordination and superordination with some results for multivalent analytic functions, *Journal of Al-Qadisiyah for Computer Science and Mathematics*, 2020, 12, 96–107.

W.G. Atshan, A.H. Battor, A.F. Abaas, G.I. Oros, New and extended results on fourth-order differential subordination for univalent analytic functions, *Al-Qadisiyah Journal of Pure Science*, 2020, 25, 1–13.

W.G. Atshan, I.A.R. Rahman, A. Alb Lupaş, Some results of new subclasses for bi-univalent functions using quasi-subordination, *Symmetry*, 2021, 13, 1653. https://doi.org/10.3390/sym13091653.

E.I. Badiwi, W.G. Atshan, A.N. Alkiffai, A. Alb Lupaş, Certain results on subclasses of analytic and bi-univalent functions associated with coefficient estimates and quasi-subordination, *Symmetry*, 2023, 15, 2208. https://doi.org/10.3390/sym15122208.

D.A. Brannan, J. Clunie, W.E. Kirwan, Coefficient estimates for a class of starlike functions, *Canadian Journal of Mathematics*, 1970, 22, 476–485.

T. Bulboacă, Classes of first order differential superordinations, *Demonstratio Mathematica*, 2002, 35(2), 287–292.

A. Cătaş, On certain class of p-valent functions defined by new multiplier transformations, in *Proceedings Book of the International Symposium on Geometric Function Theory and Applications*, August 20–24, 2007. TC Istanbul Kultur University, Turkey, 2007, pp. 241–250.

N.E. Cho, A.M.K. Aouf, Some applications of fractional calculus operators to a certain subclass of analytic functions with negative coefficients, *Turkish Journal of Mathematics*, 1996, 20, 553–562.

N.E. Cho, T.H. Kim, Multiplier transformations and strongly close-to-convex functions, *Bulletin of the Korean Mathematical Society*, 2003, 40(3), 399–410.

N.E. Cho, H.M. Srivastava, Argument estimates of certain analytic functions defined by a class of multiplier transformations, *Mathematical and Computer Modelling*, 2003, 37(1–2), 39–49.

A.M. Darweesh, W.G. Atshan, A.H. Battor, A. Alb Lupaş, Third-order differential subordination results for analytic functions associated with a certain differential operator, *Symmetry*, 2022, 14, 99. https://doi.org/10.3390/sym14010099.

S.G. Gal, A.I. Ban, *Elemente de matematică fuzzy*, Editura Universităţii din Oradea, 1996.

M. Govindaraj, S. Sivasubramanian, On a class of analytic functions related to conic domains involving q-calculus, *Analysis Mathematica*, 2017, 43, 475–487.

S.H. Hadi, M. Darus, A. Alb Lupaş, A class of Janowski-type (p, q)-convex harmonic functions involving a generalized q-Mittag–Leffler function, *Axioms*, 2023a, 12, 190. https://doi.org/10.3390/axioms12020190.

S.H. Hadi, M. Darus, F. Ghanim, A. Alb Lupaş, Sandwich-type theorems for a family of non-Bazilevič functions involving a q-analog integral operator, *Mathematics*, 2023b, 11, 2479. https://doi.org/10.3390/math11112479.

D.J. Hallenbeck, S. Ruscheweyh, Subordination by convex functions, *Proceedings of the American Mathematical Society*, 1975, 52, 191–195.

J.O. Hamzat, M.O. Oluwayemi, A. Alb Lupaş, A.K. Wanas, Bi-univalent problems involving generalized multiplier transform with respect to symmetric and conjugate points, *Fractal and Fractional*, 2022, 6, 483. https://doi.org/10.3390/fractalfract6090483.

M.E-H. Ismail, E. Merkes, D. Styer, A generalization of starlike functions, *Complex Variables, Theory and Application*, 1990, 14, 77–84.

F.H. Jackson, q-difference equations, *American Journal of Mathematics*, 1910a, 32, 305–314.

F.H. Jackson, On q-definite integrals, *Quarterly Journal of Pure and Applied Mathematics*, 1910b, 41, 193–203.

M.P. Jeyaraman, T.K. Suresh, Third-order differential subordination of analytic functions, *Acta Universitatis Apulensis. Mathematics — Informatics*, 2013, 35, 187–202.

A.S. Juma, A.H.S. Mushtaq, A.H.F. Mohammed, Third-order differential subordination and superordination results for meromorphically univalent functions involving linear operator, *European Journal of Scientific Research*, 2015, 132, 57–65.

S. Kanas, D. Răducanu, Some class of analytic functions related to conic domains, *Mathematica Slovaca*, 2014, 64, 1183–1196.

M. Lewin, On a coefficient problem for bi-univalent functions, *Proceedings of the American Mathematical Society*, 1967, 18, 63–68.

B.K. Mihsin, W.G. Atshan, S.S. Alhily, A. Alb Lupaş, New results on fourth-order differential subordination and superordination for univalent analytic functions involving a linear operator, *Symmetry*, 2022, 14, 324. https://doi.org/10.3390/sym14020324.

S.S. Miller, P.T. Mocanu, Second order-differential inequalities in the complex plane, *Journal of Mathematical Analysis and Applications*, 1978, 65, 298–305.

S.S. Miller, P.T. Mocanu, Differential subordinations and univalent functions, *Michigan Mathematical Journal*, 1981, 28, 157–171.

S.S. Miller, P.T. Mocanu, On some classes of first-order differential subordinations, *Michigan Mathematical Journal*, 1985, 32(2), 185–195.

S.S. Miller, P.T. Mocanu, *Differential Subordinations. Theory and Applications*, Monographs and Textbooks in Pure and Applied Mathematics, Vol. 225. Marcel Dekker Inc., New York, 2000.

S.S. Miller, P.T. Mocanu, Subordinants of differential superordinations, *Complex Variables*, 2003, 48(10), 815–826.

S.S. Miller, P.T. Mocanu, Briot-Bouquet differential superordinations and sandwich theorems, *Journal of Mathematical Analysis and Applications*, 2007, 329(1), 237–335.

E. Netanyahu, The minimal distance of the image boundary from the origin and the second coefficient of a univalent function in $|z| < 1$, *Archive for Rational Mechanics and Analysis*, 1969, 33, 100–113.

G.I. Oros, Strong differential superordination, *Acta Universitatis Apulensis*, 2009, 19, 101–106.

G.I. Oros, On a new strong differential subordination, *Acta Universitatis Apulensis*, 2012, 32, 243–250.

G.I. Oros, G. Oros, Strong differential subordination, *Turkish Journal of Mathematics*, 2009, 33, 249–257.

G.I. Oros, G. Oros, The notion of subordination in fuzzy sets theory, *General Mathematics*, 2011, 19, 97–103.

G.I. Oros, G. Oros, Fuzzy differential subordination, *Acta Universitatis Apulensis*, 2012a, 30, 55–64.

G.I. Oros, G. Oros, Dominants and best dominants in fuzzy differential subordinations, *Studia Universitatis Babes-Bolyai Mathematica*, 2012b, 57, 239–248.

G.I. Oros, G. Oros, Briot-Bouquet fuzzy differential subordination, *Analele Universitatii din Oradea. Fascicola Matematica*, 2012c, 19(2), 83–87.

S. Ponnusamy, O.P. Juneja, Third-order differential inequalities in the complex plane, in *Current Topics in Analytic Function Theory*. World Scientific Publishing Company, Singapore, 1992.

St. Ruscheweyh, New criteria for univalent functions, *Proceedings of the American Mathematical Society*, 1975, 49, 109–115.

G.S. Sălăgean, *Subclasses of Univalent Functions*, Lecture Notes in Mathematics, Vol. 1013. Springer Verlag, Berlin, 1983, pp. 362–372.

S.A. Shah, K.I. Noor, Study on q-analogue of certain family of linear operators, *Turkish Journal of Mathematics*, 2019, 43, 2707–2714. https://doi.org/10.3906/mat-1907-41.

H.M. Srivastava, Univalent functions, fractional calculus and associated generalized hypergeometric functions, in H.M. Srivastava, S. Owa (Eds.), *Univalent Functions, Fractional Calculus, and Their Applications*. Halsted Press (Ellis Horwood Limited), Chichester, UK; John Wiley and Sons, New York, USA, 1989, pp. 329–354.

H.M. Srivastava, Operators of basic (or q-) calculus and fractional q-calculus and their applications in geometric function theory of complex analysis, *Iranian Journal of Science and Technology, Transaction A, Science*, 2020, 44, 327–344.

S.D. Theyab, W.G. Atshan, A. Alb Lupaş, H.K. Abdullah, New results on higher-order differential subordination and superordination for univalent analytic functions using a new operator, *Symmetry*, 2022, 14, 1576. https://doi.org/10.3390/sym14081576.

B.A. Uralegaddi and C. Somanatha, Certain classes of univalent functions, in *Current Topics in Analytic Function Theory*. World Scientific Publishing, River Edge, 1992, pp. 371–374.

A.K. Wanas, A. Alb Lupaş, Applications of Laguerre polynomials on a new family of bi-prestarlike functions, *Symmetry*, 2022, 14, 645. https://doi.org/10.3390/sym14040645.

A.K. Wanas, F.M. Sakar, A. Alb Lupaş, Applications Laguerre polynomials for families of bi-univalent functions defined with (p, q)-Wanas operator, *Axioms*, 2023, 12, 430. https://doi.org/10.3390/axioms12050430.

Index

S

www.ingramcontent.com/pod-product-compliance
Lightning Source LLC
LaVergne TN
LVHW051053210926
848486LV00002B/90